Devices Based on Low-Dimensional Semiconductor Structures

NATO ASI Series

Advanced Science Institutes Series

A Series presenting the results of activities sponsored by the NATO Science Committee, which aims at the dissemination of advanced scientific and technological knowledge, with a view to strengthening links between scientific communities.

The Series is published by an international board of publishers in conjunction with the NATO Scientific Affairs Division

A Life Sciences **B Physics**	Plenum Publishing Corporation London and New York
C Mathematical and Physical Sciences **D Behavioural and Social Sciences** **E Applied Sciences**	Kluwer Academic Publishers Dordrecht, Boston and London
F Computer and Systems Sciences **G Ecological Sciences** **H Cell Biology** **I Global Environmental Change**	Springer-Verlag Berlin, Heidelberg, New York, London, Paris and Tokyo

PARTNERSHIP SUB-SERIES

1. Disarmament Technologies	Kluwer Academic Publishers
2. Environment	Springer-Verlag / Kluwer Academic Publishers
3. High Technology	Kluwer Academic Publishers
4. Science and Technology Policy	Kluwer Academic Publishers
5. Computer Networking	Kluwer Academic Publishers

The Partnership Sub-Series incorporates activities undertaken in collaboration with NATO's Cooperation Partners, the countries of the CIS and Central and Eastern Europe, in Priority Areas of concern to those countries.

NATO-PCO-DATA BASE

The electronic index to the NATO ASI Series provides full bibliographical references (with keywords and/or abstracts) to more than 50000 contributions from international scientists published in all sections of the NATO ASI Series.
Access to the NATO-PCO-DATA BASE is possible in two ways:

– via online FILE 128 (NATO-PCO-DATA BASE) hosted by ESRIN, Via Galileo Galilei, I-00044 Frascati, Italy.

– via CD-ROM "NATO-PCO-DATA BASE" with user-friendly retrieval software in English, French and German (© WTV GmbH and DATAWARE Technologies Inc. 1989).

The CD-ROM can be ordered through any member of the Board of Publishers or through NATO-PCO, Overijse, Belgium.

Series 3: High Technology – Vol. 14

Devices Based on Low-Dimensional Semiconductor Structures

edited by

Minko Balkanski
Laboratoire de Physique des Solides,
Université Pierre et Marie Curie, Paris, France

Springer-Science+Business Media, B.V.

Proceedings of the NATO Advanced Study Institute on
Devices Based on Low-Dimensional Semiconductor Structures
Sozopol, Bulgaria
September 19–30, 1995

Library of Congress Cataloging-in-Publication Data

ISBN 978-0-7923-4102-4 ISBN 978-94-009-0289-3 (eBook)
DOI 10.1007/978-94-009-0289-3

Printed on acid-free paper

TABLE OF CONTENTS

PREFACE

The NATO-ASI on Devices Based on Low-Dimensional Structures was planned as a part of a three-year program on Low Dimensional Semiconductor Structures. The first year was focused on the fundamental properties and applications. The second year was devoted to Devices Based on Low-Dimensional Semiconductor Structures, whose lecture notes are assembled in this volume. The third year will be on systems based on Low-Dimensional Quantum Semiconductor Devices.

This subject is a major development in high technology that bears considerable industrial potential. It is important to enable young scientists and engineers to be rapidly updated by the best experts in the field. The meeting of western world specialists with scientists from the newly freed eastern countries is considered an eminent priority because it concerns the basic training of future engineers and the modern industrial development of these countries.

This ASI has been mainly devoted to the fundamentals of quantum structures and devices in the two major fields: Electro-Optical devices and Pseudomorphic High Electron Mobility Transistors.

Fabricationprocedures

Low-dimensional quantum structures have become available due to the development of thin film technologies and in particular of Molecular Beam Epitaxy (MBE).

Molecular Beam Epitaxy is a very powerful technique for growing thin layers of semiconductors , metals or insulators. The molecular or atomic beam of the constituent elements crystallizes on a substrate maintained at an elevated temperature in an ultra-high vacuum environment. The composition of the grown epilayer and its doping level depend directly on the evaporation rate of the elemental sources.

In comparison to other epitaxial techniques like liquid phase epitaxy or chemical vapor deposition, MBE has two important advantages:

(a) The growth of the film is carried out under conditions far from thermodynamic equilibrium and is governed by kinetics of the surface process that occur when the impinging beam reacts with the outermost atomic layers of the substrate crystal. Under proper conditions, this permits the deposition of very pure crystal layers.

(b) The ultra-high vacuum environment allows the use of in situ surface sensitive diagnostic methods such as reflection high-energy electron diffraction (RHEED), Auger or ellipsometry.

The success of any technology depends crucially on the quality of the materials of which components are fabricated. High-performance electronic, optoelectronic and photonic devices usually involve low-dimensional semiconductor heterostructure layers, quite often complex, and they have to be grown by advanced thin-film growth techniques, such as chemical vapor deposition (CVD) vapor phase epitaxy (VPE) or molecular beam epitaxy (MBE). The vapor-phase technique includes hydride, chloride and metal-organic VPE (MOVPE), and the beam techniques include solid-source MBE (SSMBE), gas-source MBE (GSMBE), and metal-organic MBE (MOMBE), which is also called chemical beam epitaxy (CBE). Using gaseous sources in molecular beam epitaxy (MBE) extends the versatility of solid-source MBE.

Fundamentals of Quantum Structures for Electro-Optical Devices

The fundamentals of Quantum structures have been treated in extenso.

The study of the optical properties of heterostructures under an electric field is very important not only for the fundamental knowledge of these structures but also for applications, since a voltage is applied to any device structure, for example to control a switch (electro-optical modulator), generate a light beam (light-emitting diode, laser) or convert a light signal into an electric current (photo-detector) .

The role of spontaneous emission in laser diode operation has been the subject of one of the basic lectures.

When considering the operation of a diode laser, it is natural to focus attention on the process of amplification of light by stimulated emission that is the basis of laser action. Nevertheless, even though this stimulated emission dominates the coherent optical output of the laser, for lasers operating at wavelengths below about 1 μm it is the somewhat less glamorous process of spontaneous emission which is responsible for the intrinsic current at threshold. The spontaneous recombination current therefore represents the lowest threshold current that can be achieved when all extrinsic processes such as non-radiative recombination and carrier leakage have been eliminated, and it thus determines other intrinsic characteristics of the device, particularly the temperature sensitivity of threshold current.

Results on MBE growth of quantum dots in GaAs matrix leading to an injection lasing at room temperature have been presented as world premiere in this ASI.

Quantum dot (QD) arrays are considered to be very promising in semiconductor laser applications. A single quantum dot represents a potential quantum well in which the carrier motion is quantised in all three directions. The zero-dimensional electronic system can offer an improvement in threshold current density and its temperature stability owing to a delta-function-like density of states. The novel approach of forming a dense QD array in wider band-gap material matrix is a so-called 'self-assembled growth'. The method is based on a spontaneous transformation of strained epitaxial layer into and array of quantum sized islands due to the Stranski-Krastanov growth mode. Another important aspect of the quantum structure devices is the photodetector. The physics of quantum well infrared detectors has been discussed thoroughly.

The increasing performances of Quantum Well Infrared Photodetectors (QWIPs), mainly due to the maturity of III-V growth and processing technologies, allow them to be considered as an alternative to HgCdTe detectors. Two different models, namely photoconductive or photoemissive, have been proposed to describe the electro-optical behavior of QWIPs, depending on the electrical approach of the device (extrinsic photoconductor, Schottky diode or mixed system). All these models lead to roughly satisfactory results but despite intensive work, this domain still suffers from a lack of deep physical understanding. Modeling the complex photodetection and noise processes requires a good knowledge of different elementary mechanisms: injection efficiency of the contacts, electric field distribution, balance between photoionization and capture in the quantum wells, transport and scattering the barriers.

The purposes of the lectures presented in this ASI are: i) to describe the experimental procedures that have been used to determine the physical parameters which appear in the different photoresponse formalism;

ii) to discuss the physical origin and the connections between the different determined parameters (P_e, P_c, τ), iii) to show how those parameters and related physical mechanism shed light on the QWIP photoconduction mechanisms and iv) to propose a new QWIPs model.

Other interesting types of quantum structures are the semimagnetic quantum wells and superlattices.

During last ten years, there has been increasing interest and considerable experimental and theoretical activity focused on Semimagnetic Semiconductors (SMSC) or Diluted Magnetic Semiconductors (DMS), which have some unique properties that enhance their potential for use in a wide range of optoelectronic device applications.

SMSC or DMS could be II-VI, IV-VI, II-V or III-V compounds in which a small fraction of non magnetic cations has been replaced by magnetic transitions metal of rare earth ions.

Most of the research performed so far on these materials has been devoted to Mn-based II-VI compounds such as $Cd_{1-x}Mn_xTe$, $Zn_{1-x}Mn_xTe$, $Hg_{1-x}Mn_xTe$ and the corresponding Selenides and Sulfides. However, more recently, results have been reported on II-VI semiconductors containing other transitions magnetic metal ions, such as Fe or Co. Concerning the IV-VI compounds, they have been studied for about a decade. Nevertheless, the amount of knowledge that has been accumulated is much less than that concerning II Mn VI or II Fe VI alloys. The complication arises from their rather complex band structure. They are of interest because the values of their band gap are in the I.R. region.

For many purposes one may regard a SMSC (DMS) as being composed of two interacting subsystems:

- the electronic system of the carriers,
- the magnetic system of magnetic moments localized on magnetic ions.

In the lectures given in this ASI attention was focused on the II M VI and IV T VI systems, where M could be Mn, Fe or Co and T: Eu, Gd. An overview of the specific properties of such DMS is given. The properties of quantum wells and superlattices of DMS II-VI and DMS IV-VI systems respectively are also examined.

Nonlinear Optical Devices

The nonlinear optical devices constitute another important part of this ASI.

The passage of laser light through a semiconductor can produce a rich variety of nonlinear optical phenomena. The most sensitive effects occur when the photon energy is close to resonance with the band gap energy of the semiconductor. This has provided a fertile area of research for both new physics and device applications over the past 15 years.

In low dimensional semiconductor structures such as multiple quantum wells (MQW), excitonic resonances are especially important. Device applications currently include laser mode-locking elements, waveguide routing switches, optical soliton control and logic arrays for communications, signal processing and computing. Of primary interest here are the interactions of absorption coefficient and refractive index as a function of excess carrier density. Pump-probe and transient grating techniques employing mode-locked lasers can be used to monitor the dynamics of the carriers on ultrashort time-scales via these nonlinearities. The discussion presented in this ASI is restricted to room temperature nonlinear optical phenomena that are useful for devices. A new method of measuring in-well electron mobilities using transient gratings is described and some device applications are reviewed.

Linear and nonlinear coupling in waveguide based devices is discussed in detail.

It is now universally accepted that in the near future practically all the information received by us , or sent to someone else, will be carried on globally distributed optical fibre networks. In the somewhat longer term , there will be no exceptions to this technology either for long distance communications [TOLL] or local communications to and from our homes and businesses [EXCHANGE] as information is drawn off the information superhighways. The problem today is that the perceived requirement for all-optical switching has not yet been met. This form of technology is critically needed by the type of fibre systems anticipated, which will want to work at terabit data rates. Such a requirement creates a strong pressure for sub-picosecond speeds and will be beyond the ability not only of today's equipment, but of that currently anticipated for the future. Hence, the search is on for all-optical (i.e. photonic) devices. If this search fails then an interesting impasse will develop between the ability to deliver information (data) optically to a certain location (a computer, for example) and the ability to process it. The photonic devices required will need the right sort of materials, so the search for them is generating a huge amount of material assessment to see whether they have desirable optical properties, and these must include nonlinearity. This will be required in order to permit both signal and controls to be set up using onli light beams. In other words, the power of a light beam is needed to change a property like transmission of a waveguide to achieve, for example, a switching effect.

After a short introduction to the motivations and the usual material possibilities for all-optical switching, the sort of configurations that might be considered are reviewed and the foundations of how to set about designing the desired kind of all-optical coupling are set forth. The interesting ideas surrounding cascaded nonlinearity are examined in some detail and preliminary comments on this phenomenon are made in the section dealing with the examples of switches. The term cascading concerns basically the use of second-order nonlinearity to achieve an effective nonlinear index change through a mixing of a second-harmonic signal, at frequency 2ω, with the (fundamental) generating wave at frequency ω. Enhanced values of this kind on nonlinearity can be obtained from symmetric GaAlAs quantum wells. Thus, low-dimensional structures seem to be a real possibility and the promise for all-optical switching seems to be high at the present time.

High Electron Mobility Transistors

The other important category of devices based on low-dimensional quantum structures are the High Electron Mobility Transistors (HEMTs).

Today high electron mobility transistors are the most important analog electronic devices for the frequency range between 10 and 100 GHz. They achieved this position by their enormous versatility: HEMTs are used in low-noise as well as power circuits, offering outstanding performance in both applications. Particularly as low-noise devices, HEMTs are rivaled by no other technolgy. For power purposes, two additional types of III-V compound transistors might be considered , but they are potential competitors for the HEMT only in the lower part of its frequency range. The first alternative is the well established GaAs MESFET, usable up to approximately 30 GHz, but with inferior power gain. Recently, heterobipolar transistor (HBT) technology has established a second possibility. It is presently maturing rapidly, with prospective applications up to, say, 50 GHz. For higher frequencies, however, HEMTs will stay the only available solid-state power device for the foreseeable future.

Consequently, HEMTs are the only choice if a universal technology for microwave and millimeter wave ICs is required. This situation is reflected by the widespread use of HEMTs - particularly GaAs-based pseudomorphic HEMTs - in analog communication and information systems by the major companies in this field worldwide.

The lectures presented here look at HEMT device physics and epitaxial layer design from a practical point of view.

The technology used for fabrication of high performance 200 -100 nm gate length Pseudomorphic HEMTs (PsHEMTs) is examined. Most of the fabrication sequence is fairly standard and gives a good overall impression for the basic technological steps involved in the fabrication of modern PsHEMTs.

Different numerical techniques that are suitable for simulation of PsHEMTs are introduced, starting with a one-dimensional (1D) simulation of the vertical layer structure, based on a numerical solution of Poisson's equation. The carrier transport and the geometry effects are introduced later in two dimensional (2D) simulation of the whole device, based on a self consistent solution of Poisson's equation and the Boltzmann transport equation in drift-diffusion (DD) hydrodynamic (HD) or Monte Carlo (MC) approximation. The underlying physics and the limitations of the different approximations are discussed in relation to the PsHETM operation.

This volume represents a fairly broad coverage of different aspects of fundamentals and applications of devices based on low dimensional quantum structures.

M. Balkanski

Acknowledgments

The NATO Advanced Study Institute on "Devices Based on Low-Dimensional Semiconductor Quantum Structures", held in Sozopol, Bulgaria, September 19-30, 1995, was made possible by an award from the Assistant Secretary General for Scientific and Environmental Affairs. We are particularly grateful to Dr. J.A. Rausell-Colom, Program Director for Priority Area and High technology, for his constant interest and helpful attitude during the preparation of the ASI.

Of great value for the present success of this school and the future development of a Center for Scientific Culture in Bulgaria is the personal involvement of Dip. Eng. Peter Kimenov, General Manager of Administration Department of Bulbank and that of Dichko Fotev, Administrator of the Sozopol Pochivna Basa of Bulbank.

We owe special thank to the rector of Sofia University, Professor Ivan Lalov, who not only supported very generously the whole process of the organization of the school, but also came to Sozopol to attest with his presence his personal interest in the development of scientific culture in Bulgaria. Many colleagues from Sofia University have generously helped the organization of the school and we are grateful for their involvement.

We also wish to thanks Dr. Stoyan Tanev for giving so much of his energy and enthusiasm to the enterprise. Lucy Nedialkova has also generously helped the preparation of the School. Veneta Tuncheva, Michail Gyulmezov and Angel Issousov have contributed to the preparation of the book.

NATO ASI held in Sozopol, Bulgaria, September 19-30, 1995
DEVICES BASED ON LOW-DIMENSIONAL SEMICONDUCTOR STRUCTURES

Sponsored by
NATO Scientific and Environmental Affairs Division
Ministère de l'Enseignement Supérieur et de la Recherche, France
Université Pierre et Marie Curie, Paris, France
Ministry of Higher Education and Science, Sofia, Bulgaria
University St. Kliment Ohridski, Sofia, Bulgaria
Bulgarian Academy of Sciences, Sofia, Bulgaria
Institut des Hautes Etudes pour le developpement de la Culture, de la Science et de la Technologie en Bulgarie

Principal School Support
NATO Program for Priority on High Technology
Bulbank, Bulgarian Bank for Foreign Trade, Sofia, Bulgaria
Université Pierre et Marie Curie, Paris, France
Ministry of Higher Education and Science, Sofia, Bulgaria
University St. Kliment Ohridski, Sofia, Bulgaria
National Science Foundation, USA

SOLID SOURCE MOLECULAR BEAM EPITAXY

J.F. ROCHETTE
I.C.S.
c/o UMIST Ventures Ltd
PO Box 88, Manchester M60 1QD, England

1. Introduction

Molecular Beam Epitaxy (MBE) is a very powerful technique for growing thin layers of semiconductors, metals or insulators. The molecular or atomic beams of the constituent elements crystallize on a substrate maintained at an elevated temperature under ultra-high vacuum environment. The composition of the grown epilayer and its doping level depend directly on the evaporation rate of the elemental sources. The smoothness of the surface of the film is ensured by a relatively low growth rate, generally about 1μm per hour, allowing proper surface migration of the atomic species. The source beams can be interrupted very quickly by simple mechanical shutters so that changes in composition and doping can be abrupt on an atomic scale.

In comparison to other epitaxial techniques like liquid phase epitaxy or chemical vapor deposition, MBE has two important advantages :

(a) The growth of the film is carried out under conditions far from thermodynamic equilibrium and is governed by kinetics of the surface process occurring when the impinging beams react with the outermost atomic layers of the substrate crystal. Under proper conditions, this permits the deposition of very pure crystal layers at a much lower substrate temperature than what would have been possible under thermodynamic equilibrium conditions. MBE unique ability to grow quality layers at lower temperature than other epitaxial techniques is essential to achieve the deposition of very thin films with abrupt interfaces.

(b)The ultra-high vacuum environment allows the use of in situ surface sensitive diagnostic methods such as reflection high-energy electron diffraction (RHEED), Auger or ellipsometry. These powerful facilities for control and analysis eliminate much of the guesswork in the growth process and enable the fabrication of sophisticated structures reproducibly by MBE.

M. Balkanski (ed.), Devices Based on Low-Dimensional Semiconductor Structures, 1–16.

The maturity which the MBE technique has now achieved is reflected in the demand for high throughput, high yield machines. This chapter will review some of the most relevant developments in solid source MBE equipments and characterization, laying emphasis on the industrial aspect of this technique.

2. A Short History

Historically, MBE technology developed in response to the increasing attention paid by the semiconductor community to GaAs devices of increasing complexity and more generally to low dimensional structures. MBE has been first introduced by Arthur at Bell Lab's in 1969 and popularized by Cho and many other researchers during the next two decades in laboratories all over the industrial world. This technology was found very attractive because of the flexibility it offers and the ability to use in situ surface analysis. MBE has been perceived for a long time as a technology oriented towards fundamental studies and research rather than production. Recently many industrial companies have started to use intensively wafers grown by MBE in their own process lines for microwave applications. The perception of MBE has rapidly changed and is seen today as the ultimate technology to produce efficiently sophisticated structures for the electronic industry. This evolution has its origin in the remarkable development done on the equipment itself. The uniqueness of this technology is that the system can be easily scaled up without altering the intrinsic growth mechanism. It was possible to increase the size and the capacity of MBE system while keeping its important and unequaled characteristics.

The principal phases of industrial MBE development were :

- 1980 : introduction of the load lock chambers to facilitate the loading and unloading of grown samples without breaking the growth chamber vacuum. It was a major breakthrough that has increased considerably the quality of the grown material by reducing the carbon and oxygen contamination.

- 1985 : introduction of the so called "Indium free" substrate holder for 2" and 3" substrate diameter. This step has established MBE as a viable production system. In the past, substrates were bound on the molybdenum holder by melted Indium to ensure proper thermal conductivity. The back side of the wafer was irremediably damaged by this procedure and a non negligible contamination by indium migration was noticeable in the grown layers. After this date, all wafers were mechanically maintained and the back side of the substrate and its flatness was kept unchanged. In addition to a huge gain in productivity, the material grown was thus acceptable to industrial process lines where the aspect of the whole wafer is critically inspected.

- 1991 : Multi-wafer MBE systems are available from the three principal manufacturers (VG, Riber and Anelva). The design of these new machines is based on an extrapolation of existing smaller system, all 3-dimensions been increased. All multi-wafer systems have a vertical geometry with the cells pointing up at the substrates and the epi-side of the wafers looking down. The multi-wafer MBE is capable of growing five 3" wafers at the same time or three 4" wafers. These bigger machines show better uniformity, are very reliable and produce wafers with better overall characteristics. One important improvement is the noticeable reduction of the surface defects compared to single-wafer MBE.

- 1994 : Very large cells are designed to allow longer duration of the growth campaign. It is now possible to load 2 to 4 kg of Arsenic in only one cell, i.e. ten times more than in standard cells. Generally speaking, the cell design has been considerably improved recently and a lot of work is still carried on in that direction.

The availability of multi-wafer systems with large capacity cells has been the trigger to a wider usage of solid source MBE for industrial applications. This technique is now well accepted within the electronic companies and many new device developments are based on MBE material.

3. MBE System Description

3.1. GENERAL VIEW

The general design of a MBE system is shown on figure 1 and includes the following elements : the growth chamber, one or more load-lock chamber to allow easy entry/exit of substrates and a preparation chamber that is usually placed just before the growth chamber to preheat the substrates. All the different chambers are under ultra-high vacuum and linked together by a transfer mechanism that can load or unload the substrate holders. The chambers are isolated by large diameter gate valves. Nowadays, powerful computer and controllers are integral part of MBE systems and handle all the growth steps from the structure description to the transfer of the wafers. The enormous potential of MBE to produce structures of unequaled complexity can only be fully realized by utilizing effective automatic control. Stringent specifications for the accuracy and reproducibility of structures such as superlattices with layer thicknesses down to a few atomic layers necessarily involve the use of real-time computer control system.

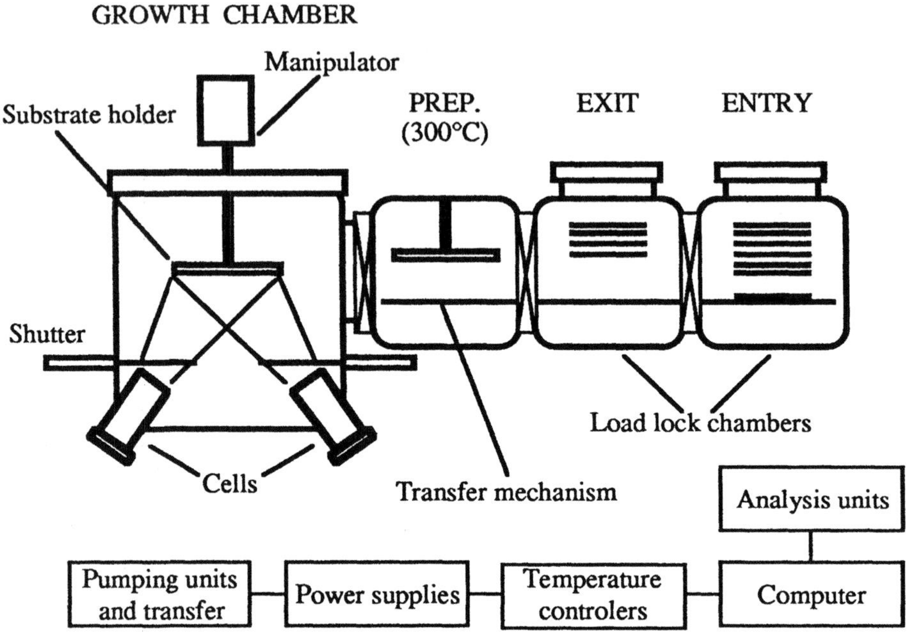

Figure 1 : View of a multi-wafer MBE system (from VG V-100 system design). The platen can accommodate 12 x 2",5 x 3" or 3 x 4" substrates.

3.2. ENTRY / EXIT CHAMBERS

The entry/exit stage consists of one or two small chambers that can be quickly pumped to a vacuum of about 10^{-8} Torr. A rapid entry lock allows cassette holder to be loaded or unloaded from the MBE machine. In multi-wafer systems eight to ten platens, each one loaded with five 3' or three 4' substrates can be introduced or removed during the operation. These simple chambers are usually pumped down by turbo-molecular pumps. All the chambers are linked together by a transfer system.

The transfer mechanism is a very crucial part of the MBE system as it is the only way to move wafers or platen from one chamber to the other. If it fails, there is no other option than opening the system for maintenance. It was the Achilles' heel of modern MBE for a while because the sophisticated mechanism has to work under high vacuum condition and support large temperature variation while transporting relatively heavy load very smoothly. The two methods that are generally employed are the magnetically coupled rods (Riber) or the wired trolley (VG). The reliability of these systems is now very satisfying.

3.3. GROWTH CHAMBER

The growth chamber is clearly the most important part of a MBE system. There are three central components : the cells with their shutters, the manipulator and the analysis units.

3.3.1. *The Cells*

The cells are the core of the MBE system. This component requires the best design skills, the ultimate purity in the material constituents and the most careful attention in mounting. The effusion cells are made of refractory materials (PBN, Mo, and Ta). The heating element is a large area foil heater which provides a very uniform and efficient radiant heating. Most of the cells have two independent heating coils to build a temperature gradient from the mouth of the cell (hotter part) to the bottom of the cell. The crucibles made from pyrolytic Boron Nitride (PBN) are tapered with an increase in diameter towards the mouth of the cell and incorporate a large flange plate at the top to prevent the substrate having a direct line of sight to any of the high temperature refractory metal components. The temperature is measured by one or more thermocouples located close to the source material.

Associated with each cell is a shutter made of refractory metal. In recent system design the shutter is usually controlled by a linear mechanism with a smooth, although fast, action. The position of the shutter should be close enough from the cell opening to prevent direct line of sight from the substrate but not too close to minimize the variation in the thermodynamic equilibrium of the cell when the shutter is opened.

In recent multi-wafer system the cell can be loaded with much larger quantity of source element as shown on table 1.

TABLE 1. Typical charge loading in Multi-wafer MBE. The capacity of R&D single wafer system is about ten times less.

Source	Nbr of Cells	Unit capacity (cc)	Total charge (g)
Ga	2	200	1 000
Al	1 (2)	200	200
In	1	200	300
As	2	500	3 000
Si	2 (1)	70	2

The number of growths between two reloading sequences can be increased, keeping the down time of the system operation as low as possible. Typically a multi-wafer system can grow between 1 500 and 2 000 3" wafers, during one campaign, depending on the structure.

Even with large effusion cells, the flux of atoms is following closely an Arrhenius law for limited temperature range of interest :

$$F = F_0 e^{-E_a / kT} \qquad 1$$

Where F is the measured flux arriving on the surface of the substrate, F_0 is a constant depending on the cell configuration and the source material and E_a is an energy constant depending on the source material itself.

3.3.2. *The Substrate Manipulator*

The manipulator is actually a substrate platen holder and heater rather than a real manipulator. In single wafer systems, the manipulator allows x-y-z and rotation motions for moving the substrate in front of the fluxes and allowing precise surface analysis. Today, the manipulator is a much more simple holder and heater which is rotating to improve uniformity. However the large quantity of calories that are produced to heat the substrates to the growth temperature, about 600°C, requires very efficient heat screening and careful design. The heating coil, slightly larger than the diameter of the substrate platen, can heat the substrate up to about 800°C. The manipulator is usually mounted with two thermocouples to measure the temperature close to the back of the wafer. This temperature measurement has always been and is still a difficulty in the MBE process. Because there is no direct contact with the surface, it is difficult to get the actual temperature of the surface of the wafer with a good accuracy. Fortunately, the RHEED transitions can be used very effectively to calibrate the manipulator thermocouples. A good manipulator should provide uniform heating to the substrates, a good confinement of the heat to prevent warming up of the wafer environment and a smooth rotation with accurate positioning to allow RHEED measurements.

3.3.3. *Analysis Units*

The three important analysis equipments in a MBE system are : the ion gauge, the mass spectrometer and the RHEED gun. The ion gauge is a very reliable method to measure accurately high vacuum in a range from 10^{-10} to 10^{-4} Torr. The vacuum background of MBE systems can go below 10^{-10} Torr but there is no real necessity to have an absolute value in such a high vacuum. Most of the growth process occurs in vacuum range from 10^{-8} to 10^{-6} Torr and the ion gauge is thus perfectly adapted. Usually, a ion gauge is mounted on a movable support that can be placed in direct view of the cells to measure the beam fluxes during the calibration procedure and retracted during the growth process. This method is straightforward and gives very good results in monitoring the growth parameters. Recent MBE systems offer the possibility to achieve this measurement automatically through a computer control procedure.

The quadrupole mass spectrometer is used mostly just after reloading the system with new sources or to check the quality of the residual background vacuum. It is not utilized during the growth itself. The mass spectrometer can accurately measure the constituent of the background like hydrogen, carbon oxide or dioxide, water or nitrogen which are naturally found in the air. It is a practical method to verify that there is no leak in the system that could degrade the growth quality. For instance, oxygen or carbon are very active impurities for semiconductor intrinsic quality.

The most interesting equipment is the RHEED because it gives the surface reconstruction of the substrate when the native oxide desorbs and also importantly the rate of atom deposition in the early stage of the growth. The surface reconstruction is necessary to start the deposition of the semiconductor and a detail study of the fundamentals of electron diffraction can be found in [1]. Let us say simply that the RHEED pattern can show As-stabilized (2x4) or Ga-stabilized (4x2) surface on (100) oriented GaAs substrates. The As-stabilized condition is usually chosen and gives a smooth, mirror-like surface. Recent work [2] on the RHEED oscillation has given MBE a considerable advantage because it is possible to monitor in real time the growth rate of the semiconductor on the substrate. The oscillations in RHEED intensity can be simply explain by the variation in the roughness of the surface during the completion of one monolayer. When the monolayer is finished, the surface is smooth and the reflection is maximum. When a new monolayer is starting to cover the smooth surface, the roughness increases as well as the diffusion, making the reflection of the RHEED beam less intense. At 50% atomic coverage, the surface gets smoother and the beam intensity increases until next maximum when the atomic layer is completed. Thus, a period in beam intensity oscillation corresponds exactly to the growth of one monolayer.

Very accurate growth rate measurements can be achieved for GaAs or AlAs on GaAs and InAs on InP for instance. The drawback of this simple and powerful technique is that the substrate has to be maintained along the same azimuth during the measurement to prevent intensity fluctuation of the RHEED beam. Because the substrate is not rotating, RHEED oscillation measurement cannot be performed during the growth sequence itself.

4. Fundamentals of MBE Growth and Doping

4.1. GROWTH CONDITIONS

MBE growth is a dynamic phenomenon. It is governed by the kinetics of atoms reacting at the surface and this kinetic process makes the thermodynamic equilibrium possible at a local level. Joyce [3] and

Madhukar [4] have given a comprehensive review of the different growth processes. The critical parameters for growing a crystal by MBE are : the temperature of the substrate, its crystalline orientation and the relative beam fluxes of the constituents

In addition, two important conditions have to be considered :

The first one is to ensure that the deposited crystal and the substrate crystal have the same or almost the same lattice parameter. If the lattice parameters are different, it is still possible to grow a film but with the limitation of the critical thickness [5]. The structure is then called a pseudomorphic growth. The figure 2 shows the deformation mechanism of a pseudomorphic layer. The energy that is accumulated in the layer can be relaxed by crystal dislocations when the thickness of the film exceeds the critical thickness of the material.

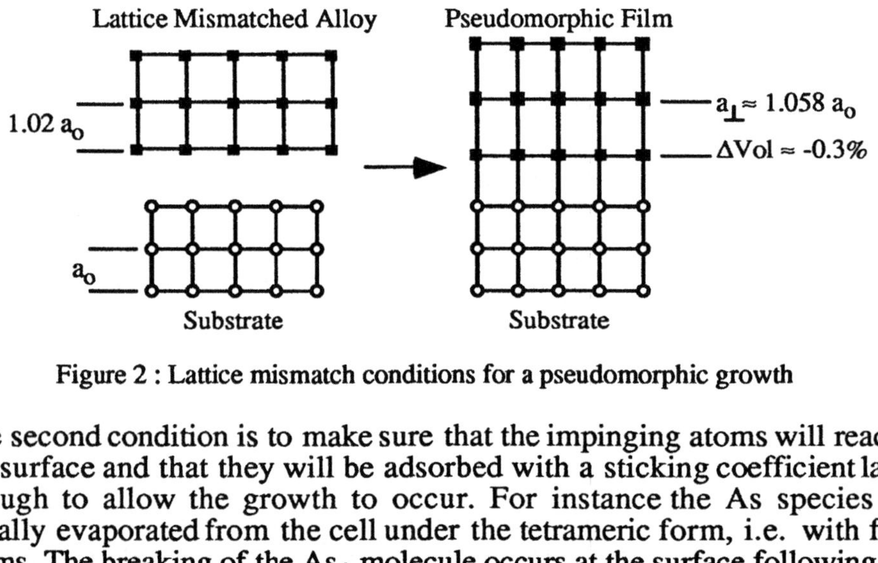

Figure 2 : Lattice mismatch conditions for a pseudomorphic growth

The second condition is to make sure that the impinging atoms will react at the surface and that they will be adsorbed with a sticking coefficient large enough to allow the growth to occur. For instance the As species are usually evaporated from the cell under the tetrameric form, i.e. with four atoms. The breaking of the As_4 molecule occurs at the surface following the figure 3 a) scheme. If a cracking cell is used, the breaking of the As_4 in As_2 is taking place into the hot zone of the cell leading to a different growth process with a better sticking coefficient as shown on figure 3 b).

The same kind of mechanism is observed for Phosphorus. This element has a sticking coefficient even smaller than Arsenic, making the growth of P-based material more tricky by solid source MBE. The sticking coefficient of atoms is strongly dependent on the substrate temperature, the nature of the surface, the orientation of the substrate and the co-incident beam. For instance, the sticking coefficient of As is close to zero on an As-stabilized surface at temperature above 200°C but is close to 0.5 between 400 and 600°C if a Ga beam is co-evaporated onto the surface.

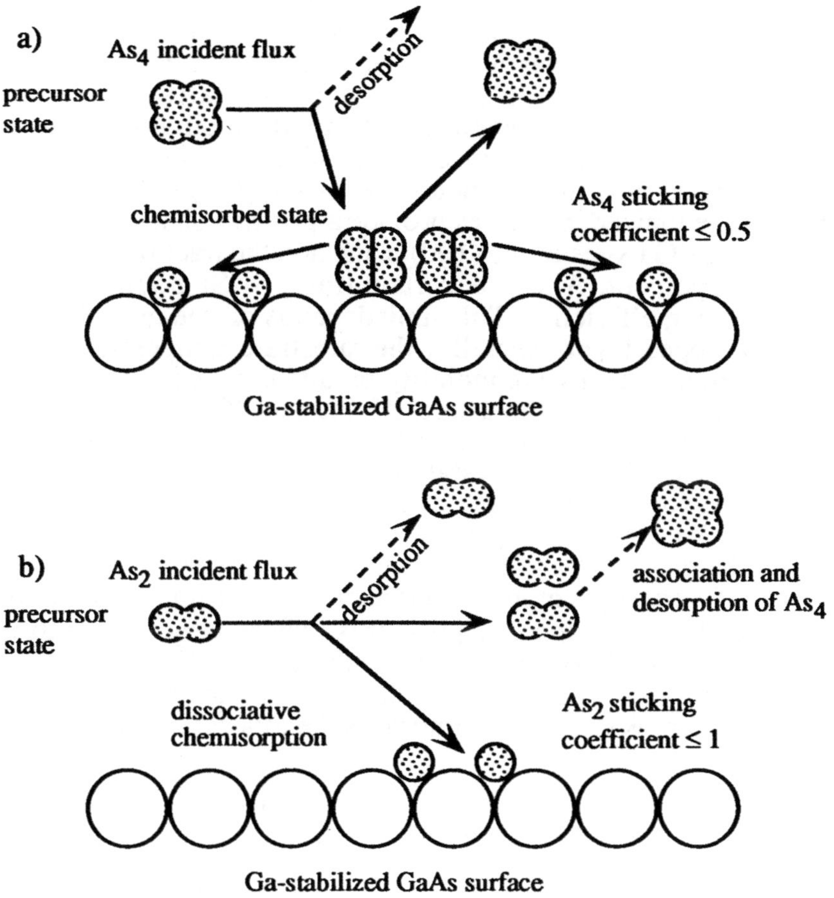

Figure 3 : Schematic illustration of the MBE growth models of GaAs on a GaAs(100) surface at a temperature around 580°C. (a) growth from Ga and As_4. (b) growth from Ga and As_2. [6]

4.2. DOPING PROCESS

Dopant incorporation is an important problem from the viewpoint of application of this growth technique for the fabrication of semiconductor devices. The intentional doping (as opposed to the unintentional or background impurities doping) of a semiconductor film can be achieved under the bulk form or delta-doping form. The bulk doping method consists in the co-deposition of doping atoms during the growth of the film. These dopants are usually Si and Be for n and p type doping in GaAs or InP. The doping efficiency is good with a dynamic range going from $1x10^{15}$ to $5x10^{18}$ cm^{-3} for Si and $1x10^{16}$ to $5x10^{19}$ cm^{-3} for Be. These values have to be compared to the residual background impurity in GaAs for instance which is about $5x0^{14}$ cm^{-3}.

Delta-doping (also known as atomic-plane doping or spike doping [7]) was originally proposed to improve the doping profile of Ge-doped n-type GaAs layers [8]. Later it was used with Si donors and Be acceptors [9] to generate symmetric V-shaped potential wells in GaAs with a quasi-2D electron (hole) gas and to generate a sawtooth doping superlattice. The basic concept of delta-doping in GaAs is illustrated in figure 4. The Si donors are located in an atomic monolayer of the (100)-oriented GaAs host material and the ionized impurities provide a positive sheet of charges. The Si atoms occupy Ga sites in the (100) plane and the fractional coverage can reach several percent. Above a certain limit, Si-Si bounds appear that reduce the doping efficiency. This limit depends on the growth conditions of the delta-doped layer. Usually, the substrate temperature is slightly reduced to prevent excessive mobility or diffusion of the Si atoms. To achieve this delta-doping, the group III shutters (Ga, Al, In) have to be closed and the group V shutter (As, P) remains open to maintain a group V stabilized surface while the crystal growth is stopped. Then the dopant shutter is open for a certain time interval, up to several minutes for high doping level. To resume the growth of the crystal, the dopant shutter is closed and the group III shutter opened again.This technique has allowed the growth of very high mobility heterostructures with 2DEG mobility in excess of $5x10^6$ cm^2/Vs at 4.2 K [10]. It is utilized very frequently now to grow structures for microwave devices with very small gate length because it allows better transfer efficiency of charges into the channel compared to the conventional bulk doping and improves the aspect ratio of the device.

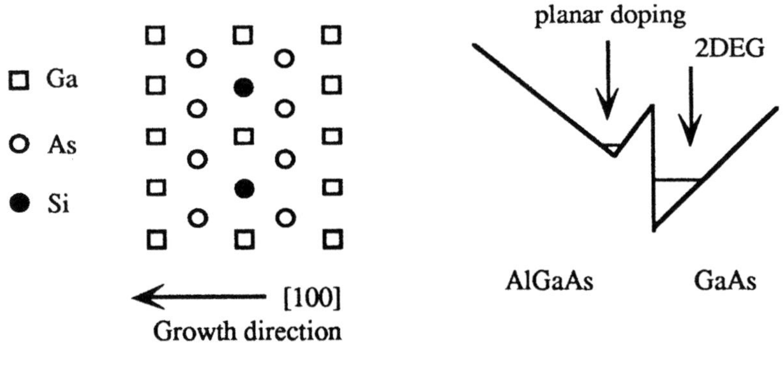

Figure 4 : Schematic illustration of delta-doping in GaAs. The Si donors are located in an atomic plane of the (100) host crystal (left). The conduction band of an AlGaAs/GaAs structure is given (right) : The electrons of the V-shaped potential formed by the delta-doping are transfered into the 2-dimension gas.

Once again, solid source MBE is the technique of choice to achieve this kind of doping with the most accurate control and the best reproducibility.

The figure 5 gives an indication of the doping efficiency of this technique for n and p-type doping in GaAs. A very good agreement of measured doping level with the number of supplied dopant atoms is obtained up to very high doping density.

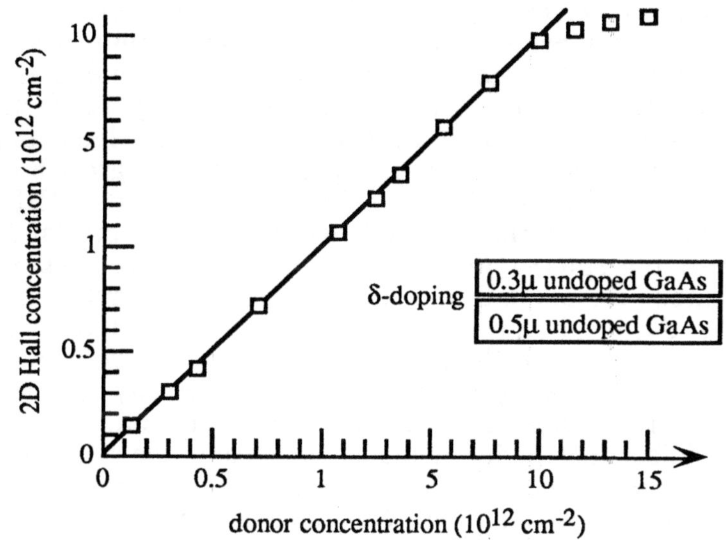

Figure 5 : 2-dimensional Hall concentration versus supplied dopant atoms in single n-type delta-doped GaAs. Doping efficiency decreases above $1x10^{13}$ at. cm^{-2}. [11]

5. Material Characteristics of Grown Materials

The material characteristics of grown epilayers are usually described by three different factors. The first one is the general aspect : smoothness, surface defect density, flatness. The second one is the structural aspect : thickness, alloy composition. The third one is the electrical and optical assessment of the grown structure. All three conditions have to be fulfilled to qualify the wafer for device processing. The Quality Control of the grown structure is a very critical part of the epitaxial process. However this discussion would go far beyond the scope of this paper. Only some basic aspects will be tackled to give a quick overview of the characteristics of MBE grown material : the surface defect density, the uniformity of thickness and doping and the electron mobility in a pseudomorphic heterojunction.

5.1. SURFACE DEFECT

Surface defects have been a real problem for a long time for MBE grown material. The progress made in designing the systems and more particularly the effusion cells have almost overcome this situation. Today, MBE grown wafers have surface density defects more than one order of magnitude lower than five years ago. Typical defect density (area $\geq 0.8\ \mu^2$) now ranges from 20 to 50 $/cm^2$ for a heterojunction structure of about 1μ thick. The defect density is even further reduced on large, multi-wafer systems because the distance between the cells and the substrate is larger. The defects at the surface of epilayer have many origins [12]. The factors which cause the surface defects are still controversial, and it is likely that there is not just a single cause but several, depending on the growth conditions employed. In opposition to CVD or LPE growth methods, MBE does not remove or alter the outermost monolayers of the host crystal during the initial step of the epitaxy process. Thus MBE is more sensitive to any pollution (carbon contamination, particulates) of the substrate surface which may happen during the preparation and the loading of the wafer. It is therefore essential to have all the substrate preparations done in very clean environment. The figure 6 give a statistic of defect density obtained over the last 10 years on typical MBE structures. There is a continuous decrease of the average defect density during the last ten years. the availability of new multi-wafer systems has accelerated the defect density improvement from 1992. The upper curve of figure 6 is an average over large number of wafers, the bottom curve shows the best values obtained. Today, MBE grown wafers have defect density very much comparable to CVD techniques. It is expected that improved design in effusion cells along with ultra-clean environment will permit further improvement in surface defects.

5.2. UNIFORMITY

The uniformity of structure parameters like thickness, alloy composition and doping level is essential to achieve high fabrication yields on device processing lines. MBE has made considerable progress and it is acknowledged as the epitaxial growth technique that produces the best overall uniformity and reproducibility for III-V and II-VI compounds. The figure 7 gives an example of the uniformity of the sheet resistivity (a function of doping density, thickness and carrier mobility) across a 3" diameter wafer grown on a R&D system and on a multi-wafer machine.Both epitaxial structures are identical. It is clear that the min. to max. and the standard deviation are better on the industrial system, although the results on the R&D MBE are very satisfying. The other advantage is that the gradient of uniformity is much smoother in the case of the multi-wafer. The uniformity improvement is a key factor to validate MBE as a reliable source of material for device foundries.

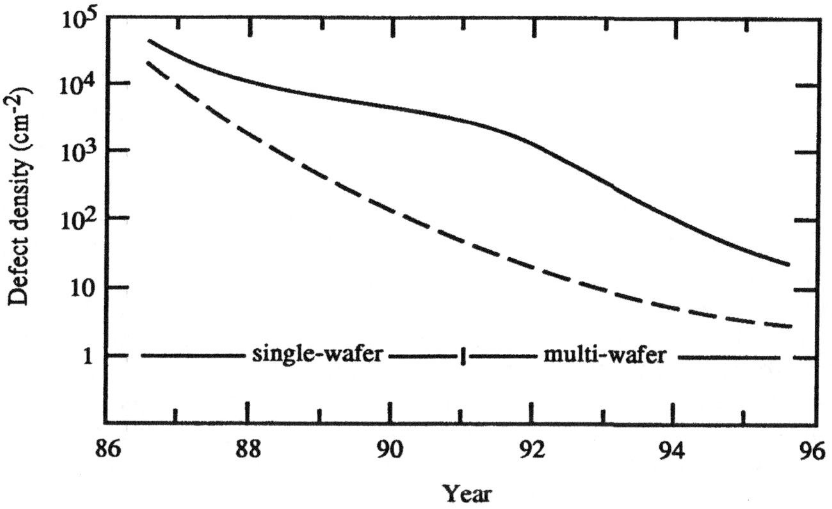

Figure 6 : Average surface defect density measured over a large number of wafers, different structures (solid curve) and best values obtained on some atypic wafers. Indication shows the beginning of production by multi-wafer MBE.

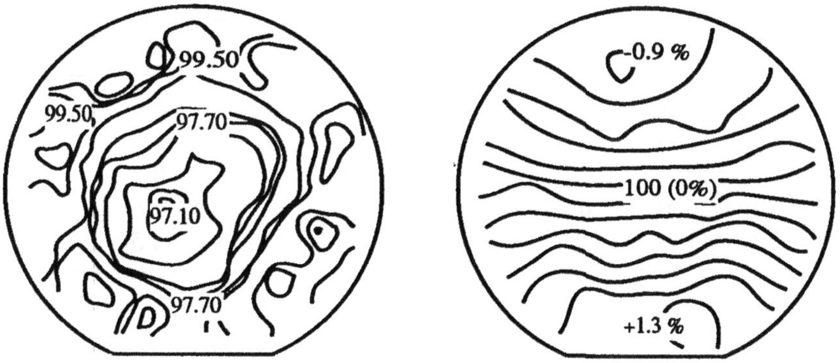

Average = 98.8 Ω / sq	Average = 100 Ω / sq
Std deviation = 1.6 Ω / sq (1.7%)	Std deviation = 0.66 Ω / sq (0.66%)
Range = 97.1 to 100.5 Ω / sq (3.5%)	Range = 98.8 to 101.2 Ω / sq (2.3%)

Figure 7 : Mapping measurement of the surface resistance across two 3" wafers grown on a single-wafer MBE system (left) and on a V-100 multi-wafer MBE machine (right).

5.3. ELECTRON MOBILITY

The other important parameter in material assessment is the quality of the crystal and the electronic properties that follow. The easiest and fastest way to check the transport properties is to perform Hall effect on a sample of the grown material. The drawback of this technique is that it is a destructive method. The table 2 gives an overview of the electron mobilities at 300 K achieved on typical heterostructure for AlGaAs HEMT on GaAs, AlGaAs/InGaAs pseudomorphic HEMT on GaAs with one and two Si planar doped layers, AlInAs/GaInAs HEMT and AlInAs/GaInAs pseudomorphic HEMT on InP. Because these structures all exhibit an· heterojunction with a charge transfer into the undoped channel, the Hall mobility shows an enhancement at low temperature due to the two-dimension electron gas [13]. This kind of structure based on a quantum effect is used very extensively in the micro-electronic industry to fabricate low noise or power devices for microwave and millimeter wave applications.

TABLE 2. Typical Hall mobility and sheet carrier density at room temperature for usual structures grown by MBE. The Pseudomorphic AlGaAs/InGaAs HEMT (P-HEMT) are delta-doped on one or both side of the InGaAs channel. The AlInAs/GaInAs HEMT and P-HEMT are grown on InP substrate.

Structure	Hall mobility (cm^2/Vs)	Sheet carrier density ($x10^{12}\ cm^{-2}$)
AlGaAs/GaAs HEMT	7 000	0.8
P-HEMT (1 side doped)	6 000	1.5
P-HEMT (2-side doped)	6 500	3.0
AlInAs/GaInAs HEMT	11 000	2.5
P-HEMT on InP	12 000	3.5

6. Outlook

The preceding paragraphs have presented a quick overview of MBE systems and some of the aspects of MBE growth method. Many of these have been skipped because of the difficulty to avoid such shortcomings in presenting a subjects so intensively investigated and so quickly developing as MBE. An exhaustive review of MBE fundamentals is given in [14]. In this presentation, the two applications comprising the growth of Si/Ge structures on Si substrates and the growth of HgCdTe and other II-VI compounds by MBE have not been investigated. The growth of III-V will probably remain the larger application of this growth technique. The industry needs will continue to guide the future development of solid source MBE. The two major directions are : a) increase in wafer size and/or wafer throughput and b) growth of new materials using the basic assembly of solid source MBE.

The increase of wafer capacity is dictated by the need to decrease the cost of epitaxial wafers. Today the cost of a 3" GaAs simple HEMT structure is about seven times more expensive than a ion implanted 6" Si wafer. Even if the competition with the silicon industry is not a realistic goal, the development of MBE technique will rely on its ability to follow market requirements. System manufacturers have the choice between two options : continue to scale up the dimension of existing systems or draw a different sustem architecture, similar to what has been done by the silicon industry with cluster design. The former option will increase the cost of MBE system because the size of the components (gate valves, manipulator, etc.) will go astronomical. However there should be no technological problem doing so. The latter option seems attractive because it opens a wide range of future developments. The risk is also very high because the new architecture, based on a cluster of small, specialized growth chambers grouped around an automatic transfer chamber, has to be entirely rethought. The most important problem will be to keep interfaces out of pollution during the transfer to the different growth chambers. Fortunately the need for 6" diameter epitaxial GaAs wafer is not immediate, giving the time to evaluate the different options and the opportunity for such a drastic change.

Closer to today's needs is the development of large effusion cells with better efficiency, i.e. a better usage of the material contained in the effusion cell. The most important development that is currently carried on is the production of very large arsenic cells that can accommodate many kilos of material in one big cylinder. A cracking stage allows the evaporation of dimeric instead of tetrameric arsenic, increasing greatly the sticking coefficient and as a consequence the efficiency of the cell.

The other alternative in solid source MBE development is the necessity of growing new materials like Phosphorus based alloys or having new doping sources. For instance Carbon is a good candidate for p-type doping to replace Be that has been used for a long time but shows a relatively strong diffusion coefficient in III-V alloys. This work will require the design of new effusion cells or components because P or C cannot be used directly in current MBE systems. Carbon evaporation needs cell temperatures greater than 2000°C and phosphorus has a sticking coefficient close to zero at current growth temperature. It is clear that cell design will concentrate a great effort in the future to overcome these problems.

Since its invention, MBE has played a major role in research and the preparation of heterojunctions and quantum well structures as well as surface physics understanding. During the last ten years, MBE has made a major breakthrough in production of sophisticated structures for low dimensional physics. It has become a mature technology that will continue to play a predominant role in microelectronic with applications tending to consumer markets like direct satellite TV and mobile phones. The competition with other growth technology is fierce and MBE has a lot of potential to offer new features which will make it the technology of choice in device fabrication.

References

1. Larsen, P.K. (1985) "RHEED and Photoemission Studies on Semiconductors Grown by MBE", in *Dynamical Phenomena at Surfaces, Interfaces and Superlattice*, Springer Ser. Surf. Sci., Vol. 3, Springer-Verlag, Berlin, p. 196
2. Neave, J.H., Joyce, B.A., Dobson, P.J., Norton, N. (1983) Appl. Phys. A **31**, 1
3. Joyce, B.A. (1985) Rep. Prog. Phys. **48**, 1637
4 Madhukar, A. (1983) Surf. Sci. **132**, 344
5. Ball, C.A.B. and van der Merwe, J.H. (1983) "The Growth of Dislocation-Free Layers", in *Dislocations in Solids*, Vol. 6, North-Holland, Amsterdam, Chap. 27
6 Foxon, C.T. (1978) Acta Electron. **21**, 139
7 Ploog, K. (1987) J. Cryst. Growth **81**, 304
8 Wood, C.E.C. (1985) "MBE III-V Compounds: Dopant Incorporation, Characteristics and Behaviour" in *Molecular Beam Epitaxy and Heterostructures*, NATO ASI Ser., Ser. E n° 87, Martinus Nijhoff, Dordrecht, p. 149
9 Ploog, K., Fisher, A., Shubert, E.F. (1986) Surf. Sci. **174**, 120
10 English, J.H., Gossard, A.C., Störmer, H.L., Baldwin, K.W. (1987) Appl. Phys. Lett. **50**, 1826
11 Ploog, K., Hauser, M., Fisher, A. (1988) Appl. Phys. A **45**, 233
12 Fujiwara, K., Nishikawa, Y., Tokuda, Y., Nakayama, T. (1986) Appl. Phys. Lett. **48**, 701
13 Störmer, H.L. (1983) Surf. Sci. **132**, 519
14 Herman, M.A., Sitter, H. (1989) *Molecular Beam Epitaxy* , Springer-Verlag Berlin

Using Gaseous Sources in Molecular Beam Epitaxy

C.W. Tu

Department of Electrical and Computer Engineering
University of California, San Diego
La Jolla, CA 92093-0407, U.S.A.

Using gaseous sources in molecular beam epitaxy (MBE) extends the versatility of solid-source MBE. We first describe the use of gaseous group V hydrides, which makes the growth of phosphides possible (before the very recent development of valved crackers). New issues in growth of phosphides arise, however; for example, As/P exchange at arsenide/phosphide interfaces and controlling the group V composition in a mixed As+P alloy. Then we discuss the use of gaseous organometallic group III sources in selective-area epitaxy, either by laser irradiation or on patterned substrates. Finally the use of gaseous dopant sources are described, in particular, carbon doping with halomethanes.

1. INTRODUCTION

The success of any technology depends crucially on the quality of the materials in which components are fabricated. High-performance electronic, optoelectronic, and photonic devices usually involve low-dimensional semiconductor heterostructure layers, quite often complex, and they have to be grown by advanced thin-film growth techniques, such as chemical vapor deposition (CVD)/vapor phase epitaxy (VPE) or molecular beam epitaxy (MBE). The vapor-phase technique includes hydride, chloride, and metal-organic VPE (MOVPE), and the beam techniques include solid-source MBE (SSMBE), gas-source MBE (GSMBE), and metal-organic MBE (MOMBE), which is also called chemical beam epitaxy (CBE). This chapter deals with the more recent development of MBE with gaseous sources, i.e., GSMBE and MOMBE/CBE. Fig. 1 summarizes these variants of MBE according to the sources used, and shows their relations to each other and to the CVD/VPE techniques according to the operating pressure[1].

MBE is basically a finely controlled evaporation process, which is carried out in a stainless-steel ultrahigh vacuum (UHV) chamber with a base pressure in the 10^{-10} - 10^{-11} Torr range to minimize impurity incorporation. To keep the growth chamber in UHV, a loadlock and a buffer chamber are usually connected to it. This very versatile thin-film growth technique was pioneered by Cho and Arthur in the late 60's[2]. It has been proven to be capable of controlling layer thickness and composition precisely, and of growing uniform, ultra-thin layers with abrupt interfaces. Many books on MBE have been published[3,4,5,6,7,8,9], including one by Panish and Temkin on MBE with gaseous sources[8].

Figure 2 shows a schematic diagram of an MBE growth chamber with either solid or gaseous sources[10]. Cryopump, turbomolecular pump and diffusion pump all can be used. The MBE process is conceptually very simple. Conventional solid-source MBE has been extensively studied and used for growing various arsenide heterostructures, e.g., GaAs/$Al_xGa_{1-x}As$, GaAs/$In_xGa_{1-x}As$, and $In_xGa_{1-x}As/In_yAl_{1-y}As$. The growth of mixed-group V compounds, such as $In_xGa_{1-x}As_yP_{1-y}$, is important for long-wavelength optical communication, but it can not be achieved easily and controllably by conventional solid-source MBE due to the high vapor pressures

M. Balkanski (ed.), Devices Based on Low-Dimensional Semiconductor Structures, 17–36.

of solid arsenic and phosphorus. Furthermore, red phosphorus is allotropic, and random allotrope transformation results in unstable beam flux output. Therefore, gaseous sources, such as phosphine and arsine, through a high-temperature cracker, were first used by Panish[11] and Calawa[12], respectively, to obtain a well controlled flux of phosphorus and arsenic species, mainly P_2 and As_2. This technique is called gas-source MBE. Since the group III elements are derived from solid sources, as in MBE, the growth mechanism is similar. In normal operation the arsine and phosphine bottles are open only for a few seconds to charge the tubing between the bottle and the first valve, and the gas in this space is enough for growing several microns of materials. The bottles are closed most of the time, making gas-source MBE much safer than vapor phase techniques, where much of arsine and phosphine, continuously flowing through the reactor, are unused and vented.

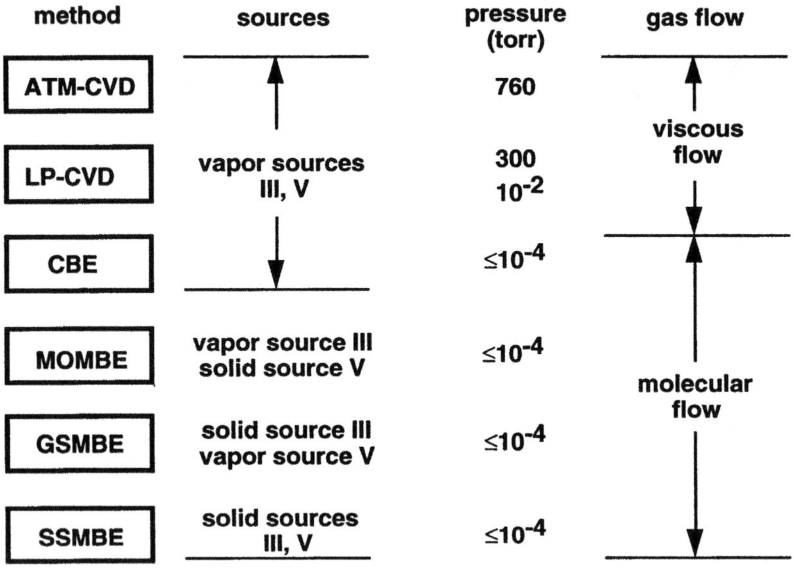

Fig. 1 Comparison of the beam techniques and the vapor-phase techniques.[1]

Recently a breakthrough in the solid-source MBE technology is the development of a valved cracker with three temperature zones (oven, reservoir and cracking zone) and a needle valve for controlling the P_4 flux into the cracking zone[13,14]. The environmentally safe red phosphorus P_4 is first heated in the oven section to about 360 °C, and the P_4 vapor is condensed on a cooled surface (room temperature) in the reservoir as white phosphorus P_4. The valve of the cell is closed during this *in situ* generation of white phosphorus. The third temperature zone is the cracking cell to dissociate P_4 at 920 °C to form dimers P_2, which has higher reactivity than P_4. Since this development is very recent, most of the reported work on phosphides grown by the beam techniques is still related to the use of gaseous group V sources.

Veuhoff *et al.*[15] first reported the replacement of the group III elemental sources by gaseous metal-organic compounds, such as trimethylgallium (TMGa), triethylgallium (TEGa), and trimethylindium (TMIn). In their experiment they used cracked arsine and phosphine as well, and they called their technique MOMBE. However, Tsang *et al.* called the all-gaseous approach CBE and demonstrated that high-quality InP and related compounds could be grown and high-performance devices could be achieved[16]. The setup of MOMBE/CBE is more complicated, involving the expensive stainless steel UHV system of MBE and the complicated gas manifold of MOVPE, but it offers advantages in lower defect density, uniformity, and selective-area growth, either on a patterned substrate or with the use of an external energy source.

Gaseous dopant sources can also be used in MBE to advantage. Particularly useful are halomethanes that are used for *p*-type carbon doping in GaAs and $In_xGa_{1-x}As$.

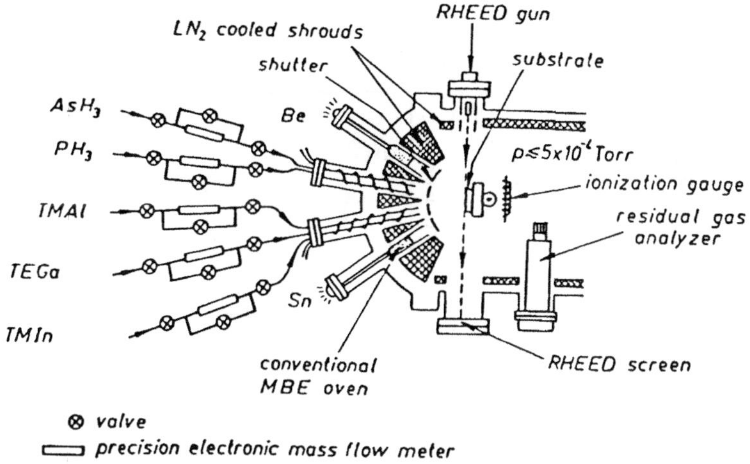

Fig. 2 A schematic of an MBE system with solid and gaseous sources[10]

In the following, we first describe the issues in using only gaseous group V sources, i.e., GSMBE. Since the main reason for using them is to grow phosphides, the issues of controlling arsenic and phosphorus flux or incorporation rate, As/P interdiffusion at an interface, and mixed group V compounds are discussed. Then we shall discuss selective-area growth in CBE, either by argon-ion-laser irradiation or on patterned substrates. Finally we describe the use of gaseous dopant sources, in particular, halomethanes for carbon doping in GaAs and $In_xGa_{1-x}As$.

2. Gaseous Group V Sources

2.1 Group V incorporation rate

Figure 3 shows the As incorporation rate in InAs as a function of the AsH_3 flow rate calibrated at the substrate temperature of 460 °C.[17] It appears that the As incorporation rate is proportional to the AsH_3 flow rate. This incorporation rate also has an Arrhenius dependence on the substrate temperature.[18] The inset of Fig. 3 shows group III- ($t_1<t<t_2$) and group V-induced ($t_3<t<t_4$) reflection high-energy electron diffraction (RHEED) oscillations during InAs growth on InAs. The InAs was grown on a 3 μm-thick, totally strain-relaxed InAs layer on InP. During $t_1<t<t_2$, the growth rate was limited by the arrival rate of the In flux as normally applied. However, the growth rate was limited by the As arrival rate during $t_3<t<t_4$. Therefore, the incorporation rates of In and As, R_{In} and R_{As}, at the given growth temperature can be determined from these oscillations.

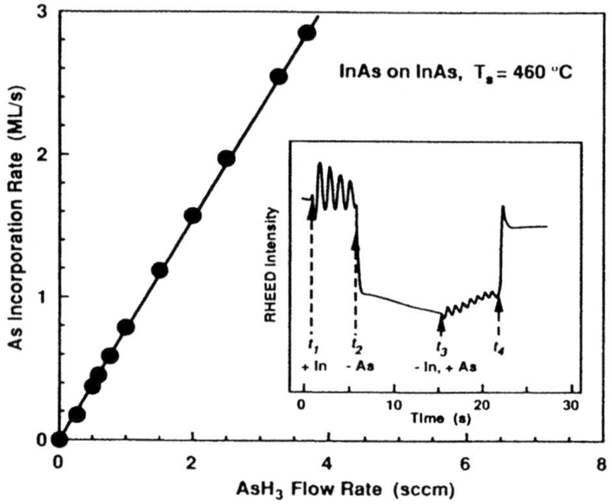

Fig. 3 The As incorporation rate as function of the AsH_3 flow rate calibrated at the substrate temperature of 460 °C for InAs grown on a thick InAs layer. The inset shows RHEED oscillations induced by In ($t_1<t<t_2$) and As ($t_3<t<t_4$).[17]

Figure 4 shows how one can establish the V/III incorporation ratio, an important growth parameter that determines the material quality. In Fig. 4 the growth rates of (a) GaP, (b) AlP, and (c) InP are plotted as a function of the reciprocal group III cell temperatures (T_{III}).[18] The phosphorus-controlled growth rates, as measured by the procedure just mentioned, are shown with thick lines. They are independent of group-III fluxes. With the phosphine flow rate fixed, the group III-controlled growth rate is measured under various group III fluxes. As the open circles show in Fig. 4(c), the growth rate increases exponentially with an increasing In cell temperature, i.e., it increases linearly as $1/T_{III}$ decreases on the semilog plot. This linear dependence stops at a certain point, and then the growth rate becomes constant even as the In flux increases further. There are now more In atoms than available P atoms in the P beam flux that can be incorporated. The growth rate in the plateau region is therefore limited by the phosphine flow rate, independent of the In flux. The point at which the growth rate saturates indicates a unity V/III incorporation ratio. It should be emphasized that this approach of studying group V-controlled growth rate is for calibrating the group-V flux and establishing the V/III ratio. Of course, during GSMBE of a heterostructure, a V/III flux incorporation ratio >1 should be used to prevent the formation of a hazy surface from In or Ga droplets. It is found that samples with the highest purity should be grown as close to stoichiometry as possible[19]. The best InP layer with 77K mobility of 112,000 cm^2/V-sec and background carrier concentration N_d-N_a of $2x10^{14}$ cm^{-3} have been achieved[20].

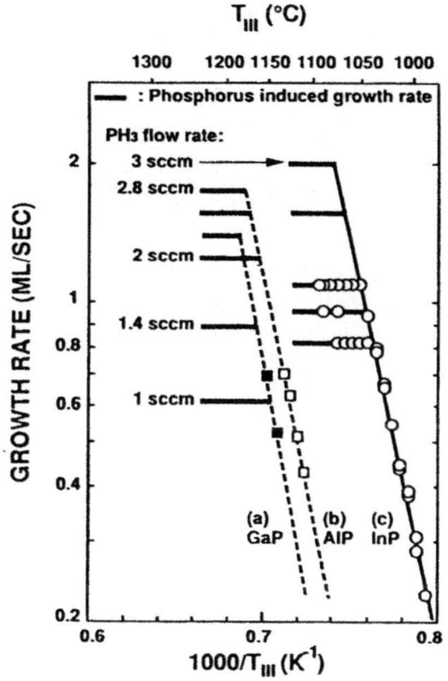

Fig. 4 Growth rate vs. reciprocal group III cell temperatures for (a) GaP, (b) AlP, and (c) InP.[18]

2.2. As/P Interfaces

Many low-dimensional structures and devices involve arsenide and phosphide in adjacent layers, e.g., GaAs/$In_xGa_{1-x}P$, $In_xGa_{1-x}As$/InP, etc. An important issue is the possibility of degraded arsenide/phosphide interfaces. Two types of problems can be present during growth: memory effect[21,22,23,24] and As/P exchange[25,26,27,28]. The memory effect occurs when residual As or P atoms from the previous layer are incorporated into the subsequent layer, forming a compositionally graded interfacial layer. This effect can be reduced by first pumping out the residual group V gas and then supplying the new group V flux during growth interruption. However, switching fluxes will cause As/P exchange: the new group V element will replace the old one in the underlying layer, leading to a compositionally graded and strained transition layer. Such interfacial layers deform the quantum well (QW) shape[29]; in severe cases, the interfaces become rough and cause increased non-radiative recombination. For our discussion here, we shall use the $In_xGa_{1-x}As/In_yGa_{1-y}P$ as an example since $In_xGa_{1-x}P$ is an attractive alternative to $Al_xGa_{1-x}As$ in a pseudomorphic high-electron-mobility transistor (PHEMT) structure.

Like a GaAs surface, the $In_{0.3}Ga_{0.7}As$ surface at the normal growth temperature (520°C) shows an initial (2 × 4) reconstruction with a streaky RHEED pattern, which gradually degrades to a spotty pattern after 40 seconds of P_2 flux exposure. For a P_2-exposed $In_{0.53}Ga_{0.47}As$ surface at the same temperature, the spotty pattern appeared quickly, within 10 seconds, as shown in Fig. 5.[30]

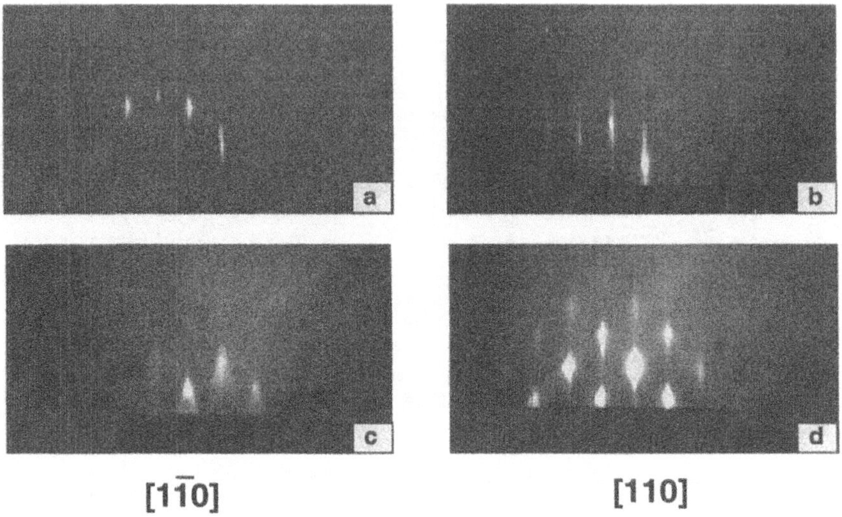

Fig. 5 RHEED patterns of an $In_{0.53}Ga_{0.47}As$ (on InP) surface exposed to a flux of As_2 and P_2 at T_s=520 °C. (a) and (b) under an As_2 flux; (c) and (d) under P_2 flux for 10 seconds. (a) and (c) are along the [$1\bar{1}0$] azimuth; (b) and (d) are along the [110] azimuth.[30]

On the other hand, at 400°C, the streaky pattern can last more than 10 minutes. For an As_2-exposed $In_{0.48}Ga_{0.52}P$ surface, the streaky pattern can last more than 10 minutes at 520°C. Figure 6 plots the critical P_2 exposure time as a function of substrate temperature (T_S) for a GaAs and an $In_{0.53}Ga_{0.47}As$ surface under a phosphine flow rate of 3.2 sccm. The figure shows clearly that surface roughening, indicated by a spotty RHEED pattern, depends on both T_S and P_2 exposure time, and an $In_xGa_{1-x}As$ surface is less stable than a GaAs surface under a P_2 flux at the same substrate temperature.

The As/P substitution at a heterointerface must be prevented in order to achieve high-quality heterostructures. One technique is to reduce the surface exposure time under the new group V flux during growth interruption, but often several seconds are still needed to stabilize the group V flow rate and to reduce the memory effect. A more effective way to prohibit As/P exchange is to grow at relatively low temperatures.[31,32] In the $In_xGa_{1-x}As/In_yGa_{1-y}P$ system, the T_s should be less than 400°C to ensure the surface stability for both $In_xGa_{1-x}As$ and $In_yGa_{1-y}P$. The quality of materials grown at below their optimal growth temperature can be usually improved by *ex situ* annealing. Figure 7 shows low-temperature photoluminescence (PL) spectra of a 10-period strain-compensated $In_{0.2}Ga_{0.8}As$ (90 Å)/$In_{0.3}Ga_{0.7}P$ (150 Å) multiple quantum well (MQW) structure grown at 350 °C before and after annealing.[29]

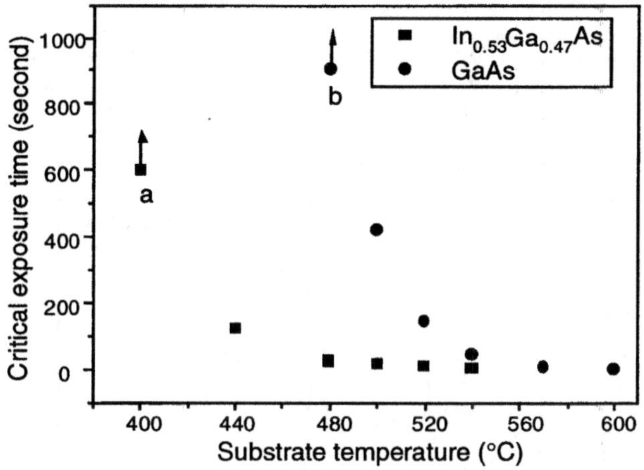

Fig. 6 Critical exposure time (P_2 flux exposure time until RHEED patterns change from being streaky to being spotty) on GaAs and $In_{0.53}Ga_{0.47}As$ under a P flux as a function of substrate temperature T_s. At points a and b, the pattern is still streaky.[30]

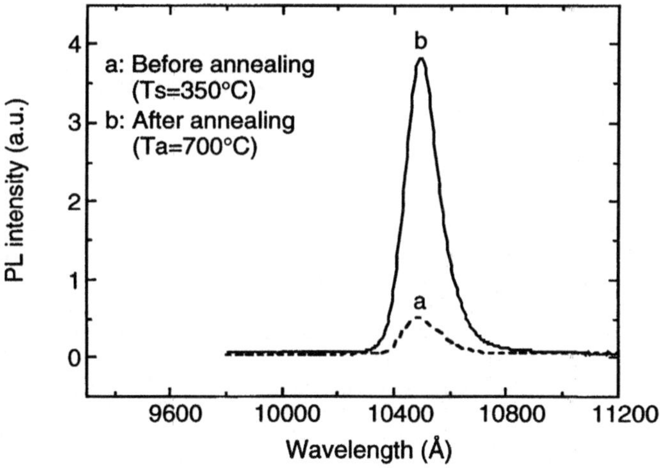

Fig. 7 10K PL spectra of a 10-period $In_{0.2}Ga_{0.8}As$ (90 Å)/$In_{0.3}Ga_{0.7}P$ (150 Å) MQW grown at 350 °C before and after annealing at 700 °C for 20 seconds.[30]

An third alternate method which ensures an abrupt interface as well as high crystal quality is to protect the $In_xGa_{1-x}As$ surface with a more stable material.[33] A protective layer must have higher stability and must not introduce extra strain into the whole stack. A thin GaAs layer meets all these requirements and this is an ideal protective layer for $In_xGa_{1-x}As$ grown on a GaAs substrate.[34] Figure 8 shows low-temperature PL spectra of an $In_{0.4}Ga_{0.6}P/In_{0.2}Ga_{0.8}As/In_{0.4}Ga_{0.6}P$ SQW with and without a 20 Å GaAs between $In_{0.2}Ga_{0.8}As$ and the upper $In_{0.4}Ga_{0.6}P$ layer. With the GaAs protective layer the PL is narrower and the intensity is almost twice as strong. We conclude that a thin GaAs protective layer is another viable approach and can be possibly better than the low T_s-and-annealing method.

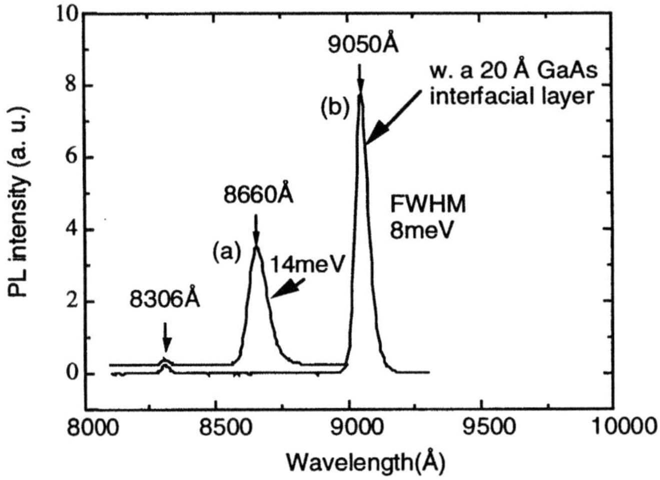

Fig. 8 10K PL spectra of (a) an $In_{0.4}Ga_{0.6}P$ (50 Å)/$In_{0.2}Ga_{0.8}As$ (50 Å)/$In_{0.4}Ga_{0.4}P$ (50 Å) SQW; and (b) an $In_{0.4}Ga_{0.6}P$ (50 Å)/GaAs (20 Å)/$In_{0.2}Ga_{0.8}As$ (50 Å)/$In_{0.4}Ga_{0.6}P$ (50 Å) SQW. Both were grown at 500 °C.[34]

2.3. Mixed Group V Compounds

Another advantage of gas-source MBE is the ability to grow mixed As+P compounds. We and others have shown previously that the ternary compound $InAs_xP_{1-x}$ is a viable alternative to the conventional quaternary compound InGaAsP for 1.3 μm modulator and laser application.[35,36,37,38] The growth rate, or the thickness, is controlled by the In flux in MBE, and the wavelength, or the composition, is determined *independently* by the arsine-to-phosphine flow-rate ratio.[39] Both parameters can be determined *in situ* by monitoring the group III- and group V-induced intensity oscillations of reflection high-energy electron diffraction (RHEED),[18] so the material composition and thickness are easily more reproducible. Once the As and In incorporation rates, R_{As} and R_{In}, respectively, at a desired growth temperature are calibrated by As- and In-induced RHEED oscillations, respectively, Hou and Tu show that when the ratio R_{As}/R_{In} is less than unity, all of As_2 dimers react with In atoms on the surface and the remaining In atoms react with P_2 dimers. The excess P_2 molecules desorb from the surface. Thus, the As composition x in $InAs_xP_{1-x}$ is just the ratio R_{As}/R_{In}[39]. Based on this idea of *in situ* controlling the As composition in $InAs_xP_{1-x}$, a series of $InAs_xP_{1-x}$/InP strained-layer superlattices (SLS's) were grown at 2 sccm

PH_3 flow rates and different R_{As}/R_{In} ratios. Shown in Fig. 9 is the *ex situ* determined As composition, by x-ray rocking curves of SLS's, versus the R_{As}/R_{In} ratio. An excellent agreement can be seen when $x<0.5$. The discrepancy for $x>0.5$ is similar to the case of $GaAs_{1-x}P_x$ discussed above. A possible reason is that the large surface strain may affect the incorporation behavior of As and P. Another possibility is that partial strain relaxation in $InAs_xP_{1-x}$/InP SLS samples may lead to an underestimate of the As composition.

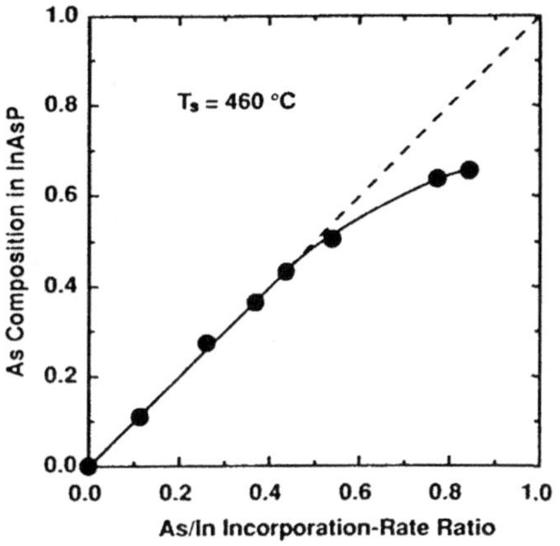

Fig. 9 Ex situ determined As composition in InAsP from X-ray rocking curves vs. in situ determined As composition from RHEED measurements.[17]

Of course, there is an upper limit to the P_2 flux for this simple in situ method of composition determination. When the PH_3 flow rate is much higher than the AsH_3 flow rate (by about a factor of five), more P_2 will be incorporated. As a result, the As composition in InAsP would start to decrease as the P_2 flux increases[39]. Therefore, to keep this simple *in situ* method valid, it is important to use as small a V/III incorporation ratio as possible. Again, one should note that at higher substrate temperatures desorption of As_2 and P_2 is significant and the incorporation of one species can be influenced easily by the presence of the other. Therefore, the *in situ* method is applicable only for substrate temperatures lower than about 500°C[39].

Since we are interested in multiple-quantum-well (MQW) light modulators based on the quantum confined Stark effect (QCSE), a relatively large number of quantum well periods (>20) is desirable for either a normal-incidence or a waveguide modulator.[40] Since the $InAs_xP_{1-x}$ quantum well layers ($x = 0.4$ for 1.3 μm and a strain of 1.3%), grown on InP, are under compression, the whole $InAs_xP_{1-x}$/InP MQW structure has a large net strain and the number of MQW period is limited by the onset of strain relaxation. Figure 10 shows double-crystal rocking curves of $InAs_{0.4}P_{0.6}$/InP (90 Å/130 Å) MQWs with 11, 7 and 5 periods, corresponding to curves (a), (b), and (c), respectively. Curve (d) is a simulation, based on the dynamical theory, of curve (c). It is clear that the lattice already starts to relax for the 7-period MQW as the superlattice satellite peaks have become slightly broader. For the 11-period MQW, the superlattice satellite peaks are hardly present, indicating almost complete lattice relaxation.

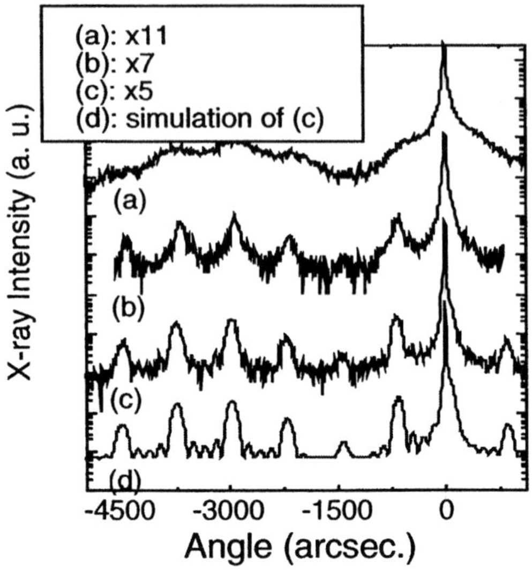

Fig. 10 Double-crystal X-ray rocking curves of strain-uncompensated $InAs_{0.4}P_{0.6}/InP$ (90 Å/130 Å) MQW structures with (a) 11, (b) 7, and (c) 5 periods. (d) is a simulation of (c) based on the dynamical theory.[41]

If we use a $Ga_yIn_{1-y}P$ barrier layer, which has a tensile strain, a strain-compensated MQW structure with a large number of periods can be obtained. The Ga composition y and the thickness of the $Ga_yIn_{1-y}P$ layer are design variables. The critical number of periods increases rapidly with increasing Ga composition until when the net strain is zero. We have grown $InAs_{0.4}P_{0.6}/Ga_{0.17}In_{0.83}P$ (86 Å/120 Å) MQWs up to 50 periods without degradation of crystallinity. Figure 11 shows an X-ray rocking curve and a simulation for a 30-period MQW, which shows sharp satellite peaks on both sides of the substrate peak, indicating excellent crystallinity.[41] The zeroth -order peak is hidden in the substrate peak, indicating excellent a nearly perfect strain compensation.

Figure 12 shows surface-normal electroabsorption measurements at room temperature for a 21-period $InAs_{0.4}P_{0.6}/Ga_{0.13}In_{0.87}P$ (93 Å/130 Å) MQW, sandwiched between n^+ and p^+ contact layers, under reverse bias. A beam of monochromatized light was focused onto the center window of the ring diode, and a photodetector was laced on the backside of the sample. The reverse electric bias was applied via the ring-shaped electrode and the back contact. Excellent QCSE is obtained. At zero bias the half width at half maximum (HWHM) of the $n = 1$ electron-heavy hole exciton peak is as small as 6 meV, which is excellent for this highly strained material system. Besides strain-compensation the $Ga_yIn_{1-y}P$ barrier layer has an added benefit of larger conduction and valence band offsets with respect to InAsP, as compared to InP/InAsP. As a result, the exciton peak is still well defined under high electric field (a HWHM of 8.4 meV at a field of 1.6×10^5 V/cm). The details of the analysis of QCSE are presented elsewhere.[42]

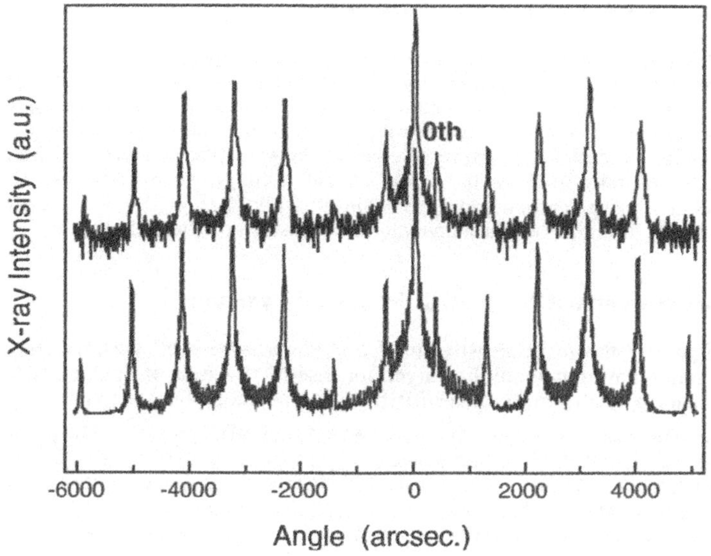

Fig. 11 X-ray rocking curve (top) of a strain-compensated 21-period $InAs_{0.4}P_{0.6}/Ga_{0.13}In_{0.87}P$ (93 Å/135 Å) MQW structure, and the bottom curve is a simulation based on the dynamical theory.[42]

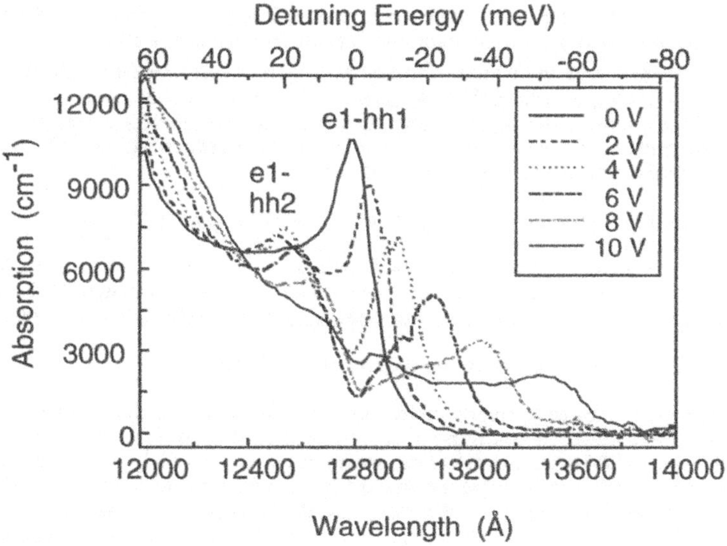

Fig. 12 Room-temperature surface-normal electroabsorption spectra of the waveguide modulator structure.[42]

3. GASEOUS GROUP III SOURCES

Using all gaseous sources in metal-organic MBE (MOMBE) or chemical beam epitaxy (CBE) has additional advantages of low thermal budget, replacing sources without opening the growth chamber, and selective-area epitaxy by argon ion laser irradiation or on oxide- or nitride-patterned substrates. We have also investigated the use of safer, alternative precursors to arsine and phosphine, in particular, tris-dimethylamino arsenic (TDMAAs)[43]. TDMAAs has no As-H nor As-C bonds, only As-N bonds. GaAs samples were grown with triethylgallium (TEGa) and TDMAAs, and *p*-type carbon doping is achieved with CBr_4 or CI_2H_2.[44] Selective-area epitaxy can be achieved by either laser irradiation or on an insulator-patterned substrate.

3.1 Selective-area CBE growth by laser irradiation

Figure 13 shows the growth rate of GaAs as a function of substrate temperature, with and without irradiation from a multiline argon ion laser.[45] Because of the long distance between the substrate and the viewport in a CBE/MBE growth chamber, the laser spot was focused to about 300 μm. The laser power density used was about 400 W/cm^2. The growth rate data are normalized to that at the high-temperature region (between 480 °C and 510 °C), where TEGa decomposition is complete. The substantial differences between laser-assisted growth using TDMAAs and AsH_3[46] or As_4[47] are that the threshold temperature for enhanced growth is about 100 °C lower and that the temperature range is about 100 °C wider. Furthermore, there are two different substrate-temperature regions for laser-enhanced growth, 340 to 440 °C and 265 to 340 °C.[48] The former is believed to be due to laser-induced heating of about 20-25 °C because the activation energy in this temperature range is similar to that without laser irradiation, whereas the latter indicates the effect of photochemical reaction at the surface.

Fig. 14 shows the hole concentration as a function of growth temperature, with and without laser irradiation. Carbon incorporation decreases as the substrate temperature increases because of the increased production of atomic hydrogen through the β-hydride elimination process of TDMAAs on a GaAs surface.[49] With laser irradiation, the hole concentration increases, despite a laser-induced temperature rise, indicating photochemical effects. It is interesting to note that there can be two orders of magnitude difference in the hole concentration with and without laser irradiation, which may have potential applications.

When laser is irradiated during growth of InGaAs in a pseudomorphic $In_xGa_{1-x}As$/GaAs quantum well structure at a normal substrate temperature (~500 °C), TEGa molecules apparently are desorbed, resulting in selective-area composition modification of the $In_xGa_{1-x}As$ quantum well layer. There is therefore a lateral variation of exciton emission wavelength.[50] Yamada *et al.* have used this kind of behavior to fabricate a dual-wavelength InP/InGaAsP laser diode.[51]

3.2 Selective-area regrowth on patterned substrates

Many devices can be improved by selective-area regrowth on etched patterned substrates, e.g., as a current and light confining area in laser diodes or as an external base layer in HBTs. With TEGa and TDMAAs, true selective growth of GaAs:C on SiO_2-patterned GaAs, i.e., growth only on the openings, can be achieved above 470 °C, which is much lower than the 600 °C in the case of using TEGa and arsine.[52] With uncracked TDMAAs, the surface morphology is always textured and the carbon incorporation efficiency is low. Thermally cracking TDMAAs can produce a specular surface morphology and a much higher carbon incorporation efficiency. In Fig. 15 (a) the growth temperature is too low such that there is polycrystalline growth on the SiO_2 surface, and in Fig. 15 (b) true selectivity is achieved at 540 °C with good surface morphology.[53]

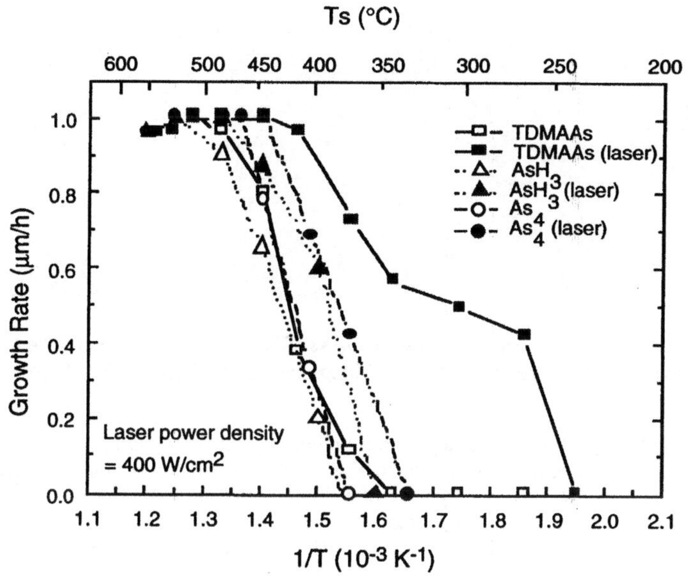

Fig. 13 The substrate temperature dependence of growth rates of both laser-irradiated and unirradiated areas. The solid lines denote the growth data obtained using TDMAAs, while the dashed and dotted lines denote the data from AsH_3 and As_4, respectively.[45]

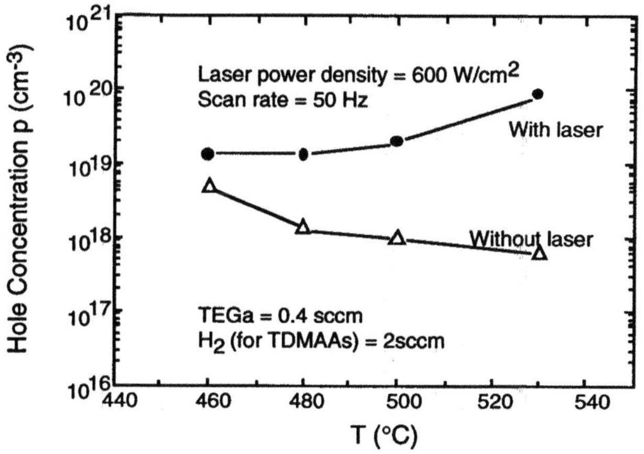

Fig. 14 Effect of Ar^+ laser irradiation on the hole concentration vs. substrate temperature.

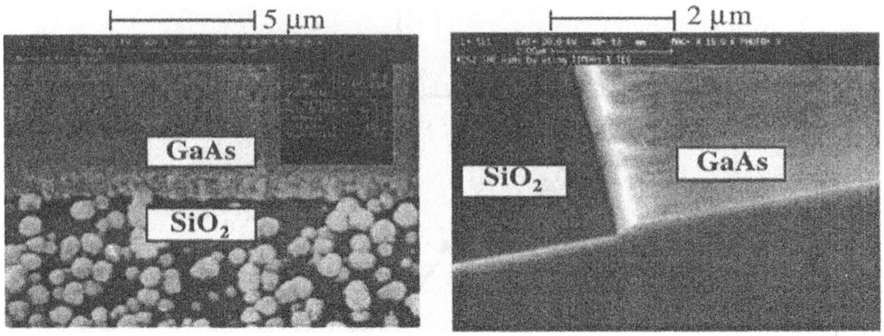

Fig. 15 Scanning electron micrographs of GaAs SAG. (a) Using uncracked TDMAAs at a substrate temperature of 440 °C; (b) Using cracked TDMAAs at 540 °C.[53]

The regrown profile can be adjusted by changing the V/III ratio, or the surface migration length of adsorbed molecules, as shown in Fig. 16.[54] To fill in the undercut area in our case, which is produced by wet etching after reactive-ion etching, so that the external base layer has a good contact with the intrinsic base region, a regrowth profile with (311) facets is used.[54] Because the external base is thicker, the resulting HBTs have lower base sheet resistance. Because the external base layer also passivates the sidewall of the internal base, the base recombination current is also reduced, resulting in higher current gain.

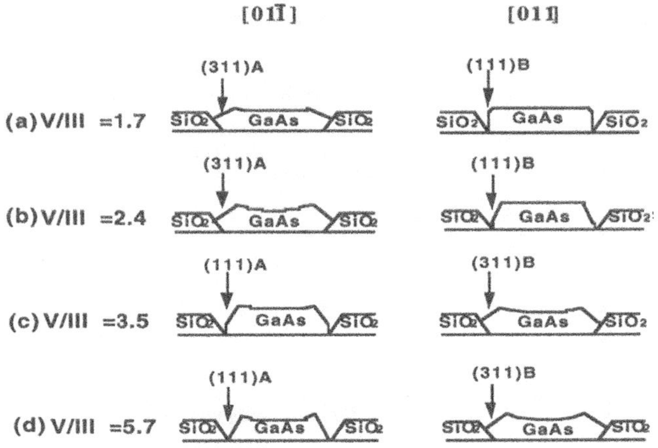

Fig. 16 Schematic cross section of GaAs grown at V/III ratios of (a) 1.7, (b) 2.4, (c) 3.5, (d) 5.7 along [01̄1̄] and [011].[54]

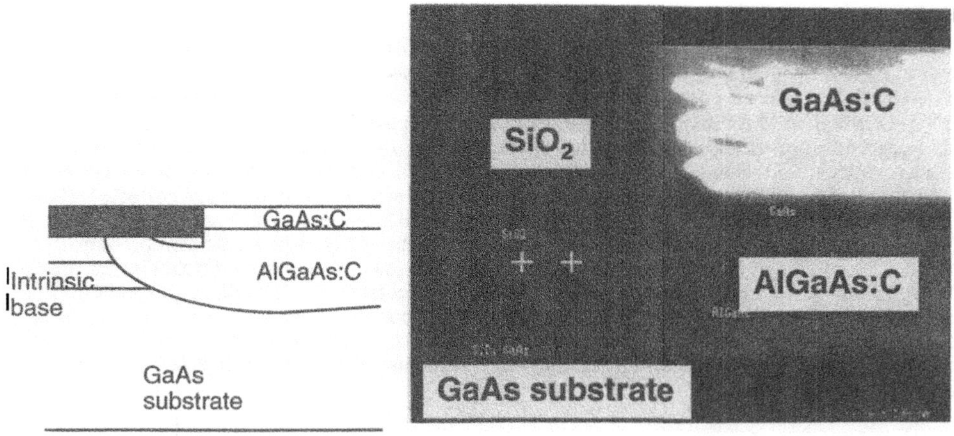

Fig. 17 An SEM photograph and a schematic cross-section of the embedded SAE $Al_{0.25}Ga_{0.75}As$:C/GaAs:C layers grown in an undercut groove at a V/III ratio of 1.3. The distance between the markers is about 0.25 μm.[54]

4. GASEOUS DOPING SOURCES

The use of carbon as a *p*-type dopant in GaAs has attracted considerable attention recently because of high achievable hole concentration[55,56,57] and much lower diffusivity than Zn or Be[58], two commonly used p-type dopants. High-performance AlGaAs/GaAs heterojunction bipolar transistors (HBT) with a base carbon doping up to 5-8 × 10^{19} cm^{-3} have been achieved.[59] InP HBT's, on the other hands, have a number of advantages over GaAs HBT's, including compatibility with long-wavelength optoelectronic components at 1.3 and 1.55 μm, higher substrate thermal conductivity, lower turn-on voltage V_{be}, and higher electron velocity in the $In_{0.53}Ga_{0.47}As$ base region. It is important to study carbon doping in $In_{0.53}Ga_{0.47}As$ because the diffusivity of carbon in $In_{0.53}Ga_{0.47}As$ is estimated to be as low as in GaAs, <3.3 × 10^{-14} cm^2 s^{-1} at 800°C.[60]

Obtaining a high *p*-type doping level from carbon in $In_{0.53}Ga_{0.47}As$, however, has not been as successful as in GaAs. First, the carbon incorporation is shown to decrease with increasing indium concentration in $In_xGa_{1-x}As$.[61,62] Second, the amphoteric nature of carbon makes it an *n*-type dopant in InAs and InP, and *n*-type or *p*-type in $In_xGa_{1-x}As$, depending on growth conditions. The means by which carbon is introduced and incorporated into the film has a pronounced effect. Table A compiles a list of carbon doping results for $In_{0.53}Ga_{0.47}As$, lattice-matched to InP. Except for MOVPE used for the results in the last row, MBE with different kinds of sources is used in the work reported in Table A.

From Table A we can see that TMGa, CCl_4 or CBr_4 can be a good *p*-type dopant for $In_{0.53}Ga_{0.47}As$. The disadvantage of TMGa is that the growth rate and doping are not independent of each other so that one can not employ techniques such as delta doping, graded doping or composition without affecting the other. Therefore, CCl_4 and CBr_4 are the most promising sources, but CCl_4 is banned for environmental reasons. The use of halomethanes, such as $CHCl_3$, CCl_4, $CHBr_3$, CBr_4, as a carbon doping source for GaAs has been reported by Buchan *et al.* for MOVPE[63] and de Lyon *et al.* for GSMBE[64]. The average bond strength is of considerable utility in predicting relative carbon incorporation. The carbon incorporation efficiency is shown to be $CBr_4 > CHBr_3 > CCl_4 > CHCl_3$.[65] The use of CCl_4 or CBr_4 in MOVPE growth of $In_xGa_{1-x}As$, however, has additional problems. The first is hydrogen passivation. High hole concentration can be obtained only after a post-growth annealing.[66] This behavior was first observed by Chin *et al.*[67] in GSMBE and MOMBE, but it was associated with post-growth cool down while AsH_3 was still supplied to protect the surface. Once the AsH_3 supply is turned off at a higher temperature, such hydrogen passivation effect is not present in GSMBE, even when the growth temperature is 420°C.[68] In MOVPE hydrogen passivation is not observed in the base region of InGaP/GaAs HBT's grown at ~625°C, but it is significant in C-doped $In_{0.53}Ga_{0.47}As$ base region grown at ~525°C because of the large amount of hydrogen in the reactor. A more significant problem with using CCl_4 or CBr_4 in MOVPE is that they can affect the In content of the epitaxial layers due to etching. Uniformity in composition may be difficult to achieve. In GSMBE such problem does not occur because of the more efficient utilization of sources in general, i.e., a much smaller amount of CCl_4 or CBr_4 is needed. Because carbon is incorporated about 100 times more efficiently with CBr_4 than with CCl_4, CBr_4 is more compatible with an MBE system. There is no noticeable memory effect.

Table A Comparison of carbon-doped $In_{0.53}Ga_{0.47}As$ reported in the literature. SPSL stands for GaAs:C/InAs short-period superlattices.

Sources	(carrier gas)		Dopant	Growth temp. (°C)	Carrier conc. (cm^{-3})	Ref.
TMGa (He)	TMIn (H_2)	As_4		500	$n=7\times10^{16}$	69
TMGa (He)	In	As_4		500	$p=1\times10^{19}$	
Ga	In	As_4	graphite	~500	$p=1\times10^{17}$	70
Ga	In	As_4	graphite	430	$p=3\times10^{18}$	71, SPSL
TMGa (He)	In	As_4		450	$p=1\times10^{18}$	10
TMGa (He)	In	As_4		420	$p=4\times10^{18}$	17
TEGa (H_2)	In	Arsine	CCl_4 (H_2)	420	$p=5\times10^{19}$	67, SPSL
TEGa (H_2)	In	Arsine	CCl_4 (H_2)	420	$p=3\times10^{19}$	67
Ga	In	Arsine	CCl_4 (H_2)	460	$p=3\times10^{19}$	
Ga	In	Arsine	CCl_4 (H_2)	420	$p=3\times10^{19}$	
TMGa (He)	In	Arsine		525	$p=7\times10^{19}$	72, SPSL
Ga	In	Arsine	CBr_4	480	$p=4\times10^{19}$	68
Ga	In	Arsine	CBr_4	420	$p=9\times10^{19}$	68
Ga	In	As_4	CBr_4	450	$p=7.5\times10^{19}$	73
TMGa	TMIn	Arsine	CCl_4	430	$p=1\times10^{20}$	74
TMGa (H_2)	TMIn (H_2)	Arsine	CCl_4 (H_2)	550	$p=1\times10^{19}$	66, MOVPE

Recently silicon tetrabromide ($SiBr_4$) is also found to be an efficient source for *n*-type doping.[75] Net electron concentrations ranging from $n = 2\times10^{17}$ to 6.8×10^{19} cm^{-3} and from 9×10^{16} to 3×10^{19} cm^{-3} have been obtained for InP and $In_{0.53}Ga_{0.47}As$, respectively. Comparison of these data with those for Si_2H_6 indicate that the Si incorporation efficiency with $SiBr_4$ is more than 10,000 times greater than with Si_2H_6 for substrate temperatures in the range of $475 \leq T_s \geq 500$ °C. During GSMBE growth, an increased Si background concentration ((Si) approximately 2×10^{17} cm^{-3}) was observed after extended use of the $SiBr_4$ source for these heavy doping concentrations, but MOMBE has no such memory effect.

5. CONCLUSION

The use of solid and gaseous sources has extended the capability of the beam technique to a variety of materials, not only compound semiconductors but also metals, silicon, silicides, high T_c superconductors and ferroelectric materials. For ~10 years multiple-wafer production by solid-source MBE for a variety of devices, e.g., $GaAs/Al_xGa_{1-x}As$ field effect transistors and lasers and InAs Hall sensors, has been a commercial reality. Although the techniques of MBE with gaseous sources are relatively recent developments with less accumulated amount of effort, compared to solid-source MBE and MOVPE, they have also produced excellent device-quality materials. Compared to vapor phase techniques, the beam techniques have the advantages of a simple growth mechanism and an in situ monitoring and control of the group III and V compositions by RHEED intensity oscillations, mass spectroscopy, etc. MOMBE/CBE with all gaseous sources, while more complicated and costly to set up, is suitable for selective-area growth for novel device applications. As the demand on layer structures becomes more complex and stringent for advanced applications, MBE and related techniques will remain the premier choice for thin-film growth.

ACKNOWLEDGMENT

I would like to acknowledge my current and former students or associates, T.P. Chin, H.K. Dong, H.Q. Hou, N.Y. Li, B.W. Liang, X.B. Mei, W.S. Wong, and C.H. Yan for many of the results presented here; the Air Force Wright Laboratory, the Office of Naval Research, National Science Foundation, and the Advanced Research Project Agency's Optoelectronic Technology Center for supports; various publishers and authors for their permission to use their figures.

REFERENCES

1 G.J. Davies, P.J. Skevinton, E.G. Scott, C.L. French, and J.S. Foord, *J. Cryst. Growth* **107**, 999 (1991).

2 For a review of the early development of MBE, see A.Y. Cho and J.R. Arthur, *Prog. Solid-State Chem.* **10**, 157 (1975).

3 B.R. Pamplin, ed., *Molecular Beam Epitaxy*, Pergamon Press, Oxford, 1980.

4 K. Ploog, ed., *Molecular Beam Epitaxy of III-V Compounds: A Comprehensive Bibliography*, 1958-1983, Springer-Verlag, Berlin, 1984.

5 L.L. Chang and K. Ploog, eds., *Molecular Beam Epitaxy and Heterostructures*, Nijhoff, Dordrecht, 1985.

6 E.H.C. Parker, ed., *The Technology and Physics of Molecular Beam Epitaxy*, Plenum Press, New York, 1985.

7 M.A. Herman and H. Sitter, *Molecular Beam Epitaxy: Fundamentals and Current Status*, Springer-Verlag, Berlin, 1989.

8 M.B. Panish and H. Temkin, *Gas-Source Molecular Beam Epitaxy*, Springer-Verlag, Berlin, 1993.

9 J.Y. Tsao, *Materials Fundamentals of Moleculr Beam Epitaxy*, Academic Press, Boston, 1993.

10 W.T. Tsang, *J. Cryst. Growth* **81**, 261 (1987).

11 M.B. Panish, *J. Electrochem. Soc.* **127**, 2729 (1980).

12 A.R. Calawa, *Appl. Phys. Lett.* **38**, 701 (1981).

13 G.W. Wicks, M.W. Koch, J.A. Varriano, F.G. Johnson, C.R. Wie, H.M. Kim, and P. Colombo, *Appl. Phys. Lett.* **59**, 342 (1991).

14 J.N. Baillargeon, A.Y. Cho, R.J. Fischer, P.J. Pearah, and K.Y. Cheng, *J. Vac. Sci. Technol. B* **12**, 1106 (1994).

15 E. Veuhoff, W. Pletschen, P. Balk, and H. Luth, *J. Cryst. Growth* **55**, 30 (1981).

16 W.T. Tsang, *Appl. Phys. Lett.* **45**, 1234 (1984).

17 H.Q. Hou and C.W. Tu, *Appl. Phys. Lett.* **60**, 381 (1992).

18 T.P. Chin, B.W. Liang, H.Q. Hou, M.C. Ho, C.E. Chang, and C.W. Tu, *Appl. Phys. Lett.* **58**, 254 (1991).

19 H. Rothfritz, G. Tränkle, R. Müller, F. Herrmann, and G. Weiman, *J. Cryst. Growth* **120**, 130 (1992).

20 M. Lambert, A. Perales, R. Vergnaud, and C. Stark, *J. Cryst. Growth* **105**, 97 (1990).

21 M. E. Sherwin, F. L. Terry, Jr., G. O. Munns, J. S. Herman, E. G. Woelk, and G. I. Haddad, *J. Electron. Mater.* **21**, 269 (1992).

22 A. Antoline, P. J. Bradley, C. Cacciatore, D. Campi, L. Gastaldi, F. Genova, M. Iori, C. Lamberti, C. Papuzza, and C. Rigo, *J. Electron. Mater.* **21**, 233 (1992).

23 T. Y. Wang, E.H. Reihlen, H. R. Jen, and G. B. Stringfellow, *J. Appl. Phys.* **66**, 5376 (1989).

24 A. M. Moy, A. C. Chen, S. L. Jackson, X. Liu, K. Y. Cheng, G. E. Stillman and S. G. Bishop, *J. Vac. Sci. Technol.* **B11**, 826 (1993).

25 K. Mahalingam, Y. Nakamura, N. Otsuka, H. Y. Lee, M. J. Hafich, and G. Y. Robinson, *J. Electron. Mater.* **21**, 129 (1992).

26 N. Kobayashi, Y. Kobayashi, *J. Cryst. Growth* ***124***, 525 (1992).

27 R. Averbeck, H. Riechert, H. Schlotterer, and G. Wiemann, *Appl. Phys. Lett.* **59**, 1732 (1991).

28 T. Anan, S. Sugou, K. Nishi, and T. Ichihashi, *Appl. Phys. Lett.*, **63**, 1049 (1993).

29 G. J. Shiau, C. P. Chao, P. E. Burrows and S. R. Forrest, *J. Appl. Phys.* **77**, 201 (1995).

30 C.H. Yan and C.W. Tu, *J. Vac. Sci. Technol. B* (1996).

31 S. Nagao, M. Takashima, Y. Inoue, M. Katoh and H. Gotoh, *J. Cryst. Growth* **111**, 521 (1991).

32 Y. Nakamura, K. Mahalingam, N. Otsuka, H. Y. Lee, M. J. Hafich. G. Y. Robinson, *J. Vac. Sci. Technol.* **B9**, 2445 (1991).

33 T. Anan, S. Sugou, N. Kenichi, and I. Toshinari, *Appl. Phys. Lett.* **63**, 1047 (1993).

34 C.H. Yan and C.W. Tu, *Inst. Phys. Conf. Ser.* **141**, 161 (1995).

35 H.Q. Hou, A.N. Cheng, H.H. Wieder, W.S.C. Chang, and C.W. Tu, *Appl. Phys. Lett.* **63**, 1833 (1993).

36 M. Yamamoto, N. Yamamoto, and J. Nakano, *IEEE J. Quantum Electronics* **QE-30**, 554 (1994).

37 T. Fukushima, A. Kasukawa, M. Iwase, T. Namegaya, *et al.*, *IEEE Photonics Technol. Lett.* **5**, 117 (1993).

38 H. Sugiura, M. Mitsuhara, H. Oohashi, T. Hirono, *et al.*, *J. Crystal Growth* **147**, 1 (1995).

39 H.Q. Hou, C.W. Tu, and S.N.G. Chu, *Appl. Phys. Lett.* **58**, 2954 (1991).

40 M.K. Chin, P.K.L. Yu, and W.S.C. Chang, IEEE J. Quantum Electronics **QE-27**, 696 (1991).

41 X.B. Mei, K.K. Loi, H.H. Wieder, W.S.C. Chang, and C.W. Tu, *Mater. Res. Soc. Symp. Proc.* (1995).

42 X.B. Mei, K.K. Loi, H.H. Wieder, W.S.C. Chang, and C.W. Tu, *Appl. Phys. Lett.* **68** (1996).

43 H.K. Dong, N.Y. Li, C.W. Tu, M. Geva, and W.C. Mitchel, *J. Electronic Mater.* **24**, 69 (1995).

44 N.Y. Li, H.K. Dong, C.W. Tu, and M. Geva, *J. Cryst. Growth* **150**, 245 (1995).

45 H.K. Dong, N.Y. Li, C.W. Tu, M. Geva, and W.C. Mitchel, *Mater. Res. Soc. Symp. Proc.* **340**, 173 (1994).

46 H. Sugiura, R. Iga, and T. Yamada, *J. Cryst. Growth* **120**, 389 (1992).

47 H.K. Dong, B.W. Liang, M.C. Ho, S. Hung, and C.W. Tu, *J. Cryst. Growth* **124**, 181 (1992).

48 H.K. Dong, N.Y. Li, S.W. Wong, and C.W. Tu, *Appl. Phys. Lett.*, to be published.

49 D.A. Bohling, C.R. Abernathy, and K.F. Jensen, *J. Cryst. Growth* **136** 118 (1994).

50 H.K. Dong, N.Y. Li, and C.W. Tu, *J. Electronic Mater.* **24**, 827 (1995).

51 T. Yamada, R. Iga, and H. Sugiura, *Appl. Phys. Lett.* **61**, 2449 (1992).

52 N. Furuhata and A. Okamoto, *J. Crystal Growth* **112**, 1 (1991).

53 N.Y. Li, Y.M. Hsin, H.K. Dong, T. Nakamura, P.M. Asbeck, and C.W. Tu, *J. Crystal Growth* **150**, 562 (1995).

54 N.Y. Li, H.K. Dong, Y.M. Hsin, T. Nakamura, P.M. Asbeck, and C.W. Tu, *J. Vac. Sci. Technol. B* **13**, 664 (1995).

55 M. Konagai, T. Yamada, T. Akatsuka, K. Saito, and K. Takahashi, *J. Crystal Growth* **98**, 167 (1989).

56 P.M. Enquist, *Appl. Phys. Lett.* **57**, 2348 (1990).

57 R.J. Malik, J. Nagle, M. Micovic, T. Harris, R.W. Ryan, and L.C. Hopkins, *J. Vac. Sci. Technol. B* **10**, 850 (1992).

58 B.T. Cunningham, L.J. Guido, J.E. Baker, J.S. Major, Jr., N. Holonyak, Jr., and G.E. Stillman, *Appl. Phys. Lett.* **55**, 687 (1989).

59 G.-W. Wang, R.L. Pierson, P.M. Asbeck, K.-C. Wang, N.-L. Wang, R. Nubling, M.F. Chang, J. Salerno, and S. Sastry, *IEEE Electron Device Lett.* **12**, 347 (1991).

60 S.J. Pearton, W.S. Hobson, A.P. Kinsella, J. Kovalchick, U.K. Chakrabarti, and C.R. Abernathy, *Appl. Phys. Lett.* **56**, 1263 (1990).

61 J. Shirakashi, T. Yamada, M. Qi, S. Nozaki, K. Takahasi, E. Tokumitsu, and M. Konagai, *Jpn. J. Appl. Phys.* **30**, L1609 (1991).

62 S.A. Stockman, A.W. Hanson, and G.E. Stillman, *Appl. Phys. Lett.* **60**, 2903 (1992).

63 N.I. Buchan, T.F. Kuech, G. Scilla, and F. Cardone, *J. Crystal Growth* **110**, 405 (1991).

64 T.J. de Lyon, N.I. Buchan, P.D. Kirchner, J.M. Woodall, G.J. Scilla, and F. Cardone, *Appl. Phys. Lett.* **58**, 517 (1991).

65 N.Y. Li, H.K. Dong, C.W. Tu, and M. Geva, *J. Crystal Growth* **150**, 245 (1995).

66 A. Stockman, A.W. Hanson, S.M. Lictenthal, M.T Fresina, G.E. Höfler, K.C. Hsieh, and G.E. Stillman, *J. Electronic Mater.* **21**, 1111 (1992).

67 T.P. Chin, P.D. Kirchner, J.M. Woodall, and C.W. Tu, *Appl. Phys. Lett.* **59**, 2865 (1991).

68 C.W. Tu, T.P. Chin, and B.W. Liang, *J. Crystal Growth* **136**, 191 (1994).

69 C.R. Abernathy, S.J. Pearton, F. Ren, W.S. Hobson, T.R. Fullowan, A. Katz, A.S. Jordan, and J. Kovalchick, *J. Crystal Growth* **105**, 375 (1990).

70 H. Ito and T. Ishibashi, *Jpn. J. Appl. Phys.* **30**, L944 (1991).

71 E.F. Schubert and R.F. Kopf, *J. Crystal Growth* **127**, 1037 (1993).

72 C.R. Abernathy, P.W. Wisk, S.J. Pearton, W.S. Hobson, P.H. Fuoss, F.J. Lamelas, S.N.G. Chu, and F. Ren, *Appl. Phys. Lett.* **60**, 1339 (1992).

73 W.Y. Hwang, D.L. Miller, Y.K. Chen, D.A. Humphrey, *J. Vac. Sci. Technol. B* **12**, 1193 (1994).

74 C.J. Palmstrom, B.P. Van der Gaag, J.-I. Song, W.-P. Hong, S.A. Schwarz, and S. Novak, *Appl. Phys. Lett.* **64**, 3139 (1994).

75 S.L. Jackson, M.T. Fresina, J.E. Baker, and G.E. Stillman, *Appl. Phys. Lett.* **64**, 2867 (1994).

OPTICAL PROPERTIES OF HETEROSTRUCTURES UNDER AN ELECTRIC FIELD

P. TRONC
ESPCI. Laboratoire d'Optique
10 rue Vauquelin - 75005 PARIS - France

The study of the optical properties of heterostructures under an electric field is very important not only for the fundamental knowledge of these structures but also for applications since a voltage is applied onto any device structure, for example to control a switch (electro-optical modulator), generate a light beam (light-emitting diode, laser) or convert a light signal into an electric current (photo-detector).

1. Theory

1.1. WAVEFUNCTIONS AND ENVELOPE FUNCTIONS

It would be very difficult to exactly calculate the wavefunctions of carriers in any quantum well (QW) or superlattice (SL). Fortunately the envelope function approximation[1,2] provides a lot of accurate results concerning the carrier energy, minibands and selection rules for optical transitions. This approximation is relevant as far as the bulk material lattice parameter can be considered as small when compared to the well and barrier width. Nevertheless, in the case of some very narrow QW's, it provided rather accurate results.[3]
The $(GaAs)_m(AlAs)_n$ SL's grown along the [001] direction (m and n are the numbers of monolayers of constituent materials) have been extensively studied. They are type I or type II depending on m and n values. The type (I or II) of $(GaAs)_m(AlAs)_n$ [001] SL's has been experimentally studied by different optical techniques : absorption[4], optically detected magnetic resonance[5], photoluminescence under hydrostatic pressure[6], etc. The estimated values of m for SL's with m=n at which the type I to type II transition takes place present discrepancies from one author to another, even if it is often claimed that the transition occurs between m=10 and m=14. On the other hand, the energy band structure of GaAs/AlAs short period SL's has been calculated using various methods. Most of the results converge to the m=8-13 interval for the type I to type II transition to occur when n=m[6]. For the $(GaAs)_m(AlAs)_n$ [001] SL's, the folding of the bulk material Brillouin zone (BZ) over the superlattice BZ (SLBZ) puts in correspondence the X_Z point of the bulk material with the Γ or Z point of the SLBZ depending on m+n is even or odd (the X_x and X_y points of the bulk BZ are always related with points at the SLBZ surface). It is generally admitted that the top of the valence band is located at the Γ point of the SLBZ whereas the position of the bottom of the conduction band in the SLBZ depends on the thickness of the GaAs and AlAs slabs. Within the envelope function approximation, this leads to two types of SL's :

M. Balkanski (ed.), Devices Based on Low-Dimensional Semiconductor Structures, 37–65.

i) if the eigenstate with the minimal energy in the SL conduction band originates from the Γ point in the GaAs BZ, it is located at the Γ point of the SLBZ ; the electrons and holes are then localised in GaAs regions and the SL is type I.

ii) if the eigenstate with the minimal energy originates from the X_Z point in the AlAs BZ, it is located at the Γ point of the SLBZ if m+n is even and at a point on the surface of the SLBZ if m+n is odd. In both cases electrons are localised in AlAs regions, i.e. the electrons and holes are spatially separated ; in the former case the optical transition is direct in the k space and indirect in the real space and the SL is pseudodirect type II ; in the latter case the transition is indirect in both k and real space and the SL is type II. Last, if the eigenstate with the minimal energy originates from the X_{XY} point in the AlAs BZ, it is located on the surface of the SLBZ and the electrons are localised in AlAs regions ; the transitions is indirect both in k and real space and the SL is type II.

It is clear that optical transitions are allowed at the Γ point for direct type I SL's. When they are allowed from symmetry in the pseudodirect ones, they should have a weak intensity since the overlap of electron and hole envelope functions is small. The relative positions is energy of the states originating from respectively the X_Z and the X_{XY} point in the AlAs BZ have been explained[7,8] by a simple model involving the competition between the weak lattice mismatch which causes a

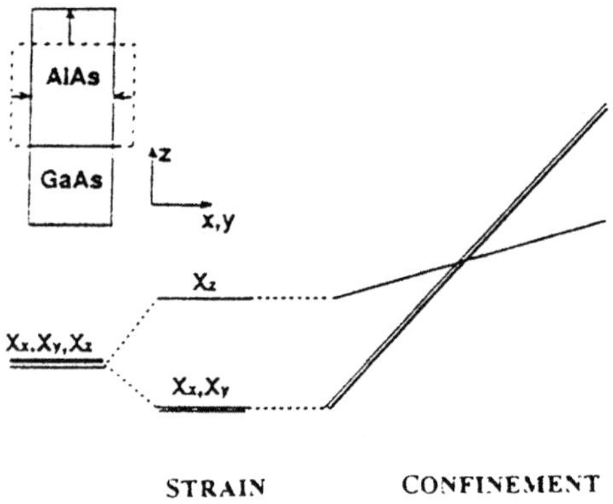

Figure 1. Lifting of the degeneracy between the X_Z and X_{XY} valleys arising from the strain and the confinement in an AlAs layer deposited on a GaAs substrate (8).

uniaxial deformation of AlAs layers, thus pushing up the X_Z valley and the anisotropy of the X valley effective mass which makes the localization energy of X_Z valley smaller, hence pushing down the X_Z state (Figure 1). Figure 2 sumarizes the results. A partition of the region II is proposed into regions R_Z and R_{XY} where the lowest electron states originate respectively from the X_Z and X_{XY} valleys. Δ_O is the computed energy difference between X_{XY} and X_Z minima of unstrainded SL's. W_1 and W_2 are the widths of respectively the GaAs and AlAs layers in the SL's.

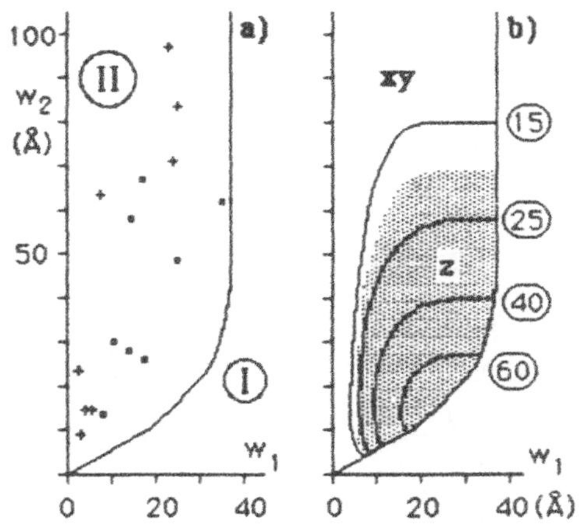

Figure 2. a) Boundary between type I and type II SL according to samples studied, classified as belonging to R_{xy} (+) or R_z (-). b) Map of energy difference Δ_0 (meV) from theory. The shaded area is the R_z region assuming a splitting of 19 meV caused by lattice mismatch (7).

The symmetry of the first conduction subband envelope function at the Γ point of the SLBZ is very important in the case of pseudodirect SL's. If the envelope function is invariant under a mirror symmetry (the mirror being parallel to the plane of the layers and located at the center of any layer) the direct optical transition in the k space is allowed. It is forbidden if the envelope function changes its sign. Going beyond the conventional envelope function approximation Sham and Lu assigned[9], in their subband structure calculations, the $(-1)^n$ parity value along the growth axis to the first electron subband, arguing that, in bulk AlAs, the X_z wavefunction changes its sign over one unit cell. The direct optical transition would then be allowed for even values of n and forbidden for odd ones. To my knowledge, no experimental proof of this prediction has been reported[10]. This lack may be due just to the difficulty to grow SL's with the exact numbers of monolayers in each slab.

The possible symmetries for electron wavefunctions and the character (allowed or forbidden) of any optical transition can be predicted from group theory analysis in SL's having a perfectly defined structure, i-e built from two stochiometric materials like for example GaAs and AlAs or GaSb and AlSb. It consists in considering these SL's as new crystals with an unit cell enlarged along the growth direction when compared to that of the parent materials. This analysis would obviously not be relevant for SL's involving an alloy. As an example of SL's having an exact symmetry, let us consider again the $(GaAs)_m(AlAs)_n$ [001] SL's. They belong to two families, corresponding to even and odd values of m+n, which have respectively the D_{2d}^5 and D_{2d}^9 space group. As a consequence their BZ's are different (Figure 3).

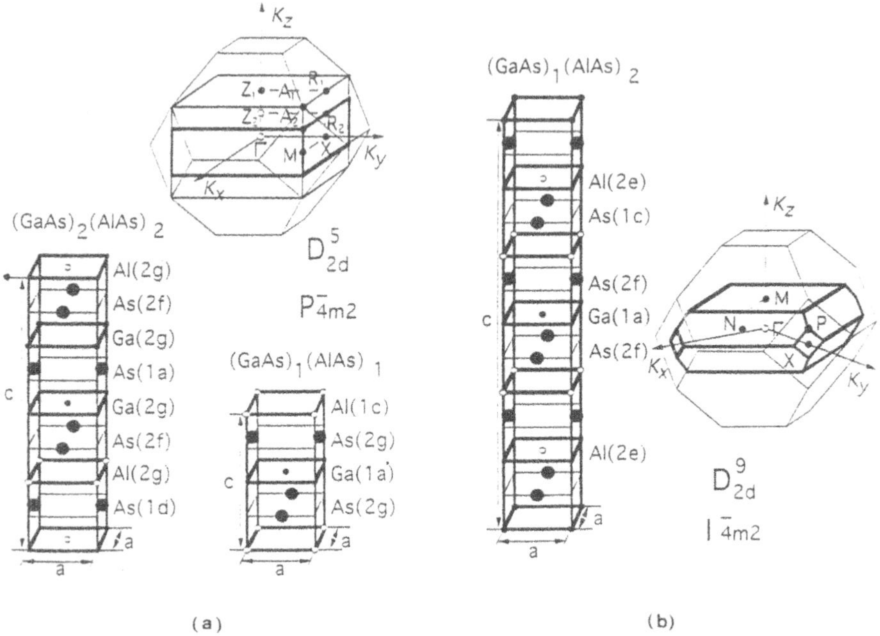

Figure 3. Unit cell and Brillouin zone of some $(GaAs)_m(AlAs)_n$ SL's belonging respectively to the (a) D^5_{2d} and (b) D^9_{2d} space group. The symmetry group of each site is indicated between parentheses (11).

A group theory analysis of the symmetry of the wavefunctions which takes into account the spin-orbit interaction and of the optical transitions in these new crystals has recently been performed[11]. This study shows that optical absorption experiments performed with linearly polarized light could help to decide among the numerous and sometimes contradictory results which have been published on band structures[11] : as a matter of fact, weak direct optical transitions, allowed only from spin-orbit interaction can provide informations about the symmetry of the valence and conduction bands. Table 1 lists the selection rules for direct optical transitions at the Γ point in the D^5_{2d} and D^9_{2d} space groups. It can be seen that several transitions are allowed only from spin-orbit interaction.

Table 1. The modification of selection rules for direct optical transitions when including spin-orbit interaction (the space groups D_{2d}^5 and D_{2d}^9). The labels of both the irreps and additional light polarizations (x, y or z) taken in parentheses refer to the case where spin-orbit coupling is taken into account.

		conduction-band states			
		$\Gamma_1(\Gamma_6)$	$\Gamma_2(\Gamma_7)$	$\Gamma_5(\Gamma_6)$	$\Gamma_5(\Gamma_7)$
valence-band states	$\Gamma_1(\Gamma_6)$	(x,y)	(x,y)z	x,y	x,y(z)
	$\Gamma_2(\Gamma_7)$	(x,y)z	(x,y)	x,y(z)	x,y
	$\Gamma_5(\Gamma_6)$	x,y	x,y(z)	(x,y)	(x,y)z
	$\Gamma_5(\Gamma_7)$	x,y(z)	x,y	(x,y)z	(x,y)

Phonon-assisted optical transitions have also been considered in the former space groups both neglecting and taking into account the spin-orbit interaction.[11]

Going back to the envelope function approximation, to which we will limit ourselves in the rest of the present paper, it is worth pointing out that the envelope functions have no longer a defined parity when the QW's are not symmetrically designed. This is also the case of a single hetorojunction when the band bending induces an asymmetric QW (Figure 4(a)). As a result the selection rules on the indexes of the minibands are lifted both for the interband and intraband optical transitions. This effect can also be achieved by applying any perturbation which is not purely even under the mirror symmetry. It is for example the case of an electric field applied along the growth axis.
The study of the hole envelope function is much more complicated that of the electron one since, in bulk III-V materials, the heavy and light holes are degenerate at the Γ point. The confining potential lifts the degeneracy[1,2] but, when the $k_\perp$ wavevector in the plane of the layers is non-zero, heavy- and light-hole envelope functions are mixed and the label heavy- or light-hole for a given subband has little significance. The mixing can be taken into account through the so-called Γ_8

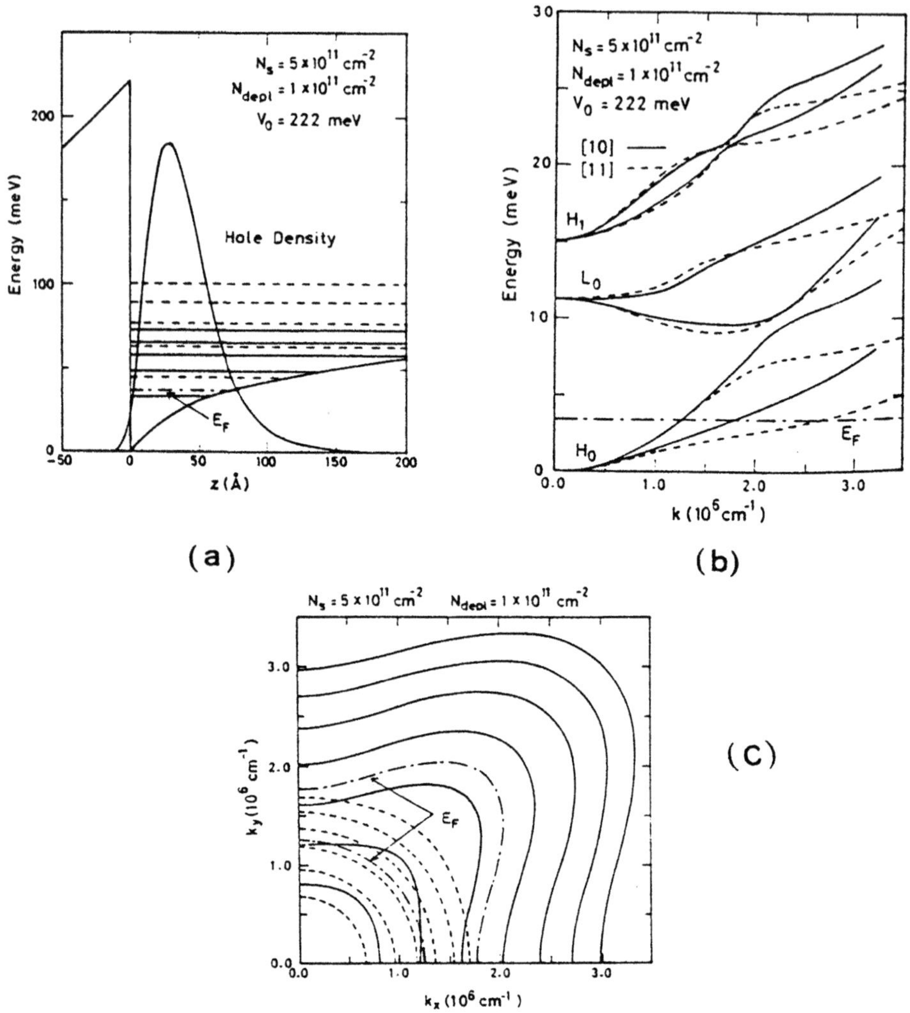

Figure 4. Hole subbands in a GaAs/AlGaAs single heterojunction. N_S is the hole concentration per unit area, N_{depl} the concentration of fixed space charges in the GaAs layer and V_o the barrier potential.

a) Calculated density distribution of holes, self-consistent potential, subband energies at $k_x=k_y=0$. The solid horizontal lines represent "heavy-hole" subbands and the dashed lines "light-hole" subbands

b) Calculated subband energy as a function of k in the [10] and [11] directions. The solid and dashed lines represent subband energies in the [10] and [11] direction, respectively. The subbands doubly degenerate at $k_x=k_y=0$ split into two due to the removal of spin degeneracy.

c) Equi-energy lines in the k_xk_y space of the lowest two subbands. The solid and dashed lines represent equi-energy contours of the lowest and spin-split first excited subband, respectively. The energy interval for adjacent contours is chosen to be 1.0 meV. The dot-dashed lines give the Fermi lines (13).

Hamiltonian[1,2] which results from considering the conduction band and spin-orbit split-off valence band as remote bands in bulk materials. This model is accurate at small $k_\perp$. It provides at large $k_\perp$, parabolic dispersion curves since it takes into account the periodic potential within the plane of the layers only through the in-plane effective masses at the Γ point. This can induce discrepancies when calculating the Fermi energy at room temperature in very heavily p-doped QW's. As a matter of fact, the real in-plane dispersion curves $E=E(k_x, k_y)$ have a slope equal to zero at the edge of the BZ whereas the parabolic dispersion curves increase monotonically with $k_\perp$.

Moreover it can happen that several subbands have to be considered when calculating the Fermi energy. For example the energy difference at the Γ point between two consecutive hole subbands taken among the three first ones in a 50 Å wide GaAs/$Al_{0.3}Ga_{0.7}As$ QW is approximatively 25 meV. This value is equal to k_BT at 300 K, where k_B is the Boltzmann constant. Tight-binding calculations of electronic states in CdTe/ZnTe SL's, which take into account the full symmetry of the system, have recently been performed by Jouanin et al.[12] This kind of calculations could allow one to introduce the states with a large $k_\perp$ value in a more accurate way. Ando[13] used the Γ_8 Hamiltonian formalism to study GaAs/AlGaAs single heterojunctions and multiple quantum wells (MQW's). To avoid difficulties arising from current conservation conditions at the interfaces, he neglected the slight differences in Lüttinger parameters values between both materials. For the sake of numerical convenience he imposed, in the case of a single heterojunction, boundary conditions such that the envelope functions should vanish at 200 Å from the interface in the GaAlAs slab and at 1500 Å in the GaAs one. In the case of MQW's the set of orthogonal envelope functions was chosen as periodic plane waves exp $(2\pi ilz / d)$, where d is the period of the SL and l is integer. Figure 4 and 5 show the calculated density distribution of holes along z, the in-plane dispersion curves in the [10] and [11] directions and equi-energy lines in the $[k_x, k_y]$ space, for the single heterojunction and MQW's respectively. One can notice the electron-like effective mass of the first light-hole dispersion curve near the BZ center. The equi-energy lines in the plane of the layers have a four-fold symmetry.[13]

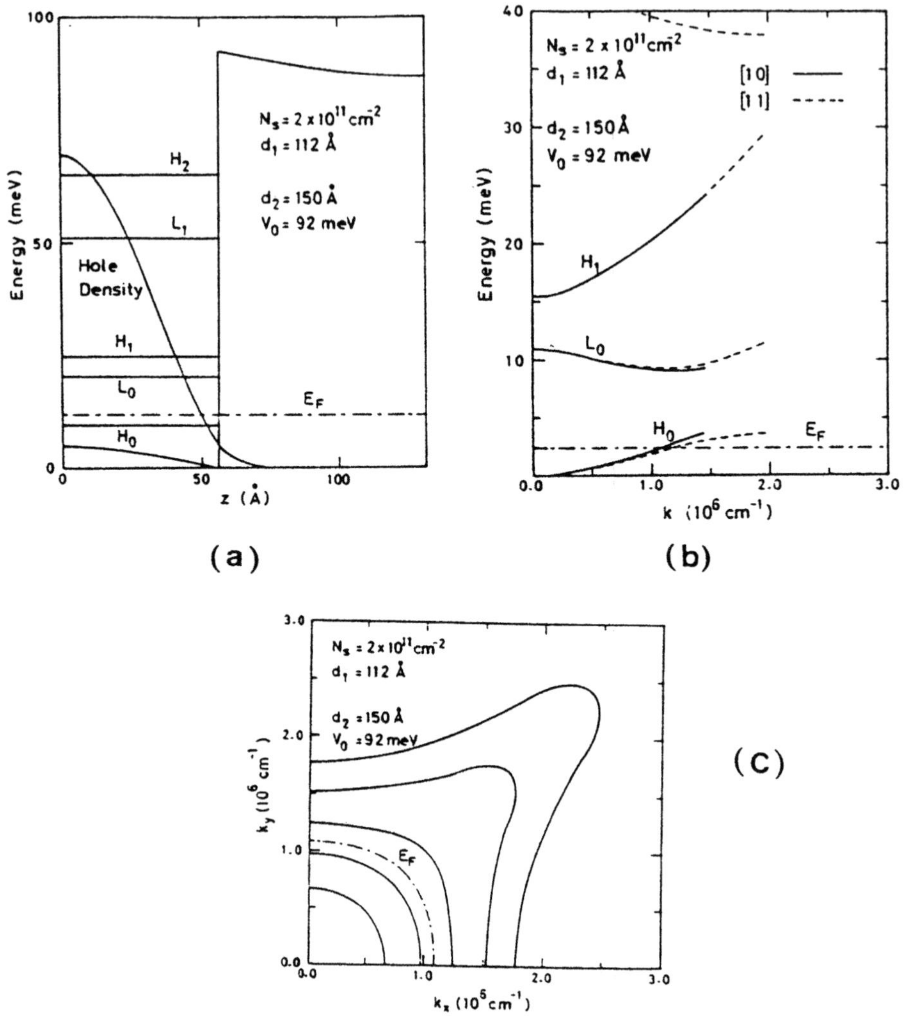

Figure 5. Hole subbands in a modulation-doped $GaAs/Al_{0.51}Ga_{0.49}As$ multiple quantum well. $d_1(d_2)$ is the thickness of the GaAs (AlGaAs) slab, Ns the hole concentration per unit area and V_0 the barrier height.
a) Subband energies at $k_x=k_y=0$, density distribution of holes, and the self-consistent potential. The origin of z is chosed at the center of the GaAs layer. Only the region $0<z<d/2$ is shown in the figure.
b) Subband energies as a function of k in the two directions [10] (solid lines) and [11] (dashed lines).
c) Equi-energy lines in the $k_x k_y$ space of the lowest two subbands. The solid and dashed lines represent equi-energy contours of the lowest and spin-split first excited subband, respectively. The energy interval for adjacent contours is chosen to be 1.0 meV. The dot-dashed lines give the Fermi lines (13).

The small difference in energy between holes subbands rises difficulties to calculate acceptors binding energies. Masselink et al[14] used a variational method to calculate the acceptor binding energies in GaAs/AlGaAs QW's. Their results are in good agreement with the measurements by Miller and al.[15] (Figure 6). They also calculated[14] the variation of the acceptor binding energy versus the strength of an

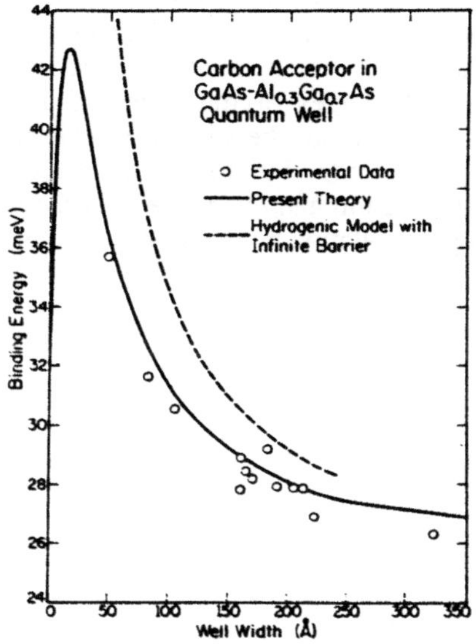

Figure 6. Calculated dependance of the on-centre carbon binding energy versus GaAs well width in GaAs-$Ga_{1-x}Al_xAs$ quantum wells. $Vb(x)=0.15x\ [\varepsilon_g(Ga_{1-x}Al_xAs)-\varepsilon_g(GaAs)]$ is the assumed hole confining barrier height. The open circles are the experimental values obtained by Miller et al. (15), (14).

electric field applied parallel to the growth axis in a 100 Å wide GaAs/$Al_{0.25}Ga_{0.75}As$ QW. At low and middle field strength the energy is quite unaffected whereas, at high field, it linearly increases with the field strength. In Figure 7, E_o is the acceptor energy in bulk GaAs when no electric field is applied.

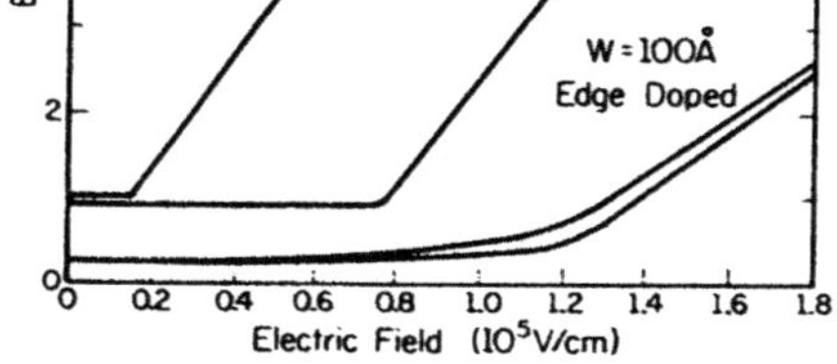

Figure 7. The absolute values of the acceptor energy measured from the bulk valence band as functions of electric field for an acceptor in GaAs and acceptor in an $Al_{0.25}Ga_{0.75}As$-GaAs quantum well. The edge doped acceptor is split because the two edges are not degenerate in an electric field (14).

When heterostructures are heavily n- or p-doped or when they are shined by an intense light beam the many-bodies effects can no longer be neglected. In bulk semiconductors, for the many-bodies ground state, the Hartree, exchange and correlation potentials can be expressed[16] as functions of the electron density n(r). Unless an external perturbation is applied to an homogeneous crystal, n(r) is constant. The results have been established for the case of a Fermi sea, i-e when k_BT is negligible when compared to the Fermi energy measured from the bottom of the band. This allows one to use a step function to model the density of states occupation. If n(r) is no longer constant, the validity of the model requires that the electronic density variation be small over distances of the order of a Fermi wavelength. Many authors[17] used the same model and the same formulae for the ground state of heterostructures, just replacing n(r) by the local density n(z) calculated from the envelope function approximation and even extended the use of such formulae to excited states. The exchange-correlation potential V_{xc} then is[17] :

$$V_{xc}(z) = -\left[1 + 0.7734 X \mathrm{Ln}\left(1 + X^{-1}\right)\right]\left(2 / \pi\alpha r_s\right) R_b$$

$$\text{where} \quad \alpha = (4/9\pi)^{1/3} \quad X = X(z) = r_s(z)/21 \quad r_s = r_s(z) = \left[\frac{4}{3}\pi a^{*3} n(z)\right]^{-1/3} \tag{1}$$

$$R_b = e^2 / 8\pi\varepsilon_0 K a^*$$

$$a^* = 4\pi\varepsilon_0 K \hbar^2 / m^* e^2$$

K=K(z) is the local dielectric constant, $m^*=m^*(z)$ the local effective mass, R_b the Rydberg (equal approximately to 5 meV in GaAs). The $V_{xc}(z)$ potential is introduced into the Schrödinger's equation which is coupled to the Poisson's one to take into account the Hartree potential and solved iterately. The Hartree energy is obviously much larger in the case of doping in the barriers that in the case of doping in the well. In the model n- and p-type heavily doped heterostructures are treated in the same manner. In many cases the carrier density does not vary slowly on the scale of the Fermi wavelength, possibly invalidating the use of the model[17]. For example in a GaAs/AlGaAs heterojunction with an electron density of 10^{12} cm^{-2}, the Fermi wavelength is about 25 nm whereas the typical envelope function width is about 10 nm for the lowest subband[17].
The above procedure for heterostructures remains controversial. Other authors[18] prefer to use Feynman diagrams. Calculations for the holes would then introduce terms different from those which appear for the electrons. These papers are too recent to allow one to perform a complete comparison between both theories.

1.2. EFFECT OF A STATIC ELECTRIC FIELD APPLIED ALONG THE GROWTH AXIS.

1.2.1. QW's (confined Stark effect)

The potential profile for electrons in such a structure is shown in Figure 8. In principle no stationary bound state does exist since the potential -eFz arising from

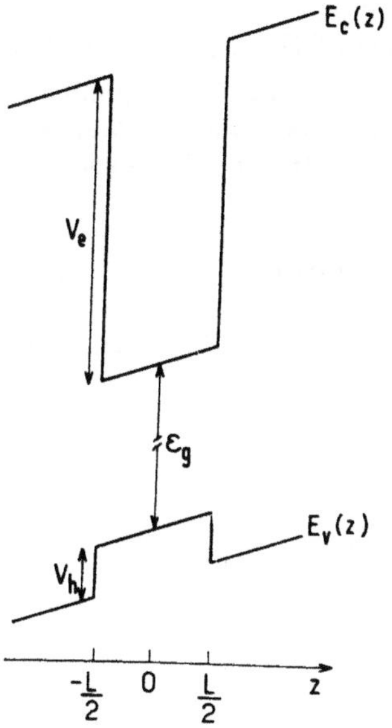

Figure 8. Position dependences of the conduction and valence band edges of a type I quantum well subjected to a longitudinal electric field F (F//z). ε_g denotes the bandgap of the well-acting material. V_e and V_h denote the conduction and valence barrier heights respectively.

the electric field F diverges when z goes to $-\infty$. Nevertheless the tunnelling effect across the low energy side barrier remains negligible for the ground bound state if its characteric penetration length K_c^{-1} remains such that[1] :

$$eFK_c^{-1} << V_b - E_1$$

$$K_c = \left(\frac{2m^*(V_b - E_1)}{\hbar^2} \right)^{\frac{1}{2}} \tag{2}$$

V_b is the barrier height and E_1 the ground state energy at zero field. The conduction electron Hamiltonian in a QW reads

$$H = H_{\perp} + H_{11}$$
$$H_{\perp} = \frac{p_x^2 + p_y^2}{2m^*} \qquad H_{11} = \frac{p_z^2}{2m^*} + eFz + V_b Y\left(z^2 - \frac{L^2}{4}\right) \tag{3}$$

where L is the well width and Y the step function.
$H_{\perp}$ decouples from H_{11}. The exact eigenstate of H_{11} is, within each slab, a linear combination of the two independent solutions of the Airy equation[19]. Matching conditions are used at each interface to obtains the complete eigenstate. Figure 9 shows eigenstates in a QW structure (an infinite barrier has been added on the left to

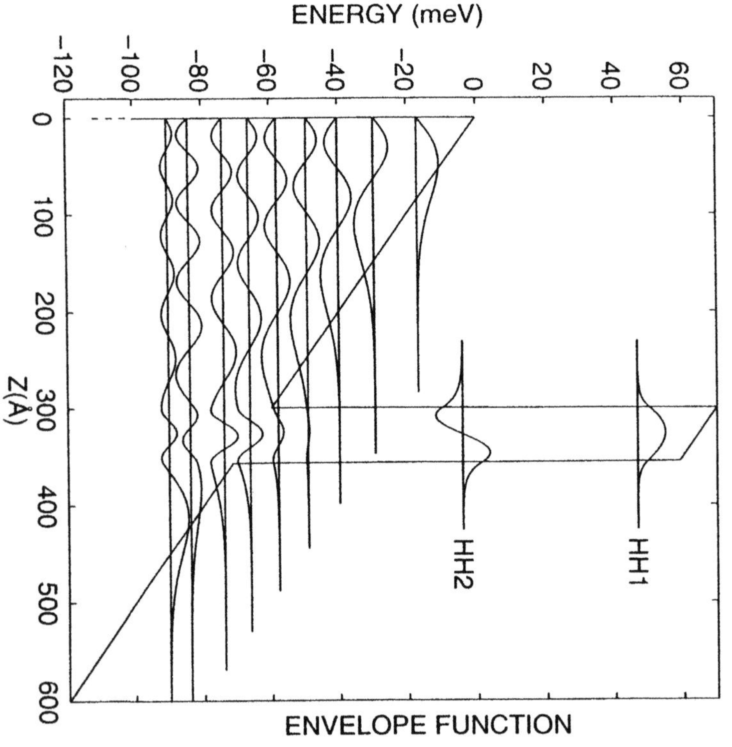

Figure 9. Heavy-hole envelope functions in a structure made of a (56 Å) GaAs/$Al_{0.3}Ga_{0.7}As$ QW and an infinite barrier on the left side under a 20 kV cm^{-1} electric field. HH_1 and HH_2 refer respectively to the ground and first excited state bound in the well. Courtesy R. Melliti.

allow to normalize the envelope functions. In such a case, no assumption has to be made about the strength of the electric field and the limitation expressed in (3) is no longer necessary). Such calculations are easily performed with the help of a computer. It can be seen that, at moderate electric field strength, the envelope functions of states bound within the well keep an approximately defined parity. An advantage of using Airy functions is that they do take into account the tunnelling effect through the left barrier. The probability of presence out of the well for a given bound state increases with its energy (increase of the probability of tunnelling).
The ground state has been analytically studied using a variational technique by Bastard et al.[20] The ground state energy decreases with increasing field strength (Figure 10).

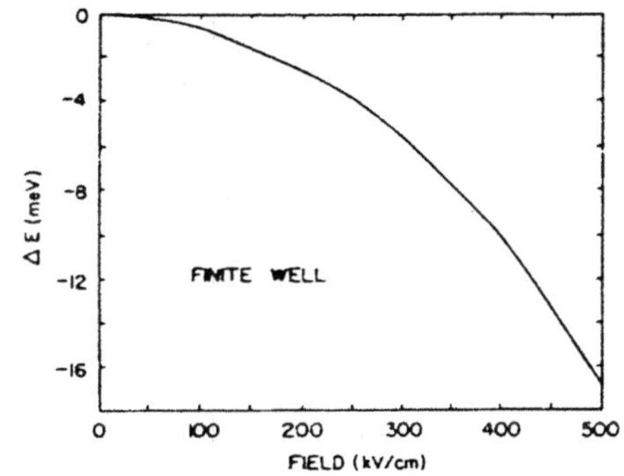

Figure 10. Energy shift vs electric field of the eigenvalue of a conduction electron in a finite GaAs QW, 30 Å. thick, with a barrier height of 0.4 eV (20).

The electron and hole envelope functions are shifted to opposite directions reducing their overlap (Figure 11) and therefore the matrix element for their optical recombination. Correlatively the binding energy of the exciton is decreased. The magnitude of this latter effect increases with the well width (Figure 11) but remains in general widely smaller than the red-shift ot the carriers energies.

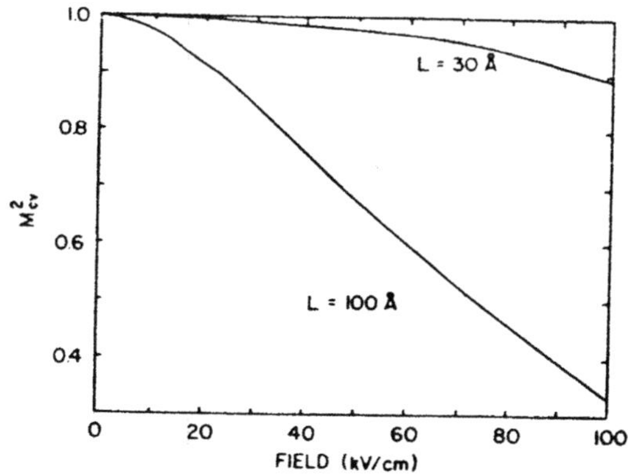

Figure 11. Square of the overlap integral between electrons and heavy holes as a function of electric field for a GaAs QW of 30 or 100 Å with a barrier height of 0.4 eV (20).

1.2.2. SL's (Wannier-Stark effect)

Airy functions can also be used to compute the eigenfunctions in a SL. States centered on a given well can be identified (in Figure 12 an infinite barrier has be added on the left side in order to normalize the eigenfunctions). Except the few wells located close to the infinite barrier, whose bound states are notably influenced by the latter, the envelope function and eigenenergy (measured from the bottom of the well) of bound states centered on a given well very weakly depend on this well. Again envelope functions built from Airy functions are exact solutions of the SL Hamiltonian.

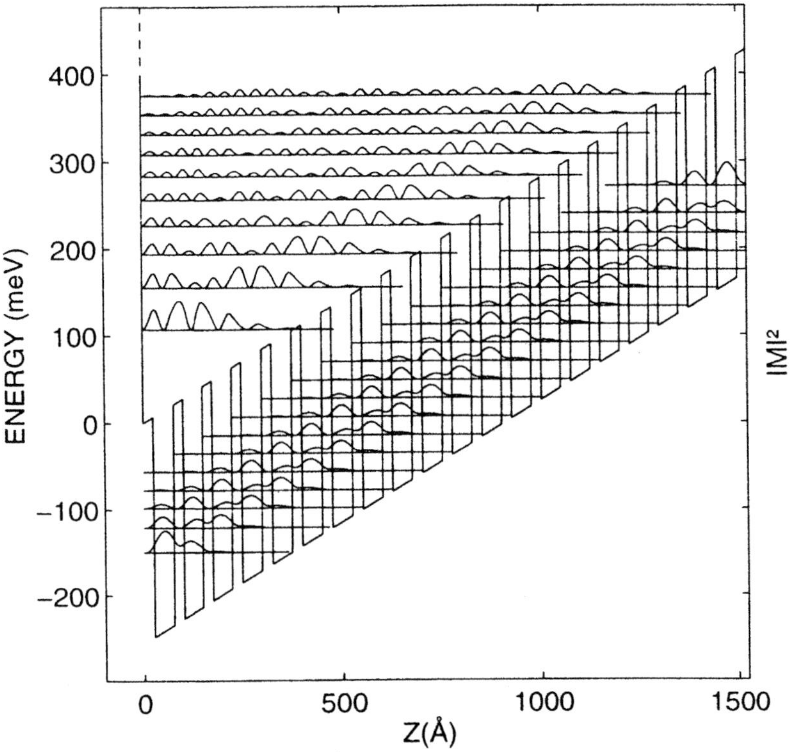

Figure 12. Square of the M modulus of the electron envelope function in a (50 Å)/(25 Å) $Ga_{0.47}In_{0.53}As/Ga_{0.24}Al_{0.24}In_{0.52}As$ SL with an infinite barrier on the left side under a 28 kV cm-1 electric field. Courtesy R. Melliti.

Carrier and exciton states and strengths of the optical transitions in a SL under an electric field have also been analytically studied in various approximations.
Due to the lack of resonant coupling between adjacent quantum wells when an electric field is applied along the growth axis of a SL, the miniband spectrum

converges towards the evenly spaced "Stark ladder". In a type I SL, the peak corresponding to the transition between a hole and an electron with a wavefunction whose maximum is p periods away from the hole state is labelled +p or -p (Figure 13), the sign indicating that its energy is respectively greater or smaller than the energy of the peak corresponding to an electron and a hole localized in the same well (peak labelled o).

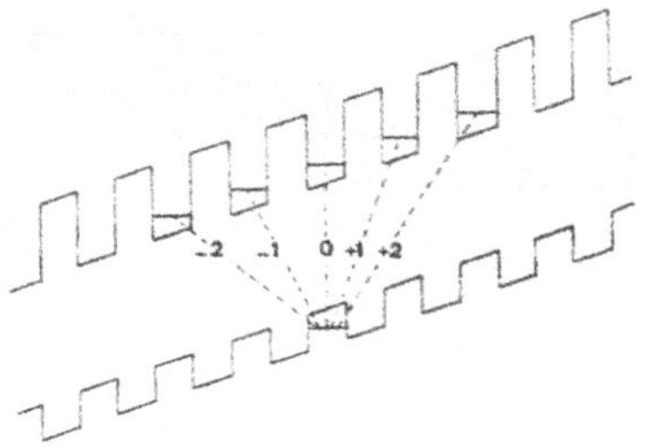

Figure 13. Sketch of the conduction - and valence - band potential profiles in a type I SL under an electric field.

The applied electric field F localizes the carriers in the SL. In III-V SL's the heavy hole is strongly localized due to its large effective mass. If N is the number of wells over which the electron wavefunctions is spread, N is limited because the potential drop over the whole spread of the electron wavefunction cannot exceed the electron first minibandwidth Δ_e (if not the probability for an electron to tunnel over N wells would be zero) :

$$NeFd \approx \Delta_e \qquad (4)$$

i) Bleuse et al.[21] used a simple tight-binding approximation and derived the absorption spectrum of an long SL. In this model the envelope function of a carrier centered on the q-th well is :

$$\eta_q = \sum_n C_{n,q}\varphi(z - nd) = \sum_n C_{n,q}\varphi_n(z) \qquad (5)$$

where n is expanded over all the υ wells of the SL, d being the period and $\varphi(z)$ the envelope eigenfunction, when no electric field is applied, of the ground state of an isolated quantum well centered at the origin. When the potential drop over the whole SL is larger than the miniband Δ, $C_{n,q}$ is given by :

$$C_{n,q} = J_{n-q}(\Delta/2eFd) \qquad (6)$$

where $J_m(x)$ is the Bessel function of integer index m, with :

$$J_{-m}(x) = (-1)^m J_m(x) \quad \text{and} \quad \sum_{m=-\infty}^{m=+\infty} J_m^2(x) = 1 \qquad (7)$$

In the theory, the approximation is made that $\varphi_n(z)$ functions are orthogonal. The same approximation is made if two functions corresponding respectively to an electron and a hole located in two different wells are considered. A result of the model is that the oscillator strengths of the -p and -p optical transitions are equal, which is not expected since the total potential has no defined parity. It can be noticed that the intra-well Stark effect is not taken into account in the model which, perhaps, could explain the latter result.

ii) Dignam et al.[22] used a tight binding model which takes into account the intra-well Stark effect through a variational procedure. This model fits the photocurrent (PC) spectra and photocurrent versus voltage under monochromatic illumination (PCV) curves recorded by Agullo-Rueda et al.[23] (Figure 14).

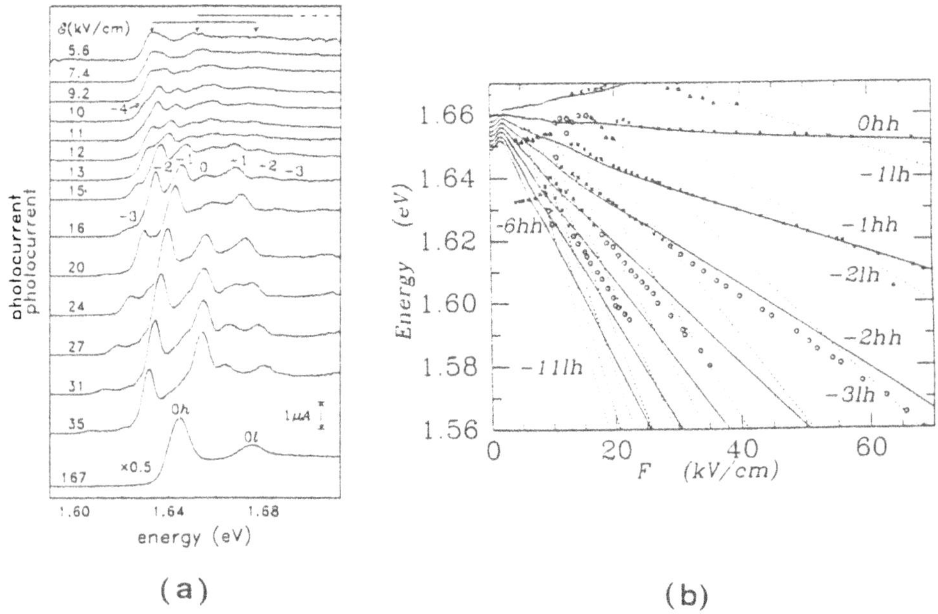

Figure 14. a) Photocurrent spectra for a (40 Å)/(20 Å) GaAs/$Ga_{0.65}Al_{0.35}As$ superlattice at 5 K, for selected values of the electric field applied perpendicular to the layers. Spectra have been offset vertically for clarity. The peak labels at low field give the Stark ladder index p of electrons for heavy-hole to conduction-band transitions (23).

b) The calculated heavy-hole (solid lines) and light-hole (dashed lines) exciton energies for the states - 6hh $\rightarrow$ +1hh and - 11lh $\rightarrow$ -11h as a function of the applied electric field. Also shown are the experimental photocurrent peak data of (a). The crosses (triangles) denote what appear to be due to the heavy-hole (light-hole) excitons ; the circles denote data points that the authors of ref. 22 feel are either of mixed light-heavy character of were improperly characterized in ref. 23 (22).

iii) Tronc proposed a perturbative model both for interband[24] and intraband[25] transitions. This model can be based on different sets of unperturbed envelope functions : Bloch states for long SL's, states taking into account the SL edge effects, ... It provides an asymmetry between the +p and -p transition oscillator strengths and predicts that the values of p with the maximal asymmetry are the two integers just superior and inferior to N/4, N being defined in (4). These results are in agreement with the measurements by Agullo-Rueda.[23]

As further property of heterostructures under an electric field is worth to be noticed[26] : in the (x,y) plane, an electric field applied along the growth direction lifts the two-fold degeneracy arising from the mirror symmetry of square QW's. Figure 15 shows in plane dispersion curves of valence subbands in QW's subjected to an external electric field parallel to the growth direction.

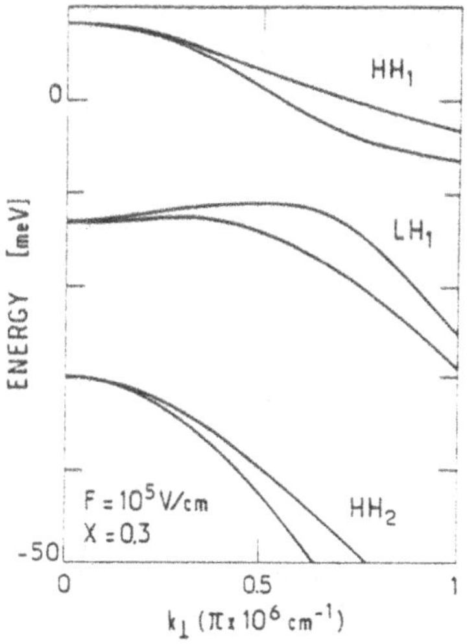

Figure 15. In-plane dispersions of the valence subbands of a GaAs-$Ga_{0.7}Al_{0.3}As$ quantum well (L=100 Å) subjected to an external electric field F//z ($F=10^5$ V/cm). The twofold spin degeneracy prevailling at $k_\perp=0$ is lifted at $k_\perp \neq 0$ (1).

1.3. OPTICAL TRANSITIONS UNDER AN ELECTRIC FIELD

When no electric field is applied to the structure, the matrix element at k=0 for an interband optical transition in proportionnal to the overlap of the initial and final envelope functions whereas it is proportionnal to the momentum matrix element for an intraband one[1,2]. There are selection rules over the indexes of the initial and final minibands ; they are based on the parity of their indexes. There are also selection rules over the direction of the light. In particular the intraband transition are forbidden with light propagating along the growth direction. Due to the heavy- and light-hole mixing, this selection rule is lifted at $k_\perp \neq 0$ for the holes[27]. This effect allows to design infrared photodetectors made of heavily p-doped MQW's. For electron it can be shown[28], using a 8-bands Kane model, that the same effect does exist, even though it is weaker, since it is induced through the coupling between electron and hole states.

An electric field parallel to the growth axis breaks the mirror symmetry of square QW's and SL's. The mirror is parallel to the layers and located at the center of the well (QW) or at the center of any QW or barrier (infinite SL's). The potential of such nanostructures is even under the above symmetry transformation, whereas the applied electric potential is odd. The total potential has then no longer a defined parity. As a result the even an odd unperturbed eigenstates are mixed and the optical transitions which where formerly forbidden from parity become weakly allowed.

2. Some optical probes for the electron states under an electric field

2.1. THE BUILT-IN FIELD AND THE PHOTOVOLTAIC FIELD

To apply an electric field onto an heterostructure one many times includes it as the intrinsic part of a pin diode or evaporates a metallic contact onto the surface to achieve a Schottky diode. When no external electric field is applied, the Fermi level remains constant throughout the structure, inducing a built-in-voltage within the intrinsic part of the structure. The actual situation is generally more complicated. For example the Wannier-Stark electro-optical modulator structure shown in figure 16 involves two non-intentionnaly doped cladding layers of 750 Å $(GaInAs)_{1-Q}(AlInAs)_Q$ material, the Q index varying linearly versus z from 0.5 to 0.91.

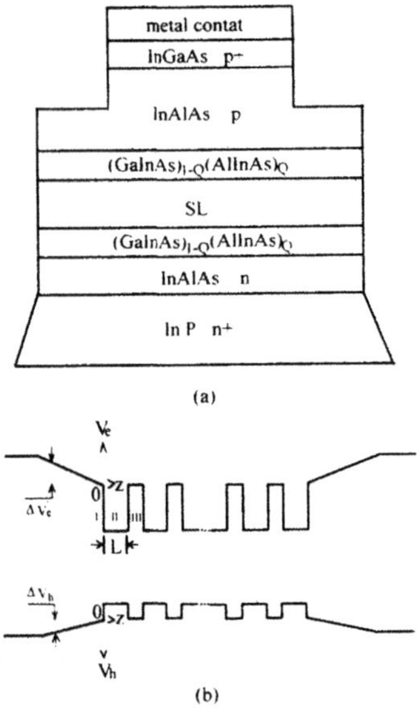

Figure 16. (a) Cross sectional view of a Wannier-Stark electro-optical modulator.
b) Potential profile of the SL in flat band condition. The SL is that presented in Figure 12. Courtesy G. Wang.

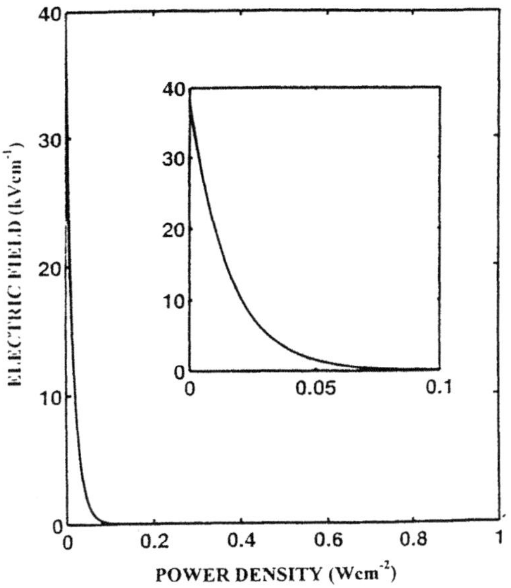

Figure 17. Electric field versus the exciting light power density P in the SL of the structure shown in Figure 16 (29).

When an external electric field is applied, the voltage drop through the above cladding layers depends on their residual doping. Moreover, when the structure is shined, the photovoltaic effect induces an electric field in a direction opposite to that of the built-one. This effect increases with the intensity of the light beam. Figure 17 displays, when no external voltage is applied, the variation of the electric field in the SL of the structure shown in figure 16 versus the exciting light power density P. The curve is obtained by resolving simultaneously the Poisson equation and the current continuity one[29]. The residual doping in the cladding layers has been assumed to be 10^{15} cm^{-3}. Last, in structures involving a large number of heterojunctions, as SL's do, carriers accumulated at the heterojunctions induce a field in a opposite direction to that of the applied one. This effect, which increases with P, can be strong and reduce the effective voltage in the SL by a factor of several units.[30,31]

2.2. ANTICROSSINGS INDUCED BY AN APPLIED ELECTRIC FIELD

Many optical experiments dealing with, for example, absorption, photoluminescence (PL), photoluminescence excitation (PLE), time-resolved photoluminescence (TRPL), PC, PCV, electroreflectivity (ER), photoreflectivity (PR) have been widely used for studying electron states in heterostructures (see for example references 1 and 2). We will here only recall few examples of anticrossing between two energy levels which occurs from an applied electric field.

i) When the two QW's of a double quantum well (DQW) structure have different widths (Figure 18(a)), the electron ground state in both QW's can be aligned using an electric field applied parallel to the growth axis. The tunneling probality accross the barrier between both QW's is then dramatically increased for the electrons. Unless the barrier is very thin, this effect is very weak for the heavy hole, due to its large effective mass. Ferreira et al. studied[32] by PL and PLE experiments such a $GaAs/Al_{0.3}Ga_{0.7}As$ structure with a thick (55 Å) barrier (Figure 18(b)). They concluded that the heavy-hole exciton anticrossing arises from the transfert of electrons from one well to the other.

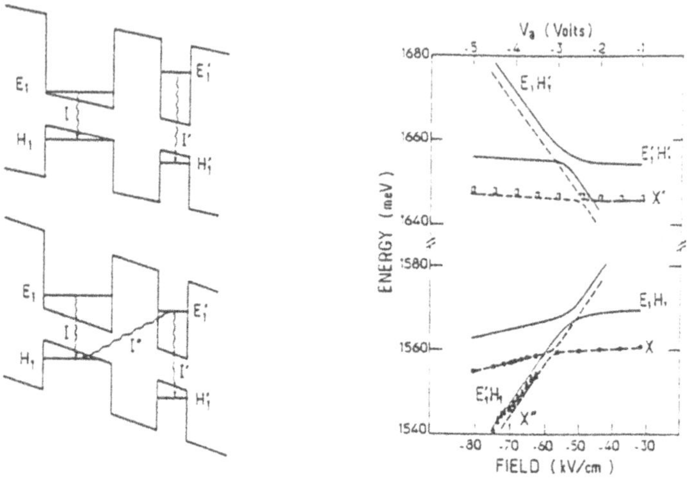

(a) (b)

Figure 18. Photoluminescence study at 2 K under an electric field of a structure made of two GaAs QW's (thicknesses 78 and 35 Å) separated by a $Ga_{0.7}Al_{0.3}As$ barrier of width 55 Å (32).
(a) Schematic level schemes of the double quantum well structure. Upper part : low reverse voltage. The ground electron and hole states are localized in the thick QW (E_1 and H_1). Lower part : large reverse voltage. The ground electron state becomes localized in the thin QW (E'_1).
(b) Curves : theoretical calculations of band-to-band transitions (solid lines) and associated excitonic transitions (dashed line) versus the electric field F near the E_1-E'_1 anticrossing. The squares, points and triangles are the experimental findings. The determination of the experimental F (i.e. of the correspondence with applied voltage V_S) is made by a best fit of the X exciton energy.

Fox et al. studied by PC experiments[33] the anticrossing between the heavy-hole exciton levels associated respectively with the ground and the first excited electron states in symmetric $GaAs/Al_{0.3}Ga_{0.7}As$ DQW's with a thin barriers (15 to 35 Å) (Figure 19). They concluded that the exciton anticrossing arises from the bonding and antibonding electron orbitals based on the electron ground state orbital in one well and the electron first excited state orbital in the other (Figure 20).

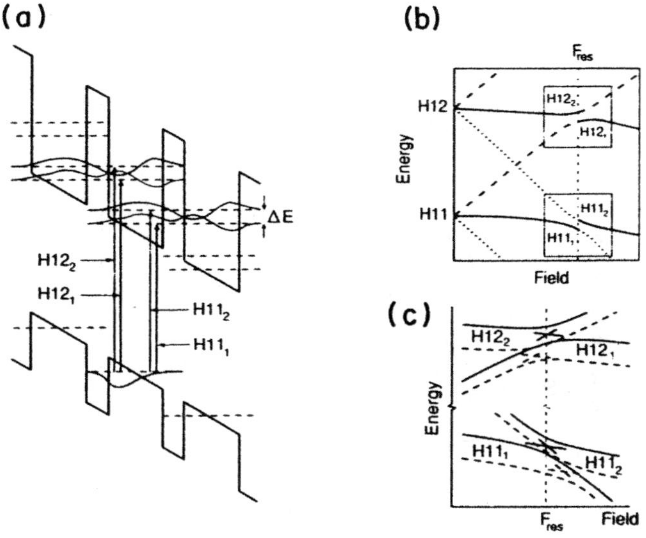

Figure 19. (a) Schematic band diagram of three wells from within a resonantly coupled multiple-quantum-well structure. The H11 and H12 intrawell doublet transitions are indicated, together with the hh1 and mixed e1-e2 wave functions. The e1-e2 splitting ΔE is exaggerated for clarity. (b) Field dependence of the H11 and H12 optical transition energies. Intrawell transitions are indicated by the solid lines, while interwell transitions are shown by broken and dotted lines. (c) Expanded version of (b) showing the detailed field-dependent optical transition energies close to resonance with (dashed lines) and without (solid lines) the electron-hole Coulomb interaction (33).

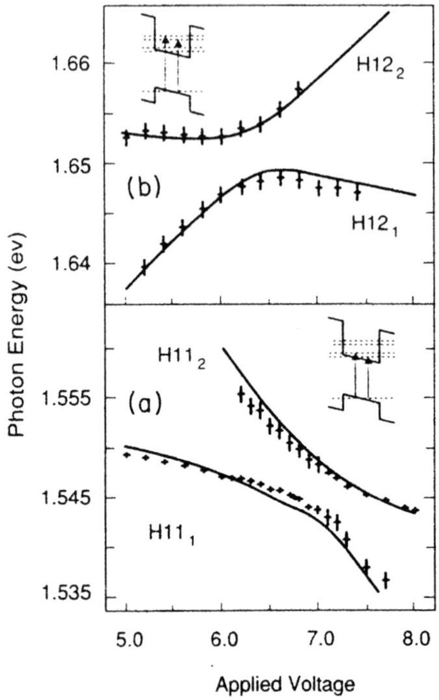

Figure 20. Photocurrent measurements of the exciton energies at 30 K in a 80 periods (95 Å)/(35 Å) $GaAs/Al_{0.33}Ga_{0.67}As$ SL for (a) the H11 and (b) H12 optical transitions defined in Figure 19. The solid lines show the results of a variational calculation of the exciton transition energies (33).

ii) From PLE experiments performed on a $GaAs/Al_{0.35}Ga_{0.65}As$ QW under an electric field, Viña et al.[34] put into evidence an anticrossing between the 1s light-hole exciton ground state and the 2s and 2p excited states of the heavy-hole exciton (Figure 21). The interaction between the exciton states is allowed from parity. It arises from the valence-band hybridization between heavy- and light-hole states and is induced by the electric field. The 2p exciton state is optically active due to the electric field.

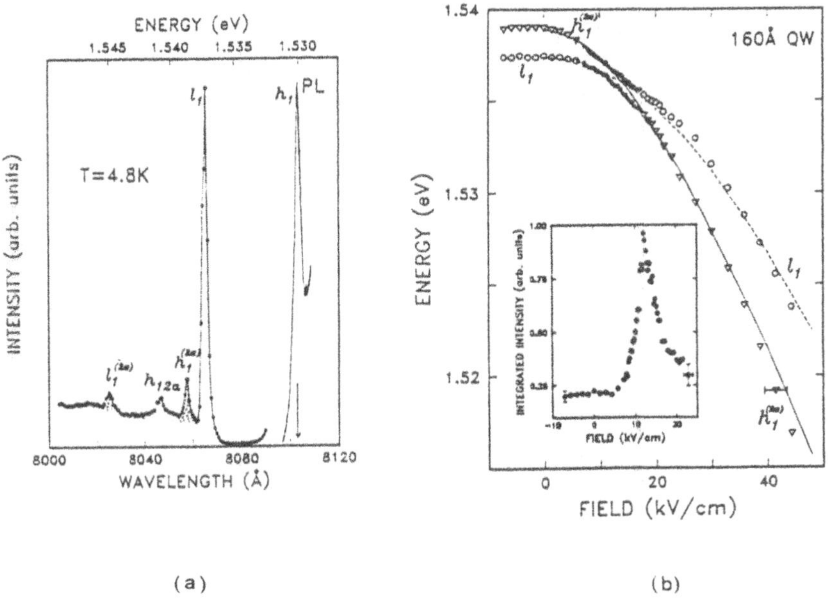

Figure 21. Excitonic coupling in a GaAs/$Ga_{0.65}Al_{0.35}As$ QW of width 160 Å in an electric field.
(a) Low-temperature excitation spectrum at 7.1 kV/cm. h_l and l_l denote the ground states of the heavy and light excitons, respectively. The shaded structures $h_l^{(2x)}$, and $l_l^{(2x)}$, correspond to excited states of the heavy- and light-hole excitons, respectively. h_{12a} is an exciton related to the n=1 conduction and n=2 heavy-hole valence band. Also shown is the photoluminescence (PL) spectrum of the heavy-hole exciton. The arrow indicates the setting of the spectrometer for the excitation spectrum.
(b) Energy thresholds of the ground state of the light-hole exciton (l_l, open circles) and excited states of the heavy-hole exciton ($h_l^{(2x)}$, triangles) as functions of electric field strength. The data are shown as solid circles in the region of strong coupling. The lines correspond to calculations of Stark shifts with use of the effective-mass approximation and neglect of changes in binding energies and coupling among excitons. Inset : the integrated intensity of the $h_l^{(2x)}$ structure normalized to that of h_l as a function of electric field. Error bars show the uncertainties in field and intensities (34).

iii) Meynadier et al. have put into evidence[35] an anticrossing between the X and Γ transitions in GaAs/AlAs [001] SL's when an electric field is applied along the growth axis (Figure 22).

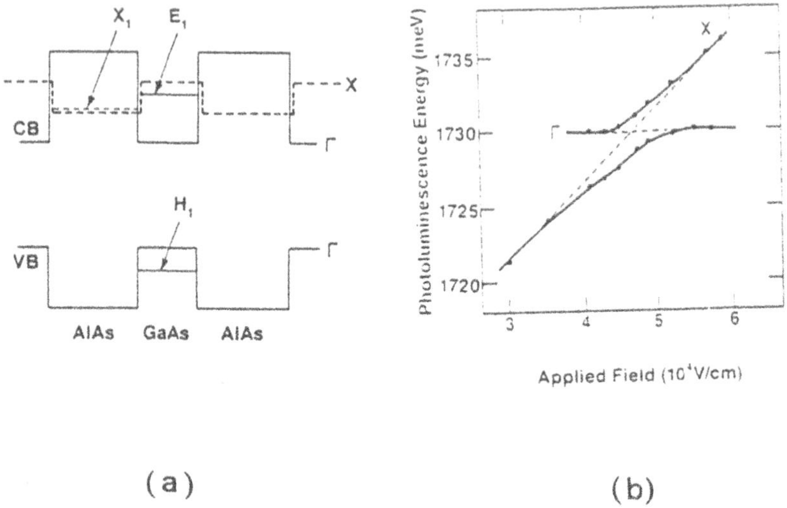

Figure 22. Photoluminescence study at 4 K of a 43 periods (35 Å)/(80 Å) GaAs/AlAs SL.
(a) X (dashed lines) and Γ (full line) conduction- and valence-band profiles in a type-II GaAs-AlAs superlattice. X1, E1 and H1 refer to the lowest quantum states of the X electron, Γ electron , and Γ heavy hole, respectively.
(b) Energies of the direct and indirect transitions at high fields showing the anticrossing between the X and Γ transitions. The linear blue shift of the X transition far from the anticrossing follows the drop of voltage across a half period. The red quantum confined Stark shift is saturated in this field regime (35).

iiii) Schneider et al. have shown[36] anticrossings between their PCV curves measured on GaAs/AlAs [001] SL's (Figure 23). One can notice that anticrossings are resolved only at large electric field strengths, for example between the OC_1-H_2 and -$1C_2$-H_2 transitions. The notations nC_i-$H(L)_j$ refers to the transition between the ith electron and the jth heavy (light) -hole subbands with a spatial separation of n SL periods.

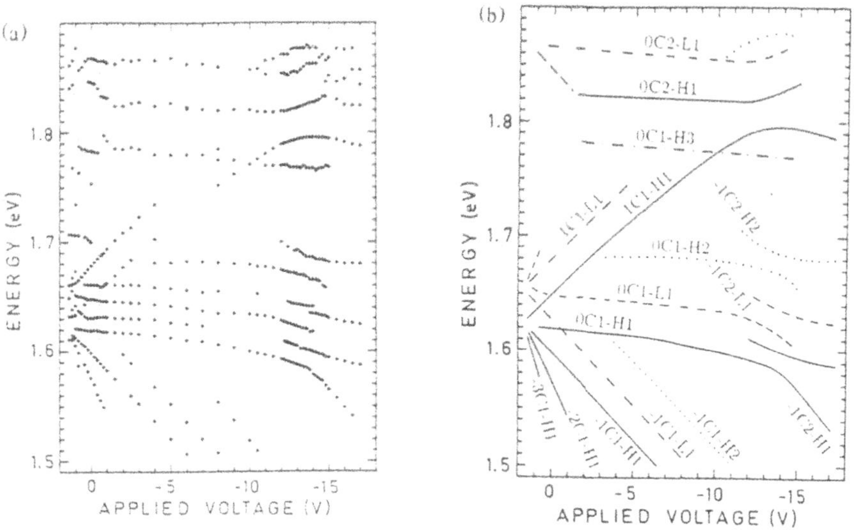

Figure 23. Photocurrent versus voltage study of a 60 periods (70 Å)/(9 Å) GaAs/AlAs SL(36).
(a) Peak energies of the spectra vs applied voltage. (b) Assignment of these peak energies to excitonic transitions. The lines refer to excitonic transitions due to the first heavy-hole (solid lines), first light-hole (dashed), second heavy-hole (dotted), and third heavy-hole (dash-dotted) subbands. The notation nCi-H(L)j is explained in the text.

3. Some devices made of heterojunctions

Many devices are based on heterostructures, for example transistors (heterojunctions), low threshold and long life-time lasers (MQW's). It is possible by doping in the barriers to achieve very high mobility quasi-2d electron gas in the wells since there are no impurities to scatter the carriers. We will consider here only two particular devices which are the MQW's infrared photodetectors and the electro-optical modulators based on confined Stark effect in MQW's or on Wannier-Stark effect in SL's.

3.1. MQW'S INFRARED PHOTODETECTORS.

These infrared photodetectors are generally made of heavily doped GaAs QW's separated one from the other by thick AlGaAs barriers. The heavy doping increases the number of carriers available for absorption. In the case of p-type doping holes with large $k_\perp$ values are created which lift the selection rule on the direction of the incident light, making allowed the intraband transitions with a light vector perpendicular to the plane of the layer. In n-type photodetectors the relaxation of the selection rule on the direction of the light is obtained by using a grating. The most widely studied detectors are n-type, even though p-type ones have been proposed[37] to take advantage of the large effective mass of the heavy hole along the growth direction. This is expected to reduce their contribution to the dark current. Indeed the dark current arises both from the carriers with an energy over the barriers and from the carriers in the well which tunnel through the left barrier (Figure 8). The dark current increases both with temperature since high energy states become more populated and with the applied voltage which, in particular, decreases the effective thickness of the left barrier and enhances the tunneling probability. Due to the heavy doping many-bodies effects (Hartree, exchange and correlation energies) have generally to be taken into account to calculate the carrier energy and envelope function. The photocurrent is generated by absorption of a photon between a state bound in a well and generally an unbound one. LO phonons seen to play a role in the process since the longitudinal lattice vibrations, contrary to the transverses ones, generate a strong local electric field by moving apart the ions and the carriers. The coupling of carriers with LO phonons is well described by the Fröhlich Hamiltonian[38]. The photogenerated free carriers undergo multiple scattering, in particular on the GaAs/AlGaAs interfaces which very quickly lower their kinetic energy.

3.2. ELECTRO-OPTICAL MODULATORS

Such devices can be achieved using the confined Stark effect in QW's since the electric field lowers the energy of both the electron and the hole, therefore red-shifting the absorption line of the structure.

The devices can also be based on Wannier-Stark effect in SL's. The blocking state is achieved using the -1 transition of the Stark ladder whereas the on-state corresponds to zero voltage applied to the structure (Figure 24). When compared to the confined Stark effect modulator, the Wannier-Stark one has an advantage : its control voltage[39] is of the order of 0.8 V instead of several V. It has also a disadvantage since the maximal power beam it can operate is limited to 10 meV approximately. This limitation arises probably from carriers which are trapped at the numerous interfaces and whose number increases with the beam power[30,31]. These carriers generate an electric field in the direction opposite to that of the external applied field. If one considers the long range telecommunication systems operating at 1.55 µm both kinds of modulator are very fast (of the order of 20 Ghz).

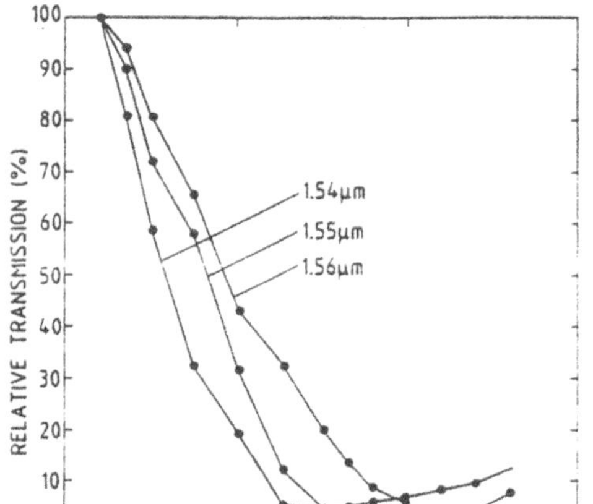

Figure 24. Relative transmission of the TE-guided mode vs reverse bias voltage for different wavelengths in a 10 periods (60 Å)/(10 Å) InGaAs/InAlAs SL. Waveguide length is 560 µm (24).

References

1. Bastard, G. (1990) Wave Mechanics Applied to Semiconductor Heterostructures, Les Editions de Physique, Les Ulis, and references therein.
2. For a brief review of electronic properties of heterostructures see : Tronc, P. (1995) Electrons and Holes in Quantum Structures, in M. Balkanski and I. Yanchev (eds.), Fabrication, Properties and Applications of Low-Dimensional Semiconductors, Kluwer Academic Publishers, Dordrecht, pp. 153-172.
3. Leonelli, R., Tran, C.A., Brebner, J.L., Graham, J.T., Tabti, R., Masut, R.A., and Charbonneau, S. (1993) Optical and Structural Properties of Metalorganic-vapor-phase-epitaxy grown InAs Quantum Wells and Quantum Dots on InP, Phys. Rev. B 48, pp. 11135-11143.
4. Violotis, V., Grousson, R., Lavallard, P., Ivchenko, E.L., Kiselev, A.A., and Planel, R. (1994) Absorption Coefficient in Type-II GaAs/AlAs Short-period Superlattices, Phys. Rev. B 49, pp. 2576-2584.
5. Romanov, N.G., Baranov, P.G., Mashkov, I.V., Lavallard, P., and Planel, R. (1994) Optically detected magnetic resonance study of the transition from pseudodirect type-II to type-I GaAs/AlAs superlattices, Solid-States Electron. 37, pp. 911-914.
6. Li, G.H. (1992) Type-I-type-II transitions in GaAs/AlAs superlattices, in Z.C. Feng (ed.), Semiconductor Interfaces and Microstructures, World Scientific, Singapore, pp. 120-148.
7. Scalbert, D., Cernagora, J., Benoit a la Guillaume, C., Maaref, M., Charfi, F.F., and Planel, R. (1989) Nature of the lowest electron states in short period GaAs/AlAs superlattices of type-II, Solid. State Commun. 70, pp. 945-949.
8. Tessier, R., (1992) PhD Thesis, Paris.
9. Sham, L.J. and Lu, Yan-Ten (1989) Theory of electronic structures in superlattices, J. Lumin. 44, pp. 207-221.
10. Ge, Weikun, Schmidt, W.D., Sturge, M.D., Pfeiffer, L.N. and West, K.W. (1994) Electronic states in GaAs-AlAs short-period superlattices : energy levels and symmetry, J. Lumin. 59, pp. 163-184.
11. Kitaev, Yu. E., Panfilov, A.G., Tronc, P. and Evarestov, R.A. (1995) Electron states symmetries and optical transitions in $(GaAs)_m(AlAs)_n$ short-period superlattices, in Russian Academy of Sciences (ed.), Nanostructures : Physics and Technology, pp. 130-132.
12. Jouanin, C., Bertho, D. and Benoit, C. (1993) Orthogonalized-moment method and the study of the electronic structure of heterostructures : Application to CdTe/ZnTe superlattices, Phys. Rev. B 47, pp. 3706-3713.
13. Ando, Tsuneya (1985) Hole subband at GaAs/AlGaAs heterojunctions and quantum wells, J. Phys. Soc. Jpn. 54, pp. 1528-1536.
14. Masselink, W.T., Chang, Yia-Chung and Morkoç, H. (1985) Acceptor spectra of $Al_xGa_{1-x}As$-GaAs quantum wells in external fields : Electric, magnetic and uniaxial stress, Phys. Rev. B 32, pp. 5190-5201.
15. Miller, R.C., Kleinman, D.A., Tsang, W.T. and Gossard, A.C. (1982) Extrinsic photoluminescence from GaAs quantum wells, Phys. Rev. B 32, pp. 3871-3877.
16. Sham, L.J. and Kohn, W. (1966) One-particle properties of an inhomogeneous interacting electron gas, Phys. Rev. 145, pp. 561-567.

17. Stern, F. and Das Sarma, S. (1984) Electron energy levels in GaAs-$Ga_{1-x}Al_xAs$ heterojunctions, Phys. Rev. B 30, pp. 840-848.
18. Betbeder-Matibet, O., Combescot, M. and Tanguy, C. (1994) Coulomb energy of a quasi-2d electron gas in a quantum well, Phys. Rev. Lett. 72, pp. 4125-4128.
19. Antosiewicz, H.A. (1970) Bessel functions of fractional order, in M. Abramowitz and I.A. Stegun (eds.) Handbook of mathematical functions, Dover Publications, New York, pp. 435-478.

20. Bastard,G., Mendez, E.E., Chang, L.L. and Esaki, L. (1983) Variational calculations on a quantum well in a electric field, Phys. Rev. B 28, pp. 3241-3245.
21. Bleuse, J., Bastard, G. and Voisin, P. (1988) Electric-field-induced localization and oscillatory electro-optical properties of semiconductor superlattices, Phys. Rev. Lett. 60, pp. 220-223.
22. Dignam, M.M. and Sipe, J.E. (1991) Exciton Stark ladder in semiconductor superlattices, Phys. Rev. B 43, pp. 4097-4112.
23. Agullo-Rueda, F., Mendez, E.E. and Hong, J.M. (1989) Quantum coherence in semiconductor superlattices, Phys. Rev. B 40, pp. 1357-1360.
24. Tronc, P. (1992) Oscillator strengths of the optical transitions in a semiconductor superlattice under an electric field, J. Phys. I France 2, pp. 487-499.
25. Tronc. P. (1994) Oscillator strengths of intraband transitions in a semiconductor superlattice in an electric field, Phys. Stat. Sol. (b) 182, pp. 393-389.
26. Bastard, G. in ref 1, p. 111.
27. Chang, Yia-Chung and James, R.B. (1989) Saturation of intersubband transitions in p-type semiconductor quantum wells, Phys. Rev. B 39, pp. 12672-12681.
28. Yang, Rui Q., Xu, J.M. and Sweeny, M. (1994) Selection rules of intersubband transitions in conduction-band quantum wells, Phys. Rev. B 50, pp. 7474-7482.
29. Wang, G., Demory, R., Tronc, P., Depeyrot, J., Melliti, R., Harmand, J.C., Palmier, J.F., Kochereshko, V.P. and Platonov, A.C. (1995) Observation of the Wannier-Stark ladders associate to the light-hole ground state and to the heavy-hole first excited state in GaInAs/AlGaInAs superlattices, to appear in Nuovo Cimento D.
30. Tronc, P., Cabanel, C., Palmier, J.F. and Etienne, B. (1990) Stark localization in GaAs-GaAlAs superlattices under a low electric field, Solid. State Commun. 75, pp. 825-829.
31. Devaux, F. (1993) PhD Thesis, Paris.
32. Ferreira, R., Delalande, C., Lin, H.W. Bastard, G., Etienne, B. and Palmier, J.F. (1990) Resonances in the excitonic transfert in biased double quantum wells, Phys. Rev. B 42, pp. 9170-9173.
33. Fox, A.M., Miller, D.A.B., Livescu, G., Cunningham, J.E. and Jan, W.Y. (1990) Excitonic effects in coupled quantum wells, Phys. Rev. B 42, pp. 1841-1844.
34. Viña, L., Collins, R.T., Mendez, E.E. and Wang, W.I. (1987) Excitonic coupling in GaAs/GaAlAs quantum wells in an electric field, Phys. Rev. Lett. 58, pp. 832-835.
35. Meynadier, M.H., Nahory, R.E., Worlock, J.M., Tamargo, M.C. and de Miguel, J.L. (1988) Indirect-direct anticrossing in GaAs-AlAs superlattices induced by an electric field : Evidence of Γ-X mixing, Phys. Rev. Lett. 60, pp. 1338-1341.

36. Scheider, H., Grahn, H.T., V. Klitzing, K. and Ploog, K. (1990) Resonance-induced delocalization of electrons in GaAs/AlAs superlattices, Phys. Rev. Lett. 65, pp. 2720-2723.
37. Levine, B.F., Gunapala, S.D., Kuo, J.M., Pei, S.S. and Hui, S. (1991) Normal incidence hole intersubband absorption long wavelength GaAs/$Al_xGa_{1-x}As$ quantum well infrared photodetectors, Appl. Phys. Lett. 59, pp. 1864-1866.
38. Thibaudeau, L. and Vinter B. (1994) Phonon-assisted carrier capture into a quantum well in an electric field, Appl. Phys. Lett. 65, pp. 2039-2041.
39. Bigan, E., Allovon, M., Carre, M. and Voisin, P. (1990) Very low drive voltage optical waveguide modulation in an InGaAs/InAlAs superlattice, Appl. Phys. Lett. 57, pp. 327-329.

THE ROLE OF SPONTANEOUS EMISSION IN LASER DIODE OPERATION

PETER BLOOD
Department of Physics and Astronomy
University of Wales Cardiff
PO Box 913
Cardiff CF2 3YB
UK

1. Introduction

When considering the operation of a diode laser it is natural to focus attention on the process of amplification of light by stimulated emission which is the basis of laser action. Nevertheless, even though this stimulated emission dominates the coherent optical output of the laser, for lasers operating at wavelengths below about 1μm it is the somewhat less glamorous process of spontaneous emission which is responsible for the intrinsic current at threshold. The spontaneous recombination current therefore represents the lowest threshold current which can be achieved when all extrinsic processes such as non-radiative recombination and carrier leakage have been eliminated, and thus it determines other intrinsic characteristics of the device, particularly the temperature sensitivity of threshold current.

Rather than attempt to reproduce all my lectures given at Sozopol on quantum well lasers in some condensed form, I have chosen to devote this chapter to a discussion of the fundamental limits to the operation of diode lasers which are imposed by the spontaneous emission process. This was the theme of my final lecture. In so doing I will often ignore significant practical difficulties, such as effects due to defects and impurities, but this is not to say that I regard these as unimportant: indeed control of these effects often makes the difference between failure and success. In discussing these limits I also review the historical development of diode lasers by considering the role of spontaneous emission in their operation, and I assess recent developments such as microcavity devices. This article is therefore a survey of the diode laser field from a particular viewpoint which provides valuable tutorial insight into the manner of their operation, and places the development of the device into a unified context.

M. Balkanski (ed.), Devices Based on Low-Dimensional Semiconductor Structures, 67–90.

While the discussion of spontaneous emission is applicable to all lasers, other intrinsic processes, such as Auger recombination and inter-valence band optical absorption, contribute to the threshold current in lasers at wavelengths beyond 1μm. Consequently, remarks made regarding the impact of spontaneous emission on threshold behaviour have less relevance for these lasers.

The chapter begins by reviewing the fundamental origins of spontaneous emission in semiconductors, contrasting them with the textbook two-level atomic model. I describe the technique we have use to observe spontaneous emission from working lasers, and in section 4 I describe the results of some of those investigations with particular emphasis on the implications for the lasing process. I then give a critical review of the fundamental limitations of diode lasers, and show how key developments in the device have been directed to gaining control over the spontaneous emission process to achieving more effective utilisation of radiative recombination.

2. Recombination and gain in semiconductors

Most textbook treatments of spontaneous and stimulated emission begin by considering optical transitions between two discrete energy states of non-interacting atoms [1]. The electrons are localised within each atom and the behaviour of the system is determined by the distribution of electrons between the upper and lower states. This atomic system is considered to be in equilibrium with the radiation of a classical optical cavity having physical dimensions many orders greater than its wavelength so that the allowed optical modes are sufficiently closely spaced to constitute a continuum distribution in frequency. Even though all frequencies are allowed by the cavity, the absorption, emission and lasing spectra are intrinsically sharp due to the nature of the atomic levels, though there is some natural broadening of order 10^{-8}eV arising from the finite lifetime of electrons in the excited state. In practice the line width is usually dominated by Doppler broadening and by collisions which occur in a real gas. Since the atoms are largely non-interacting the electrons are localised, so pumping of the system to achieve population inversion must be done optically, using radiation which interacts with all the atoms.

In a crystalline solid the atomic spacing is sufficiently small that the wavefunctions associated with the atomic energy levels overlap and interact to form bands of allowed electronic states. The wavefunctions extend through the whole crystal and the electrons are no longer localised. These bands each have quasi-continuous distributions of allowed electron states in energy and, provided electron-electron and electron-phonon scattering is sufficiently rapid, the electron occupancy within each band is given by a thermal Fermi distribution. Optical transitions occur between occupied and empty states in conduction and valence bands (figure 1), and the spontaneous recombination spectrum is determined by a convolution of the energy distributions of (i) electron allowed states, (ii) their thermal occupancy probability, and (iii) the allowed optical modes of the cavity [2]. Typically this spectrum has an intrinsic width of a few (kT), much greater than that of an atomic transition, with only a modest contribution from lifetime broadening at all but the highest recombination rates.

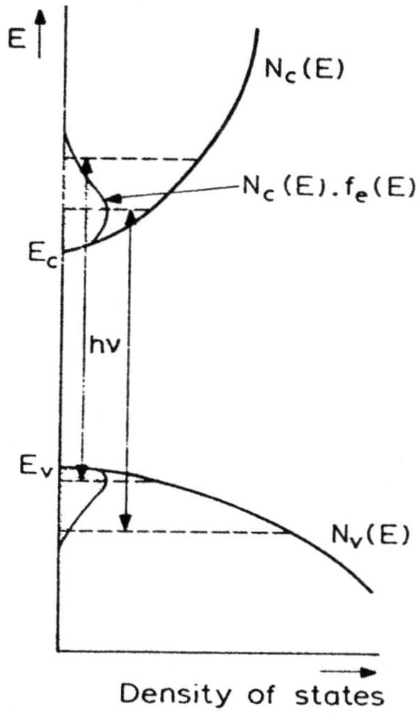

Figure 1. Thermal distributions in energy of electrons and holes injected into the conduction and valence bands of a semiconductor. The carrier distributions are the products of the density of allowed states $N_c(E)$, $N_v(E)$, and the Fermi distribution function $f_e(E)$, $f_h(E)$. Optical transitions occur between full and empty states at the same value of wavevector, **k**.

The optical gain spectrum is also broad, with gain being produced by all transitions with photon energies below the quasi-Fermi level separation, according to the Bernard and Durraffourg condition. Nevertheless the laser spectrum is narrow because, due to the feedback in the cavity formed by the laser waveguide and end-mirrors, photons experience net amplification at the photon energy where the optical gain first matches the losses. As this threshold is passed the stimulated recombination rate builds up rapidly, in proportion to the photon density, and the Fermi level separation and the carrier density become pinned at fixed values (figure 2). Threshold is defined as the condition where the gain first matches the optical losses, just before the photon density in the lasing mode has built up. At this stage the stimulated recombination rate remains small and the dominant recombination process is spontaneous emission, consequently the threshold current density, J_{spon} , is identical to the spectrally integrated spontaneous recombination rate, $r_{spon}(h\nu)$, multiplied by the electronic charge [2]:

$$J_{spon} = e \int r_{spon}(h\nu) dh\nu \tag{1}$$

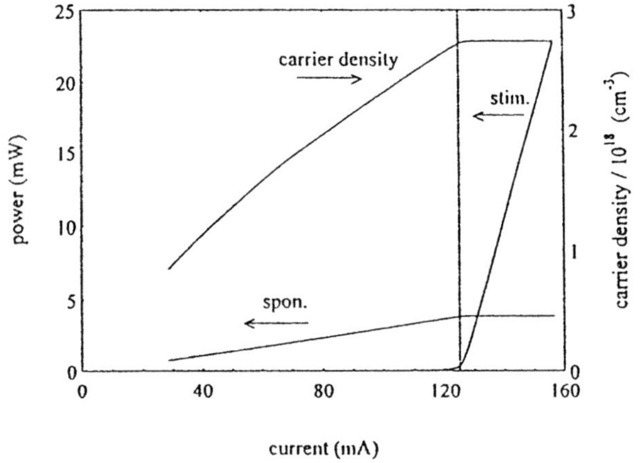

Figure 2. Calculation of the carrier density, spontaneous emission rate and stimulated emission rate as functions of the total recombination current. As the stimulated emission rate builds up the carrier density and the spontaneous emission rate "pin" at fixed values, and it can be seen that at the onset of lasing, ie threshold, the dominant recombination process is spontaneous emission. The calculation follows ref[3] (p 239).

Thus, although the lasing process is itself confined to a very narrow spectrum, in contrast to an atomic system the inverted carrier population density has to be maintained against a spontaneous recombination process which spans a wide energy spectrum, corresponding to the thermal distribution of occupied electron and hole states in the two bands (see figure 1). The laser diode is in fact a voltage controlled device: the peak gain is determined chiefly by the quasi-Fermi level separation and the threshold current is the total current which flows at that voltage. The task of minimising the threshold current is therefore one of minimising the recombination current at this voltage. Because the spontaneous emission spectrum in semiconductors is so wide relative to the width of the laser line, the spontaneous recombination process is critical to the performance of diode lasers. To minimise J_{spon} the strategy should therefore be to reduce the width of the spontaneous emission spectrum, concentrating all the recombination in the vicinity of the laser line.

The laser diode is not only spectrally inefficient but, in common with some other laser systems, it is also wasteful of spontaneous recombination in terms of direction and polarisation. The lasing mode is in a direction along the laser waveguide, and has a specific polarisation, whereas spontaneous emission in a classical cavity can occur in all directions and polarisations. Overall the fraction of spontaneous emission coupled into the lasing mode in an edge emitting diode laser is only about 10^{-3} or 10^{-4} so in principle there is ample room for improvement in the threshold current.

On the basis of detailed studies of 1.3μm lasers [4,5,6], it has been suggested recently that, in addition to the spontaneous emission emitted into the "free space" cavity, below threshold there are fluctuations of the photon density in the laser waveguide modes which produce an additional current in the device. Although these ideas await confirmation, they are physically reasonable and the possibility of an additional current should not be ruled out. Nevertheless, spontaneous recombination accounts for much of threshold current and remarks below regarding reduction and control of spontaneous emission remain valid.

Significant improvements in the intrinsic threshold current of diode lasers can therefore be achieved by striving to concentrate all the spontaneous emission into the laser mode in terms of its frequency, its direction and its polarisation. With hindsight, all significant fundamental advances in laser diodes can be seen as moves toward this goal, even though they were not all originally conceived in this way. This view provides a valuable insight into the effects of quantum confinement and the physics of the temperature dependence of threshold, and it is the theme of the this paper.

Before pursuing this line of thought in more detail, I describe the method we have used to observe spontaneous emission spectra from working lasers, then in section (4) I describe some studies of quantum well lasers, highlighting the influence of spontaneous emission on the working of the device.

3. Observation of Spontaneous Emission

The spectrum observed from the end facet of an edge emitting laser is the convolution of the spontaneous emission spectrum at various points along the length of the device with the spectrum of absorption or gain which this light experiences along its path to the facet. Below threshold this end-emission spectrum is concentrated at low photon energies where the absorption is negligible and, as threshold is approached, low energy photons are amplified, enhancing this part of the spectrum, and light at higher energies is lost by absorption. Above threshold a narrow laser line emerges on a very weak broad background of amplified spontaneous emission. Thus the spectrum of incoherent light emerging from the facet does not represent the true spontaneous emission spectrum which is intrinsic to the gain medium. If, however, the light is observed in a direction perpendicular to the axis of the waveguide then, because the gain generating region in this direction is very short compared to the absorption or gain "length", the emission is not distorted by absorption or amplification.

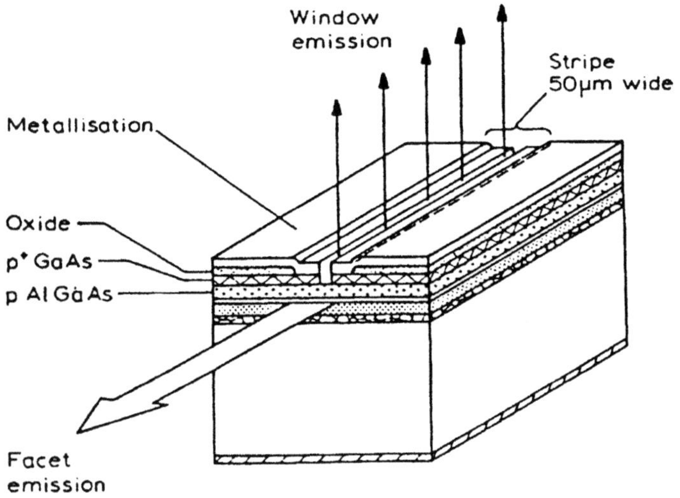

Figure 3 Diagram of an oxide stripe laser with a narrow window opened in the top contact metallisation and contact layer to allow spontaneous emission to be observed in a direction normal to the axis of the laser waveguide. See ref [7].

In long-wavelength lasers grown on InP substrates the substrate is transparent to the radiation from the active region so the spontaneous spectrum can be observed simply through the substrate of the laser. However in shorter wavelength devices grown on GaAs the substrate is absorbing so the window structure shown in figure 3 has been developed for spontaneous emission studies. It comprises an oxide-isolated stripe geometry laser with a narrow window about 4μm wide opened centrally in the metallisation and absorbing upper contact layer along the length of the device. The region below such a narrow window is uniformly pumped across its width by current spreading in the highly doped cladding layer, even though the injected carrier

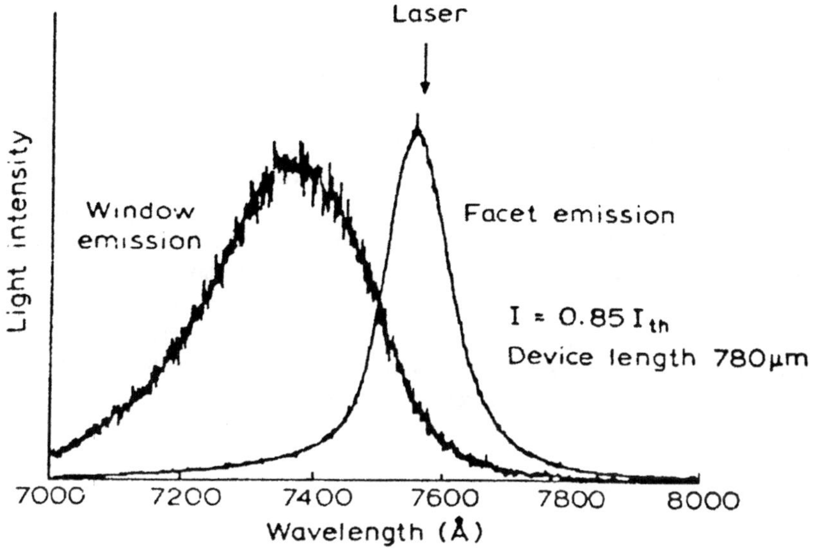

Figure 4. Emission spectra recorded from the end facet and the top window of a GaAs quantum well laser operating at 85% of its threshold current; the two spectra are recorded at different sensitivities and their amplitudes are unrelated. The vertical arrow indicates the wavelength at which laser emission is observed. The effect of absorption and gain along the cavity in concentrating the facet spectrum at long wavelength is very clear.

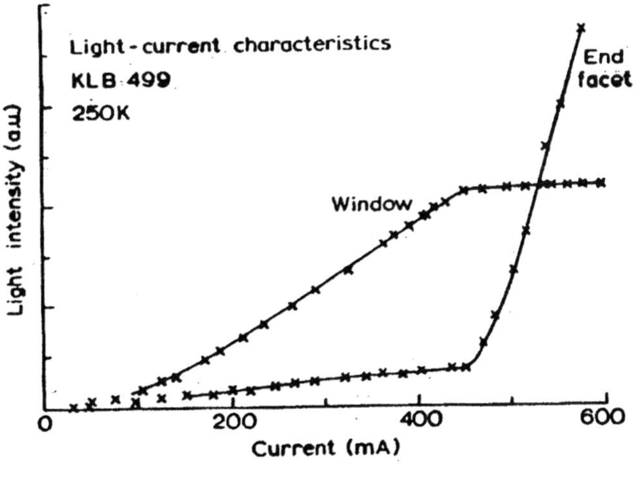

Figure 5. Intensity of spontaneous emission from the window and of "laser" emission from the facet measured as functions of the drive current for a GaAs quantum well laser. [P.Blood, unpublished]

population falls off toward the edges of the stripe. Figure 4 shows a spectrum measured in this way at 85% of threshold compared with the spectrum observed from the facet; it can be seen how the spontaneous emission from the end of the laser is seriously distorted, being concentrated at low energies by absorption during its passage along the length of the device. Details of the device structure and the measurement technique are given in ref [7].

Figure 5 shows the spectrally integrated intensity of spontaneous emission from the window and of "laser" emission from the facet measured as functions of the drive current for a GaAs quantum well laser. It can be seen that the spontaneous emission intensity saturates when the laser emission builds up, providing experimental confirmation that the carrier density, and hence the Fermi level separation, does indeed "pin" at threshold as is commonly assumed and as indicated by the calculation in figure 2.

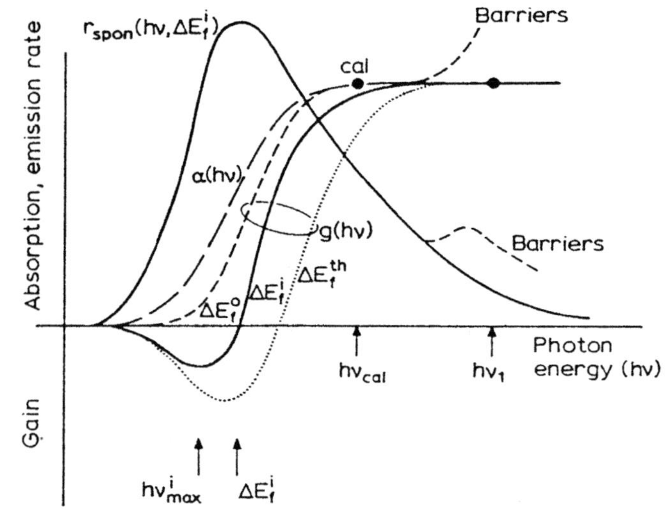

Figure 6. Schematic diagram of the relation between the absorption spectrum in the absence of any injected carriers, α(hν), and the emission and gain spectra when the material is pumped to a Fermi level separation of ΔE_f^i as defined by equation (2).

Through detailed balance arguments there is a simple relation between the spontaneous emission spectrum and the absorption or gain spectrum, g(hν), which depends only upon the separation of the quasi-Fermi levels for electrons and holes, ΔE_f.

$$g(h\nu,\Delta E_f)=(\text{const})(h\nu)^{-2}\left\{1-\exp\left(\frac{h\nu-\Delta E_f}{kT}\right)\right\} r_{spon}(h\nu,\Delta E_f) \tag{2}$$

so the emission spectrum measured as a function of carrier density contains information concerning the non-linear behaviour of the absorption and gain. Figure 6 illustrates the relation between the absorption spectrum in the absence of any injected carriers, $\alpha(h\nu)$, and the emission and gain spectra when the material is pumped to a Fermi level separation of ΔE_f^i as defined by equation (2). As the Fermi level separation is increased so the peak gain increases. In view of the difficulties measuring gain directly these measurements offer an attractive alternative for study of gain media [7,8], though in practice the principal difficulty is determining the Fermi level separation to the required degree of accuracy [8].

4. Studies of Spontaneous Emission

4.1 EMISSION FROM QUANTUM WELLS

We recall that in a quantum well the separation of allowed values of wavevector $\underline{\mathbf{k}}$ across the well is large compared with (kT) and the conduction and valence bands become split into a small number of sub-bands, each having a step-like density of states function [9]. Unlike bulk materials, with their $E^{1/2}$ energy dependence of the density of states, these structures should have an abrupt low-energy edge to their emission spectra corresponding to this truncated distribution of allowed states. It was surprising therefore to find that quantum well lasers operated at photon energies smaller than the measured energy gap separating the abrupt distributions of conduction and valence band states [10]. This suggested that gain was possibly being generated at energies below that of the smallest radiative transition, possibly by simultaneous emission of a phonon. [11]. This question was resolved by measurements of spontaneous spectra from lasers operating at threshold. These showed that at the high injected carrier densities necessary for population inversion, the spectra were broadened on their low energy side (see figure 4) and shifted to lower energy, and recombination was observed at the photon energy of laser emission. The shape of the spectra on the high energy side is due to the thermal distribution of carriers at increasing energy in the bands. By invoking the relation between gain and emission, equation (2), it was shown that band-to-band recombination accounted for the lasing process without the need to invoke phonon participation [12]

Thus, under lasing conditions, the observed emission spectra differ significantly from initial, simple-minded expectations. Further studies of emission spectra as a function of current showed that the energy broadening was made up of two components: one which increased with increasing carrier density and is associated with carrier-carrier scattering (ie the finite lifetime of electrons in the quantum states) and another which was independent of carrier density but varied from sample to sample [13]. This latter component is due to broadening of the density of states distribution by variations in the width of the quantum well along the laser waveguide, and may also include a contribution from carrier-phonon scattering. It is inevitable that there are monolayer steps along the structure and, remembering that the well has two interfaces, we cannot therefore expect its width to be controlled to better than ±1 monolayer. In a 2.5nm wide well the transition energy changes by about 25meV for a monolayer change in width so the broadening can be significant even compared to thermal effects. Moreover, the effect on the transition energy of a monolayer change in well width becomes greater as the well width becomes smaller so, in the quest to increase the sub-band separation by decreasing the well width, the effect of quantum confinement in producing an abrupt density of states distribution is lost.

These observations of broadening of spontaneous emission spectra by well width variations are important because they show there is a limit to the extent to which quantum confinement effects can be exploited in real structures. These ideas would not have emerged from simple measurements of laser emission and threshold current.

The practical consequences of both of these broadening processes is that the intrinsic recombination current at the Fermi level separation necessary to produce a specific value of gain is increased causing an increase in threshold current by as much as a factor two in thin well devices[13]. To illustrate this, calculated broadened and unbroadened gain spectra for the same value of peak gain from a 2.5nm wide GaAs well are shown in figure 7.

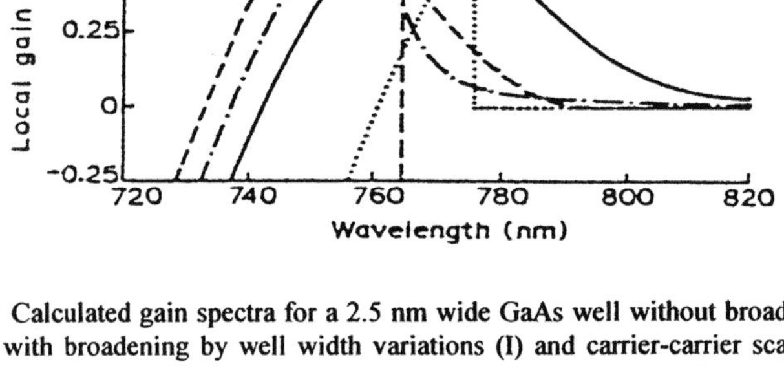

Figure 7 Calculated gain spectra for a 2.5 nm wide GaAs well without broadening, dotted line (V), and with broadening by well width variations (I) and carrier-carrier scattering (II). Line (IV) includes both broadening processes and band-gap narrowing. All curves are calculated to give approximately the same peak gain and the carrier density is given alongside the symbol. [13].

The carrier-carrier scattering process is just one manifestation of "many body effects", others being band gap narrowing (the effects of which are also seen in spontaneous emission spectra), and Coulomb Enhancement. While broadening of the sharp one-electron transitions with an appropriate lineshape function gives a physically intuitive way of treating carrier-carrier scattering, a full many-body theory is necessary to give a rigorous and consistent treatment of these effects [14].

4.2 TEMPERATURE DEPENDENCE OF THRESHOLD CURRENT

One of the most convincing manifestations of the relation between spontaneous emission and threshold current comes from studies of the temperature dependence of threshold. The intrinsic recombination current is predicted to vary as

$$J_{spon} = J_0\, T^{n/2} \qquad (3)$$

according to the dimensionality, n, of the system [15]. This relation follows from the temperature dependence of the effective density of states and the radiative recombination coefficient, and it can easily be shown that in a quantum well this varies linearly with temperature rather than the $T^{3/2}$ dependence characteristic of bulk material. Thus for a quantum well the intrinsic current should vary linearly with temperature. Most measurements of threshold show a linear dependence at low temperatures but an exponential-like increase sets in as T increases above 200-300K [16] so the question naturally arises whether this theory, and with it the fundamental attractions of quantum wells, is really valid. The difficulty in verifying these predictions arises because the externally measured threshold current may include additional contributions from non-radiative and leakage processes which mask the intrinsic behaviour. These matters have been clarified by spontaneous emission studies.

Figure 8 shows a series of spontaneous emission spectra from a GaAs quantum well laser measured at threshold at each temperature [17] The relative areas under each spectrum represent the intrinsic threshold current (equation 1) and the plot in figure 9 shows that this does indeed vary linearly with temperature over the full range as theoretically predicted. The measured current follows this linear behaviour at low temperatures but a contribution in excess of a linear extrapolation appears above about 300K. In this case the excess current is due to recombination in the barrier alongside the well, which explains its strong temperature dependence, and detailed studies show that in these GaAs devices the excess current is due to non-radiative recombination from the X-minimum in the barrier material [18].

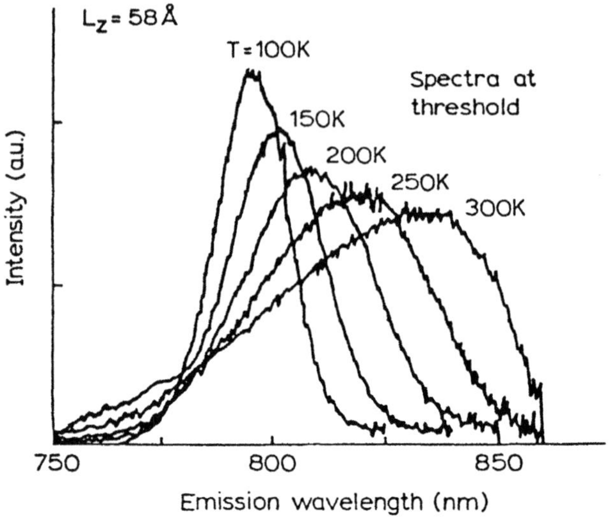

Figure 8. Spontaneous emission spectra measured through a top-contact window at threshold at each of the temperatures indicated for a quantum well laser diode with two 5.8nm wide GaAs wells [17].

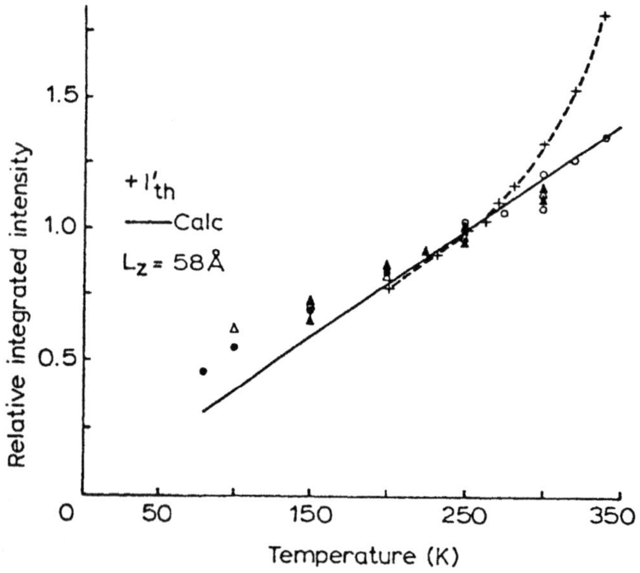

Figure 9. Spectrally integrated spontaneous emission intensity at threshold as a function of temperature for a quantum well laser, obtained from data such as that in figure 8. The crosses are the measured values of threshold current normalised to the intensity data at low temperatures, and the line is obtained from a calculation of the spontaneous recombination current [17].

These experiments showed, for the first time, that the intrinsic recombination current in a quantum well at threshold does indeed vary linearly with temperature as predicted. The significance of the spectra is that they show clearly that the linear increase in intrinsic threshold current is due to the increase in width on the high energy side of the spectra caused by the increasing thermal spread of carriers among the allowed states. To match the optical losses (which are assumed independent of temperature) it is necessary to achieve the same *peak* gain, but because the area under the emission spectrum increases with temperature the current required to sustain this gain increases due to the increased thermal spread of carriers in the bands.

The linear temperature dependence occurs only when the lowest pair of sub-bands are occupied. When the gain requirement is large such that the Fermi level separation is significantly greater than the n=1 transition energy, it is possible for the n=2 sub-bands to become occupied even though the laser line remains within the n=1 gain spectrum. This effectively broadens the spontaneous emission spectrum and leads to a super-linear increase in *intrinsic* current [19] Thus to realise the full intrinsic benefit of quantum confinement it is necessary that the sub-band separation exceeds many (kT) and that the n=1 transitions can meet the gain needed for laser operation. As the sub-band separation is reduced by increasing the well width, so the density of states function approaches that of a 3D material and the attractive linear temperature dependence of threshold is lost.

4.3 EQUILIBRIUM BETWEEN WELLS AND BARRIERS

The processes of capture and subsequent thermalisation of carriers in the wells are possible fundamental limitations to the operation of quantum well lasers; these have to be sufficiently fast to sustain the stimulated recombination process (sub-ns) going on near the bottom of the well. Capture could be a serious problem because the well widths are similar to, and even smaller than, the typical mean free paths of carriers between collisions, so the probability of a carrier in the barrier experiencing an energy-loss collision while in the vicinity of the well could be quite small. The use of a separate confinement heterostructure to provide a waveguide aids the capture process by localising the injected carriers in the vicinity of the well, and this idea was developed further in the graded confinement structure which "funnels" carriers into the well [ref[9] section 23, and [20]] Once in the well, the indications are that carrier-carrier and, more importantly carrier-phonon scattering, is sufficiently rapid ($\approx$10ps) for carriers to loose energy and approach a thermal distribution under lasing conditions. Such a distribution is usually assumed in calculations of laser gain and recombination.

Spontaneous emission spectra provide a means of studying the effectiveness of the capture and thermalisation processes. Figure 10 shows a spectrum measured from a GaAs quantum well laser: in addition to the main peak due to recombination in the quantum well there is a small peak at a shorter wavelength which corresponds to the position of the absorption edge of the barrier material measured by photovoltage techniques [21]. This peak is due to radiative recombination in the barrier adjacent to the well and in this case the integrated intensity is about 1% of the total radiative intensity, showing that in the absence of non-radiative recombination the contribution of the barriers to the radiative recombination current is small. This supports the conclusion in section 4.2 that the barrier contribution to the measured current is non-radiative [22]. The barrier emission is not easily observed from the end facet of the laser because it is strongly absorbed by the quantum well as it propagates down the waveguide.

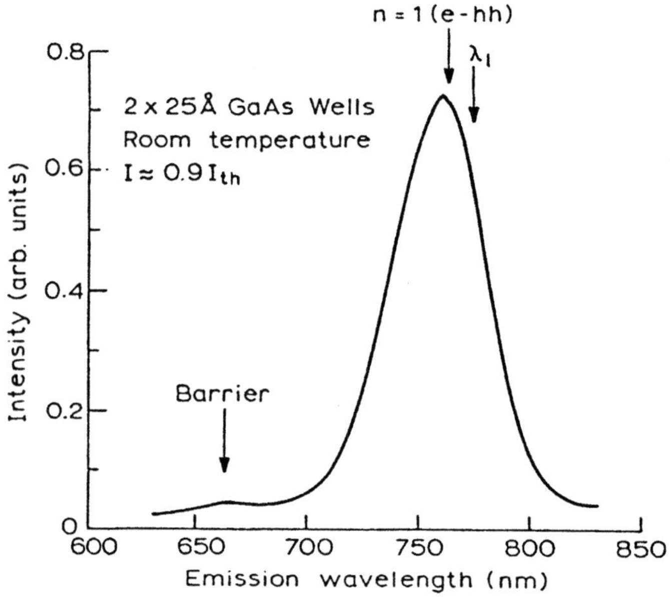

Figure 10 Spontaneous emission spectrum from a GaAs quantum well laser showing weak radiative recombination in the barrier region appearing at short wavelength [22].

If the capture and thermalisation processes are not sufficiently rapid to maintain a thermal distribution of carriers within the well, and between the well and the adjacent barrier, then carriers will "pile up" in the barrier and at high energies in the well, increasing the radiative recombination rate in the corresponding regions of the emission spectrum. This should be observable in steady-state emission spectra and causes an increase in the intrinsic recombination current.

The spontaneous emission spectrum is given by [7]

$$r_{spon}(h\nu, \Delta E_f) = (\text{const})(h\nu)^2 \alpha(h\nu) \exp\left\{\frac{\Delta E_f - h\nu}{kT}\right\} \quad (4)$$

and at high photon energies where the absorption spectrum α(hν) is flat the spectrum at a given Fermi level separation has the form

$$r_{spon}(h\nu)=(\text{const})(h\nu)^2\exp(-h\nu/kT) \quad (5)$$

Here T is the temperature appearing in the Fermi distribution function which therefore represents an effective carrier temperature. Equation (5) shows that in the appropriate spectral region a semi-logarithmic plot of the observed emission divided by $(h\nu)^2$ should be linear with a slope inversely proportional to the carrier temperature. Measurements on GaAs wells show that such a plot is linear and that the carrier temperature is close to the lattice temperature indicating efficient thermalisation within the well under lasing conditions. Furthermore, the emission from the barrier appears to have the same high-energy slope, though this is difficult to determine accurately and the interpretation is complicated by the fact that in bulk material the density of states distribution, and hence the absorption spectrum, is not constant in energy so equation (5) is not such a good approximation in this case.

A more accurate assessment of the equilibrium between well and barrier can be obtained by integrating the appropriate regions of the spectra (for example figure 10) to obtain the ratio of light emitted from the barrier to that from the well. Any "pileup" of carriers in the barrier due to inefficient capture will enhance this ratio above that expected from an equilibrium calculation. For GaAs wells as thin as 2.5nm we find that the observed ratios agree with an equilibrium model [23], which justifies the use of such a model in theoretical calculations for these structures.

We must qualify this conclusion however. Firstly, although these experiments were done under lasing conditions they only apply to operating currents near threshold (ie low power) and to steady state operation. Under the measurement conditions the stimulated time constant is ≈100ps so all we can conclude is that the equilibriating processes are somewhat faster than this. However, when driven to higher power the stimulated rate increases and the capture rate may eventually become a limitation. Similarly, attempts to modulate the lasers at frequencies in excess of a few GHz may also become limited by capture processes. Secondly, capture processes may be slower in other materials and structures, and there is considerable data to support the view that in long wavelength lasers carrier capture and transport (especially of low-mobility holes) are serious rate limiting steps, even in steady state operation [24]. The impact of rate limiting processes on the modulation response of lasers has understandably concentrated attention on dynamical studies as the means of investigating carrier capture processes. However these are not always straightforward to interpret and steady state spontaneous emission studies could be developed further to provide insight into these effects.

4.5 SUMMARY

These studies reveal several aspects of the intimate relation between the emission spectrum and operation of the laser. In particular we see that an increase in the width of the spectrum with increasing temperature produces a linear increase in intrinsic threshold current. Measurements of the total current show an excess current at high temperatures due to extrinsic recombination in the barrier. We have seen that dimensional and many-body effects, and possible carrier pile-up due to slow capture, broaden the emission spectrum and increase the threshold current because of the reduced utilisation of spontaneous emission in the lasing process.

5. Fundamental limits to diode lasers

5.1 THE PROBLEM

In the previous section we have given several examples of experimental observations of the key role played by spontaneous emission in determining the intrinsic threshold current of diode lasers. In the limiting case of negligibly small rates of non-radiative recombination and carrier leakage, spontaneous emission determines the fundamental limits to the performance of these devices. As we noted in section 2, and as the examples in section 4 demonstrate, the diode laser is very wasteful because of the wide spontaneous emission spectrum compared with the laser line. This arises because the allowed energy states in a crystalline solid form bands.

Get rid of the bands and the problem is solved! It's not quite so straightforward as that. The formation of bands allows the electrons to move throughout the material, providing the conduction path to the active region. More importantly, the bands with their energy gap enable us to make a p-n junction by which we can inject very high concentrations of minority carriers to achieve population inversion. These attributes of a semiconductor enable inversion to be achieved by direct electrical pumping which is the distinctive feature of the laser diode. Furthermore, through our ability to engineer the band gap by changing the composition of semiconductor alloys we can change the refractive index and thereby fabricate an integral slab waveguide, again a key feature of the device. We throw away many attractive features if we throw away the energy bands of a solid. Indeed we could take the view that the advantages which follow from using a semiconductor medium far outweigh its inherent fundamental limitations.

Can we get the best of both worlds: atomically sharp emission spectra and a semiconductor-based junction laser? Let us re-examine the origins of the wide emission spectrum, following the discussion in section 2. They are as follows:

(i) the allowed electron states form a closely spaced continuum,

(ii) these states are occupied according to a thermal distribution function, and

(iii) the allowed photon modes of the free-space cavity also form a closely spaced continuum, accepting radiation of any frequency.

5.2 TEMPERATURE

Taking the temperature first, cooling the device to cryogenic temperatures is not an acceptable option. In fact it does not give great improvements because many-body broadening remains and the spectrum retains a significant (though reduced) width as a consequence of this process. We have measured spectra down to 10K and find no reduction in width below about 40K for GaAs quantum well devices.

5.3 DENSITY OF ELECTRON STATES

5.3.1 *Quantum confined structures*

A more attractive prospect is to reduce the width of the bands, and this is the basis of the improvements brought about by quantum confinement: the progression from 2D well, to 1D wire, to 0D dot leads to a progression in the density of states function from a parabola, eventually to a delta function: the engineered atom! [ref [9] section 28] Increasing degrees of quantum confinement produce a reduction in the width of the emission spectrum by restricting the energy distribution of allowed electron states. Further improvements have been achieved using elastically strained quantum wells produced intentionally by growth on a mismatched substrate. The distortion of the unit cell increases the separation of light and heavy valence sub-bands in compressively strained structures which reduces the area under the spontaneous emission spectrum, thereby reducing the threshold current [25]. In long-wavelength lasers further improvements are predicted to strain-induced reductions in Auger recombination and inter-valence band absorption.

Very significant improvements have been achieved using confinement by quantum wells, and this suggests that, if the engineering problems can be overcome, following the progression to its limit of the quantum dot will get close to our goal. However, we have shown in section 4.1 that broadening masks the effects of quantum confinement and we should regard broadening by dimensional fluctuations as "fundamental" in that it is most unlikely that a fabrication process will be devised to make semiconductor structures dimensionally accurate to one monolayer. Moreover, as the number of critical dimensions in the structure is increased so the task of keeping them all correct throughout the device becomes more difficult. Self-organising structures may offer the best prospect of realising practical quantum dots and nature does it with atoms and molecules, which is surely a pointer for scientists to follow. Quantum dots made by conventional semiconductor fabrication techniques are unlikely to offer *significant* improvements over quantum wells in reward for the immense effort needed to make them. The quantum well is probably the practical optimum for this class of structure: provided the wells are not too thin they give worthwhile improvements.

5.3.2 *Structures incorporating optically active atoms*

However, as we commented above, nature offers a way achieving precise control of the energy levels by use of molecules or atoms. Suppose we fabricate a semiconductor p-n junction double-heterostructure using indirect gap material throughout to suppress intrinsic radiative recombination, and that we are able to eliminate extrinsic non-radiative processes, then in the active region we introduce a dilute concentration of optically active atoms. If the upper level of the atoms can be populated with electrons from the conduction band and the lower level emptied by loss of electrons to the valence band, then it is possible to achieve population inversion between these levels and hence laser action. In such a structure we retain the attractive features of the diode lasers, namely direct electrical pumping and an integral waveguide, we eliminate the broad emission spectrum by use of an indirect gap material, and we provide isolated ions having a narrow emission spectrum to generate the optical gain.

Such a device is not fantasy. There is considerable interest in doping silicon with optically active erbium, though driven by different motives to the scenario described here. The chief motivation is that the scheme provides possible means of generating laser light in silicon, furthermore the Er emission occurs at 1.55μm, the wavelength used in fibre communications systems. If such a device can be realised it will in fact represent a significant development in the fundamental mechanism of operation of diode lasers, with prospect of significant reductions in the intrinsic threshold current. This structure offers two other advantages as a result of using isolated atoms as the gain medium. The band gap of a semiconductor, and hence the emission wavelength of a diode laser, is temperature dependent, in contrast to the energy of an optical transition in an atom, consequently the device sketched above should have a wavelength almost independent of temperature. The threshold current should also be less dependent upon temperature.

There is one serious fundamental difficulty with this device concept in that the gain generated per unit length is low due to the dilute concentration of the atoms, and to achieve laser operation the gain has to exceed losses from the laser mode as the light passes along the waveguide, caused by imperfections in the waveguide and scattering by free carriers. Assuming that these losses can be exceeded it will nevertheless be necessary to use coated low-reflectivity facets to reduce the mirror loss, and probably a long cavity. Increasing the concentration of atoms increases the gain per unit length but also increases coupling between the wavefunctions of the atomic states, broadening the emission spectra and taking us toward a "atomic solid" structure. There is therefore an important balance to be made between maximising gain while minimising wavefunction coupling and the recombination current.

There are also serious practical problems to be overcome with the Er in silicon embodiment of this device concept. There are difficulties in achieving high concentrations of Er without precipitation, non-radiative recombination in the silicon matrix must be suppressed, and the mechanism by which the Er is electronically coupled to the conduction and valence bands must be fast enough to sustain the stimulated emission process under lasing conditions. Progress is being made [26] and both light emission from a silicon matrix and fundamental improvements in the laser diode mechanism may emerge from this research.

The concept of using an atomic recombination process in a semiconductor matrix is fundamentally sound and it offers improvements in temperature stability of wavelength and current as well as threshold reductions. This approach has inherent advantages over fabrication of quantum dots in terms of both manufacturability and control of dimensions. It represents the limit of engineering the emission spectrum through the density of states distribution. However, while the use of an atomic-like recombination process can achieve a good match between emission and laser spectra, it does not eliminate the waste of spontaneous emission in terms of its direction and polarisation. To deal with this problem we have to turn to the design of the cavity which receives the spontaneous emission.

5.4 CONTROLLING THE PHOTON MODES

Apart from subtleties relating to polarisation, the emission due to recombination in a semiconductor, as that from an atom, occurs in all directions in free space. Even in the waveguide of a semiconductor laser only a small fraction of the emission occurs in a direction and polarisation coupled to one of the guided modes of the structure. For this reason we can ignore the wavegiude in calculating the spontaneous recombination rate. This is of course the downfall of the laser diode which concerns us. The ideal structure is one in which the radiation field can couple only to the laser mode:

this gives us the so-called "threshold-less laser" in which the light output versus current characteristic has no change in slope at the transition to laser operation, threshold is simply identified by a transition to coherent emission.

Photon-gap structures offer, in principle, a means of realising a device of this kind. The idea is now well known [eg ref 27]: in a structure with a three-dimensional periodic variation in refractive index the allowed photon modes are grouped into bands separated by gaps corresponding to frequencies of light which cannot be propagated through the material. It is an optical crystal, with photonic properties analogous to the electronic properties of a crystalline solid, and the analogy extends to the formation of allowed modes with the gap by introducing departures from perfect periodicity. In principle it can be arranged that lasing occurs on such a "defect" mode, and that no other modes are available to receive spontaneous emission.. The fabrication of such structures presents serious practical difficulties because the period of the refractive index variation has to be on the scale of the wavelength of light in the medium, that is about 100nm, nevertheless they can, in principle, ensure coincidence of spontaneous emission and laser spectra in frequency, direction and polarisation. Furthermore in a device where the spectrum is controlled by the cavity the emission wavelength is temperature insensitive. Even though the dimensional challenge is difficult, the scale is an order of magnitude greater than that involved in the fabrication of electronic quantum dot structures because of the different wavelengths of the electronic wavefunctions and the light. Thus the task is to engineer a photonic crystal which provides only one allowed optical mode and to incorporate within it an electrically-pumped gain generating region which incorporates a p-n junction. It is desirable that the "free space" emission spectrum of the gain medium is not too narrow so that it is not necessary to achieve a precise matching of the allowed states and modes of the electronic and photonic components of the structure. A quantum well should be satisfactory.

The practical realisation of a device of this kind is some way off, nevertheless it is possible to achieve some cavity control of the emission spectrum by less ambitious means. For example it is possible to fabricate edge emitting structures in which the number of optical modes available for propagation of radiation in directions normal to the axis of the laser cavity is reduced. Using multilayer Bragg reflectors above and below the active region of a conventional laser, it is possible to design a "vertical" cavity in which there are no allowed modes available within the emission spectrum for propagation in this direction, thereby reducing the total emission rate while leaving the generation of gain in the lasing mode unimpaired. This approach reduces spontaneous emission in unwanted directions rather than controlling the spectrum.

We have achieved some progress with these structures, having demonstrated a reduction in threshold current of about 25% by use of Bragg reflectors in the manner described [28] to supress spontaneous emission in a vertical direction. Figure 11 shows that the structure is designed so that the reflectivity spectrum of the Fabry-Perot Bragg cavity spans the free-space emission spectrum of the quantum well and has no allowed vertical modes within this spectrum. The observed threshold reduction is larger than that predicted by our present simple theory and there may also be a contribution from photon recycling. Vertical cavity laser structures are designed *to have an allowed vertical mode* and can achieve some performance improvements by *enhancing* spontaneous emission into the lasing mode. The generic term for these structures is *microcavities* [29]: cavities with dimensions on the scale of the wavelength of light which contain only a small number of allowed photon modes.

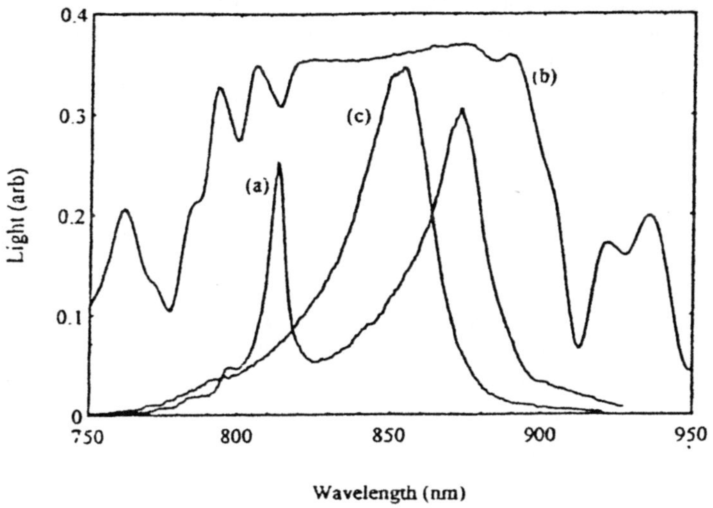

Figure 11. Measured reflectance spectrum (curve b) of a vertical Fabry-Perot cavity with no allowed cavity mode, and the spontaneous emission spectrum of a quantum well placed in this cavity, measured on a control sample without the reflectors (curve c). The cavity is designed to suppress spontaneous emission from the well in a vertical direction to reduce the threshold current for conventional edge-emitting laser operation. The spectrum from the well observed vertically through the reflectors (curve a) has significantly lower intensity and is magnified in this figure.

6. Final summary

In this article I have indicated the crucial role played by spontaneous emission in the operation of laser diodes. This is a direct consequence of the use of a crystalline solid with a continuum of allowed electronic states as the gain-generating medium. While the p-n junction and the built-in waveguide are the origin of many of the attractions of diode lasers, these bands lead to temperature dependence of both threshold current and wavelength and the broad emission spectrum in these materials make the devices fundamentally inefficient in their use of radiative emission in the lasing process. These fundamental flaws can, in principle, be overcome by designing a structure in which the recombination and lasing emission is near coincident in terms of the three characteristics of frequency distribution, direction and polarisation. Developments in diode lasers from their inception have been attempting, unknowingly, to do just that. The use of increasing degrees of quantum confinement and optically active ions all bring about reductions in the width of the emission spectrum by engineering the energy distribution of electronic states. More recent interest in microcavities and photon gap structures take the idea further by designing the total optical environment to achieve matching of all three characteristics of recombination and laser radiation. While fabrication of these structures is far from a simple matter, it is encouraging that the scale of photon gap and microcavity structures is about an order of magnitude greater than that of electronic quantum structures. In all this the challenge is to develop optically engineered materials while retaining the advantages of direct electrical pumping which endows the diode laser with many of its attractive features.

7. Acknowledgements

This article draws on research carried out over many years with a large number of colleagues and students both at Philips Research Laboratories, Redhill, and in Cardiff. To all of them my thanks are due for their contributions: experiments, calculations, questions, answers, crystal structures, devices. Their names appear in the references cited to our work. Especial thanks go to those who have grown the device structures over the years, Karl Woodbridge, Tom Foxon, John Roberts, David Westwood and David Woolf, and to Paul Hulyer who has turned them into devices. I thank the University of Wales and EPSRC for financial support.

References

1 Loudon, L. (1973) *The Quantum Theory of Light,* Clarendon Press, Oxford.
2 see for example: Casey, H.C. and Panish, M.B. (1978) *Heterostructure Lasers* Academic Press, San Diego.
3 Agrawal, G.P., and Dutta, N.K., (1993) *Semiconductor Lasers,* Van Nostrand Reinhold, New York.
4 O'Gorman, J., Levi, A.F.J., Schmitt-Rink, S., Tanbaum-Ek, T., Coblentz, D.L., and Logan, R.A., (1992) On the temperature sensitivity of semiconductor lasers, *Appl Phys Letts* **60**, 157-159.

5 O'Gorman, J., Chuang, S.L., and Levi, A.F.J., (1993). Carrier pinning in laser diodes, *Appl Phys Letts* **62,** 1454-1456

6 Chuang, S.L., O'Gorman, J., and Levi, A.F.J., (1993) Amplified spontaneous emission and carrier pinning in laser diodes, *IEEE Journ Quantum Electron* **29**, 1631-1639

7 Blood, P., Kucharska, A.I., Jacobs, J.P., and Griffiths, K., (1991) Measurement and calculation of spontaneous recombination current and optical gain in GaAs-AlGaAs quantum well structures, *Journ Appl Phys* **70,** 1144-1156

8 Rees, P. and Blood, P., (1995) Derivation of Gain Spectra of Laser Diodes from Spontaneous Emission measurements, *IEEE Journ Quantum Electron.* **QE 31**, 1047-1050

9 Weisbuch, C., and Vinter, B., (1991) *Quantum semiconductor structures.* Academic Press, San Diego.

10 Woodbridge, K., Blood, P., Fletcher, E.D., and Hulyer, P.J., (1984) Short wavelength (visible) GaAs quantum well lasers grown by molecular beam epitaxy, *Appl. Phys. Letts.* **45** 16-18

11 Holonyak, N.,Jr., Kolbas, R.M., Dupuis, R.D. and Dapkus, P.D., (1980) *IEEE Journ Quantum Electron* **QE 16** 170-185

12 Blood, P., Fletcher, E.D., Hulyer, P.J., and Smowton, P.M., (1986) Emission wavelength of AlGaAs-GaAs multiple quantum well lasers, *Appl Phys Letts,* **48** 1111-1113.

13 Blood, P., Colak., and Kucharska, A.I., (1988) Influence of broadening and high-injection effects on GaAs-AlGaAs quantum well lasers, *IEEE Journ Quantum Electron* **QE 24** 1593-1604.

14 Chow,W.W., Koch, S.W., and Sargent III, M., (1994) *Semiconductor Laser Physics,* Springer-Verlag, Berlin

15 Haug, A., (1987) Relations between the T_0 values of bulk and quantum well GaAs, *Appl Phys* **B44** 151-153

16 see for example Chin, R., Holonyak, N., Jr., Vojak, B.A., Hess, K., Dupuis, R.D., Dapkus, P.D., (1980) *Appl Phys Letts* **36** 19-21, and data and references in Blood, P. (1988) Reappraisal of GaAs-AlGaAs quantum well lasers, in *Quantum wells and superlattices in optoelectronic devices and integrated circuits,* ed. A.R. Adams *Proc SPIE no 861*, 34.

17 Blood, P., Kucharska, A.I., Foxon, C.T., and Griffiths, K., (1989) Temperature dependence of spontaneous emission in GaAs-AlGaAs quantum well lasers, *Appl Phys Letts* **55** 1167-1169.

18 Blood, P., Fletcher, E.D., Woodbridge, K., Heaseman, K.C., and Adams, A.R., (1989) Influence of the barriers on the temperature dependence of threshold current in GaAs/AlGaAs quantum well lasers, *IEEE Journ Quantum Electron* **QE 25** 1459-1467.

19 Blood, P., Fletcher, E.D., Woodbridge, K., and Vening, M., (1990) Spontaneous recombination current in InGaAs/GaAs quantum well lasers, *Appl Phys Letts* **57** 1482-1484

20 Tsang, W.T., (1981) *Appl Phys Letts* **39,** 134-136

21 Blood. P., (1986) Capacitance-voltage profiling and the characterisation of III-V semiconductors using electrolyte barriers, *Semicond Sci Technol.* **1,** 7-27.

22 Blood, P., Tsui, E. S-M., and Fletcher, E.D., (1989) Observations of barrier recombination in GaAs-AlGaAs quantum well structures, *Appl Phys Letts* **54** 2218-2220

23 Tsui, E.S-M., Blood, P., and Fletcher, E.D., (1992) Electroluminescent processes in quantum well structures, *Semicond Sci Technol.* **7** 837-844.

24 Nagarajan, R., Ishikawa, M., Fukushima, T., Geels, R.S., and Bowers, J.E., (1992) *IEEE Journ Quantum Electron.* **QE 28** 1990

25 O'Reilly, E.P., (1988) Valence band engineering in strained layer structures, *Semicond Sci Technol* **4** 121-137.

26 see for example: Efeoglu, H., Evans, J.H., Jackman, T.E., Hamilton, B., Houghton, D.C., Langer, J.M., Peaker, A.R., Perovic, D., Poole, I.,Ravel, N., Hemment, P., and Chan, C.W., (1993) Recombination processes in erbium doped MBE silicon, *Semicond Sci Technol.* **8** 236-242.

27 Yablonovitch, E., et al, 3-dimensional photonic band stucture, (1992) *Opt and Quantum Electron.* **24** S273.

28 Yang, F., Blood, P., and Roberts, J.S., (1995) Edge-emitting quantum well laser with Bragg reflectors, *Appl Phys Letts* **66** 2949-2951.

29 Yamamoto, Y., Machida, S., and Bjork, G., (1992) Microcavity semiconductor lasers with controlled spontaneous emission *Opt and Quantum Electron.* **24** S215.

MBE GROWTH OF (In,Ga)As SELF-ASSEMBLED QUANTUM DOTS FOR OPTOELECTRONIC APPLICATIONS

V.M.USTINOV, A.E.ZHUKOV, A.YU.EGOROV, N.N.LEDENTSOV, M.V.MAKSIMOV, A.F.TSATSUL'NIKOV, P.S.KOP'EV, D.BIMBERG*, AND ZH.I.ALFEROV

A.F.Ioffe Physico-Technical Institute

194021, Politekhnicheskaya 26, St. Petersburg, Russia

** Institute fur Festkorperphysik, Technische Universitat Berlin, Hardenbergstr. 36, D-10623 Berlin, Germany*

Quantum dot (QD) arrays are considered now to be very promising in semiconductor laser applications. Single quantum dot represents a potential quantum well in which the carrier motion is quantised in all three directions. The zero-dimensional electronic system can offer an improvement in threshold current density and its temperature stability owing to a delta-function-like density of states [1]. The novel approach to form a dense QD array in wider band-gap material matrix is a so-called 'self-assembled growth'. The method is based on a spontaneous transformation of strained epitaxial layer into an array of quantum sized islands due to Stranski-Krastanov growth mode [2]. In the present work we demonstrate MBE growth of (In,Ga)As QD in GaAs matrix leading to an injection lasing at room temperature [3].

Lattice mismatch between GaAs substrate and InAs epilayer is as high as 7%. Using RHEED, it was found that, when growing (In,Ga)As on GaAs (100) surface, transition from two-dimensional (layer-by-layer) to three-dimensional (island-like) growth mode occurs at (In,Ga)As effective thickness of about 12Å for 50% In content and about 5Å for pure InAs. Having been covered by GaAs layer these islands may be considered as an array of (In,Ga)As quantum dots in GaAs matrix. TEM studies revealed that InAs dots have a pyramidal shape and their typical sizes are 120Å in

M. Balkanski (ed.), Devices Based on Low-Dimensional Semiconductor Structures, 91–94.

base and 35-50Å in height (Fig.1a). These dimensions are small enough for quantum confinement for both electrons and holes.

Optical studies have shown that (In,Ga)As quantum islands are the origin of long-wavelength luminescence line which is dominant in PL spectra up to 300 K (Fig.1b). The numerical simulations of quantum levels in InAs/GaAs QDs taking into account real QD size and shape have shown that only one electron and several hole levels are located inside QD (Fig.2). The theoretical value of ground state transition energy (1.086 eV) agrees well with experimental result.

The strained dislocation-free InAs islands exist within a very tight interval of InAs effective layer thickness (up to 12Å) and MBE growth conditions, i.e. As beam equivalent pressure, growth temperature, etc. Outside these optimal regimes, InAs deposition leads to the formation of either large defected clusters or rough dislocated film. In these cases PL signal is very small to be detected due to a large amount of non-radiative recombination channels.

In conventional PL experiments large sample area covered with dots is excited by a pumping laser beam. Due to a size distribution of QD and a large number of dots inside laser spot the PL line exhibites rather wide linewidth (FWHM ~ 40 meV). The zero-dimensional nature of (In,Ga)As islands was clearly demonstrated by high-resolution cathodoluminescence (HRCL) experiments with electron beam diameter as small as approximately 500Å. Even taking spreading into account, only several dots can be excited which leads to a drastic change in luminescence spectrum. It consists of a number of well-resolved sharp lines with FWHM limited by the resolution of experimental set-up (0.15 meV) (Fig.3a). Each peak corresponds to an individual size of given dot. FWHM of individual peak is independent on temperature as expected for a zero-dimensional system (Fig.3b).

High luminescent properties of (In,Ga)As QDs can be used in laser application. AlGaAs/GaAs GRIN-SCH laser structures with recombination region based on (In,Ga)As quantum dots were grown by conventional solid source MBE, and shallow mesa stripe lasers with stripe width of 20 μm were fabricated. The threshold current densities at 77K were ~ 120 and 70 A/cm^2 for $In_{0.5}Ga_{0.5}As$ and InAs QDs, respectively. The most exciting feature of such a laser is very weak temperature dependence of J_{th} at cryogenic temperatures (Fig.4a). Characteristic temperature T_0 was measured to be ~350 and 425 K, respectively, for these two In compositions. These T_0 values are considerably higher than theoretical prediction for QW laser.

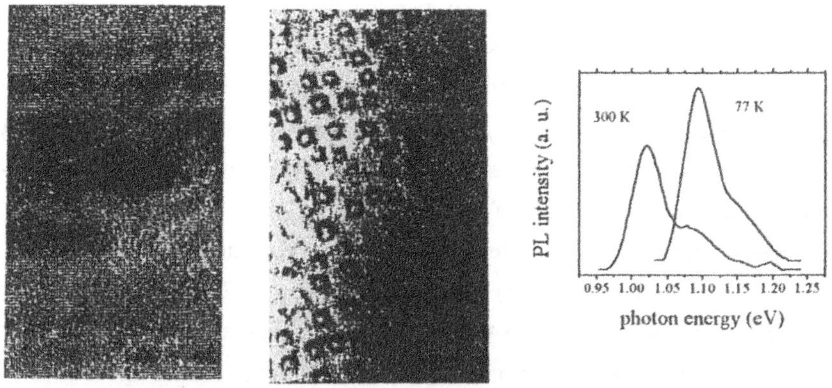

Fig.1. Cross section and plan-view (a) and typical PL spectra (b) of InAs quantum islands inserted into GaAs matrix

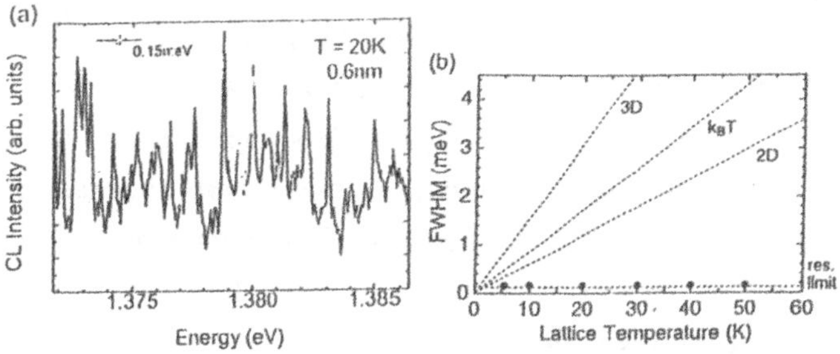

Fig.3. High resolution cathodoluminescence spectrum (a) and FWHM of an individual CL line (b) for InAs/GaAs QD array

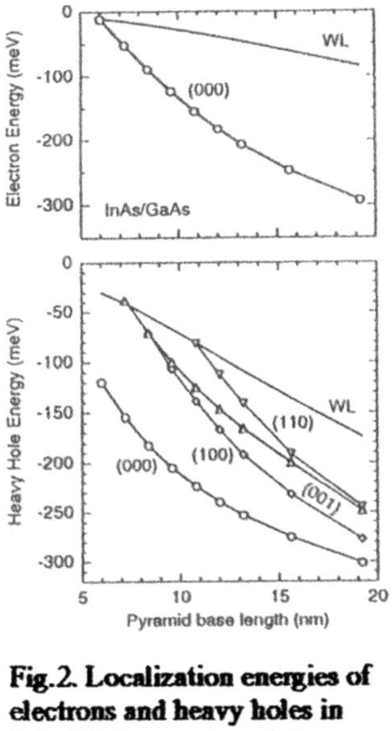

Fig.2. Localization energies of electrons and heavy holes in InAs/GaAs quantum dot

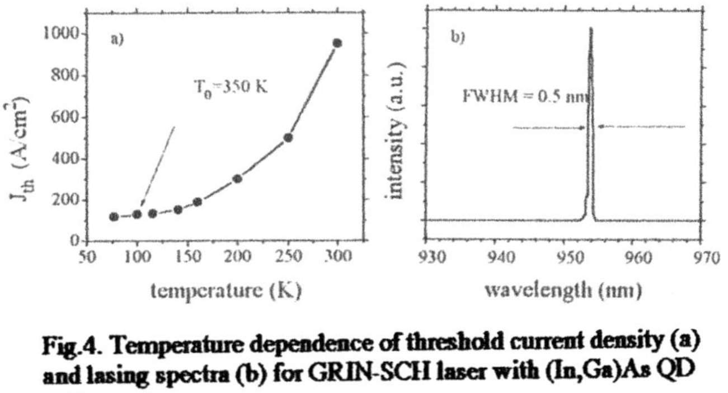

Fig.4. Temperature dependence of threshold current density (a) and lasing spectra (b) for GRIN-SCH laser with (In,Ga)As QD active region

However, the threshold current density increases rapidly when temperature exceeds ~ 120÷140 K, so RT threshold density lies in the kA/cm^2 range. Simultaneously the laser wavelength blue-shifts toward wetting layer transition. We believe that the increase in threshold current density does not result from non-radiative channels but is due to an evaporation of carriers from dots. This problem can be solved by increasing the localization energy of carriers in QD.

Laser diodes with cavity length less than 250 μm reproducibly demonstrated single longitudinal (Fabry-Perot resonator) mode operation up to 22 mW output power with FWHM of emission line of about 5Å (Fig.4b). This effect is not typical for broad-area quantum well lasers and we attribute it to the manifestation of specific density of states (DOS) which leads to very narrow gain function. We have also found that optical gain is approximately linear function of injection current with no saturation in a wide interval of current densities (up to ~ 7 kA/cm^2) which is indicative of the fact that density of states participating in laser action is sufficiently high.

Thus, the self-assembled (In,Ga)As quantum dots are suitable for laser application. The low-threshold operation through zero-dimensional states in QD has been demonstrated at LN_2 temperatures. The main goal is to reduce 300K threshold current density down to sub-kA range by extending high-T_0 region up to RT. The work in this direction is in progress now.

[1] Arakawa, Y. and Sakaki, H. (1982) Multidimensional quantum well laser and temperature dependence of its threshold current, *Appl. Phys. Lett.* 40, 939-941.

[2] Moison, J.M., Houzay F., Barthe, F., Leprince, L., Andre, E., and Vatel, O. (1994) Self organized growth of regular nanometer-scale InAs dots on GaAs, *Appl. Phys. Lett.* 64, 196-198,

Petroff, P.M. and Den Baars, S.P. (1994), MBE and MOCVD growth and properties of self-assembling quantum dot arrays in III-V smiconductor structures, *Superlat. Microstruct.* 15, 15-21.

[3] Kirstaedter, N., Ledentsov, N.N., Grundmann, M., Bimberg, D., Ustinov, V.M., Ruvimov, S.S., Maximov, M.V., Kop'ev, P.S., Alferov, Zh.I., Richter, U., Werner, P., Gosele, U., and Heydenreich, J. (1994) Low threshold, large T_0 injection laser emission from (InGa)As quantum dots, *Electron. Lett.* 30, 1416-1417.

EXCITON AND MAGNETOEXCITON LUMINESCENCE IN $Ge\text{-}Ge_{1-x}Si_x$ MULTIPLE QUANTUM WELL STRUCTURES.

N.G.KALUGIN, V.B.TIMOFEEV[+], A.V.CHERNENKO[+],
L.K.ORLOV, O.A.KUZNETSOV

Institute for Physics of Microstructures of the Russian Academy of Sciences, 46 Ulianov Str., 603600 Nizhny Novgorod, Russia,
[+]Institute of Solid State Physics of the Russian Academy of Sciences, 142432 Chernogolovka, Russia.

There has been a considerable progress in the photoluminescence (PL) properties investigations of SiGe/Si and Ge/Si structures of late. The PL investigations of $Ge\text{-}Ge_{1-x}Si_x$ structures with quantum wells in germanium layers have been started recently[1,2]. This work reports the studies of the exciton and magnetoexciton luminescence of $Ge\text{-}Ge_{1-x}Si_x$ heterostructures grown by CVD method on Ge [111] substrates.

The high- frequency shift of the PL line positions[1,2] versus decreasing Ge layer thicknesses in the investigated $Ge\text{-}Ge_{1-x}Si_x$ multiple quantum well structures indicates to the presence of size quantization of both electrons and holes in the Ge layers. The bottom of conduction band in Ge layers of investigated $Ge\text{-}Ge_{1-x}Si_x$ structures is the three (L3) valleys, located in equivalent for [111] directions, the bottom of valence band is the havy (moment projections $j_z=\pm 3/2$) holes.

The temperature dependencies of integral PL line intensities, which allow to evaluate the exciton binding energies for investigated samples, show the 2D-character of excitons in our structures.

Fig.1 and Fig.2 show what the NP exciton PL line of structures #261 (with parameters: Ge thickness d_{Ge}=210 Å, period D=530Å, layer number N=72, x=12.3% and in-plane lattice period a=5.638 Å) and #262 (d_{Ge} =120 Å, D=420 Å, N=262, x=14%, a=5.629) looks like for different values of the magnetic field. In the both cases the magnetic field is perpendicular to the structure layers.

M. Balkanski (ed.), Devices Based on Low-Dimensional Semiconductor Structures, 95–98.

The spectra of structure #261 (d_{Ge}=210 Å) exhibit just a slight blue shift of the PL lines and a higher PL intensity in the high fields due to enhanced localization of the charge carriers. In the spectra of structure #262 (d_{Ge}=120 Å) the exciton lines were observed to split beyond the point B=4-5 T due to transitions between the Landau levels. The exciton energy spectrum of Ge-$Ge_{1-x}Si_x$ structures in a magnetic field is quite complicated since the Landau levels are mixed up in the valence band. However, the calculations[3] for structures with Ge thicknesses d_{Ge}~100 Å in the fields B≥10 T show that

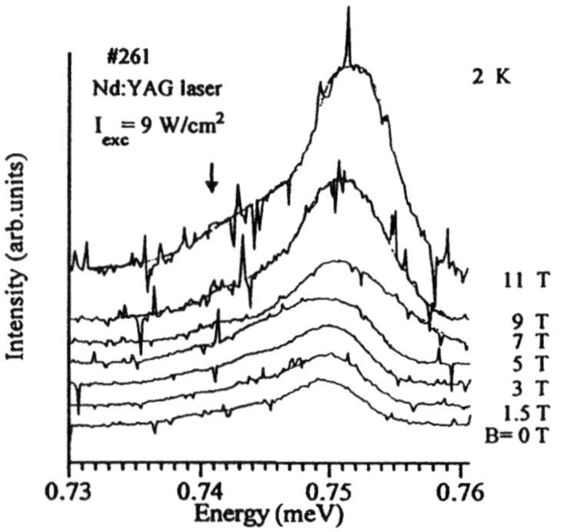

Figure 1.

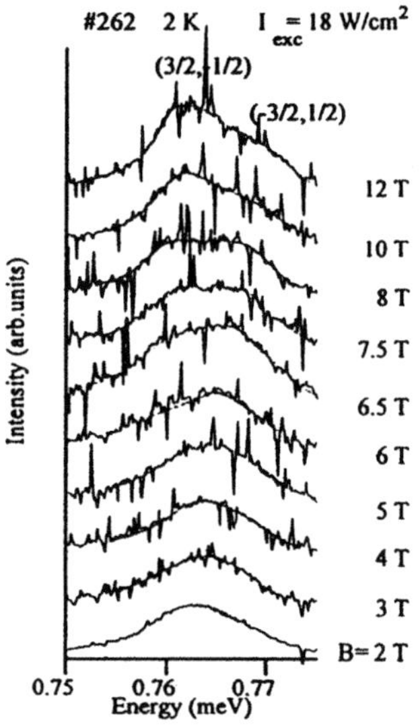

Figure 2.

even for rather high (~2-3· 10^{11} cm^{-2}) hole concentrations it is possible to reach the ultraquantum limit when only the zero Landau level (split into 2 sublevels with j_z=3/2 and j_z=-3/2) of holes is filled by carriers. The lines observed in the spectrum of structure #262 under the fields B≥10 T must correspond to the luminescence of excitons formed by the zero Landau level holes with moment projections j_z and the electrons with spin projections s_z (j_z, s_z):(3/2,-1/2) and
(-3/2,1/2). Estimation of the g-factors (the Lande factors) from the NP line splitting value at B≥10 T yelds: $|g_h-g_e/3|\approx 4$ (where g_h and g_e are g-factors for holes and electrons, respectively). No splitting in the spectra of

structure #261 in the magnetic field were observed because of a weaker size quantization effect. According to the calculations[3] , in the structures with d_{Ge}~200 Å the energy levels in the valence band, even in fields B~10 T, are located in the close proximity to each other and it is more difficult to resolve the different levels experimentally.

We investigated the PL spectra in a magnetic field versus the exciting radiation intensity I_{exc}. The observed effect of the Ge wells filling[4] (Fig.3)

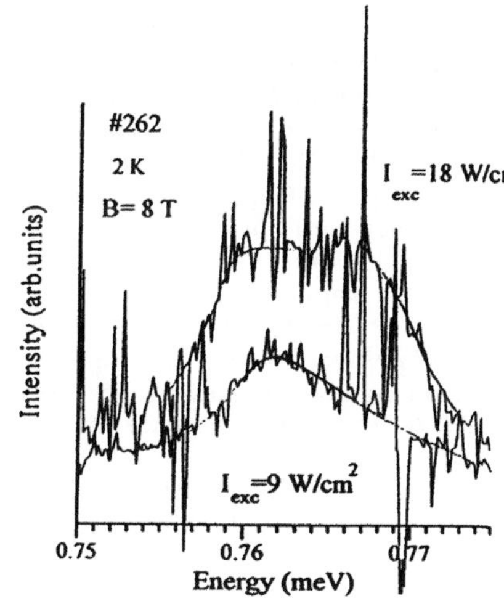

Figure 3.

allowed to determine the excitons concentration: for structure #262 in the field B~10 T a high frequency PL peak appears at I_{exc}~10 W/cm^2, when the exciton concentration

n_{ex}=eB/hC~2·10^{11} cm^{-2}. Due to that the recombination time was evaluated at ~10^{-8} s, which implies that the main recombination channel is the non-phonon radiative transitions.

The magnetoluminescence measurements provide a tool for investigations into the nature of PL lines. The low-frequency wing of the line (marked by an arrow in Fig.1), observed at B=0, remains in the magnetic fields at B>2-3 T. This leads us to a conclusion that this line has no relation to biexcitons[5,6], and its existence depends on the quality of structures under study.

The experimentally observed in this work gas of spin-aligned quasi-2D excitons is an interesting object for a study of phase transitions, for example, the Bose-Einstein condensation of excitons[7].

References

1. N.G.Kalugin, L.K.Orlov, O.A.Kuznetsov. (1993) Observation of 2D-exciton luminescence in germanium layers of Ge-$Ge_{1-x}Si_x$ periodical heterostructures, *JETP Letters* Vol.58, 200-205.
2. L.K.Orlov, V.Ya.Aleshkin, N.G.Kalugin, N.A.Bekin, O.A.Kuznetsov, B.Dietrich, G.Baquet, J.Leotin, M.Brousseau, F.Hassen.(1995) Exciton luminescence in Ge-$Ge_{1-x}Si_x$ multiple quantum well structures, *Submitted to J.Appl.Phys.*
3. Yu.G.Arapov, N.A.Gorodilov, V.N.Neverov, M.V.Yakunin, A.V.Germanenko, G.M.Minkov, O.A.Kuznetsov, L.K.Orlov, R.A.Rubtsova, A.L.Chernov.(1994) The quantum Hall effect in the multilayer p- Ge-$Ge_{1-x}Si_x$ heterostructures and energy spectrum of the two-dimensional hole gas in magnnetic field, *JETP Letters*, Vol.59, 268-272.
4. L.V.Butov, V.D.Kulakovskii, G.E.W.Bauer, A.Forchel, D.Grutzmacher.(1992) Excitons in dense two-dimensional electron-hole magnetoplasmas, *Phys.Rev.B46*, 12765-12768.
5. V.D.Kulakovskii, V.G.Lisenko, V.B.Timofeev.(1985) Exciton molecules in semiconductors, *Uspekhi Fizicheskih Nauk*, v.147,n.1, 3- 48.
6. M.L.W.Thewalt.(1993) Photoluminescence processes in $Si_{1-x}Ge_x$/Si heterostructures, *Extended Abstracts of the 1993 International Conference On Solid State Devices and Materials*, Makuhari, 494-496.
7. V.B.Timofeev, V.D.Kulakovskii, I.V.Kukushkin.(1983) Spin aligned exciton gas in uniaxially compressed Ge, *Physica 117B&118B*, 327-332.

THE PHYSICS OF QUANTUM WELL INFRARED DETECTORS

E. ROSENCHER, Ph. BOIS and J. Y. DUBOZ
Laboratoire Central de Recherches / Thomson-CSF
Domaine de Corbeville F-91404 Orsay (France)

We present our current understanding of electron unipolar emission and capture processes in multiquantum wells. All the peculiarities of QWIPs mechanisms (contact injection, capture and emission, QW depletion, ...) are taken into account in a new QWIP device model.

INTRODUCTION

The increasing performances of Quantum Well Infrared Photodetectors (QWIPs), mainly due to the maturity of III-V growth and processing technologies, allow them to be considered as an alternative to HgCdTe detectors (1). Two different models, namely photoconductive (1,2) or photoemissive (3), have been proposed to describe the electro-optical behavior of QWIPs, depending on the electrical approach of the device (extrinsic photoconductor, Schottky diode or mixed system). All these models lead to roughly satisfactory results but despite an intensive work, this domain still suffers from a lack of deep physical understanding. Modeling the complex photodetection and noise processes requires a good knowledge of different elementary mechanisms: injection efficiency of the contacts, electric field distribution, balance between photoionization and capture in the quantum wells, transport and scattering in the barriers, ...

The purposes of this paper are thus: i) to describe the experimental procedures that we have used to determine the physical parameters which appear in the different photoresponse formalisms, ii) to discuss the physical origin and the connections between the different determined parameters (p_e, p_c, τ_c...), iii) to show how those parameters and related physical mechanisms shed light on the QWIP photoconduction mechanisms and iv) to propose a new QWIPs model.

PHOTOCONDUCTIVE vs PHOTOEMISSIVE MODEL OF QWIPs

A good picture of a QWIP is that of an energy adjustable extrinsic photoconductor, where the active impurities are quantum wells, *i.e.* emission-recombination centers with a huge optical cross section, spatially localized but 2D extended, like heterostructures or Schottky interfaces. QWIPs can be considered either as extrinsic but non homogeneous photoconductors (photoconductive model), or as Schottky photoemitters mounted in series (photoemissive model). The link between the two approaches is of course achieved when all peculiarities of QWIPs are taken into account, *i.e.* when it is described as a stack of 2D extended optical and thermal traps.

Let us consider a series of N $GaAs/Al_xGa_{1-x}As$ quantum wells of width d and barrier thickness L_B with a donor concentration ρ_s per unit area in each QW. The barrier energy height E_B is a well known function of the Al concentration x (4). In most of this paper,

M. Balkanski (ed.), Devices Based on Low-Dimensional Semiconductor Structures, 99–113.

the parameters d and x have been chosen such that only one bound state of energy E_1 is present in the QW, the excited state E_2 being close to resonance with the bottom of the barrier conduction band. For this design (quasibound) the QWIP performances have been shown to be optimum (5). The formalism developed elsewhere for the absorption is in excellent agreement with experimental spectra (6). The electrons excited in the final continuum state have a given emission probability p_e not to be recaptured into the well, thus to contribute to the photocurrent. For instance, if the second level E_2 were deep in the QW, p_e should be mainly the (very low) tunneling probability through a triangular barrier. This leads to a surface optical generation rate G_{op} ($cm^{-2}.s^{-1}$) at each QW given by:

$$G_{op} = p_e \alpha_{1QW} \Phi \qquad [1]$$

where Φ is the photon flux, and α_{1QW} the dimensionless absorption coefficient per QW.

Photoconductive model

In the photoconductive model, the steady state *surface* density n_{2d} of the photoexcited carrier in the barrier is given by the balance equation *in each period* between the surface generation G_{op} and the capture R rates. The latter is given by:

$$R = \frac{n_{2d}}{\tau_c} \qquad [2]$$

where τ_c is the electron capture time from the AlGaAs conduction band into the QW. However, the photocurrent J_{op} is related to a *volume* density of excited carriers n_{3d} through $J_{op} = n_{3d}\, q\, v_d$. For small applied electric fields, the carrier velocity v_d is given by $v_d = \mu F$ where μ is the electron mobility and F is the electric field. Thus one needs a relationship between n_{2d} and n_{3d}. Assuming an homogeneous medium, it is standard to choose(2):

$$n_{3d} = \frac{n_{2d}}{L} \qquad [3]$$

where $L=L_B+d$ is the period of the QWs. This leads to the following expression for the QWIP number responsivity, *i.e.* the ratio (number of electrons collected / number of photons in):

$$R_{op} = \frac{J_{op}}{q\,\Phi} = p_e \alpha_{1QW} g_{1QW} \qquad [4]$$

In the latter expression, g_{1QW} is the photoconductive gain of a single QW, *i.e.* the ratio of the capture time τ_c to the drift time through the QW period:

$$g_{1QW} = \frac{\tau_c}{L/v_d} \qquad [5]$$

Let us note that, still assuming an homogeneous system, we could have considered the total absorption $\alpha_{tot} = N\alpha_{1QW}$ and the total photoconductive gain $g_{tot} = g_{1QW}/L$ corresponding to the drift of the electron through the entire structure. This would have led to an expression similar to Eq. 4, *i.e.* $R_{op} = p_e\, \alpha_{tot}\, g_{tot}$. However, this blurs one prediction of Eq. 4 that the responsivity of the QWIPs is independent of the number of wells, which is experimentally verified up to a single QW (7, 8). The quantities α_{1QW} and g_{1QW} are thus more physically sound. Finally, in this approach, the role of the contacts is to provide enough carriers to ensure electrical neutrality.

Photoemissive model

In the photoemissive approach (3), the injected photocurrent at the contact J_{op} is imposed by the constraint that the flux of photoexcited electrons out of each QW ($p_e \alpha_{1QW}\Phi$) be equal to the flux of captured electrons $p_c J_{op}/q$ in this well, where p_c is the capture probability, *i.e.* the ratio between the captured and the incident electron flows.

This approach leads to the same equation (Eq. 4) as the photoconductive approach, but with a photoconductive gain given by:

$$g_{1qw} = \frac{1}{p_c} \qquad [6]$$

A complete model including the dark current must include the thermal part J_{th} of the current (injected or thermoexcited out of the wells):

$$J_{inj}\frac{p_c}{q} = (J_{opt}+J_{th})\frac{p_c}{q} = p_e\alpha_{1QW}\Phi + G_{th}$$
$$J_{dark}\frac{p_c}{q} = J_{th}\frac{p_c}{q} = G_{th} \qquad [7]$$

where G_{th} ($cm^2.s^{-1}$) is the surface thermal generation rate.

Considering the whole well as a single 2D trap, one can write the emission-capture equilibrium as (6):

$$\frac{\partial \rho_s}{\partial t} = \frac{c_n}{A} n_{3D} - e_n \rho_s \qquad [8]$$

where A is the sample area, e_n (s^{-1}) is the emission rate from the QW while c_n ($cm^3.s^{-1}$) is the classical capture coefficient in the QW. In our case of 2D trap, the capture coefficient c_n is proportionnal to the sample area so that the significant capture parameter is the capture velocity v_{QW} defined as $v_{QW} = c_n / A$. Table 1 summarizes the correspondences between the classical 0D impurity trap and the 2D QW trap. At this stage, it is important to recall that the capture probability is the ratio between the captured electrons flux to the total electron flux:

$$p_c = \frac{J_{cap}}{J_T} = \frac{n_{3D}.q.v_{QW}}{n_{3D}.q.v_d} = \frac{v_{QW}}{v_d}$$
$$v_{QW} = \frac{L}{\tau_c} \qquad [9]$$

In the latter approach, the QW period L plays no role, so that, for the two models to be equivalent, the equality

$$\frac{1}{p_c} = v_d \frac{\tau_c}{L} \qquad [10]$$

must hold. As long as the distance between wells is large enough to make p_c a property of a single well, this means that the *ratio* $\tau_c/L = 1/v_{QW}$ *is independent from the QW period* for a given applied field. We have checked without any ambiguity the proportionality between the recombination lifetime τ_c and the QW period (15). This establishes the link between the photoconductive and photoemissive models.

	POINT DEFECT (0D)	QUANTUM WELL (2D)
Emission-Capture Equations	$\frac{\partial N_{0D}}{\partial t} = c_n n_{3D} - e_n N_{0D}$	$\frac{\partial \rho_s}{\partial t} = \frac{c_n}{A} n_{3D} - e_n \rho_s$
Capture Parameters	capture rate c_n ($cm^3\ s^{-1}$)	capture velocity $v_{QW} = \frac{c_n}{A}$ ($cm.s^{-1}$)
Ratio of Captured Parameters	Capture cross section $\sigma = \frac{c_n}{v_d}$	Capture probability $p_c = \frac{v_{QW}}{v_d}$

Table 1 : Equivalence between the relevant emission-capture equations and parameters in point defects and in quantum wells in semiconductors.

INJECTION AT CONTACTS

Generally, it is assumed in the preceeding models, that the injecting contacts of QWIPs are perfectly ohmic, *i.e.* that they provide as many electrons as QWs require. However, in classical QWIP structures, this is obviously false (8): the GaAs n^+ /AlGaAs heterojunction is blocking at low temperature. We have investigated this problem with a single QW quasi-bound detector. The band diagram of the device is sketched in Figure 1. The current provided by the cathode can be estimated under the WKB approximation for temperature assisted tunneling:

$$J_1(F_1) = \frac{qm^*kT}{2\pi^2\hbar^3}\int_{E_c}^{\infty} T(E,F_1)\ln\left[1+\exp\left(\frac{E_{Fc}-E}{kT}\right)\right]dE \qquad [11]$$

where F_1 is the electric field across the injecting AlGaAs barrier, T is the quantum transparency of the barrier due to electric field assisted tunneling, E_c and E_{Fc} are respectively the bottom of the GaAs conduction band and the Fermi energy of the cathode. The reverse current is neglected because $\Delta E_c \gg kT$. We assume that the current emitted by the quantum well depends only of the local electric field F_2. Following Levine and Liu (1,9), the dark current from quantum well toward the anode is given by a similar expression, changing F_1 to F_2 and E_{Fc} to E_{Fqw}.

If the QW emits electrons (thermally or optically), an electric field builds up in the cathode barrier until it counterbalances the current flowing from the QW. Under steady state conditions, the whole system is described by i)the total bias $V_T = L_B.(F_1 + F_2)$, ii)the current conservation $J_T(V_T) = J_1(F_1) = J_2(F_2) + q.p_e\alpha_{1qw}\Phi.\mu\tau_c.F_2/L_B$ and iii) the induced charge change in the QW given through Gauss' law. Basically, the current flowing from the QW is far larger than the current available from the cathode because $\Delta E_c - E_{Fc} > \Delta E_c - E_{Fqw}$. This leads to a reconfiguration of the electric field, which is even higher under illumination. It is important to keep in mind that in multiquantum well QWIPs, a part of the total bias, illumination dependent, is used to maintain a high field domain at the cathode. In the last section of this paper, the electric field dependence of the capture probability (here taken constant) will be included.

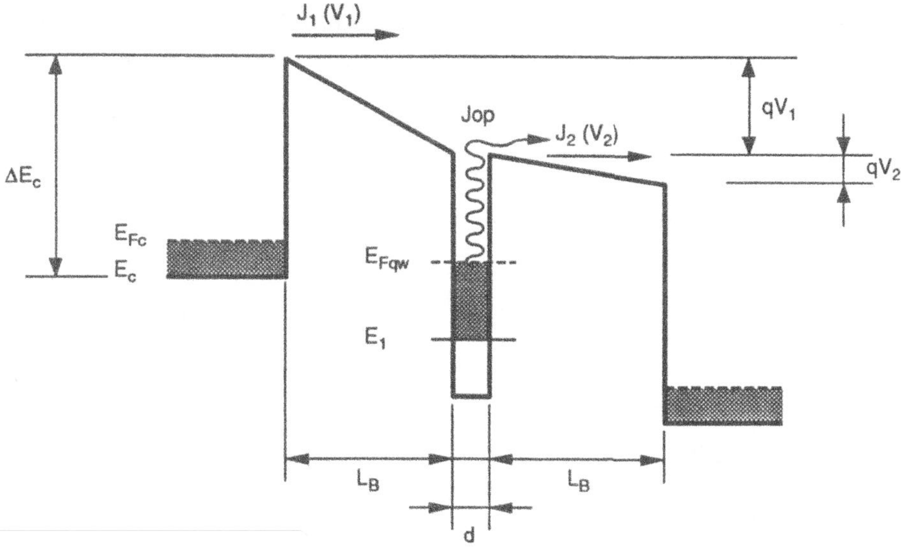

Fig. 1 : Band diagram of a biased single QW detector. The electric field at the cathode is higher than at the anode so that the contact provides enough electrons.

EMISSION PROBABILITY

In order to determine the emission probability p_e, we have measured (10) the transient photocurrent fom a multiquantum well structure in a Schottky diode. The technique used is very similar to the Deep Level Optical Spectroscopy, classically applied to the study of the optical cross section of defects in semiconductors. The Schottky diode ensures that no refilling current flows through the device under reverse bias: it is then possible to measure exactly the rate at which electrons will be ionized out of the QW, without the complications due to the steady state, *i.e.* the capture of electrons into the wells. The experiment follows the usual voltage cycle of transient optical spectroscopy (11). Under infrared illumination, the structure is first forward biased so that electrons from the GaAs contact are injected in the AlGaAs barrier and captured in the wells. The diode is then reverse biased and the trapped electrons are emitted from the well to the lower contact.

Since the current has a transient behavior, it is the *sum of the contribution of all the wells:*

$$I(t) = -Aq\sum_i \frac{\partial \rho_i}{\partial t} = Aq\Phi\sum_i \sigma_p(i)\left(\rho_i - \rho_i^{eq}\right) \qquad [12]$$

where i stands for the well index, *eq* stands for equilibrium, Φ is the photon flux and A the sample area. $\sigma_P(i)$ is the photoionization cross section ($\sigma_P = p_e\, \sigma_{op}$), which may depend on the electric field at the ith QW. This is markedly different from steady state photoconduction where the current is constant through the structure (see Eq. 4). The experimental photoionization cross section σ_p is found in the 10^{-15} cm^{-2} range.The variation of the photoionization cross section with the photon energy hν is in total agreement with the quantum mechanical calculation α (hν) (12). The experimental variation of σ_p and then p_e with the electric field F is shown in Figure 2.

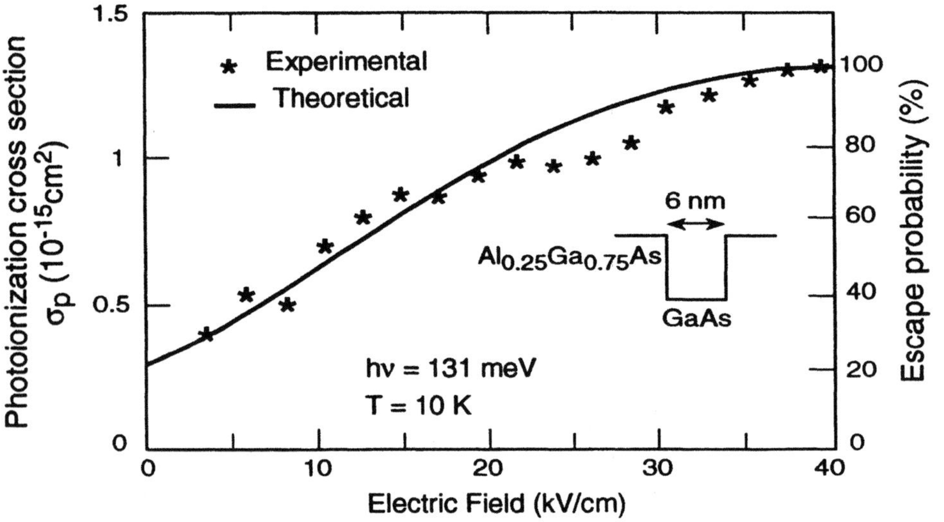

Fig. 2 : Experimental variation of σ_p and p_e as a function of applied electric field. The fit is based on the statistical Gaussian distribution of the ionization energy ($\Delta E = 5$ meV).

The escape probability is usually connected (3) to an electron escape time τ_e and the relaxation time τ_r through the relation $p_e = (1 + \tau_e / \tau_r)^{-1}$. When the electric field is enhanced, τ_r should not be significantly affected ($\tau_r \approx 1$ ps) while τ_e should vary with the tunnel barrier transparency. However, the theoretical variation of the transparency with the electric field is very steep, from practically 0 to 1 for electric field values close to F_s given by $qF_s d / 2 = E_B - E_2$. *The variation of p_e should then be described by a Heaviside function at $F = F_s$,* which is clearly not in agreement with the experimental results. We believe that this smooth variation of p_e has a statistical origin, *i.e.* E_2, and consequently the excited state ionization energy $E_B - E_2$, fluctuates in each QW with a Gaussian distribution. The final p_e vs F curve is then the convolution of the Heaviside function Y with this Gaussian distribution and is given by:

$$p_e(F) = \int_{-\infty}^{+\infty} Y(E_2 - E_B + \frac{qFd}{2}) D(E_2)\, dE_2 = \frac{1}{2}\, \mathrm{erfc}\left(\frac{E_B - \frac{qFd}{2} - E_m}{\sqrt{2}\,\Delta E} \right) \qquad [13]$$

where E_m is the mean value of ionization energy E_2 and ΔE the variance of this energy distribution. The best fit of the experimental results, obtained with $\Delta E = 5$ meV and $E_B - E_2 = E_m = 4$ meV is shown in Figure 2. The former value can be reasonably accounted for by a *one monolayer fluctuation* in the QWs on both sides.

CAPTURE INTO A QUANTUM WELL

Capture velocity:

We have performed impedance spectroscopy on a sample containing only one QW in a Semiconductor-Insulator-Semiconductor structure (13), in order to determine the capture of electrons from the continuum in the barriers into a QW. This technique is widely used to study the capture velocity of Si-SiO_2 interface traps.

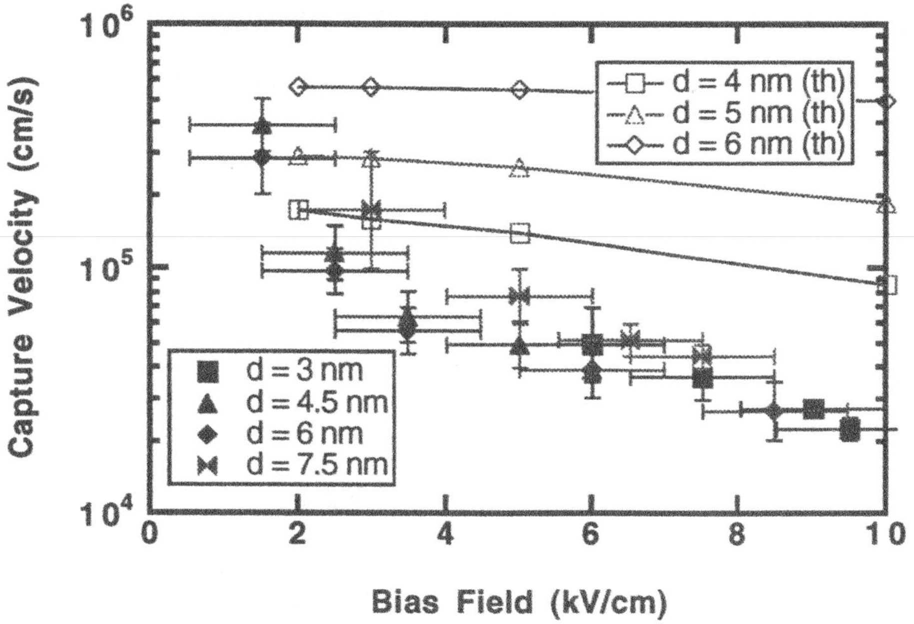

Fig. 3 : Experimental and theoretical (th) variation of the recombination velocity in a QW as a function the applied electric field for different QW thicknesses.

In Figure 3 we report experimental and theoretical determinations of the capture velocity v_{QW} as a function of the electric field (in the 80 K-120 K range) for different QW thicknesses (the barrier thickness is kept constant at 50 nm). The derivation of v_{QW} is based on a full quantum calculation using the Fröhlich polar optical phonon interaction as the scattering process (14). We suspect that hot electron effects in the continuum states are the main causes of the weaker field dependence of the capture reported by the model. Let us note that, in contrast to what has been reported from time-dependent photoluminescence experiments, we found no clear dependence of v_{QW} on well width. The results lead to a capture probability p_c (Eq. 9) in the range 0.03-0.006 for an electric field between 1 to 10 kV/cm (g_{1QW} in the range 30-150).

Electron lifetime measurements:

Optical measurements of the electron intersubband relaxation time T_1 in MQW bound to bound (BB) structures under zero bias give T_1 around 1 ps (16-18) whereas electrical measurements performed on bound to extended (BF) QWIPs give τ_c larger than 10 ps (1). We have achieved optical measurements (19) on QWIPs for BB and BF structures, in order to link the preceeding experiments. We have determined the electron lifetime in QWIPs by absorption and photoresponse saturation using picosecond micropulses with a power density up to 10 GW/cm^2 delivered by a tunable free electron laser. The beam is made of macropulses (9.5 µs) consisting of micropulses (in the ps range) separated by 32 ns.

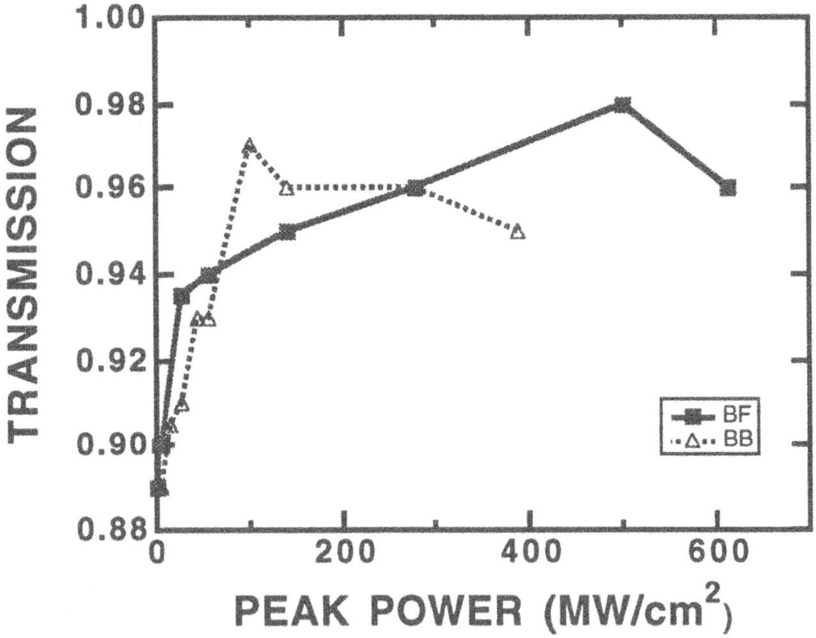

Fig. 4 : Transmission of the BF sample (λ = 9.7 µm) and the BB sample (λ = 10.8 µm) as a function of the incident power density. For both samples, the absorption is halved at the saturation power P=80 MW/cm^2.

Absorption saturation. We first compared BB and BF absorption. The transmission of both samples was measured as a function of the incident power at different wavelengths. Figure 4 reports results obtained at the peak wavelength on each sample.

We developed a model that includes the effect of T_2 on the stimulated emission and thus on the saturation power. Excited electrons are scattered with the time constant T_2 toward a different k state. Then, they cannot suffer any transition toward the ground state by stimulated emission anymore (the optical transitions must conserve the total k value) and can get back to the ground state by inelastic relaxation only, with the T_1 time constant. This model gives T_1 times of 0.45 ps and 1 ps for the BB and the BF sample respectively. This difference may be related to the overlap between the fundamental and excited levels. The overlap is smaller when the excited level is in the continuum and the relaxation time is larger.

Photoresponse saturation. The setup was almost the same as for the absorption saturation measurements. The QWIP is mounted at Brewster incidence in a LN_2 cryostat. We measured the current during the macropulse and subtracted the dark current so as to get the photocurrent as a function of the incident power. We present the results for a BF sample with 49 wells doped to 7.10^{10} cm^{-2} peaked at 9.6 µm (barrier thickness ≈ 35 nm). The laser wavelength was set to 9.88 µm. The optical cross section σ(9.88 µm) is 10^{-15} cm^2. The micropulse duration was set to be around 1 ps for this experiment.

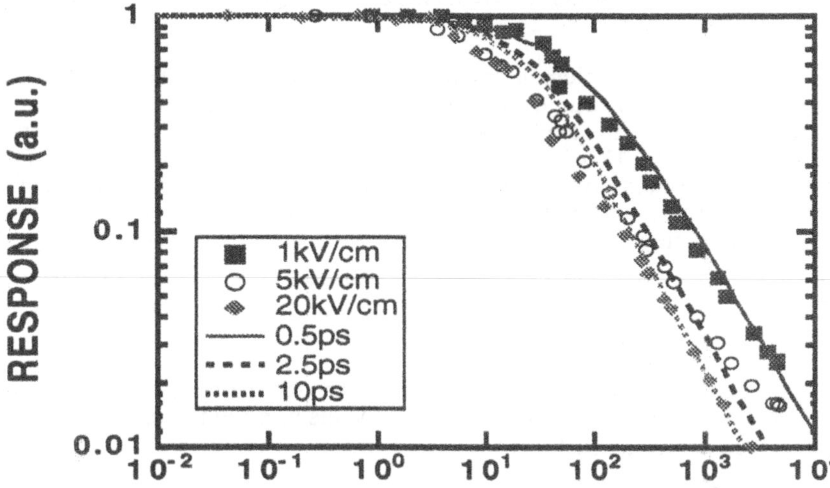

Fig. 5 : Photoresponse *vs* incident power density. It has been normalized with the low power response. Results are fitted by a theoretical model that gives τ.

In Figure 5, we present the responsivity normalized. When the incident power increases, the response is first constant and then decreases down to 0.01 or so. We observe that the saturation occurs at lower and lower power when the electric field increases. We developed a two-level model with n_1 and n_2 being the fraction of electrons in the ground and excited state respectively. As the electron scattering increases with the applied electric field, we neglected the stimulated emission from level 2 to level 1 ($\delta n_2/\delta t = \sigma\phi n_1 - n_2/\tau$). By fitting the data, we found τ times in the range 0.5 to 1 ps at low field (1 kV/cm), increasing to 1.5 to 2 ps at moderate field (5-10 kV/cm) and up to 10ps or more at higher fields (20 kV/cm). It has to be noted that we cannot determine τ times that are larger than 10 ps because the fit becomes very little dependent on τ (the optical pulse is too short compared to τ and there is almost no relaxation during the optical pumping of level 1). This increase of the lifetime with the electric field may be interpreted in the following way: First, as the field increases, the relaxation of photoexcited electrons back into the well they originate from, decreases because electrons are swept away by the electric field. Second, the excess energy above the barrier increases and the probability that an inelastic collision leads to an electron energy lower than the top of the well decreases.

As a conclusion of this part, we have measured the electron lifetimes in MQW structures in the case of interest for infrared detectors, i.e. in the case of a bound-to-free transition with an applied electric field. First, we have shown that the lifetime for a bound-to-free transition is larger than the one for a bound-to-bound transition by a factor of 2. Second, we have shown the increase of the lifetime with the electric field.

MODELING PHOTODETECTION IN QWIPS

Model:

We have studied the three elementary bricks that governs the photodetection in QWIPS, namely the contact injection, the emission and capture mechanisms. First of all, we have demonstrated that the steady state operating mode imposes the value of the electric field at the cathode. This is necessary for the injection flux to match the QWs thermal and optical generation fluxes. We now propose to extend this redistribution of the applied electric field to the whole structure, and to give a sequential procedure for the modeling. The QWIP structure used in the calculation is a stack of 10 GaAs doped wells ($d = 6$ nm, $\rho_0 = 1.5\ 10^{11}$ cm^{-2}) and 11 $Al_{.22}Ga_{.78}As$ barriers ($L_B = 35$ nm) clad between two n+ ($n_{3d} = 5.10^{17}$ cm^{-3}) GaAs contact layers. The theoretical absorption spectrum is peaked near 10 µm. The operating temperature is 77 K.

Injection current: In steady state, the current flowing through the device is the injection current J_{inj}. We assume that J_{inj} depends only on the electric field applied on the first barrier F_0 and is well described by Eq. 11.

Capture: Let us assume that we have a precise knowledge of the capture probability including hot electrons effects in a full quantum mechanical calculation, *i.e.* we are able to numerically express $p_c(F)$. In practice, this would be a tremendous task like the theoretical validation of transport Boltzmann equations, so we prefer to describe $p_c(F)$ by an empirical expression:

$$p_c(F) = a \exp\left(\frac{b}{c+F}\right) \qquad [14]$$

where a=.05, b=20 kV/cm and c=7 kV/cm are derived from experimental measurements (1,13). F is in kV/cm.

Emission: For the photoionization probability $p_e(F)$ we use expression [13].

The optical generation rate is defined by Eq.1. The absorption coefficient α_{1QW} depends on the electron density ρ_i in the ith well by :

$$\alpha^{i}_{QW} = \rho_i \, \sigma^{i}_{op} \qquad [15]$$

In addition, we include the spatial variation of the optical cross section σ_{op} in each well in order to account for the variation f(i) of the electromagnetic field with depth due to the coupling:

$$\sigma^{i}_{op} = \sigma^{0}_{op} \, f(i) = \left[A + B \exp\left(-\frac{i}{C}\right)\right] \qquad [16]$$

where A, B and C can be adjusted from coupling simulation. If the electric field is homogeneous, the corrective factor reduces to the constant term A in the case of gratings, or to the exponential term for Brewster configuration. According to 2D-gratings simulations, we take the following parameters: $A=3.33\ 10^{-14}\ cm^2$, $B=1.27\ 10^{-13}\ cm^2$ and C=14.

The thermal escape out of the QW in a 2D model must include a scattering process. Unfortunately this scattering process is still unclear and a crude approximation is often used: It is commonly assumed that all 2D electrons having the required total energy (instead of the energy in the growth direction) escape the well (20). Moreover, this model requires using the electron mobility in the barrier, which remains largely undefined. For that reason, we prefer to describe the thermal emission out of the QW as a 3D process. J_{th} is thus given by Eq.11, where we simply replace the Fermi energy of the contact E_{Fc} by the Fermi energy of the quantum well E_{Fqw}..

Algorithm: The basis of our model is the balance equilibrium in each well between the captured and emitted electron fluxes:

$$J_{inj}(F_0).p_c(F_{i-1}) = q\, p_e(F_i)\sigma^{i}_{op}\, \rho_i \Phi + J_{th}(\rho_i, F_i) \qquad [17]$$

where Φ is the incident photon flux, F_i and F_0 are respectively the electric fields in the next barrier after the ith well and at the contact. Gauss's law introduces a relation between the local variations of F_i and the electron density ρ_i of the wells (ρ_0 is the doping density):

$$\rho_i - \rho_0 = \frac{\varepsilon_0 \varepsilon_r}{q}(F_i - F_{i-1}) \qquad [18]$$

We use F_0 as a parameter. We first calculate the injected current $J_{inj}(F_0)$. Expressions [17] and [18] allow then to calculate sequentially the electric field F_i and the electron density ρ_i. The total distribution $\{F_i\}$ gives $V_T(F_0.\Phi)$, the true applied bias.

Results:

Field distribution: Figure 6 shows the field distribution, for different initial electric field F_0 and a moderate illuminating condition, $\Phi = 10^{17}\ cm^{-2}/s$. For the test structure, where the doping level in the contact is weak, most of the applied bias is used by the injection condition. We present in Figure 7 the resulting I-V curves in the dark case.

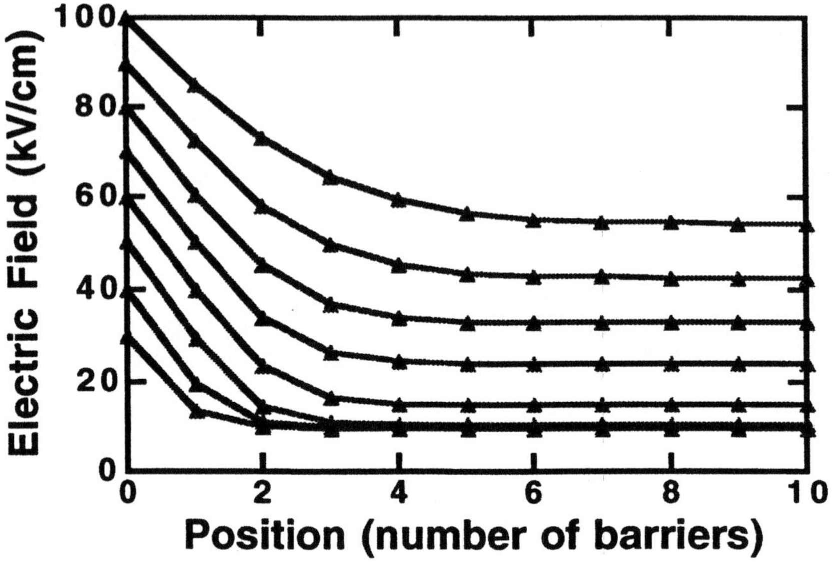

Fig. 6 : Distribution of the electric field in the test QWIP under operation, for different F_0. T=77 K, $\Phi = 10^{17}$ $cm^{-2}s^{-1}$.

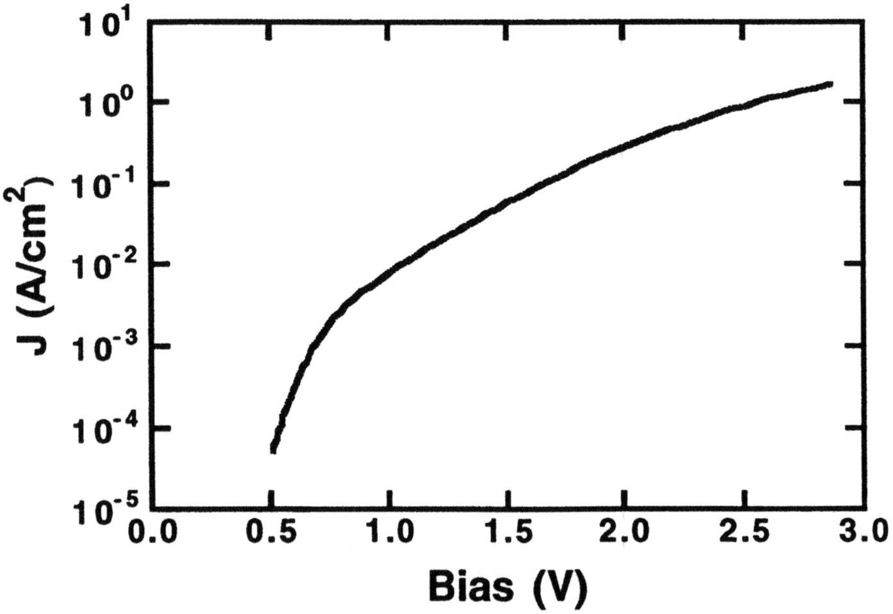

Fig. 7 : Total current flowing through the device at 77 K in dark condition.

Response: The photoresponse is given by the difference of the total current under illumination and the dark current:

$$R(V_T) = \frac{J_{inj}(\Phi, F_0^{\Phi}) - J_{inj}(\Phi = 0, F_0^{dark})}{hv.\Phi} \qquad [19]$$

where F_0^{Φ} and F_0^{dark} lead to the same bias V_T. Figure 8 shows the variation of the photoresponse with the applied bias. The linearity of the response with the flux has been checked up to $\Phi = 5.10^{17}$ cm^{-2}s^{-1}. Note that, contrary to what might be expected for a bound to quasibound QWIP, the photoresponse is not linear at small applied biases. Moreover, the signal saturates at higher biases. These two features are in accordance with experimental facts (1, 21).

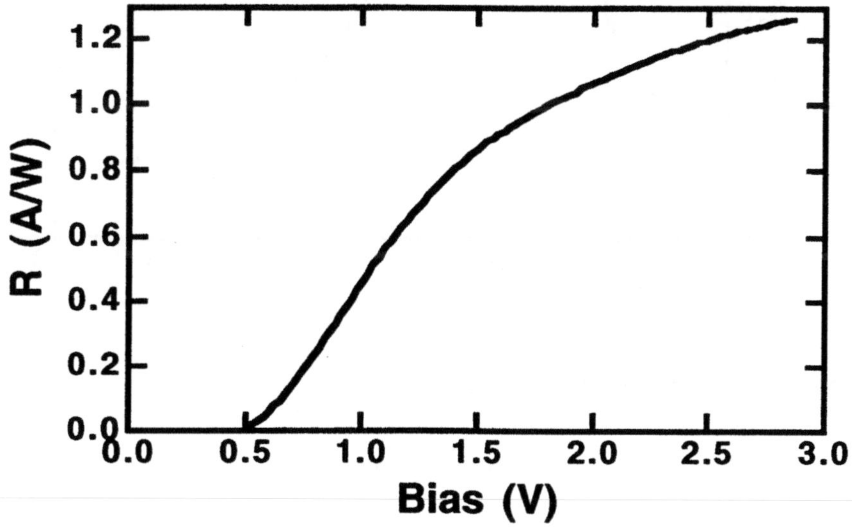

Fig. 8 : Responsitivity vs bias at 77 K.

Noise: We develop a new approach of the noise in QWIPs. Following Beck (22), the noise current generated by a single QW is given by the fluctuations of the electron flux in the external electrical circuit, $i_n^2 = 2qI_0\Delta f\,(2\text{-}p_c)/p_c$. In our general picture of QWIPs as QW periods mounted in series, we can assume that each period is a local noise generator, contributing independently to the total noise current. In this case, we have to sum up the voltage fluctuations and the total noise current is given by:

$$i_n^2 = \frac{\sum_{n=1}^{N} \left(i_n^2(i)\, Z_i^2\right)}{\left(\sum_{n=1}^{N} Z_i\right)^2} = 2\,q\,I_0\,\Delta f\, \frac{\sum_{n=1}^{N} \left(\frac{2-p_i}{p_i} Z_i^2\right)}{\left(\sum_{n=1}^{N} Z_i\right)^2} \qquad [20]$$

where $p_i = p_c(F_{i-1})$ is the local capture probability, Z_i is the impedance of each period, and $I_0 = A.J_{inj}(F_0,\Phi)$ is the total current flowing through the device under illumination. Figure 9 shows the dependence of the detectivity D* against the applied bias. The optimum bias corresponds to the saturation threshold of the photoresponse.

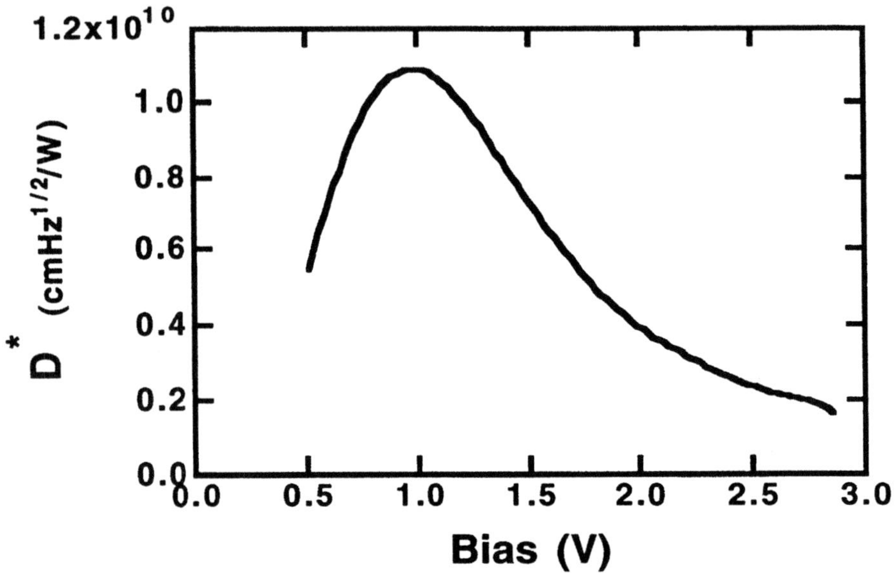

Fig. 9 : Detectivity vs applied bias. $\Phi_B = 10^{17}$ cm2/s, T=77 K.

CONCLUSION

The originality of our approach of QWIP physics has been the use of elaborate characterization tools of semiconductor deep levels (impedance measurements, DLOS, ...) for the study of emission and capture in QWs. We have developped a rather precise physical understanding of the transport mechanisms in QWIPs where the main pecularities are i) the role of the injecting contact, ii) the electric field dependence of emission and capture probabilities and iii) the electric field redistribution to maintain the current conservation in the structure. All these elements have been integrated in a new numerical model which allows us to calculate the QWIP performances under different conditions. The physical origin of some discovered effects are still unclear, such as the strong field dependence of the capture probability, that we have begun to study by saturation absorption using a free electron laser.

ACKNOWLEDGEMENTS

The authors thank the other contributors to this work: B. Vinter, L. Thibaudeau, F. Luc, E. Costard, N. Laurent and J. Nagle from Thomson CSF/LCR and D. Jaroszinski and J. M. Ortega from the free electron laser team of CLIO (Orsay). This work was partially supported by DRET.

REFERENCES

1. B. F. Levine, J. Appl. Phys. **74**, R1 (1993).
2. M. A. Kinch and A. Yariv, Appl. Phys. Lett. **55**, 2093 (1989).
3. H. C. Liu, Appl. Phys. Lett. **60**, 1507 (1992).
4. M. Missous, in Properties of Aluminium Gallium Arsenide, S. Adachi, Editors, p.73, INSPEC, London (1993).
5. H. C. Liu, A. G. Steele, M. Buchanan and Z. R. Wasilewski, in Intersubband Transitions in Quantum Wells, E. Rosencher, B. Vinter and B. Levine, Editors, p. 57, Plenum, New York (1992).
6. E. Rosencher, B. Vinter, F. Luc, L. Thibaudeau, P. Bois and J. Nagle, IEEE J. Quantum Electron. (accepted for publication) (1994).
7. G. Hasnain, B. F. Levine, S. Gunapala and N. Chand, Appl. Phys. Lett. **57**, 608 (1990).
8. E. Rosencher, F. Luc, P. Bois and S. Delaître, Appl. Phys. Lett. **61**, 468 (1992).
9. H. C. Liu, A. G. Steele, M. Buchanan and Z. R. Wasilewski, J. Appl. Phys. **73**, 2029 (1993).
10. F. Luc, E. Rosencher and P. Bois, Appl. Phys. Lett. **62**, 2542 (1993).
11. A. Chantre, G. Vincent and D. Bois, Phys. Rev. B **23**, 5335 (1981).
12. E. Rosencher, F. Luc, L. Thibaudeau, P. Bois and B. Vinter, in Quantum Well Intersubband Transition Physics and Devices, H. C. Liu, B. F. Levine and J. Y. Andersson, Editors, p. 97, Kluwer, Dordrecht (1994).
13. F. Luc, E. Rosencher and B. Vinter, Appl. Phys. Lett. **62**, 1143 (1993).
14. L. Thibaudeau and B. Vinter, Appl. Phys. Lett. (accepted for publication) (1994).
15. E. Rosencher, F. Luc, P. Bois and Y. Cordier, Appl. Phys. Lett. **63**, 3312 (1993).
16. B. Deveaud, A. Chomette, F. Clérot and A. Regreny, in Intersubband Transitions in Quantum Wells, E. Rosencher, B. Vinter and B. Levine, Editors, p. 275, Plenum, New York (1992).
17. J. Faist, F. Capasso, C. Sirtori, D. L. Sivco, A. L. Hutchinson, S. N. G. Chu and A. Y. Cho, Appl. Phys. Lett. **63**, 1354(1993).
18. F. H. Julien, J. M. Lourtioz, N. Herschkorn, D. Delacourt, J. P. Pocholle, M. Papuchon, R. Planel and G. LeRoux, Appl. Phys. Lett. **53**, 116 (1988).
19. J. Y. Duboz, E. Costard, E. Rosencher, Ph. Bois, J. M. Berset, D. Jaroszynski, J. M. Ortega, Superlattices and Microstructures (accepted for publication) (1994)
20. S. V. Meshkov, Sov. Phys. JETP **64**, 1337 (1986).
21. B. Xing, H. C. Liu, P. H. Wilson, M. Buchanan, Z. R. Wasilewski and J. G. Simmons, J. Appl. Phys. **76**, 1889 (1994).
22. W. A. Beck, Appl. Phys. Lett. **63**, 3589 (1993).

SEMIMAGNETIC QUANTUM WELLS AND SUPERLATTICES

M. AVEROUS

Groupe d'Etude des Semiconducteurs, URA 357

Université de Montpellier II,

Pl. Eugène Bataillon, F-34095 Montpellier Cedex 5, FRANCE

Introduction

During last ten years, there has been increasing interest and considerable experimental and theoretical activity focused on Semimagnetic Semiconductors (S.M.S.C.) or Diluted Magnetic Semiconductors (D.M.S.) which have some unique properties that enhance their potential for use in a wide range of optoelectronic device applications [1-5].

S.M.S.C. or D.M.S. could be II-VI, IV-VI, II-V, or III-V compounds in which a (small) fraction of non magnetic cations has been substituted by magnetic transitions metal or rare earth ions.

It was Galazka [6] who draw the attention to these compounds and first emphasized the crucial role of the interaction between the mobile carriers and the local moments as the driving mechanism behind the interesting semiconducting and magnetic properties.

Most of the researchs performed so far on these materials have been devoted to Mn-based II-VI compounds such as $Cd_{1-x}Mn_xTe$, $Zn_{1-x}Mn_xTe$, $Hg_{1-x}Mn_xTe$ and the corresponding Selenides and Sulfides. However, more recently, results have been

M. Balkanski (ed.), Devices Based on Low-Dimensional Semiconductor Structures, 115–170.

reported on II-VI Semiconductors containing others transitions magnetic metal ions, such as Fe or Co [7-9]. Concerning the IV-VI compounds, they have been studied for about a decade [10]. Nevertheless, the amount of knowledge which has been accumulated is much less than compared to the situation of II Mn VI or II Fe VI alloys. The complication arises from their rather complex band structure. They are of interest because the value of their band gap are in the I.R. region.

For many purpose one may regard a S.M.S.C. (D.M.S.) as been composed of two interacting subsystems:

- the electronic system of the carriers,
- the magnetic system of magnetic moments localized on magnetic ions.

The most striking phenomena observed in D.M.S. are those which arise from the sp-d interaction between the electronic and magnetic subsystems. Because of this interaction, a perturbation of the magnetic subsystem, e.g. by applying a magnetic field H or changing the temperature T, affects the electronic subsystem. This leads to a lot of interesting magneto-optical and magneto-transport phenomena, such as large Zeeman splitting, giant Faraday rotation, bound magnetic polarons and giant magnetoresistance anomalies.

In this first chapter we shall focus on the II M - VI and IV T VI systems, where M could be Mn, Fe or Co and T : Eu, Gd. We give an overview of specific properties of such D.M.S. In chapter II and chapter III, we shall develop the properties of quantum wells and superlattices of D.M.S. II-VI and D.M.S. IV-VI systems respectively.

1. Diluted magnetic II-VI compounds

The organization of this chapter is as follows. At first we will briefly review the crystal structure, the electronic structure, the Mn d-sp Band Exchange, the Mn d - Mn d exchange since we have choose as example the II_{1-x} Mn_x VI D.M.S., for the sake of simplicity.

1.1. CRYSTAL STRUCTURE

Almost of the II VI compounds crystallize either in the Zinc Blende or wurtzite structures. In these structures each atom of one element is tetrahedrically bounded to four atoms of the other elements. In Zinc Blende these tetrahedra are arranged in a cubic type structure while in wurtzite they are in an hexagonal type structure.

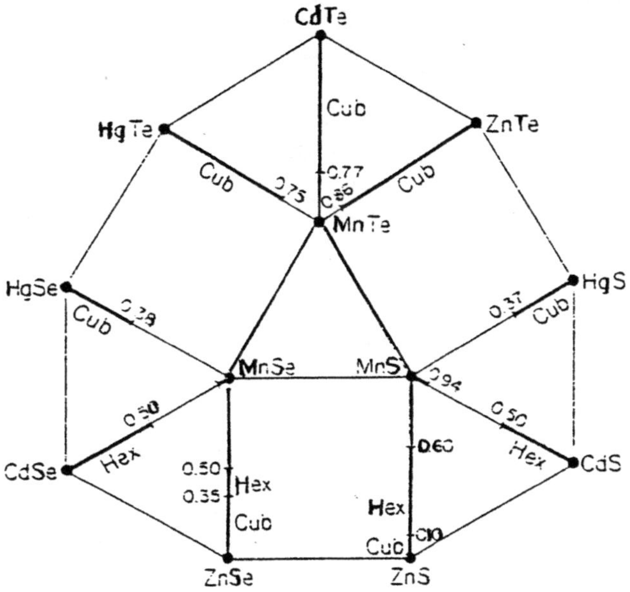

Figure 1. A^{II}_{1-x} Mn_x B^{II} alloys and their crystal structure.

The tetrahedral bondings (s-p^3), involve the six valence p electrons of the element VI and the two valence s electrons of the element II. Manganese is a transition metal with valence bound electrons corresponding to $4s^2$ orbital. So it can contribute to be s-p^3 bonding by these $4s^2$ electrons and therefore replace the element II in the wurtzite or Zinc Blende structure, despite the fact that it differs from this element by its 3 d shell half-filled. However the exactly half filled 3d orbitals play a role in the existence of very wide solid solutions. On fig. 1 we report the crystal structure of II_{1-x} Mn_x VI alloys as a function of x.

Usually for small x they have the crystal structure of the II VI parent [11]. Except for ZnMnS and ZnMnSe which are cubic Zinc Blende at low x, then, above a critical x, they are wurtzite. For each ternary compound an upper limit of x is imposed by the fact that MnTe, MnSe or MnS do not crystallize in the Zinc Blende or wurtzite structure.

Fig. 2 shows that the lattice parameters for all S.C.S.M. obey Vegard's low very closely [12].

The microscopic situation is more complicated (local structure).

Fig. 3 shows EXAFS results for CdTe and MnTe bond length. [13]. This result is not in contradiction with fig. 2 since X- ray diffraction data give the lattice parameter averaged over all anion and cation sites.

To adjust the CdTe and MnTe constant bond lengths, the anion sublattice must be locally distorded. Since the MnTe bond length is shorter than the CdTe one, the Te ion will be displaced from its equilibrium position toward Mn and away from Cd (Fig. 4). This fact might be of great importance for the exchange constants as seen later.

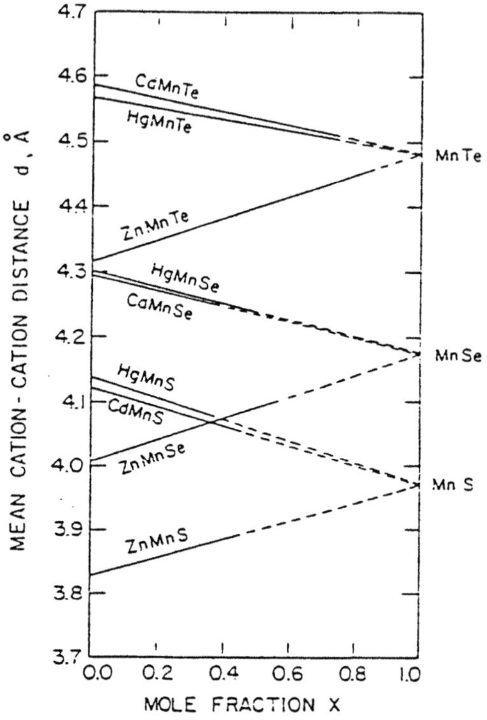

Figure 2. **Mean cation-cation distance vs Mn mole fraction.**

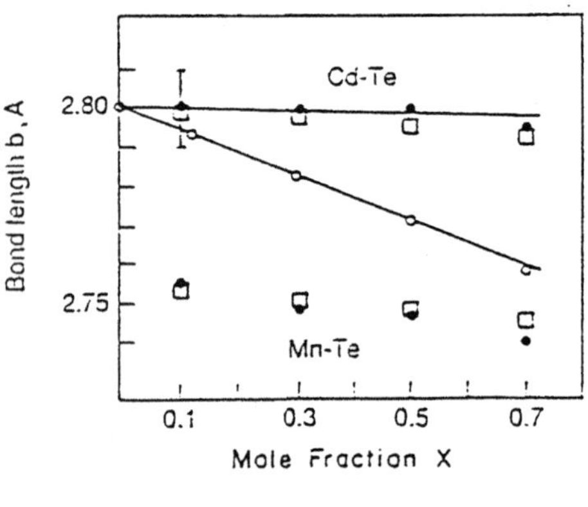

Figure 3. **EXAFS results on CdTe and MnTe bond lengths.**

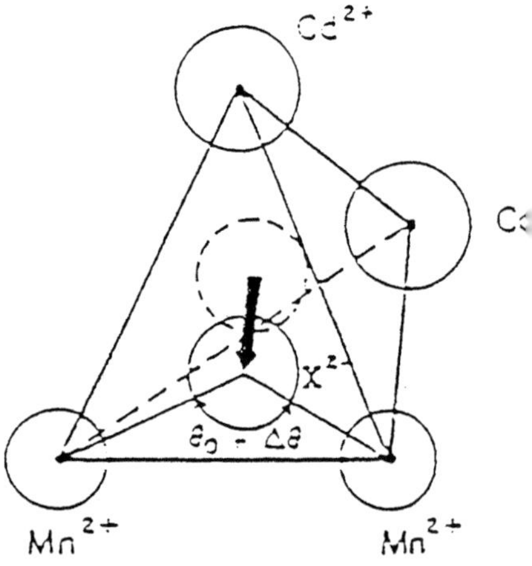

Figure 4 . **Distorsion position of the anion in mixed CdMnX.**

1.2. ELECTRONIC STRUCTURE

One of the attractive features of $II_{1-x}Mn_x$ D.M.S. is the fact that the band structures of the II-VI host compounds are well known. In the absence of magnetic field the semiconducting properties of $II_{1-x}Mn_{1-x}$ VI D.M.S. are remarkably similar to those of the II-VI alloys.

All II-VI compounds with Zn, Cd or Hg and S, Se or Te exhibit direct band gaps at k=O. This along with the wide energy range of these gaps (0 - 3.8 eV) make these materials promising for optical applications.

The band structure could be understood in terms of s-p^3 orbital bonding configuration. The valence (conduction) bands in these compounds are derived primarily from bonding (antibonding) combinations of s-p^3 hybrid orbitals on neighboring atoms [11].

In addition to having two valence s electrons, each group II cation has a fully occupied d shell that lies relatively below in energy (> 1 Ryd). We neglect the role of these d electrons to the bonding.

For example for ZnTe the highest-lying valence band is three fold degenerate (without S.O. Interaction), and the states at the center of the Brillouin zone are bonding combinations of functions which are p - like about the nuclei. The lowest conduction band is s-like (Fig. 5).

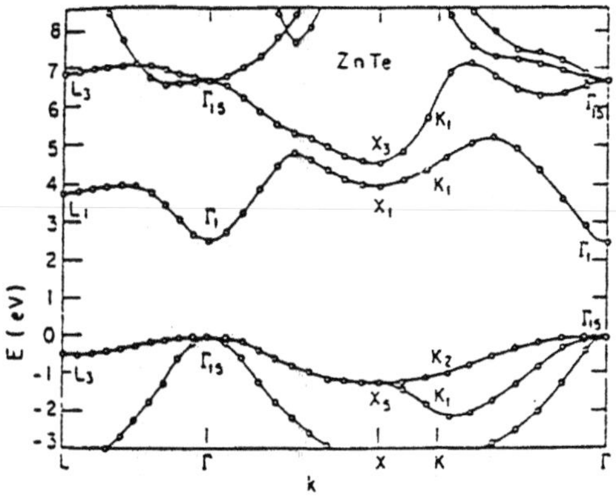

Figure 5. Band structure of Cubic ZnTe.

When S.O. interaction is include the degeneracies are lifted. The Γ_{15v} state sixfold degenerated (threefold if we don't consider the two spin states) splits into a fourfold Γ_8 (J = 3/2) state and a lower twofold Γ_7 (J = 1/2) state.

HgTe is a semi metal [14]. The six p like states are decoupled by the S.O. interaction into a quadruplet Γ_8 (J = 3/2) and a doublet Γ_7 (J = 1/2) as in the case of CdTe. The Γ_7 forming a lower lying valence band. The s - like band, transforming as Γ_6 (J = 1/2) lies below the Γ_8 quadruplet, leading to a symmetry induced zero gap semiconductor. The Γ_8 levels then form the degenerate valence and conduction band edges, separate from the Γ_6 band by the interaction gap:

$E_o = \Gamma_6 - \Gamma_8 = 0.3$ eV at 4.2 K

There exits a solid solution in the entire range of composition between HgTe and CdTe and Fig. 6 shows the band structure of $Hg_{1-x}Cd_xTe$.

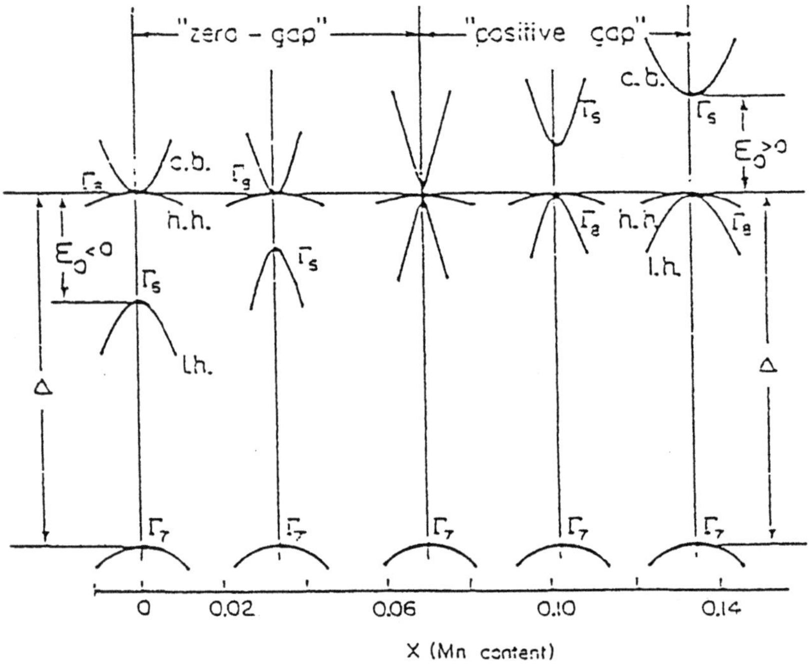

Figure 6. **Band structure of $Hg_{1-x}Cd_xTe$ or $Hg_{1-y}Mn_yTe$**

The interaction gap E_o increases with x, and passes through zero for x = 0.165 at 4.2 K giving rise to a semi-metal/semiconductor transition.

The behavior of the band structure, without magnetic field, of $Hg_{1-x}Mn_xTe$ is the same. The values of the gaps for MnTe are extrapolated in the case of cubic phase (virtual). In this case the transition occurs for x= 0.075 at 4.2K.

We have anticipated for the Mn VI compounds the same structure as in II-VI compounds according to computational results based for example on virtual crystal approximation. The intermediate D.M.S. compounds behave alike, and in fact their energy gaps follow the Vegard's low very closely (Fig. 7).

In addition, there are properties due to the presence of the Mn^{2+} ions in the tetrahedral crystal field. They not only change the band structure, but also introduce new optical transitions. These include absorption and emission features which can be correlated with the expected splitting of Mn multiplets in the crystal field. A schematic diagram of Mn^{2+} energy levels without and with crystal field is shown in Fig. 8.

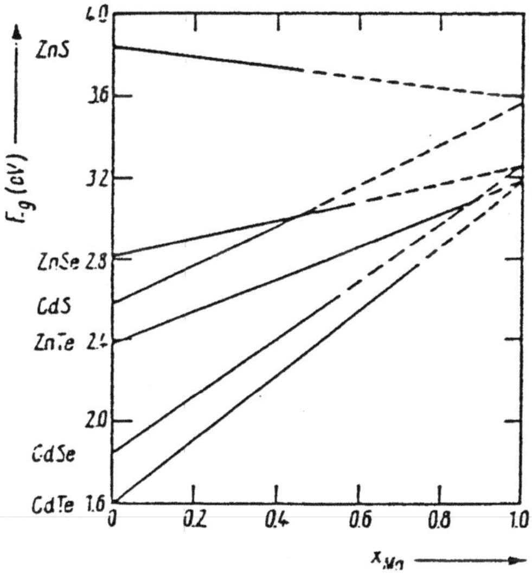

Figure 7. Energy gap Eg(x) of II_{1-x} Mn_x VI compound.

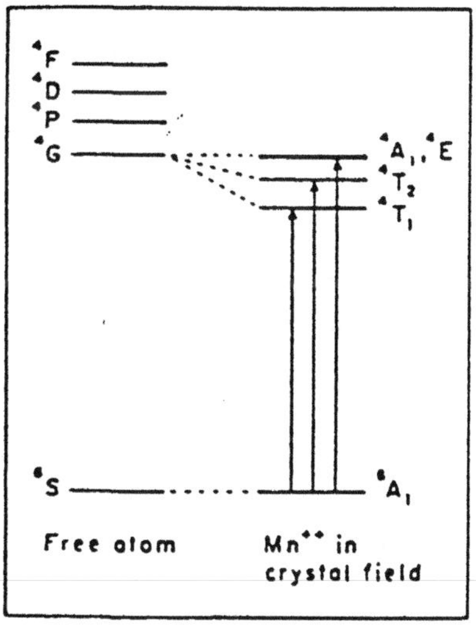

Figure 8 : Energy levels of Mn^{2+} in a tetrahedral crystal field.

This diagram is governed by the five electrons of the 3d shell of Mn^{2+} ; the ground state of the latter is characterized by S = 5/2 L = 0. The lowest Mn^{2+} electronic transition ${}^6A_1 \rightarrow {}^4T_1$ is the most important and occurs around 2.2 eV. This transition determines the absorption edge for large x (E_g (x) > 2.2 eV). In some D.M.S. as Zn MnS or ZnMnSe (X ~ 0.03) it is responsible for intense electroluminescence phenomena [15]. All the energy levels presented above including II Mn VI alloys and Mn^2 ions are obtain without magnetic field.

When a magnetic field is applied the magnetic exchange interactions must be taken into account through sp -d exchange interactions, or d - d interactions.

1.3. sp-d EXCHANGE INTERACTIONS

The largest magnetic exchange interaction in II_{1-x} Mx_x VI alloys is that between the local Mn spins (S_i) and the spin densities $\sigma(r)$ associated with "sp" valence and conduction states. The dominant (isotropic) interaction of this kind may be described by the phenomenological Kondo-like Hamiltonien [17,18].

$$H_{ex} = -\sum_{R_i} J_{sp-d}(r - R_i).S_i.S(r)$$

r and R_i the coordinates of band electrons and magnetic ion, respectively, with an exchange function J_{sp-d} that falls off rapidly with the distance from a Mn site. This interaction is responsible for the enhanced magneto-optical and magneto transport properties of S.M.S.C. that have attracted both scientific and technological interests.

Since the carriers have a spatially extended wave function they interact simultaneously with a large number of Mn^{2+} ions. Then using the molecular field approximation (M.F.A.) one replace S_i by the thermal average $< S>$. Obviously without magnetic field $< S> = O$. If the magnetic field direction is z, $< S > = < S_z>$ (the average spin per Mn site). $<S_z>$ is directly related to the total magnetization by $M = N_o\, x < S_z>$, where No is the number of unit cells per unit volume.

The summation no longer runs over the exact position of the spins (R_i) but over all cation sites (R) weighted by the concentration magnetic ions x. Now, the exchange integral follows the periodicity of the semiconducting lattice (Virtual Crystal Approximation) and provides the possibility to solve H_{ex}

$$H_{ex} = -x.s_z < S_z > \sum_{R} J(r - R)$$

In the case of parabolic conduction band and for wide gap materials further approximations are allowed. The effective mass m* is large, both Landau and spin splitting of the band states are considerably smaller, and they are neglected in the

Hamiltonian of the system so in this approximation the magnetic splitting is due mainly to H_{ex}. For a cubic Zinc Blende structure the matrices of H_{ex} are diagonal;

$$<\Psi_{\Gamma_6}|H_{ex}|\Psi_{\Gamma_6}> = \begin{vmatrix} 3A & 0 \\ 0 & 3A \end{vmatrix}$$

$$<\Psi_{\Gamma_8}|H_{ex}|\Psi_{\Gamma_8}> = \begin{vmatrix} 3B & 0 & 0 & 0 \\ 0 & B & 0 & 0 \\ 0 & 0 & -B & 0 \\ 0 & 0 & 0 & -3B \end{vmatrix}$$

and $$<\Psi_{\Gamma_7}|H_{ex}|\Psi_{\Gamma_7}> = \begin{vmatrix} -B & 0 \\ 0 & B \end{vmatrix}$$

Where $A = -\frac{1}{6} N_0 . x . \alpha < S_z >$ and $B = -\frac{1}{6} N_0 . \beta . x < S_z >$

$\alpha = <s|J|s>$ and $\beta = <x|J|x>$ are exchange integrals for conduction and valence band respectively ($\alpha >$ O, $\beta <$ O). Then large splittings appear with the magnetic field. The Fig. 9 shows a schematic illustration of the spin split valence band maxima and conduction band minima in magnetic field, for a Zinc Blende wide gap S.M.S.C..

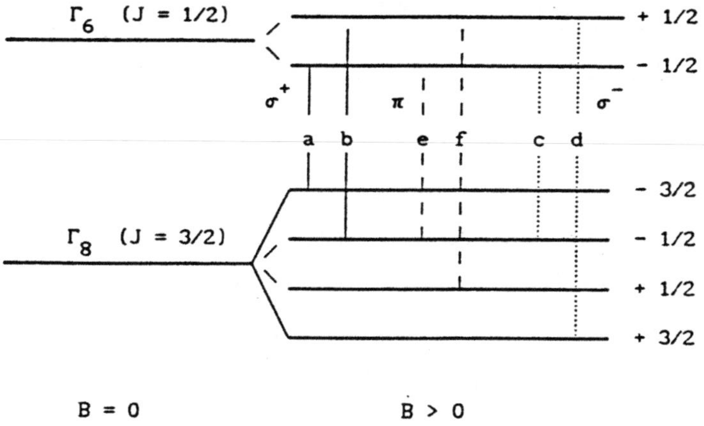

Figure 9. Allowed transitions between Γ_8 and Γ_6 states splitted by magnetic field.

Allowed transitions are indicated by arrows. In Faraday geometry, the transitions a and d are strong components where as transitions b and c are weak. In Voigt configuration the observed transitions are e and f [19].

The Zeeman splitting ΔE between the two strong components $(|-3/2_v> \rightarrow |-1/2_c>$ and $|+3/2>_v \rightarrow |+1/2_c>)$ is given by : $\Delta E = -N_0(\alpha - \beta)x < S_z >$

Table 1 gives some values of $N_o\alpha$ and $N_o\beta$ deduced from magneto reflectivity or magneto absorption.

TABLE 1. Exchange integrals for the conduction and valence bands.

Compound	Composition x	T	$N_0\alpha$ (eV)	$N_0\beta$ (eV)
CdMnTe	≤0.3	1.4	0.22 ±0.01	0.88
ZnMnTe	≤0.1	22	0.19 ±0.01	-1.09
	≤0.25	1.6	0.18 ±0.01	-1.05
CdMnSe	0.10	1.5	0.26	-1.24
	0.05	1.7	0.26	-1.30
ZnMnS	≤0.1	2.2	0.26	-1.31
ZeFeSe	0.05	2.0	0.22	-1.57
	0.043	4.5		
CdCoSe	0;01	1.8	0.32	

The difference in both sign and magnitude of α and β had been explained by Bhattacharjee at al.[20] There exist two competitive mechanisms, the first originating from direct potential interaction between bands and d electrons is rather weak and ferromagnetic (positive). The second arises from the hybridization between d levels of the magnetic ions and the band states and called kinetic exchange. In the absence of disorder, the sp - d hybridization is strictly for bidden for the Γ_6 edge, by symmetry. By contrast, the valence band edge does hybridize strongly with the d states, and in that case kinetic exchange dominates. It follows that the valence band exchange $N_o\beta$ should be AF (Negative) and larger in magnitude than N_o α (see table 1).

In the case of small gap S.M.S.C. as $Hg_{1-x}Mn_xTe$ for low x, the effective masses are small, and the Landau level splittings cannot be neglected, so the matrix elements :

$$<\Psi_{\Gamma_6}|H_{ex}|\Psi_{\Gamma_6}>, <\Psi_{\Gamma_8}|H_{ex}|\Psi_{\Gamma_8}> \text{ and } <\Psi_{\Gamma_7}|H_{ex}|\Psi_{\Gamma_7}>$$

are little more complicated. However the landau levels are obtained by solving two decoupled matrix equations corresponding to a and b ladders [21].

For each quantum level n>1, there exists four solutions in each a and b ladder that describe Γ_6, Γ_8 (heavy and light) and Γ_7 Landau levels.

In the Γ_6 band, the energies of a and b ladders corresponding to n ↑ (or a) and n ↓ (or b) levels are (parabolic approximation) :

$$E_{a,b}(n) = (n+\frac{1}{2})\frac{eH\hbar}{m_{\Gamma_6}c} \pm \frac{1}{2}g^*_{\Gamma_6}\mu_B H \mp \frac{1}{2}\alpha N_0 x <S_z>$$

$m^*_{\Gamma_6}$ and g_{Γ_6} are the band edge effective mass and gyromagnetic factor deduced from k-p interaction :

$$\frac{m_0}{m_{\Gamma_6}} = 1 + \frac{E_p}{3\varepsilon_0}\cdot\frac{3\varepsilon_0+2\Delta}{\varepsilon_0+\Delta}$$

$$g^*_{\Gamma_6} = 2(1-\frac{E_p}{3\varepsilon_0}\cdot\frac{\Delta}{\varepsilon+\Delta})$$

In eq. 1 to have an usual form, one can replace :

$$g^*_{\Gamma_6} \text{ by } \overline{g^*_6} - \frac{5}{2}\alpha N_0 x$$

then $$E_{a,b}(n) = (n \pm \frac{1}{2})\frac{eH}{m_{\Gamma_6}c} \pm \frac{1}{2}g^{\alpha}_{\Gamma_6}\cdot\mu_b H$$

For the Γ_8 bands :

$$\frac{m_0}{m^*_{\Gamma_8}(\text{light})} = -(\gamma_1+2\overline{\gamma}+\frac{2E_p}{3\varepsilon_0}) \; ; \qquad \frac{m_0}{m^*_{\Gamma_8}(\text{heavy})} = 2\overline{\gamma}-\gamma_1$$

$$g^*_{\Gamma_8(\text{light})} = 2(3\overline{\gamma}-\chi+\frac{2E_p}{3\varepsilon_0}) - \frac{1}{3}\cdot\frac{\beta N_0}{\mu_B H}<S_z>$$

$$g^*_{\Gamma_8(\text{heavy})} = -6x - \frac{\beta N_0}{\mu_B H}<S_z>$$

It is worth to note that the effective masses are not changed by the exchange interaction, but an additional term due to the exchange appears in the gyromagnetic factors, that depend on T and H through $<S_z>$. Thus the effect of exchange gives additional splittings that must be added to the previous Zeeman splitting .

Fig. 10 and Fig. 11 present respectively the Γ_6 - Γ_8 transitions for x = 4‰ at 2 K and the Γ_8 Landau level energies at 4.2 K and 2 K for various x, versus Magnetic field [2].

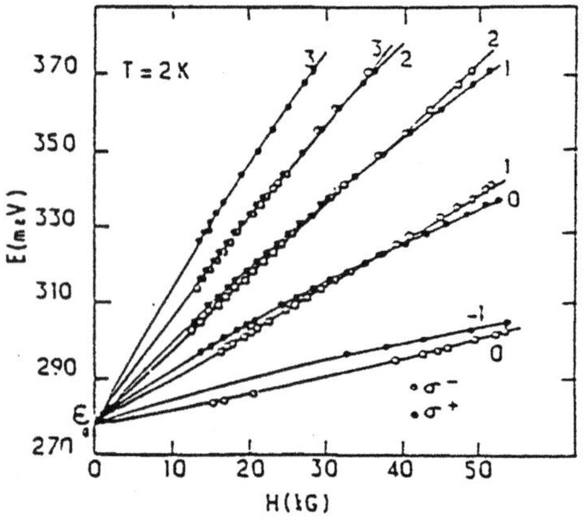

Figure 10. Γ_6 - Γ_8 transitions vs magnetic field for $Hg_{1-x}Mn_xTe$ (x = 4‰)at 2 K

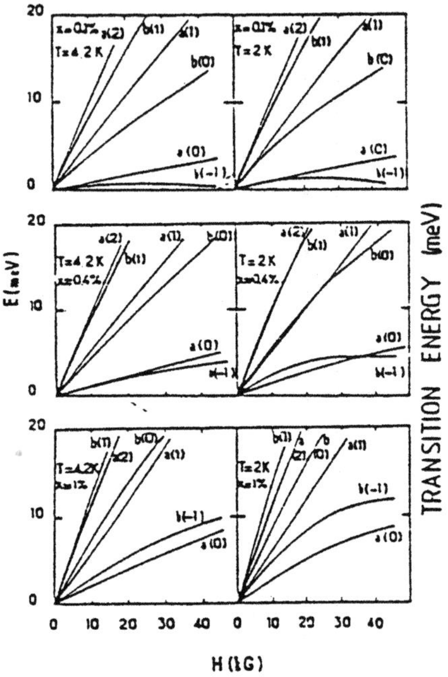

Figure 11. Γ_8 landau level energies vs H $Hg_{1-x}Mn_xTe$ at 4.2K and 2K

The above examples of wide gap and small gap S.M.S.C., show the difficulty in the case of quantum well, or superlattice of CdTe/Hg MnTe or Cd MnTe/HgTe but also, the large and rich possibilities of such heterostructure. Already, we can imagine how their properties will be strongly dependent of the magnetic field and the temperature.

1.4. Mn d - Mn d EXCHANGE

The magnetic exchange interactions (22) among the Mn spins themselves is also of great interest. From general experience with 3 d ions one expects that the dominant magnetic coupling between these ions is the isotropic exchange interaction.

$$H_{Mn-Mn} = -2\sum_{<i,j>} J_{ij} S_i . S_j$$

where J_{ij} is the exchange constant between spins S_i and S_j and the sum is on all pairs $<i,j>$ of magnetic ions, J_{ij} fall off rapidly with distance and are predominantly A.F. The nearest neighbor exchange, which will be denoted J_1 is on the order of - 10 K (-1 meV) in all II_{1-x} Mn_x VI. The second-neighbor exchange constant J_2 or J_{nnn} is an order of magnitude smaller than J_1.

Two other types of exchange interactions may exist. The usual anisotropic exchange is the part of exchange which depends on the direction of the spins with respect to the lattice. Usually it is very small and could be neglected in the case of Mn ion which is an ideal spin S = 5/2. However we have studied by E.P.R., the anisotropic exchange of the Co^{++} which may be regarded as an effective spin S = 3/2 and subjected to a weak single ion anisotropy ΔS^2_j.

A second type of anisotropic exchange is the Dzyaloshinski-Moriya interaction which is antisymmetric with respect to the inter exchange of the two interacting spins. For a pair of spins, i and j, it has the form :

$$H_{DM} = - 2\, D_{ij} \,.\, (S_i \times S_j)$$

The largest D_{ij} is expected to be for the nearest neighbors. Its magnitude D_1 for NN Mn^{2+} ions was calculated by Larson and Ehrenreich (22). As a result D_1 increases with the atomic number of the anion, i.e. largest for tellurides and smallest in sulfides.

In the presence of a magnetic field H there is a Zeeman interaction :

$$H_{Zeeman} = \sum_i g\mu_B S_i . H$$

The exchange constant J_{ij} which are the largest one in the series of exchange interactions describe above could be deduced from Curie-Weiss Temperature :

$$\chi = \frac{C}{T-\theta}$$ where the Curie constant C is given by

$$C = xNg^2\mu_B S(S+1)/3k_B$$

N is the total number of cations (magnetic or non magnetic) per unit volume and k_B is the Boltzmann constant.

The Curie Weiss temperature θ is given by :

$$\theta = \left[(2x.S(S+1)/3k_B\right]\sum_i z_i J_i$$

J_i is the exchange constant between a central magnetic ion and a magnetic ion which is on the ith coordination sphere (sphere of ith distant neighbors) and z_i the number of cations sites on that sphere. Assuming that J_1 much larger them all other exchange constants, one deduces

$\theta \cong 2 z_i x.S(S+1)/3k_B$ where $z_i = 12$ in Zinc Blende structure.

However the most accurate methods of determining J_i are based on studies of pairs, each consisting of two magnetic ions which are NN of each other. The NN cluster model is a simple model which ignored all other exchange interactions. Obviously if J_2 is larger than J_i it cannot be used and a more sophisticated model must be used taking into account the singles, the pairs, the closed triplet, the open triplet and so on. On Fig. 12, the energy level diagram for a J_1 pair of Mn^{2+} spins at H = O is given.(22) Arrows show the allowed "Stockes transitions for inelastic scattering of neutrons". The inverse "anti Stockes" between the same levels are also allowed.

In magnetization, the Zeeman splitting of these levels, gives steps corresponding to a change of the ground state with the magnetic field (Fig. 13).

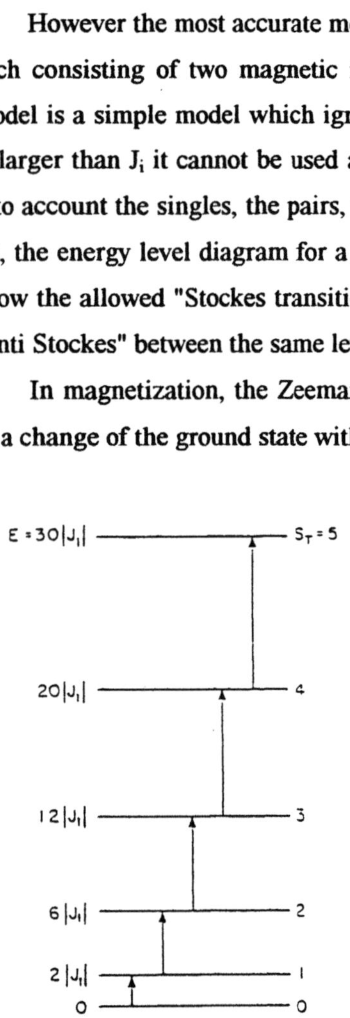

Figure 12. Energy-level diagram for a J_1 pair of Mn^{2+} spins at H = O

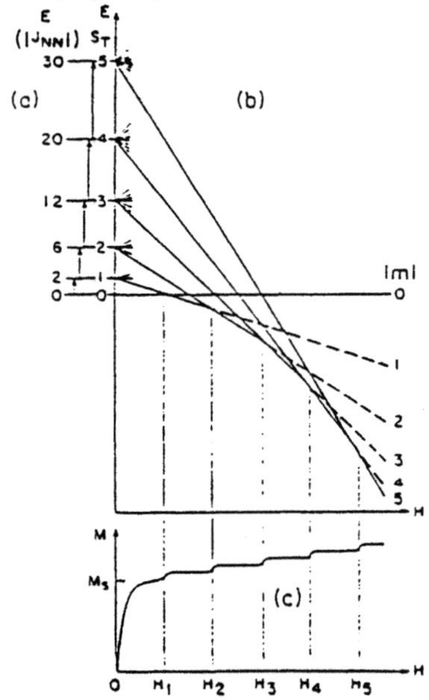

Figure 13. Energy level splitting of same diagram with H and magnetization steps.

Obviously the exact determination of M is very important to obtain $<S_z>$ which is a key parameter in the interpretation of the optical properties. For a complete treatment of the "modified cluster model". the reader can follow the very nice work of Y. Shapira in the book "Semimagnetic Semiconductors and Diluted Magnetic Semiconductors" edited by M. AVEROUS and M. BALKANSKI in the Ettore Majorana international Science Series [22].

T ABLE 2 . Values of J obtained from neutron, Raman and MST studies (M.S.T.: Magnetization steps).

J_1/k_B Material	Neutron	Raman	M.S.T (22)
ZnMnTe	-9.35		-88
	-8.79		-9.0
ZnMnSe	-12.3		-122
ZnMnS	-16.1		-16.9
ZnCoTe	-38.0		
ZnCoSe	-49.5		
ZnCoSe	-47.5		
CdMnSe		-8.1	-7.7
CdMnS		-10.6	10.4
CdMnTe			-6.1
HgMnTe			-5.1
HgMnSe			-6.0

2. Diluted magnetic IV-VI compounds

These compounds have attracted attention for two reasons : They are of interest in the I.R. wave length range as well detectors, as emitting lasers.

They permit a study of the effects of incorporation of magnetic ions with partially filled 3 d or 4 f states in a cubic structure which is not the "close packed" zinc blende type as was the case in the II - VI hosts.

The common IV - VI D.M.S. are those based on the pseudo-binary lead compounds (PbTe, PbSe, PbS) and little is known about the other possible IV - VI D.M.S.(10). The most commonly substituted ions are Mn, Eu and Gd. Some very recent results appear with Yb and U [23].

The solubility limits of the magnetic ions are generally small and vary markedly from alloy to alloy. Materials of this group of D.M.S. show interesting properties which make them useful for devices applications. For example substitution of Eu produces larger variation in "band gap" than Mn in PbTe or PbSe [24].

These IV - VI compounds, have for small concentrations of magnetic ions, a direct narrow gap at the L point of the Brillouin zone.

Fig. 14 shows the crystal structure of IV Mn VI D.M.S. as a function of manganese concentration [3].

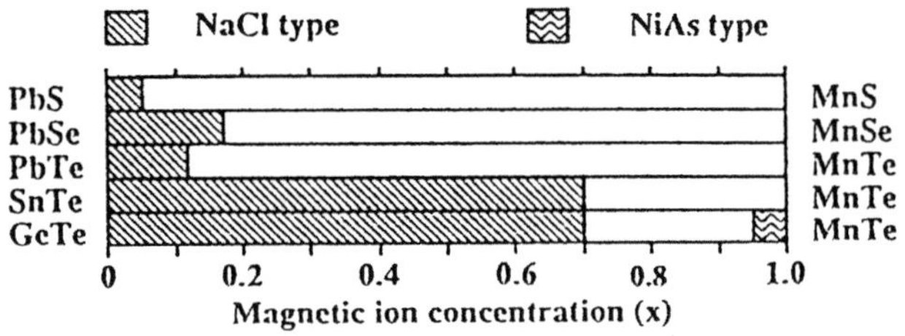

Figure 14. $A^{IV}_{1-x}Mn_xB^{VI}$ alloys and their crystal structure

2.1. ELECTRONIC STRUCTURE

Let us recall briefly the electronic structure of host crystals like PbTe, PbSe or PbS. In these cases apart from lowest lying conduction and highest valence band which form the forbidden gap, two further conduction and two valence bands have to be taken into account. The ordering of these levels which lie within an energy spread of about 3 - 5 eV, have been established by empirical pseudopotential calculation. Fig. 15 shows the sequence of energy levels in PbSe and PbTe following Bernick and Kleiman [25].

The spin orbit interaction is very strong, it not only removes the degeneracy of the L_3 levels but the originating energy gaps are of the order of the gap between L_1 and L_3 levels. The symmetry of the band edges for the conduction and valence bands given by Mitchell and Wallis are respectively :

$$L_6^-(L_2^-)_\alpha = -\sin\theta^- z\downarrow - \cos\theta^- x_+\downarrow$$

$$L_6^-(L_2^-)_\beta = \sin\theta^- z\downarrow - \cos\theta^- x_-\uparrow$$

and

$$L_6^+(L_1^+)_\alpha = i.\cos\theta^+ R\uparrow + \sin\theta^+ S_+\downarrow$$

$$L_6^+(L_1^+)_\beta = i.\cos\theta^+ R\downarrow + \sin\theta^+ S_-\uparrow$$

The relative mixing of the states by spin orbit coupling is determined by the parameters θ^+, θ^-. R, Z, X_{+-} and S_{+-} denote functions which are derived from functions with s, p, d symmetry [26].

In PbTe the lowest conduction band is dominated by the L_{3-}^{6-}. For PbSe it is the L_{2-}^{6-}. The energy gaps increase with the magnetic ion concentration Fig. 16 [10].

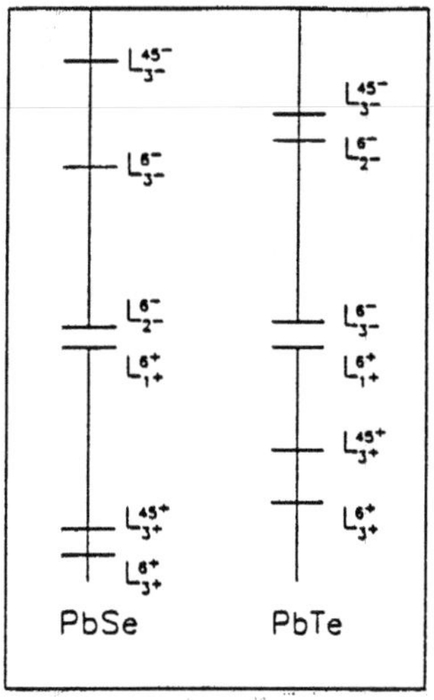

Figure 15. Energy levels in PbSe and PbTe[25]

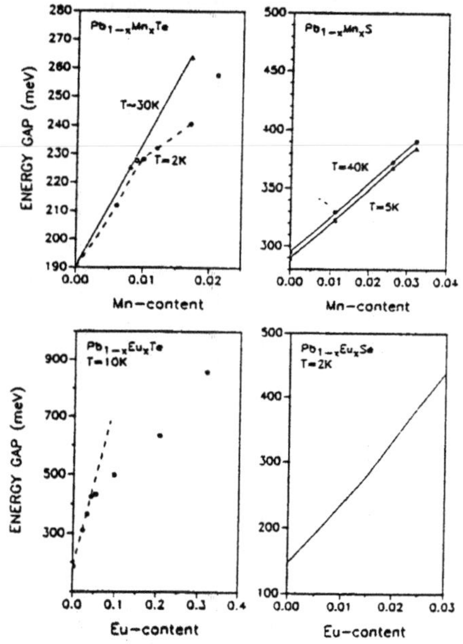

Figure 16. Energy gap versus composition for Mn nd Eu compounds.

In the case of PbEuTe a large bowing is observed. Whereas in EuTe the f-levels form a narrow band in the band gap, in PbEuTe with low Eu content the corresponding levels should be situated well below the valence band edge. Then for concentrations higher than 10% the gap does no longer behave PbTe - like.

2.2. EXCHANGE INTERACTION

For low magnetic ion concentration the band structure remains PbTe - like (PbSe, PbS). As seen already, for the description of the band extrema, situated at the L - points of the Brillouin zone, a two band model with four far bands treated in k^2 perturbation is used.

Within the frame work of the Molecular field Approximation (M.F.A.), Zawadzki had calculated the exchange included corrections to the (4 x 4) k.p Hamiltonian (in Ref 10). The exchange Hamiltonian added to the 4 x 4 matrix Hamiltonian has the general form :

$$H_{exch} = J(|\vec{r} - \vec{R}|).\vec{\sigma}.\vec{S}$$

$\vec{\sigma}$ the electron spin and $\vec{S}$ the magnetic ion spin (5/2 for Mn, 7/2 for Eu). Due to the many valley band structure H_{exch} is : (for Mn)

$$H_{exch} = \frac{1}{2} x S_0 B_{5/2}\left[\frac{5\mu_B B}{\mu_B(T+T_0)}\right] J(\sin\phi\sigma_x + \cos\phi\sigma_z)$$

J is the exchange integral, σ_x and σ_z the electron spin operators, ϕ denotes the angle between the magnetic field B and the (111) direction for $\vec{B}$ in a (x,z) ($[\bar{2}11]$ [111]) plane. This 4 x 4 exchange Hamiltonian contains off diagonal terms which cause a strong mixing of

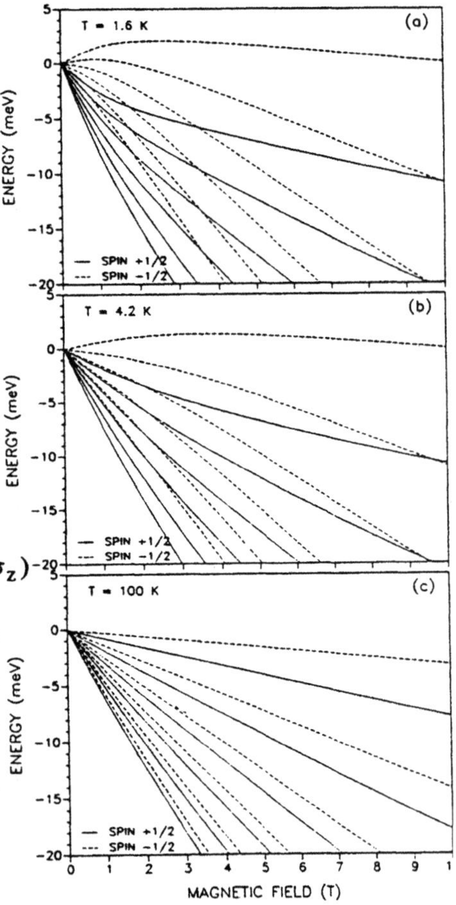

Figure 17. **Valence band Landau levels of $Pb_{1-x}Mn_xTe$ (x = 0.01) at 1.6 K, 4.2 K and 100 K**

Landau states with different spin and Landau levels.

Fig. 17 shows as an example the valence band Landau levels of $Pb_{1-x}MnTe$ (x = 0.01) where it could be seen very clearly at low temperature, that the exchange interactions alter drastically the magnetic field dependence of the Landau levels [10]. At small magnetic field the 0^- valence band level exhibits a positive slope. Obviously when T increases the exchange interactions decrease and at 100 K, the behavior is the same as the host crystal.

Obviously these exchange interactions give very rich magneto transmission spectra as shown in Fig. 18 and Fig. 19.

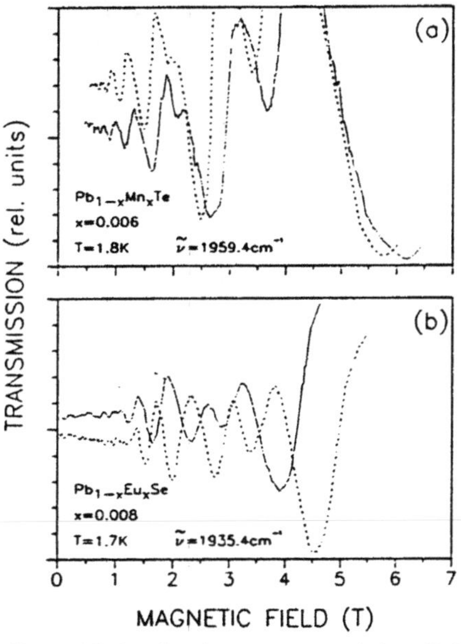

Figure 18. **Interband magneto transmission : Full curves σ^+ and broken curves σ^- circular polarization**

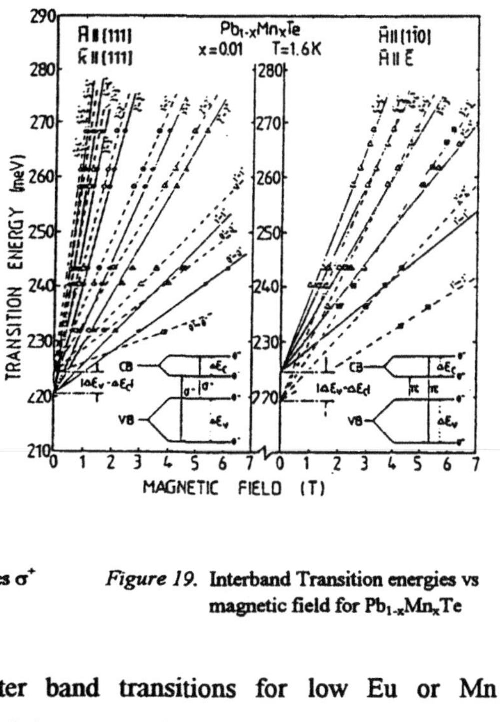

Figure 19. **Interband Transition energies vs magnetic field for $Pb_{1-x}Mn_xTe$**

The study of magneto optical inter band transitions for low Eu or Mn concentration requires radiation wavelength between 4.5 - 7.5 μn. The transmission spectra presented on $Pb_{1-x}Eu_xSe$, and $Pb_{1-x}Mn_xTe$ are obtained by Bauer et al [10].

The effective electron and hole g factors experimentally derived are plotted as a function of magnetic field for $Pb_{1-x}Mn_xTe$ and $Pb_{1-x}Eu_xSe$. For an orientation of H//110, four g factors are experimentally accessible : two for the conduction and two for the valence band. From H_{exch} an explicit expression for the longitudinal g factors can be derived :

$$\text{C.B. } g^{eff}_{C,1} = +\frac{4P_\perp^2}{m_0E_g} + g_1^- + x\alpha_1\frac{<S_z>}{\mu_0 B}$$

$$\text{V.B. } g^{eff}_{V,1} = -\frac{4P_\perp^2}{m_0E_g} + g_1^+ + x\beta_1\frac{<S_z>}{\mu_0 B}$$

Let us recall that :

$M = - x N_o g \mu_B <S_z>$.

α_1 and β_1 are ones of the two exchange parameters of the conduction and valence bands.

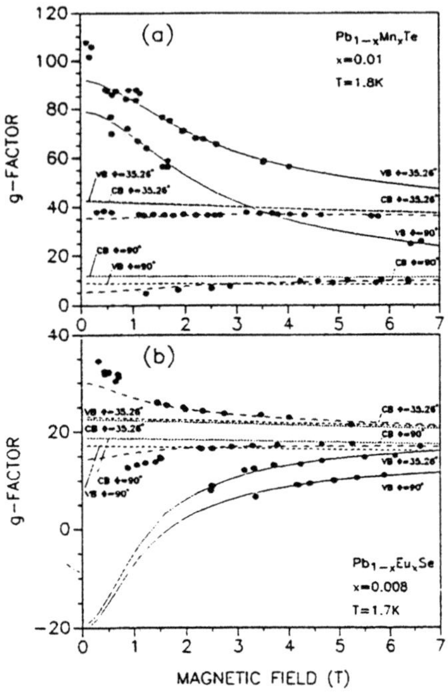

Figure 20. Magnetic dependence of g factors
full lines : calculated for holes
dashed lines : calculated for electrons.

2.3. d-d OR f-f INTERACTIONS

As in the case of II-VI compounds the exchange Hamiltonian between magnetic ions, has the same expression :

$$H_{d-d} = -2\sum_{i,j} J_{i,j} S_i . S_j$$

Again, the J_{ij} are the exchange constant between a central magnetic ion and a magnetic ion which is on the ith coordination sphere. The exchange constants which are the largest one could be deduced either by susceptibility measurements through the Curie Weiss temperature, or by magnetization measurements.

$$\chi = \frac{C}{T-\theta}$$ corrected by the lattice diamagnetic susceptibility

Finally the nearest neighbor exchange interaction is found to be :

$$\frac{J_{NN}}{k_B} = \frac{3\theta}{2\overline{x}(S+1)S_z}$$

The magnetization is fitted by :

$$M = M_s + M_p + \chi_0 H$$

where $M_S = M_0 S \overline{x}_1 B_S(\xi)$ is the single contribution,

with: $M_0 = g_M \mu_B N_0$

$B_S(\xi)$ is a modified Brillouin function

$$B_S(\xi) = \frac{2(S+1)}{2S}\coth\left[\frac{2S+1}{2S}\xi\right] - \frac{1}{2S}\coth\left[\frac{\xi}{2S}\right] \quad \text{and} \quad \xi = \frac{S g_M \mu_B H}{\mu_B (T+T_0)}$$

$$M_P = \frac{1}{2} M_0 \overline{x}_2 \cdot \frac{\sum_{s=0}^{S_{max}} \exp\left[\frac{J_P}{k_B T} S(S+1)\right] \sinh\left[\frac{2S+1}{2S}\xi_P\right] . s B_s(\xi)}{\sum_{s=0}^{S_{max}} \exp\left[\frac{J_P}{k_B T} S(S+1)\right] \sinh\left[\frac{2S+1}{2S}\xi_P\right]}$$

is the pair contribution to the magnetization.

where $\xi_P = s g_M \mu_B H / k_B T$ and $S_{max} = 2S$

the terms M_p, as seen previously, yields a step like behavior for $J_p/k_B T >> 1$.

TABLE 3. Values of J_{NN} or J_i obtained from susceptibility and magnetization.

Material	J_1/k_B (from χ)	J_1/k_B (from M)
PbMnTe		-0.88
PbMnSe		-0.93
PbMnS		-0.53
PbEuTe	-0.43	-0.43
PbEuSe (28)	-0.49	-0.32
PbGdTe (27) (29)	-0.5	-0.3
PbUTe (23)	-0.1	-0.1

It could be noted that the values of J_i are smaller than in the case of II VI compounds. This is due to additional interacting mechanisms as Blomberger Rowland interaction or RKKY interaction. For example, it has been found a ferromagnetism behavior, when the carrier concentration is higher than 3.10^{20} cm^{-3}. In fact there exists a Free Carrier Induced Ferromagnetism.

Even if this study seems to be out of the field of heterostructures, one can understood easily that a perfect knowledge of the magnetic properties is necessary to reach a good value of $<S_z>$ which is a key parameter on the energy level positions.

3. D.M.S. Heterostrcutures

3.1. INTRODUCTION

DMS Heterostructures exhibit new opportunities with respect to classical heterostructures. These new effects are the consequence of the peculiar properties of exchange interactions. In fact there are three main reasons.

- The spin splitting of electronic states in D.M.S. may easily be greater by as much as two orders of magnitude in the case of wide gap D.M.S. at low temperatures, than in adjacent non magnetic layers.

- The spin splitting of the electronic states in Diluted Magnetic Semiconductor layers is of the same order of magnitude than the ionization energy of the shallow impurities.

- The spin splitting is considerably greater in the case of wide gap or comparable in the case of small gap D.M.S. to the energy separation between consecutive Landau levels in the presence of magnetic field.

The chapter will be organized in the following. In a first paragraph different possibilities depending of the 3 above properties will be describe : i.e .- the spin superlattice-exchange effect in shallow non magnetic wells-transition from type I to type II superlattice-magnetic freeze-out and boil-off - Tunable resonant tunneling - Quantum oscillation [30].

All these effects will be described on very simple band model that means, that additional possibilities not presented here could exist. In a second paragraph, a review of II-VI D.M.S. superlattices will be given, then the same review will be presented on IV-VI D.M.S. heterostructures.

3.2. BASIC ON D.M.S. HETEROSTRUCTURES [3O]

Advances in the growth techniques like the molecular beam epitaxy has lead to high quality heterostructures incorporating layers of a dilute magnetic semiconductor.

3.2.1. *The spin superlatice*

The first idea is to create a superlattice, with a material A without magnetic ion, and a material B with diluted magnetic ions. The two components have the same energy band gap (often narrow gap) and the same effective mass. In the absence of magnetic field an electron moving along the growth direction of the superlattice see an uniform potential as shown on figure 21.

When a magnetic field is applied, due to the exchange interaction, the g* factor of the D.M.S. layer changes, and the supplementary term gives additional splittings that must be added to the previous Zeeman splitting.

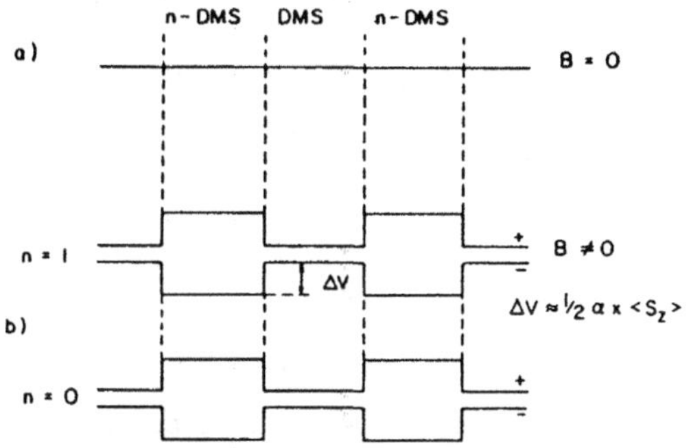

Figure 21. Schematic potential seen by an electron in the spin superlattice

Let us recall that the effective mass is unchanged by the exchange interaction. In this situation, the same electron moving along the growth axis see a potential which has the form of a sequence of square wells. The magnetic field splits the conduction band into a series of "minibands" separated from each other by "minigaps", which are proportional to the difference of spin splitting in the materials A and B. This difference is due to the exchange interaction in material B. These kind of spin superlattice can be obtained either with $Hg_{1-x}Mn_xSe$ and $Hg_{1-y}Cd_ySe$ with x=0.01 and y=0.024, or $Hg_{1-x}Mn_xTe$ and $Hg_{1-y}Cd_yTe$ with x=0.165 and y=0.075.

3.2.2. *Exchange effect in shallow non-magnetic wells*

In the case of wide band gap semiconductors, if we consider non magnetic quantum wells between D.M.S. barriers, and if without magnetic field, the potential wells seen by an electron is of the same magnitude of the exchange energy in the barrier, with magnetic field, the depth of the wells depends on the magnetic field. This means that the g factor which increases in the barrier, increases also in the non magnetic well since the energy position of the states in the well depends on the depth of the well. This phenomenon could be described as being due to the "leakage" of the wave functions of the electrons confined in the well into D.M.S. barriers, where it can interact via the magnetic ions through the sp-d exchange. This "leakage" is the greater, the closer a given electronic state is situated to the top of the well.

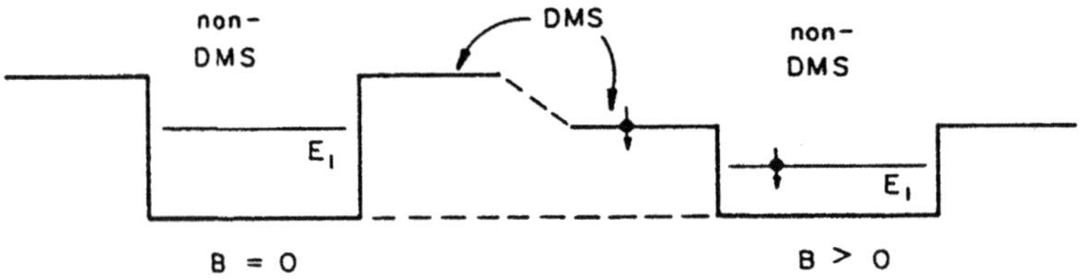

Figure 22. Shallow non D.M.S. quantum well between DMS barriers

3.2.3. *Magnetic field induced type I - type II transition*

Let us consider superlattices in which the barriers contain magnetic ions, whereas the well are non magnetic. In the particular case where the valence band offset is smaller than the conduction band one, the wells are for the electrons and holes in the non magnetic semiconductor (figure 23a).

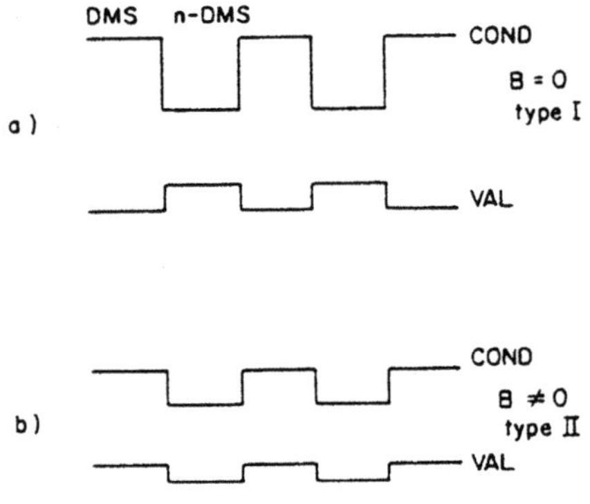

Figure 23. Schematic band offsets in $CdTe/Cd_{1-x}Mn_xTe$ superlattice without and with magnetic field

In the absence of magnetic field, this superlattice is type I.

The electrons and holes are confined in the same region (non D.M.S. material)

When a magnetic field is applied the electrons remain located in the non D.M.S. layer if the depth of the well is sufficiently large. For the valence band, the spin splitting can be compared to the valence band offset, as it is rather small. If this spin splitting is larger than the zero-field valence band offset one may expect a transition to a type II superlattice to occur (Figure 23b). This type I - type II transition will drastically change the exciton binding energy and the oscillator strength of the corresponding optical transition.

3.2.4. *Magnetic freeze-out and boil-off*

Another spectacular effect due to the exchange interaction could be magnetic freeze-out and the reverse magnetic effect the boil-off. If we start from a modulation-doped single well of non D.M.S. material embedded between two n-type D.M.S. barriers. For a suitable choice of materials, one can obtain at zero magnetic field the situation depicted on figure 24.

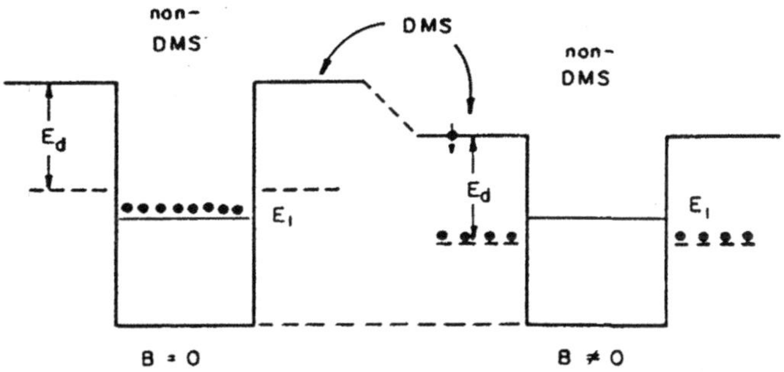

Figure 24. Non DMS quantum well between DMS barrier with a donor level in the barrier

The E_1 ground electronic state in the well is slightly below the donor state in the barrier (few meV). At low temperature the electrons of the donor level transfer to the E_1 states giving a conducting two dimensional sample. When the magnetic field is applied, one of the spin components of the donor state shifts down in energy, while the E_1 position in the well is practically unaffected. In this situation, E_1 becomes higher than the position of the donor level, and the electrons are transferred to the donor level giving a sudden increase of the resistivity. This is the magnetic freeze-out.

A reverse effect can be expected if the well are magnetic and the barrier non magnetic, with E_1 slightly higher than the donor level. This effect is described on figure 25. In that case the electrons transfer from the donor level to the well giving a two-dimensional conducting sample under magnetic field.

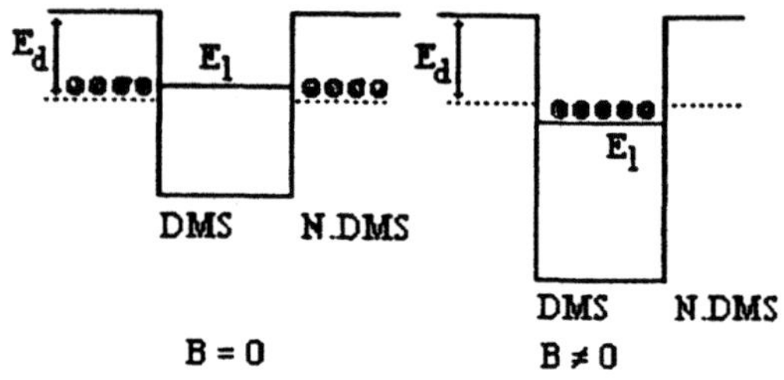

Figure 25. DMS quantum well between non DMS barriers.

This is the magnetic boil-off effect. These effects have been already observed. For the magnetic freeze-out the sequence of material was $Cd_{0.982}Zn_{0.018}Te$ and $Cd_{0.95}Mn_{0.05}Te$ (lattice matched). for the magnetic boil-off: $Hg_{0.1}Cd_{0.85}Mn_{0.05}Te$ and $Cd_{0.977}Zn_{0.023}Te$ (lattice matched system). Of course the "escape" rate of electrons depends on the temperature.

3.2.5. *Tunable resonant tunneling*

The above examples suggest the possibility to adjust the position of the energy levels in the D.M.S. quantum wells and the donor levels in the non magnetic barriers. giving a resonant tunneling of carriers through the barriers tuned by the magnetic field and the temperature.

The ground state and the first excited state in the quantum well are used. The tunneling involves both spin component of E_1 ground state spin split by the magnetic field. The electric field is used to satisfied the tunneling

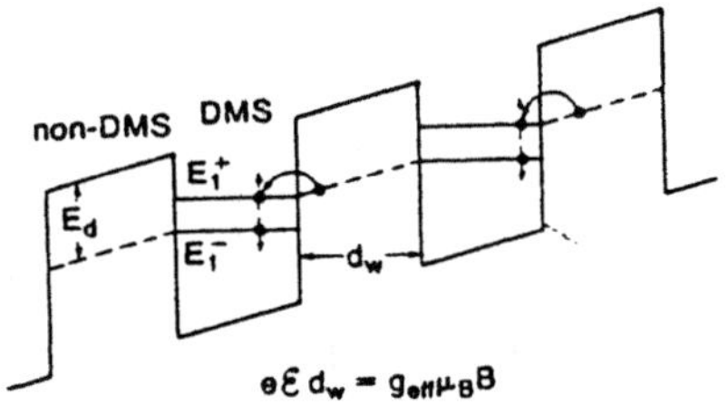

Figure 26. DMS quantum well between non DMS barriers (n type) with electric and magnetic fields.

condition ($e\varepsilon d_w = g_{eff}\,\mu_B B$). It can be adjust to ensure the equality of the E_1 level spin splitting and the voltage drop across the barrier.

If one imagine that, the electrons injected to the excited states in the well, lose their energy to a radiative process, it is possible to obtain efficiency infrared sources of radiation, with a definite polarization of the out going light.

The wave functions of the higher states (E^+_1) extend in the barrier, more than the lower ones (E^-_1). Then the probability of the tunneling process will be higher for the spin up electrons. This property could be used as a "spin filtering" mechanism with only one spin type being passed through a serie of quantum well. If the spin splitting is large this filter could be very efficient.

There exists some others systems to obtain infrared emitters based for example to the opposite sign of the dependence of the g factor with the spin interaction, and the exchange interaction for the conduction electrons.

These possibilities are well described in the excellent paper of J. Kossut and J.K. Furdyna [30] that I have used to develop here the different opportunities of the D.M.S. quantum wells.

3.2.6. *Quantum Oscillations*

Up to now, because the relatively large values of the effective masses (wide gap D.M.S.), the orbital quantization of the electron states into the Landau levels has been neglected.

If we consider that well resolved Landau levels are formed but the spin splittings far exceed the separation between consecutive Landau levels. Starting with a configuration of Fig. 24, where, with B = O, E_1 is situated slightly below the donor level in the barrier. With magnetic field E_1 splits into Landau levels whose energy increases with B. An oscillation appears when a Landau level coincides with the Fermi level. The Fermi level is determined by the donor level in the barrier, but E_F is non constant, since the donor states move to lower energies due to the exchange interaction. This effect modifies the usual B^{-1} periodicity of the oscillations and also permits to observe the oscillation associated with the lowest Landau level. This observation is usually not possible in non magnetic semiconductors (Fig. 9).

Now if we consider Fig. 24 (Boil-off effect) and if we superimposed the landau levels splitting, one obtains the situation depicted in Fig. 27. Due to the initial strong downward spin splitting of E_1, the oscillations related to the lowest Landau index n appear first in the lowest field. This is in complete contrast with the usual sequence observed, where

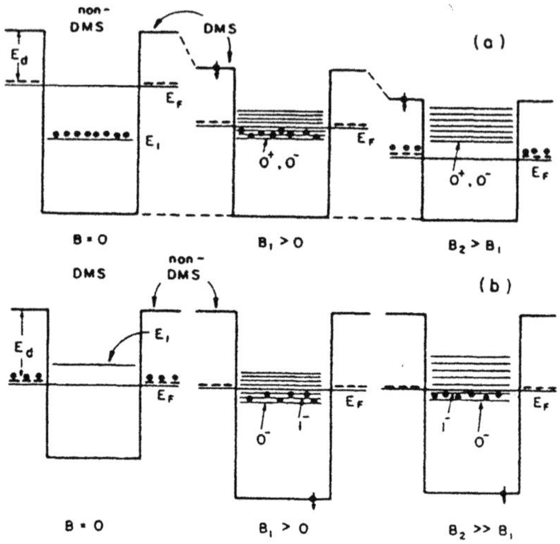

Figure 27. a) Non DMS quantum well between DMS barriers when a magnetic field is applied.
b) DMS quantum well, non DMS barriers.

the oscillations appear in the order corresponding to decreasing Landau indices when B increases. For very strong magnetic field, i.e. when the magnetization saturates, the exchange interaction reaches a saturation, and the order of the oscillations should eventually reverse, and a usual sequence should take place.

3.2. II-VI D.M.S. HETEROSTRUCTURES

We report in this paragraph the latest results obtained on quantum wells or superlattices involving D.M.S. in the barrier or in the quantum well. We restrict ourselves to two species of such heterostructures. The first concerns (CdMn)Te/CdTe and the second one (ZnMn)Se/ZnSe heterostructure. A lot of teams work on the (CdMn)Te/CdTe system. It seems that recently the part of (ZnMn)Se/ZnSe system increases.

3.2.1. *(CdMn)Te/CdTe heterostructures*

The relatively small valence band offset, the strain induced splitting of heavy and light holes, as well as the spin dependent tuning of potential depths in an external magnetic field result in many interesting phenomena observable in these heterostructures.

3.2.1.1. *Single quantum well structures*

We present, at first, results obtained by B. Kuhn et al [31] on single quantum well structures. The substrates are (100) oriented CdTe. The well is CdTe and the barriers which are thick are of (CdMn)Te. The Mn concentration of the barriers varies between 18% and 26% and the well widths between 32 and 150Å.

The thickness of the barrier is larger than the critical thickness. Then the wells are completely strained where as the barrier are unstrained.

The experiment used, was low temperature photoluminescence excitation (PLE) under magnetic field.

The Fig. 28 presents their PLE results for a quantum well with L_z = 88 Å and x = 0.22. The different structures observed are assigned to heavy hole exciton X(11 h) light hole exciton (11 h) and to excitons formed between electrons and holes of excited subbands as x(22 h) or x(31 l).

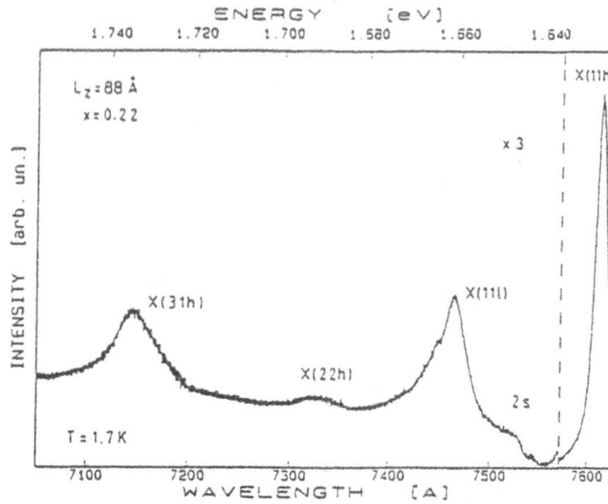

Figure 28. **Photoluminescence excitation spectrum [31] for a quantum well with L_z 88Å and x= 0.22.**

These assignments are supported by the calculation of energy transitions and also by their magnetic field dependence;

In a magnetic field, as we have shown on fig. 9, in Faraday configuration the recombination of the exciton observed splits into two strong components corresponding to a and d transitions where as b and c are weak (a for 6^+, d for σ^-)

The Fig. 29 shows the Zeeman splitting of heavy hole excitons as a function of magnetic field, for three well width 30 Å, 52 Å and 80Å. The Mn concentration in the barriers for the 3 samples is about 20%, i.e. comparable magnetic behavior of the barriers for all wells.

The splitting increases when L_z (well width) decreases. This result is as expected due to the sp-d exchange interaction of the moments of the exciton with the magnetic moment of the Mn^{2+} ions within its orbit. The increase of the Zeeman splitting when L_z decrease is caused by the increase of the wave function penetration into the barriers. It could be noted that the splitting is not symmetric. Chang and coworkers explained this unexpected behavior[32].

They suggest that the external magnetic field alters the quantum well potential heights due to the spin splittings in the (CdMn)Te barriers. Following the approach of Chang et al, B. Khun and coworkers have performed variational calculations for the exciton energies as a function of the external magnetic field. If they reproduce qualitatively the asymmetric splitting, they obtain some disagreement between their theoretical

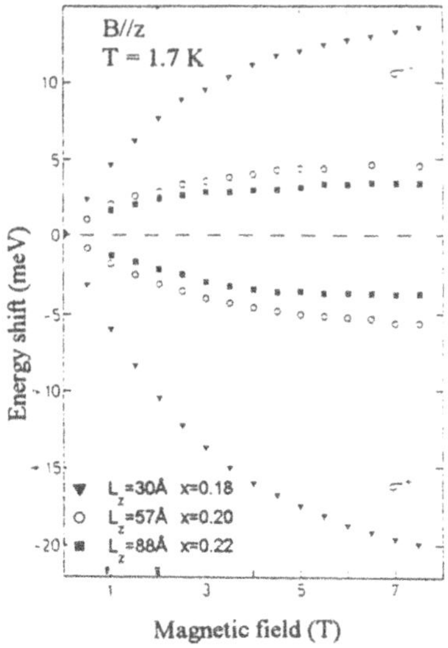

Figure 29. **Zeeman splitting of heavy-hole excitons vs magnetic field for various well widths [31].**

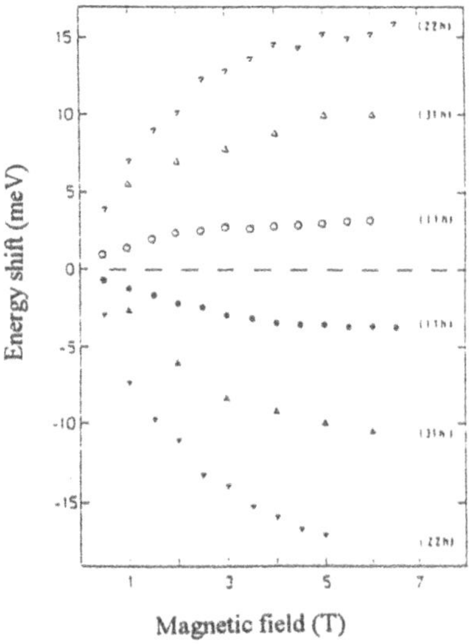

Figure 30. **Zeeman splitting of excitons formed between electrons and holes of different subbands [31] for sample of Fig. 28.**

treatment and experimental results. Several interpretations explained this discrepancy : localization of excitons at the interface, or residual Mn concentration in non magnetic well, or frustrated antiferromagnetic at the interface. This latest assumption has been supported by Chang et al [32] who have shown that interfaces influence the behavior of antiferromagnetic material, and anomalous characteristics are interpreted in terms of frustration against antiferromagnetic ordering.

On Fig. 28, it can be see several clearly resolved structures. From the knowledge of the magnetic field dependence of excitons formed between the lowest subbands, one can explain the Zeeman splitting of excitons connected to excited subbands. The Fig. 30 shows the Zeeman splitting of excitons formed between electrons and holes of different subbands.

It can be observed that the difference of splitting between the excitons (11 h) and (31 h) is larger than about 2.2 times. This difference is attributed to the electrons only, because they have calculated the ratio of the splitting with respect to the penetration of the wave function, they found 2.2.

The large and asymmetric splitting of X(22 h) is dominated by the heavy hole. The exchange interaction of the hole is about four times larger than that of the electrons (see table 1) small changes in the penetration of the wave functions are related in a more pronounced change of the splitting. Due the rather poor confinement of the second heavy hole, the tuning of the potential depth by an external magnetic field results in an asymmetric change of the excitons energy for different spin components.

On Fig. 31 the Zeeman splitting of light hole exciton has been plotted. It can be seen that it is larger and strongly asymmetric compared with the heavy hole exciton one. If we remark that the well is strained, the biaxial strain gives a potential height for the light holes in zero magnetic field which is smaller than that for heavy holes.

In addition, the mass of the $J_z = \pm \frac{1}{2}$ state is lighter, giving a pronounced penetration of the wave functions into the barriers, i.e. a larger exchanged induced splitting of the spin components. As a conclusion of this part, on single quantum well, between two D.M.S. barriers, it could be noted that :

- Well confined sub - band states close to the bottom of a quantum well are not strongly influenced by a change of the potential height caused by an external magnetic field ;

this gives a symmetric and small splitting of the components as it is the case for the heavy hole exciton X(11 h) in the broad quantum well - rather poorly confined sub - band states gives a strong and asymmetric Zeeman splitting due to the spin dependent tuning of the potential heights as it is the case for the heavy hole exciton X(22 h) and the light exciton (11l) of Fig. 30.

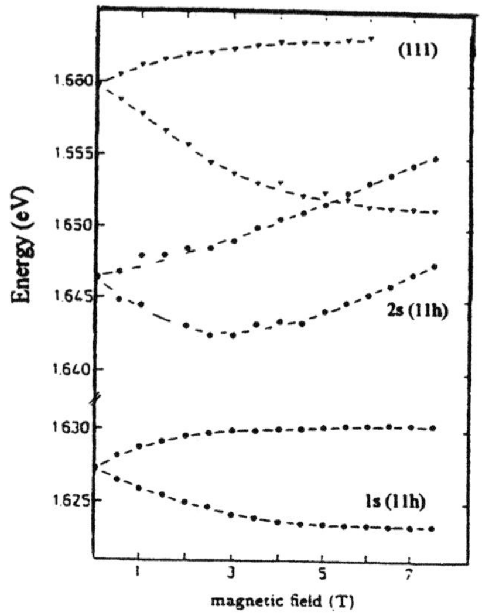

Figure 31. Zeeman Splitting of the light hole exciton, and the 1s and 2s states of the heavy hole exciton versus magnetic field.

3.2.1.2. *Multiple quantum well structures*

In this part I would like to present two works, which at a first look, seem rather near. They refer to magneto-optical study of CdTe/$Cd_{1-x}Mn_xTe$ multiple quantum wells. They are rather complete. The first one is from the team of Hull university, (33) where as the second one is from E.N.S. and I.B.M. Center.(34). The only difference is the following, one observes the type I. Type II transition, where as the other doesn't observe this transition.

At first we present the results obtained by S.R. Jackson et al (33) from the University of Hull. Their multiple quantum wells of CdTe/$Cd_{1-x}Mn_xTe$ were grown on (001) In Sb substrates by M.B.E. A 0.1 µm buffer layer of CdTe was grown before de MQW stack.

Two samples are selected to be presented here. The structures had 15 wells of CdTe and $Cd_{1-x}Mn_xTe$ barriers, with the top and bottom barriers of the same thickness as those of the stack. Their double crystal X ray diffraction (D.C. X R.D.) as well photoluminescence spectra prove the very good quality of their samples. Fig. 32 and Fig. 33.

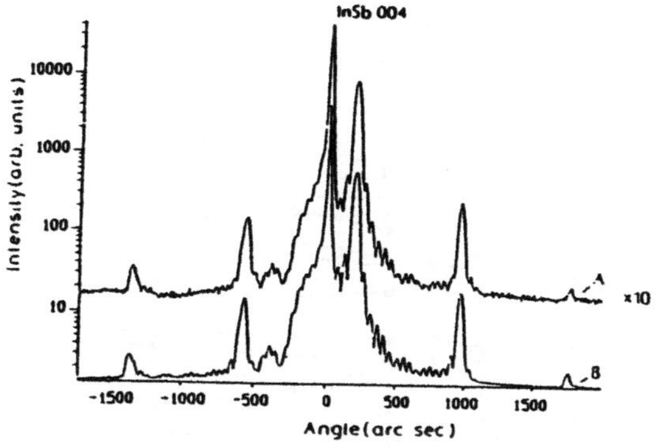

Figure 32. Recorded DCXRD of a CdTe/ $Cd_{1-x}Mn_xTe$ MQW (15 wells). The lower curve is a dynamical simulation of the spectrum [33].

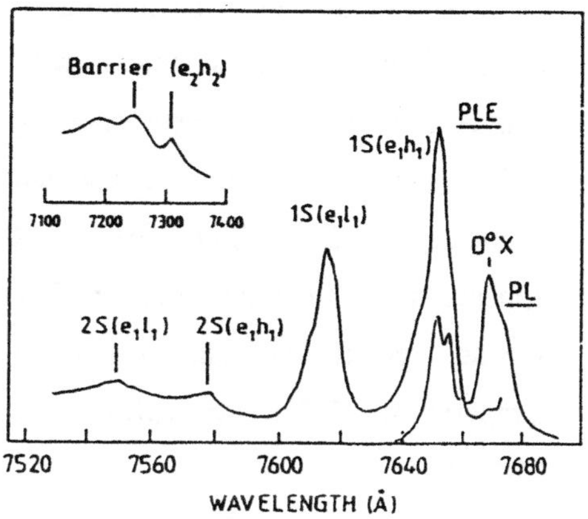

Figure 33. Photoluminescence and photoluminescence excitation spectrum [33] of sample of Fig. 32.

From Fig. 32, a determination of the layer thicknesses within the structure has been made. The wells and barriers thicknesses are respectively 75 and 158Å, with x = 0.075 for sample 1, and for sample 2, 52 and 176 Å with x = 0.077. The P.L., shows emission bands that are blue shifted through confinement, by about 25 meV from the CdTe band edge. The higher energy pair of lines are attributed to three exciton emission (e_ih_i) involving heavy holes. The broader emission of 7667Å is assigned to the recombination of donor-bound exciton D°X, by comparison with bulk CdTe. The double peak observed in the (e_1 h_1) transition is due to the interface roughness, and has been clearly explain by Gaj et al [35].

On the same Fig. P.L.E. spectrum is presented. The lowest energy peak resonance at 7651Å coincides with the P.L. (e_1h_1) exciton peak. Calculations based on the enveloppe function approximation were carried out to obtain energies of the various quantum states. On the basis of these calculations the transition around 7653 and 7613Å were assigned to the heavy hole (e_1h_1) and the light holes (e_1l_1) excitons. The two resonances at 7309 and 7240Å were attributed to the (e_2 h_2) transition and an excitonic transition in the barriers (7240Å correspond to free exciton in $Cd_{1-x}Mn_xTe$ with x = 0.075).

The Fig. 34 shows the splitting of the quantum states of sample 1, with magnetic field. The magnetic splitting of both states are similar, except that in the 2s state (LH 2S) and (HH 2S) it appears a diamagnetic behavior. The second remark is that at higher temperature (20K) the splitting is smaller.

This is easily understood because the giant Zeeman splitting as shown in Chapter I, is temperature and magnetic field dependent through the exchange interaction. The exchange interaction involves a Brillouin like function $B_{5/2}$ for the magnetization of the 5/2 Mn^{2+} spins located in the $Cd_{1-x}Mn_xTe$ barrier. The variation of the ± 3/2 heavy hole state in the valence band for example is given by :

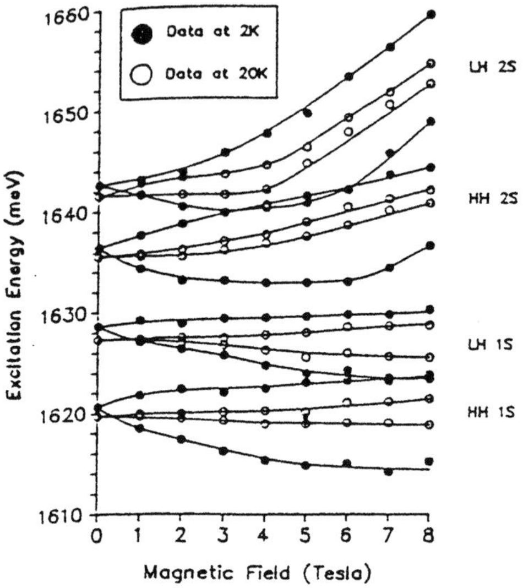

Figure 34. Zeeman splittings versus magnetic field in the Faraday configuration [33] at 2 and 20 K for same sample.

$$\Delta E_{\pm 3/2}(B) = \pm \frac{N_0\beta}{2} x S_0 B_{5/2}\left[\frac{5/2.g\mu_B B}{\mu_B(T+T_0)}\right]$$

All the parameters have been defined previously. A similar expression could be derived for the ± ½ light hole states. Obviously since the exchange interaction decreases when T increases the diamagnetic behavior of the 2S state is more apparent. The diamagnetic shift of the exciton, must be taken into account and included as a corrective term. This correction, using a first order theory perturbation, leads to :

$$\Delta E_{diam} = \frac{3}{16} \cdot \frac{e^2}{\mu_{//}} a_B^2 . B^2$$

Where $\mu_{//}$ is the reduced mass in the layer plane and a_B the Bohr radius of the exciton.

The Fig. 35 shows the exchange induced splitting in the Voigt configuration : B// to the plane of the layers and ⊥ to both incident light and the emission. This Fig., because the selection rules, confirms the identification of the (e_ih_i) and (e_il_i) states.

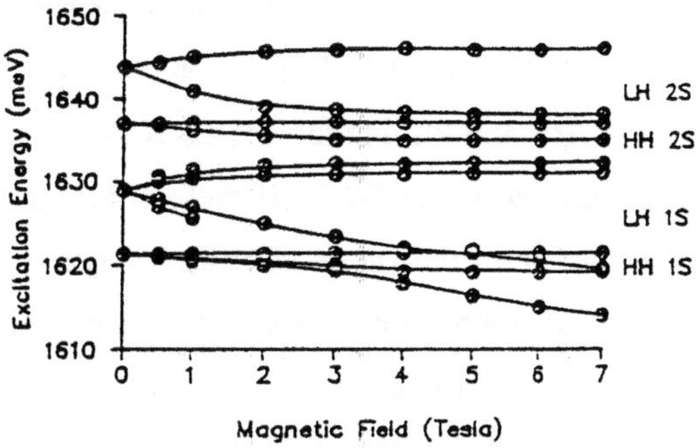

Figure 35. The same Fig. 34 in Voigt configuration [33].

3.2.1.3. *Type I → type II transition*

We expect a type I - type II transition if : $\Delta E_{-3/2}(B) - \Delta E_v = 0$ is fulfilled. Where ΔE_V is the valence band off set.

E. Deleporte et al [34] have calculated the initial magnetic field where the transition occurs, as a function of T. They took the relative valence band offset as a parameter (Q_v = 15% or ΔE_v = 18.2 meV ; Q_v = 20% or ΔE_v = 24.2 meV ; Q_v = 25% or ΔE_v = 29.7 meV). The strain effects are included in the values of the offsets (the strain effects are responsible of a decrease of the conduction band offset of 26 meV and an increase of the valence band offset of 2.1 meV.

Their sample consists of a (111) $CdTe/Cd_{0.93}Mn_{0.07}Te$ well of 25 periods with symmetric barriers and wells of a thickness of 86 Å. The multi-quantum-well is grown on a GaAs (100) substrate followed by a 1500 Å CdTe (111) buffer layer. The Fig 36 shows the result of their calculation.

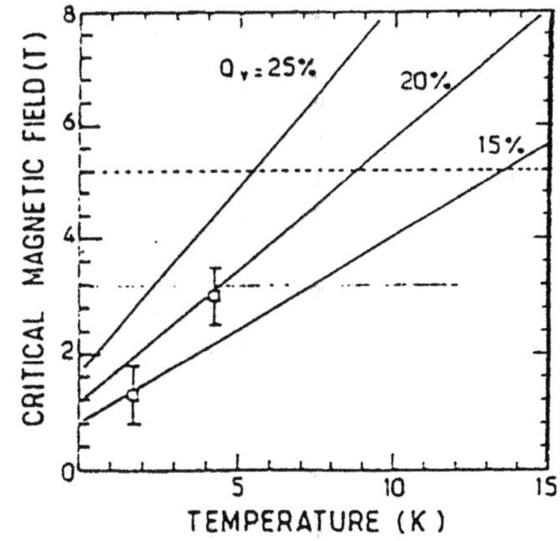

Figure 36. Calculated critical magnetic field versus temperature for three different values of Q_v [34]. Dashed line : highest available experimental magnetic field. Squares : experimental values.

The Fig. 37 shows the variation of the energies of the PL lines and the $A\sigma^+$ and $A\sigma^-$ (Faraday configuration) of the excitation spectrum. They correspond respectively to the D°X and $1s(e_1\ h_1)$, $1s\ (e_1\ l_1)$ of the Fig. 34. In this Fig., clear shoulders are observed in the $A\sigma^+$ and PL line for T = 1.7 K and T = 4.2 K, where as at T = 15 K no shoulders are observed up to 5.2 T. If these effects are due to the transition type I - type II as expected from the values of critical magnetic field, we also must have a drastic change in the exciton bending energy and in the oscillator strength of the corresponding optical excitonic transition.

E. Deleporte et al [34] have calculated the fundamental 1s hole excitons binding energy in term of a variational method taking into account the small and possibly negative valence band offset. Their results [36] are reported on Fig. 38.

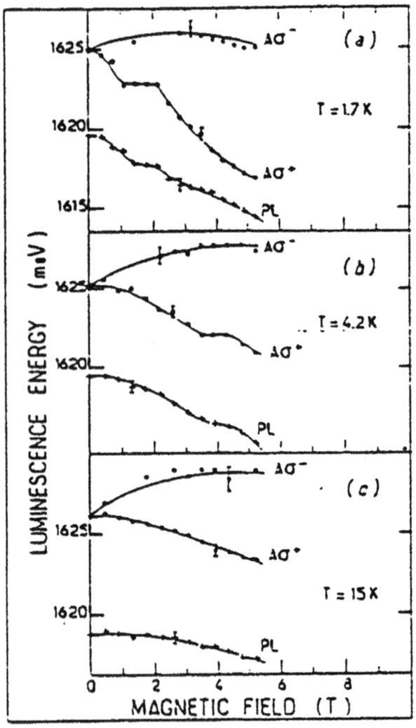

Figure 37. **Experimental variations of the PL and $A\sigma^+$ and $A\sigma^-$ lines of the excitation spectrum vs magnetic field for T = 1.7 K, 4.2 K and 15 K [34].**

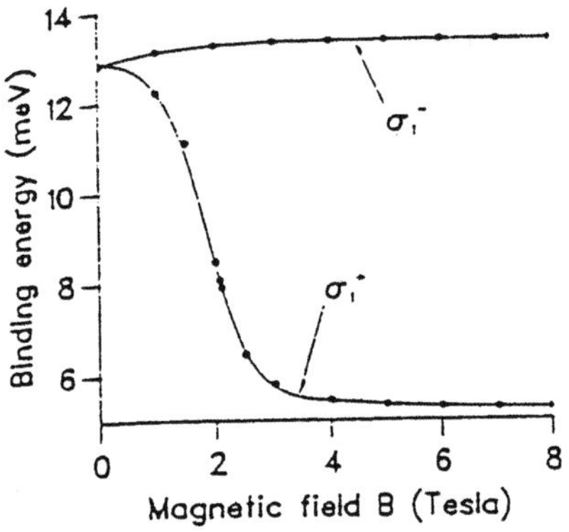

Figure 38. **Calculated variation of the binding energy of the 1s states of the $(e_1h_1)^{\pm}$ excitons, ($Q_v = 0.2$).**

The Fig. 39, shows the variation of the exciton binding energy Ex with magnetic field (dashed line) at T = 4.2 K and a relative valence band offset $Q_v = 20\%$. E_x exhibits a steep decrease near B_o. The energy levels are calculated within the framework of the three band envelope approximation, including the dependence of the valence and conduction band offsets on magnetic field (37). As for the variation of the exciton binding energy, the variation of the band to band transition $E_1H_1^B$ (dashed dotted line) exhibits a steep decrease near B_o, leading to the formation of a shoulder near the critical magnetic field in the variation of the excitonic transition energy (solid line) through the relation :

$$E_1H_1^X = E_1H_1^B - Ex$$

In fact, if we compare now theoretical results and experimental ones, it can be note that theoretically one find a peak and experimentally a plateau at critical magnetic field. Probably this discrepancy is due to inhomogeneous broadening effects like width fluctuations which tend to spread out the resonance - like - structure.

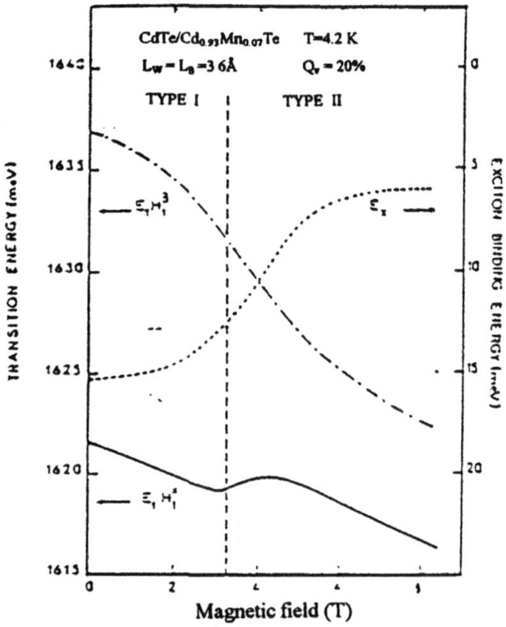

Figure 39. **Calculated fondamental : 1s heavy hole exciton binding energy E_x (dashed line), excitonic transiton $E_1H_1^x$ (solid line). Band to band transition $E_1H_1^B$ (dashed-dotted line).**

Now the question is : why these effects of type I - type II transitions are observable in the work of Deleporte et al [34], and not in the one of S. R. Jackson [33].

If we compare carefully Fig. 34 and Fig. 37, the lines corresponding to heavy hole 1S, it could be note that the splitting in σ^- and σ^+ configuration is equivalent. For ex at 4 Tesla the splitting in 7.3 meV in Fig. 37 and only 7.1 meV in Fig. 34 for about the same T (1.7 and 2 K respectively). They announced roughly the same Mn concentration 0.07 and 0.077. The giant Zeeman splitting is directly proportional to $<S_2>$, that is to say to the magnetization. The two results confirm this point. Then the only difference is the well width : 86 Å and 75 Å respectively, that changes the carriers confinement energy.

Because the band gap and binding energy variation counteract, the appearance of type I - type I transition depend markedly on the value of the assumed valence band offset and the width of the quantum well. A. Wasiela et al [38] have compared two samples in which the well widths are respectively 75 Å and 150 Å where as the barrier widths are 200 Å. These samples contained only 5% of Mn, that means that the potential barriers are lower compared to the previous studied sample. Their results, as well theoretical as experimental for the $e_1 h_1$ exciton line are given on Fig. 40.

This Fig. illustrate the effect of reducing the QW width. While the magnetic - field induced changes of the binding energy remain nearly the same, the band gaps variations of sample I (L_W = 75 Å) are much larger than those of sample II (L_W = 150 Å) as a result of the larger confinement energies. Consequently the variations of the exciton energies with applied field are also larger. For sample II the data fit well with a value of Q_v = 0.3. The magnetic splittings observed for sample I are larger than those

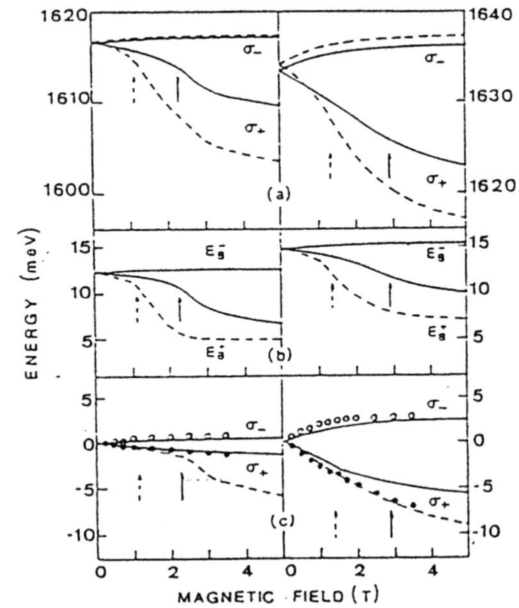

Figure 40 : Calculated variation with magnetic field of: band gap energies, binding energies, and transition energies of the heavy hole exciton for two samples (left side L_W = 150 Å, right side L_w = 75 Å) for two values of Q_v : Q_v = 0.3 solid line, Q_v = 0.2 dashed line. The points are experimental data. The arrows indicate the critical field(34).

calculated with Q_v = 0.3. They are nearer than those of Q_v = 0.2. These may be due to a slight diffusion of Mn into quantum well where interaction with the carrier is high.

The enhancement of the Zeeman splitting due to such a diffusion can be very large for sample with x = 0.7 .0.8%.

In fact on sample I, although type I - type II transition takes place, the Zeeman energy shift of the σ^+ heavy hole exciton does not show any clear signature at the transition. Calculations performed by the authors show that this is due to the exact compensation of the magnetic variations of the band gap and the exciton binding energy.

- Provide that samples with very precise specifications in terms of Mn concentration in the barrier, well widths and barriers widths, very good crystalline quality and very interfaces are available, the type I - type II transition, can be obtained, when an external magnetic field is applied. This is a very nice tool to study in a reversible manner such a transition in multiple quantum well, and (CdMn)Te/CdTe is an adequate system for these studies.

3.2.2. *ZnSe/ZnMnSe Heterostructures*

This system requires a particular attention for several reasons. The ZnSe based materials are of interest because their blue emission. The first blue laser emission has been obtained with ZnSe based compounds [39].

In terms of superlattice this system presents several advantages. The energy gap of $Zn_{1-x}Mn_xSe$ exhibits a rather striking bowing with Mn concentration x at low temperature, and then rapidly increases. It is thus possible to find a value of x at which $Zx_{1-x}Mn_xSe$ has the same energy gap as ZnSe. This occurs for x = 0.04 at 1.5 K. It turns out that this value of x also leads to zero band offsets in both the conduction and valence bands, giving consequently a schematic diagram [40] as Fig. 41.

When a magnetic field is applied the Zeeman effect lead to the formation of a spin superlattice structure.

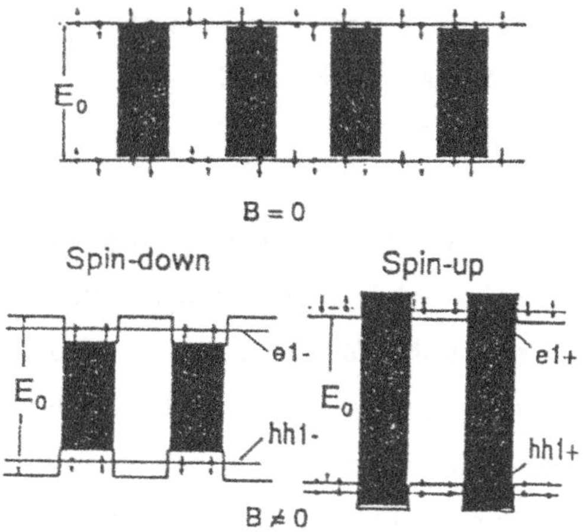

Figure 41 : Diagram of the band structure of a magnetic field-induced spin superlattice with DMS and non MS layers with equal energy gaps at B = O [40].

The band edge of $Zn_{1-x}Mn_xSe$ for spin down electrons moves down, and moves up for spin down holes. The ZnMnSe layers thus become wells for both carriers of the spin - down orientation. Similarly, for spin - up states the conduction band moves up in energy and the heavy - hole band moves down, so that for this spin orientation the ZnSe layers become the wells. This magnetic - field induced band offset has therefore the consequence of spatially separating electrons and hole states of different spin orientation, we obtain a magnetic field - induced spin superlattice. This effect can be observed in magneto absorption experiments in Faraday geometry in which only spin - conserving transitions are allowed ($\Delta S = O$). Fig. 42. The left side shows the magnetic field dependence of excitonic transitions observed with the σ^- and σ^+ polarization in ZnSe/ $Zn_{0.96}Mn_{0.4}Se$ superlattice. The spin down transitions show a rapid red shift of the absorption line since the initial and final states are both in the D.M.S. layers, and thus follow the magnetic field dependence of the valence and conduction band of D.M.S. material (Fig. 9 Chap I : transition a).

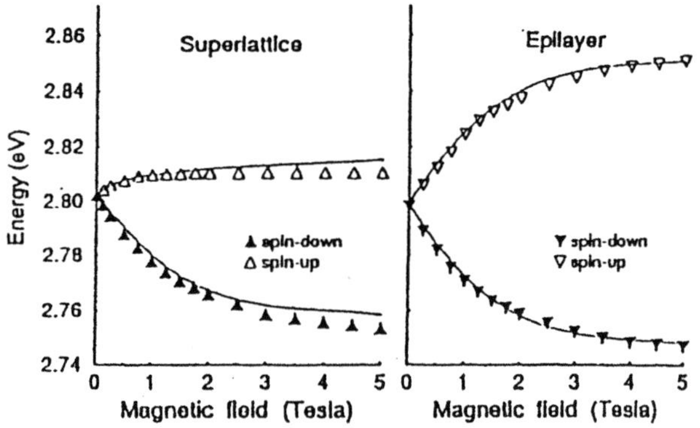

Figure 42: Zeeman splitting of the excitonic ground state transitions in a $Zn_{0.96}Mn_{0.04}Se$ (100 Å/100 Å) superlattice[40]. On the right side, results of an epilayer with $x = 0.04$.

In contrast, the spin-up transitions take place between states localized in ZnSe layers, which become the wells for this spin orientation when the field is applied. This is responsible for the flatness of the transition energy at higher fields, i.e., once the offset becomes well defined. The small initial rise in energy at low field (the offset is just beginning to form) is because the state of ZnSe well (extremely shallow well) is more sensitive to the height the 'D.M.S. barrier, which increases with the field. When the barrier continues to increase, the energy of the states confined in the well are less and less sensitive to the barrier height. For comparison, on the right side of Fig. 42, the position of free exciton lines, in an epilayer with the same Mn concentration is shown.

3.2.2.2. *Type I - to type II exciton transition in ZnSe/ZnMnSe*

These transition has been studied by E. Deleporte (37). The structure consists of two ZnSe quantum wells whose thickness are respectively 53 Å and / 106 Å separated by a $Zn_{0.73}Mn_{0.27}Se$ barrier of 350 Å thicknesses. The double quantum well is deposit on a 1200 Å $Zn_{0.73}Mn_{0.27}Se$ layer grown on a 9700 Å thick ZnSe buffer layer. A cap layer of 350 Å $Zn_{0.75}Mn_{0.27}Se$ covers the structure. The substrate is GaAs. The 1200 Å ZnMnSe layer is partially relaxed on ZnSe buffer layer. The barriers are compressed. The residual strain in the barrier is found to be $\varepsilon_{bar} = -5.510^{-3}$. An elastic accommodation of the lattice parameter of the wells with the barrier one is assumed.

Moreover from the energy difference between the heavy and light hole excitons, the percentage of the strain free band gap difference which is in the valence band has been evaluated (41) $q^{\circ}_{v} = 10\%$. The resulting band structure is shown in Fig. 43. The photoluminescence spectrum of this heterostructure is shown in Fig. 44.

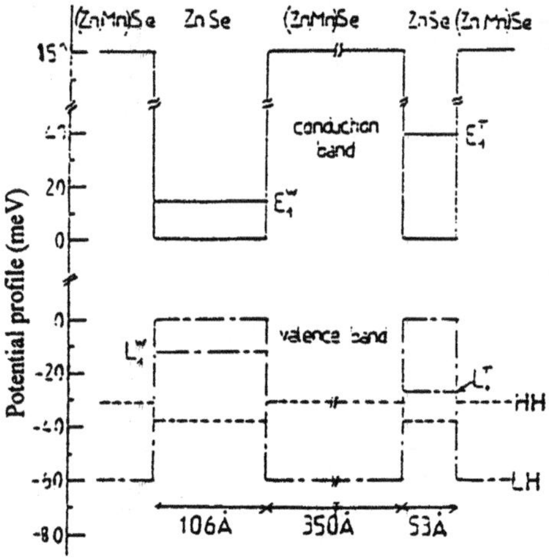

Figure 43 : **Calculated potential profile taking into account the strain effects ($Q_v^{\circ} = 10\ \%$; $\varepsilon_{bar} = -5.5\ 10^{-3}$) [41].**

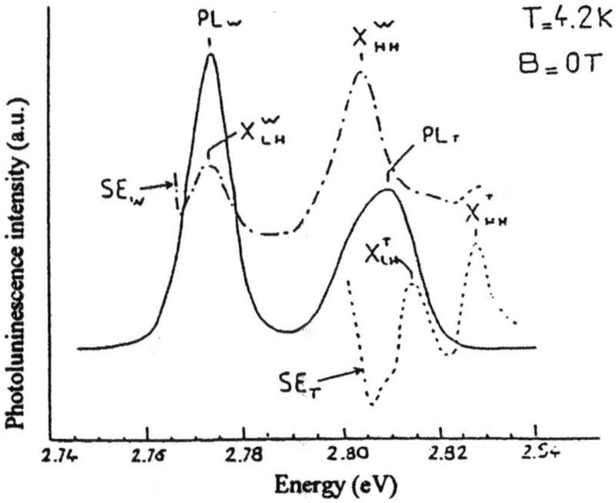

Figure 44 : **PL and excitation spectroscopy of the two quantum wells at B = O and T = 4.2 K [41].**

From this spectrum the following comments could be made : the small stoke shift (6 meV for the thin well, 2 meV for the wide well) between the PL lines and the first line of their excitation spectra indicate a good quality of the sample. Comparing the two excitation spectra it can be conclude that the two wells are not coupled and therefore treated like single quantum wells.

By the comparison of the experimental and calculated energie excitonic transitions, the different lines of the excitation spectra have been interpreted and used to determine the band structure. It has been found that the fundamental optical transitions of the two wells : X^{w}_{LH} and X^{T}_{LH} are type I. The first excited transitions X^{w}_{HH} and X^{T}_{HH} are type II.

Fig. 45a and b show the evaluation of the excitation spectra of the two wells for different magnetic field. Because the position of light holes in the non D.M.S. material and the position of heavy holes in D.M.S. barriers the X_{LH} does not change (in addition the exchange interaction is three times smaller), where as, due to the giant Zeeman effect the X_{HH} lines split into two components.

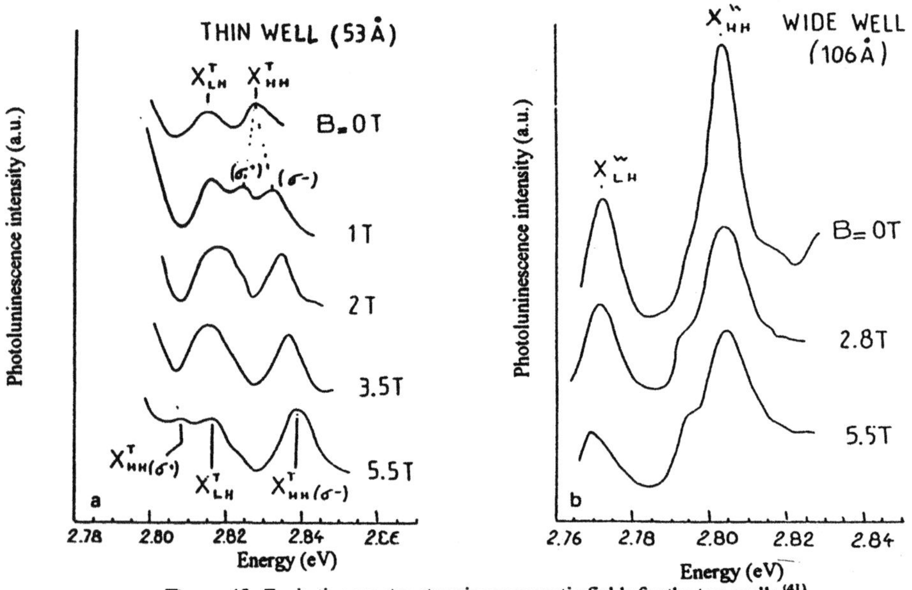

Figure 45 : Excitation spectra at various magnetic fields for the two wells [41].

The results are summarized in Fig. 46 and compared to the calculated lines with a very good agreement. This Figure shows very clearly the crossing between the X^{T}_{LH} and X^{T}_{HH} (σ^{+}) transitions and then the existence of a critical B. For $B<B_o$, the fundamental

optical transition is a type - I light exciton. For the wide well such a transition is expected at 8 T and not seen here. B_o depend strongly of the well width this property could be understood from Fig. 43 since the distance L^w_1 - HH is higher than L^T_1 - HH.

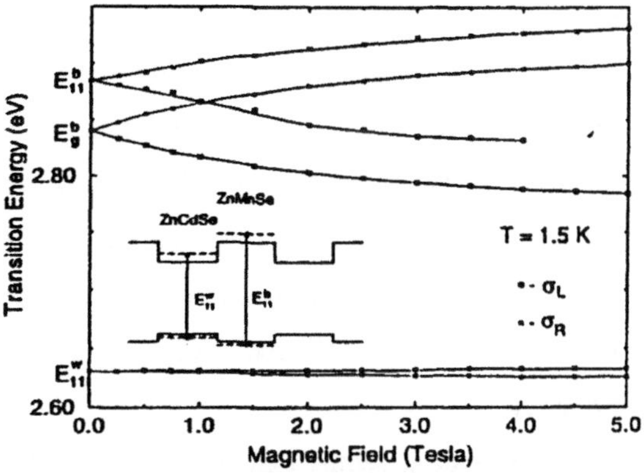

Figure 46 : Energy position of the different lines versus magnetic field deduced from Fig. 45.

3.2.3. *Conclusion*

We have focus our attention on two families of heterostructures in II VI D.M.S. heterostructures, CdTe/CdMnTe and ZnSe/ZnMnSe, which are rather reach to point out several special interest of D.M.S. in heterostructures like Spin Superlattices, type I - type II transition : type I heavy hole exciton to a type II heavy hole exciton for CdTe/CdMnTe heterostructure ; type I light hole exciton to type II heavy hole exciton for ZnSe/ZnMnSe.

Obviously, we have simplified the interpretation without any details on the models used in the calculation except for the exchange interaction, which is the original contribution of the D.M.S.

There exist a lot of other possibilities as for instance ZnCdSe/ZnMnSe superlattices (type I) or CdSe/ZnMnTe and CdMnSe/ZnTe type II super lattices. The reader will find in the ref. all the details to access to such works.

3.3. IV.-VI. D.M.S. HETEROSTRUCTURES

The development of studies in IV-VI D.M.S. heterostructures is due to the potential applications for the fabrication of tunable diode lasers in the mid-infrared range (3-6.5 µm). recently, Feit et al. (42) reported the highest cw operating temperature (203 K) for such mid-infrared lasers using a buried heterostructure concept. Partin et al. (43) have achieved up to 175 K cw operation using a side optical cavity quantum well. G. Bauer et al. (44) have shown the large refractive index enhancement in $PbTe/Pb_{1-x}Eu_xTe$ multi-quantum -well structures which is relevant for the design and fabrication of mid infrared quantum well lasers. The teams working in this area are mainly: General Motors Research Laboratories, M.I.T., Tuft University, Institut für Halbleiterphysik at Linz University, Polish Academy of Sciences and G.E.S. in Montpellier University, University of Notre Dame Indiana US, the Fraunhofer Institut für Physikalische Freiburg.

However it seems that the most complete work has been done by Bauer and his team at Linz University in Austria. So in the following section, a lot of results have been borough to their works. As in the case of II-IV heterostructures in IV-VI materials there exist several family, namely: $PbTe/Pb_{1-x}Eu_xTe$, $PbSe/Pb_{1-x}Eu_xSe$ and $PbS/Pb_{1-x}Eu_xS$.

3.3.1. *$PbTe/Pb_{1-x}Eu_xTe$ heterostructures*

$PbTe/Pb_{1-x}Eu_xTe$ multi-quantum-well were grown by MBE on (111) oriented BaF_2 substrates on which a comparatively thick $Pb_{1-x}Eu_xTe$ buffer layer (2 to 5 μm) was deposited with the same x as in the D.M.S. barriers. The Bragg (222) triple axis Ω/2θ X ray diffraction curve for a MQW grown at 350°C and at 200°C are shown on figure 47.

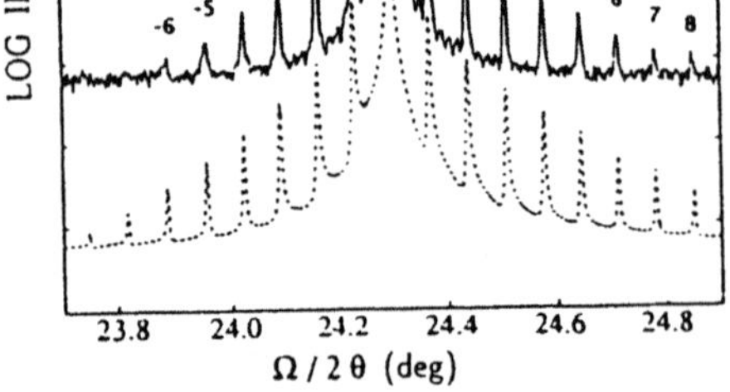

Figure 47 : Bragg diffraction of a PbTe/PbEuTe multilayer structure [45].

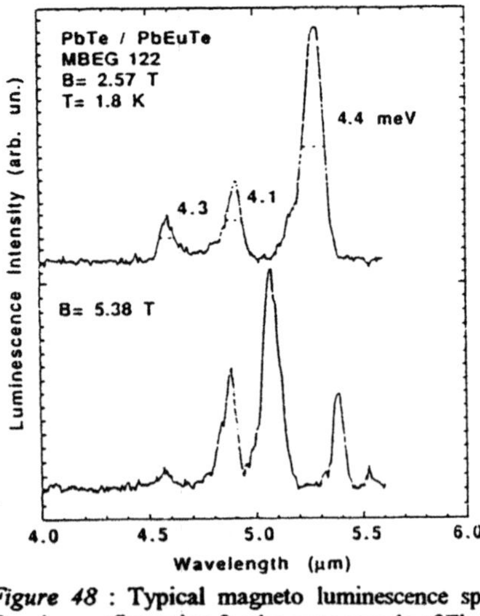

Figure 48 : Typical magneto luminescence spectra in Faraday configuration for the same sample of Fig. 47.

Besides the 0th order superlattice (222) peak (SL0), up to eight satellite peaks are observed on both sides of the central peak. Because of the small content of Eu (3%) the lattice mismatch between PbTe and $Pb_{1-x}Eu_xTe$ is only 0.19%, The PbTe wells are fully strained. The average lattice constant of the MQW stack is almost identical to that of the $Pb_{1-x}Eu_xTe$ buffer layer because the barriers are thicker than the well (d_B = 600-700 Å, d_w=100 Å; 50 periods). On the MQW of figure 47, a cap layer of PbTe of 2700 Å has been grown. The SL0 $\Omega/2\theta$ FWHM value is 20 arc/sec (width in the [111] direction, increases gradually to 33 arc/sec for SL+5 and to 31 arc/sec for the SL-4).

We have selected one photoluminescence spectrum, in order to study the transition energies and to show, as in the case of II-VI D.M.S. MQW, the variation with the magnetic field, i.e. the role of magnetic exchange interaction. On figure 48 it can be seen some typical spectra in the Faraday configuration, B//[111], which is the growth direction of the $PbTe/Pb_{1-x}Eu_xTe$ MQW for different magnetic fields. This sample consists of three different quantum well thicknesses. The luminescence peaks originate from the different quantum wells.

As in the case of II-VI D.M.S. superlattices, the transition energies , as a function of the applied magnetic field obtained by Bauer et al., are presented on figure 49.

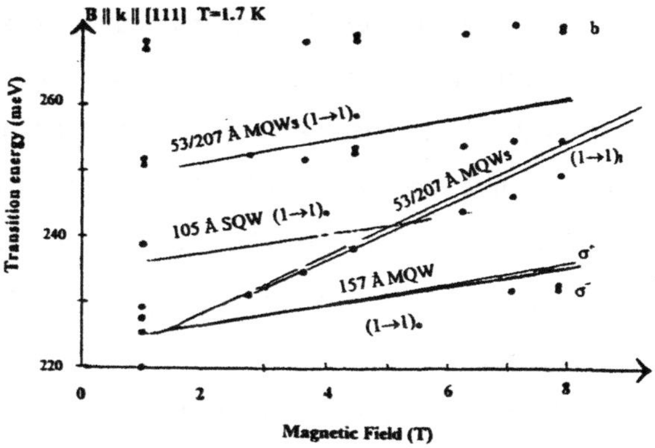

Figure 49 : Transition energies versus magnetic field for the same sample as Fig. 47 The full lines calculated in the frame work the envelop functions [46].

They are compared to the results derived from envelope function calculations, assuming a type I band structure and a band offset of $\Delta E_e / \Delta E_g = 0.5$, including the exchange interaction of carriers penetrating into the barriers. The highest energy transition (b) is a direct interband transition in the PbEuTe buffer layer. Four transitions are observed.

For the longitudinal valleys, the electric and magnetic quantizations are decoupled. The interband magnetic selection rules are bulk like. The $\Delta i=0$ selection rule is for type I SL. For the oblique valleys, electric and magnetic quantizations are coupled giving rise to more complicated selection rules. The transitions observed involved the lowest lying levels and the extrapolation at B = 0 T gives the lowest lying electric subbands. In figure 49, transitions with the oblique valleys are observed for the different MQW. For the 5.3 nm thick PbTe WQ, transitions involving the longitudinal valleys are also observed.

The calculated transition energies by E.F.A. formalism fit well with the experimental data. It could be seen that the σ^+, σ^- splitting is quite small. We have to remember that the exchange parameter of PbEuTe is about a factor of twelve smaller than in CdMnTe. It cannot be seen clearly in experimental results.

3.3.2. *Large refractive index enhancement in PbTe/$Pb_{1-x}Eu_xTe$*

It is well known that the refractive index exhibits remarkable changes at the fundamental absorption edge of bulk materials in III-V compounds, as well as for the narrow gap lead salts, where the non parabolicity affects absorption and consequently the refractive index $\eta(\varpi)$. It has been demonstrated that is GaAs/AlAs quantum wells the enhancement of $\eta(\omega)$ close to interband resonances depends on the wells and barriers thicknesses, and barrier composition. The highest reported changes of η are of the order of 5 %.

In PbTe/$Pb_{1-x}Eu_xTe$, the dependence of the refractive index on the frequency exhibits pronounced maxima with $\Delta\eta/\eta$ as high as 19 % close to the interband transitions between confined hole and electron states in PbTe/$Pb_{1-x}Eu_xTe$ MQW structure. It is necessary to take into account these large variations of the refractive index to explain quantitatively the optical transmission spectra PbTe/$Pb_{1-x}Eu_xTe$ MQWS.

In Fig. 50 the transmission spectrum for a typical sample (MQW N° 151) normalized to the transmission of the BaF_2 substrate is shown. The absorption is due to the $Pb_{1-x}Eu_xTe$ buffer layer and barrier layers, as well as to the interband transitions between hole and electron subbands in the PbTe wells. The fringes arise from multiple reflections and interference effects in the whole multilayer stack including the buffer. The drastic steplike changes in the transmission signal are due to absorption involving the two dimensional (21) confined states in the PbTe wells.

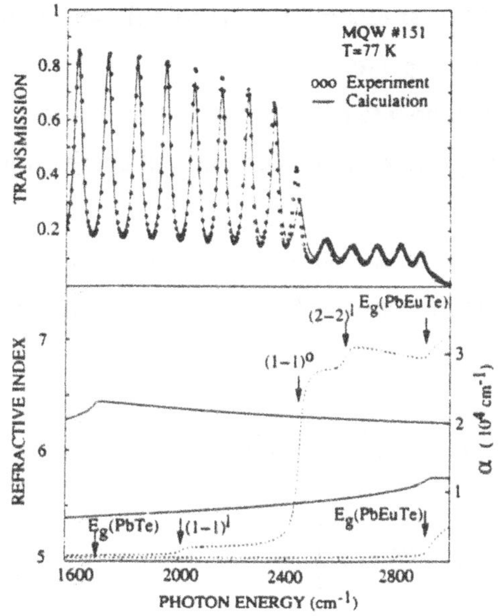

Figure 50 :Transmission spectrum of PbTe/$Pb_{1-x}Eu_xTe$ MQW with L_w = 62 Å and 46 periods. Circles : experimental data. Full lines :calculated data. Lower panel, dashed line : absorption coefficient. Full line : refractive indexes of bulklike PbTe and $Pb_{1-x}Eu_xTe$

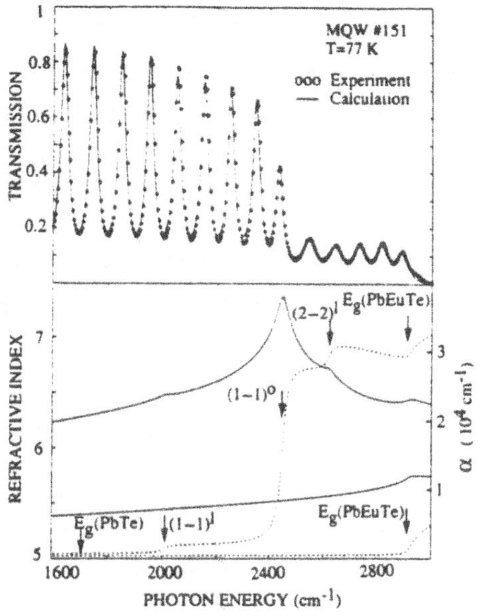

Figure 51 : The same as Fig. 50, but considering the changes of refractive index between confined carrier states in the PbTe wells.

For the calculation of the transmission, the optical constants of $Pb_{1-x}Eu_xTe$ buffer and barrier layers are need. They were determined independently from a $Pb_{1-x}Eu_xTe$ reference sample.

For the PbTe QW, a perfect 2D system is assumed : i.e the penetration of the carriers wave functions into the barrier is neglected.

The extinction coefficient in the Qw,, is then given by :

$$K(\hbar\omega)=\sum_j A_j\left[\frac{\pi}{2}+\operatorname{Arc\,tan}\left(\frac{\hbar\omega-E_j}{\Gamma_j}\right)\right]\frac{1}{(\hbar\omega)^2}$$

E_j denotes the interband transitions between the confined states in the VB and in the CB of the PbTe wells. Γ_j is a broadening parameter. j, is the interband transition index. $\hbar\omega$ is the photon energy.

Since the growth is along (111) direction, two subband systems occur. One associated with the valleys at the L point of the Bz along the [111] axis (l) which have circles as surfaces of constant energy in the k_x, k_y plane.

The 3 other valleys with their main axis orientated obliquely (O) with respect to the [111] growth direction give rise to 3 ellipses as surfaces of constant energy in the 2D case.

Consequently Aj which depends both on the 2D density of states as well as the transition matrix elements is different for transitions involving subbands of (l) valley (O) valleys

$$A_j^{(l)} = \frac{e^2 E_g P^2/P_\perp^2)}{8\eta\varepsilon_o \pi d_w}\left[\frac{1}{3}\sqrt{1+8P^2/P_\perp^2}\right]^{-1} A_j^{(l)} = \frac{e^2 Eg P_{//}^2/P_\perp^2}{8\eta\ \varepsilon o\ \pi\ d_w}$$

and $$A_j^{(o)} = \frac{e^2 g_v Eg\left(\frac{5}{9}+\frac{4}{9}P_{//}^2/P_\perp^2\right)}{8\eta\varepsilon_0\pi d_w}\left[\frac{1}{3}\sqrt{1+P_{//}^2/P_\perp^2}\right]^{-1}$$

E_g is the energy gap of PbTe, g_r (=3) is the oblique valley degeneracy index. $P_\perp$ and $P_{//}$ are the transverse and longitudinal inter band momentum matrix elements, d_w is the well width.

The corresponding variation of the absorption coefficient $\alpha(\omega) = 4\pi\, K(\omega)/A$ versus photon energy is shown on the lower panel of Fig. 50 .

The step in $\alpha(\omega)$ for the oblique valley interband transition $(1 - 1)^{o}$ is 5 times larger than the longitudinal one $(1 - 1)^{l}$, $(2, 2)^{l}$ because of the three fold degeneracy, the differences in point density states masses, and the transition matrix elements.

It could be observed that the transmission spectra calculated reproduces the main features of the experimental data.

However, in particular close to the i = 1 (VB) $\rightarrow$ i = 1 (CB) transition in the oblique (O) subband ladder, the position of the interference fringes does not coincide with the measured values.

Since the extinction coefficient $K(\omega)$ and the refractive index $n(\omega)$ are related via the Kramers-Kronig relation ship.

$$\eta(\omega) = \eta_{\infty} + \frac{1}{\pi} P \int_{-\infty}^{+\infty} \frac{K(\omega') - K(\infty)}{\omega' - \omega} d\omega'$$

One can expect that especially in the IV-VI Qw, in the 2D case, interband transitions will lead to substantial changes in $\eta(\omega)$. This is shown in the Fig. 51 where $\eta(\omega)$ has been calculated from the above relation (error < 3%). At the frequencies where $\alpha(\omega)$ exhibits a step like increases, $\eta(\omega)$ shows maxima. The stronger the change in $\alpha(\omega)$, the more pronounced are the peaks in $\eta(\omega)$. For the (1-1)° transition, the increase is 19% with respect to bulk PbTe at 2450 cm^{-1}.

Using the calculated $\eta(\omega)$ for the PbTe well, and the experimental data for $\eta(\omega)$ for $Pb_{1-x}Eu_xTe$ buffer and barrier layers the theoretical transmission spectra are in better agreement with the experimental data. In particular the positions of the transmission extrema which are sensitive to the dispersion of $\eta(\omega)$ are well reproduced.

The interband transition energies between confined states in C.B. and V.B. of the PbTe wells are calculated in the envelope function approximation formalism (E.F.A.). It can be conclude that for relatively wide $Pb_{1-x}Eu_xTe$, the band offset in $Pb_{1-x}E_xTe$ is quite symmetric for low x (x < 5%) $\Delta Ec/\Delta Eg = 0.55$.

The refractive index changes considerably at photon energies close to the interband transitions between confined states. This fact is important for the optical confinement of radiation in Mid Infra QW lasers.

3.3.3. *PbSe/Pb$_{1-x}$Eu$_x$Se and PbS/Pb$_{1-x}$Eu$_x$S MQW systems*

PbEuSe is a valuable ternary compound for the fabrication of laser heterostructures in the 3-8 μm wavelength region. The PbSe based heterostructures present an alternative to the $Pb_{1-x}Eu_xTe_{1-y}Se_y$/PbTe lattice matched heterostructures ($y \leq x$) investigated by Partin et al (43) G. Springholz et al (45) have produced quantum well lasers of $PbSe/Pb_{1-x}Eu_xSe$. In figure 52 the energy gap and the lattice constant as a function of the Eu concentration x is presented.

Compared to PbTe, the lattice constant of PbSe is smaller (6.460 Å for PbTe, 6.128 Å for PbSe). Also different are the cation and anion radii

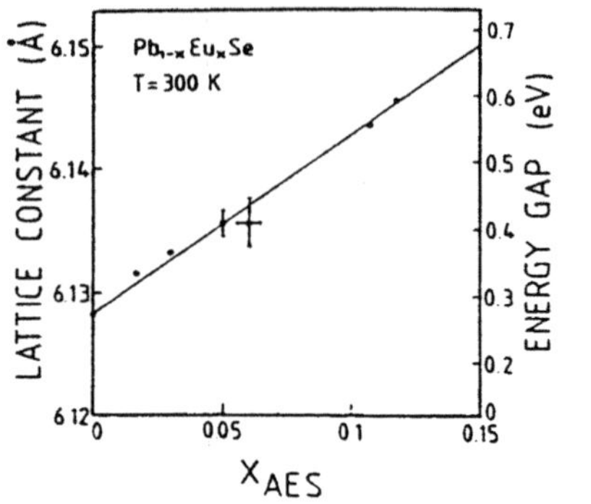

Figure 52 : **Energy gap and lattice constant of $Pb_{1-x}Eu_xSe$ as a function of x** [47].

$$\left.\begin{array}{ll} \text{for: PbTe} & 1.29 \text{ and } 1.94\,\text{Å} \\ \text{for: PbSe} & 1.22 \text{ and } 1.93\,\text{Å} \end{array}\right\} \text{ determined from charge density calculations}$$

Volume filling, 58.72 % for PbTe and 65,58 % for PbSe.

There is less free space in the PbSe unit cell than in PbTe. The ionic radius of Eu^{2+}(1.09 Å) agrees quite well with the Pb^{2+} ionic radii in PbTe (1.29 Å) and even better in PbSe (1.22 Å). On can expect that the Eu interdiffusion in PbSe is less than in PbTe. Obviously the interdiffusion coefficient D, depends also on the crystalline perfection at the interfaces.

As a reference PbSe and $Pb_{1-x}Eu_xSe$ single layers were p-type with $p = 3.10^{17}$ cm^{-3} and $p = 5.10^{17}$ cm^{-3} at 77 K. This is due to the excess Se flux during MBE growth which creates metal (Pb, Eu) vacancies in the crystal and each induces 2 acceptor levels in the valence band.

In typical MQW, carrier concentration is 5.10^{16} cm^{-3} and the mobility àf 20.000 cm^2/V.s which is a factor of 2-3 lager than PbSe single layer, and 8 times the mobility in $Pb_{1-x}Eu_xSe$ single layer. This is an indication for confinement of the holes in PbSe layers.

Multi-quantum-well structures of not lattice matched $Pb_{1-x}Eu_xSe$/PbSe (x=0.02) were grown on (111) BaF_2 substrates by MBE. The θ/2θ X ray diffraction pattern obtained for the (222) Bragg reflection is presented on figure 53. The diffraction pattern exhibits superlattice satellite peaks around the (222) Bragg reflection of PbEuSe up to the order i = 8, that demonstrate the high quality. After annealing for all together 1024 min at 370 °C no change of the satellite intensities in the diffraction patterns can be detected. In principle, from a decrease of the intensity of the high indexed satellite peak, information on the mearing out of the Eu from the barriers of the

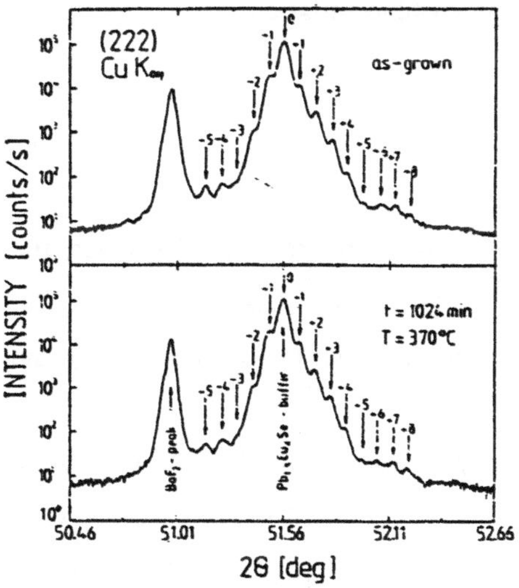

Figure 53 :θ - 2θ X-ray diffraction (222) Bragg reflection of the $Pb_{1-x}Eu_xSe$/PbSe MQW structure (15 double layers with a QW period of d = 109 nm; D_{PbSe} = 27.5 nm ; d_{PbEuSe} = 81.5 nm Upper part : As Grown Lower part : After a nnealing

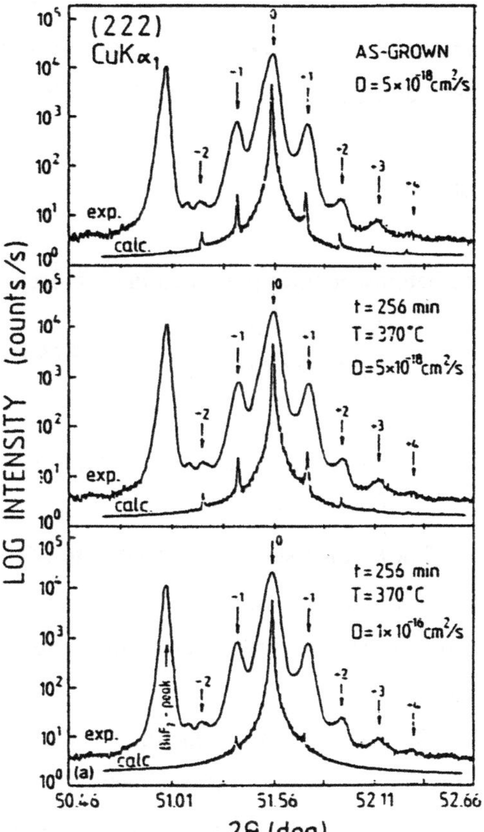

Figure 54 : The same as Fig 53 with d = 58 nm ; d_{PbSe}=27 nm d_{PbEuSe}= 31 nm. Calculated diffraction patterns based on Eu profiles deduced from D = $5.10^{-8}cm^2/s$ and 1.10^{-16} cm^2/s, are added.:

quantum wells accross the heterointerface can be obtained. These results show rather clearly that the Eu interdiffusion coefficient in this system must be rather small at these temperatures. Springholz et al. have studied the interdiffusion of Eu in such structures and shown that the interdiffusion coefficient of Eu is comparatively small with respect to the Mn in PbMnSe/ PbSe or to Eu in PbEuTe/ PbTe [45].

This system then offers the advantage that relatively high substrate temperatures up to 420 °C can be chosen for the growth of MQW structures. This ensures high mobilities in the p-n layers and good crystalline quality. They show that the interfaces are very abrupt. Therefore the PbEuSe/PbSe system seems to be superior compared to the PbEuTe/PbTe one, as far as the fabrication of heterostructures is concerned.

Conclusion

The heterostructures presented in these lectures are of interest for emitting systems.

CdMnTe / CdTe ZnMnSe / ZnSe	large energies green-blue emission
PbMnTe / PbTe PbEuTe / PbTe PbEuSe / PbSe	Mid I R Region

These interests are not related to Magnetic exchange interactions. However, the magentic exchange interactions are a powerfull tool to study these superlattices. It enable to assign without error the observed transitions, and to know perfectly the structure.

Obviously, in addition, the magnetic exchange opens a new field of interest in basic physic as well of magnetism and individual (by opposition to collective) interaction as to study the transition between type I - type II in a smooth and reversible way.

Finally it enables to determine the penetration of the wave function of state of the well, in the barrier, in that way we come back to the first point.

References

1. *Diluted Magnetic (Semimagnetic) Semiconductors*, (1987). Edited by Aggarwal, R.L., Furdyna, J.K., and Von Molnar, S. Mat. Research Society Symposia Proceeding, Pittsburgh, **vol. 89**.
2. *Semimagnetic Semiconductors and Diluted Magnetic Semiconductors*, (1991). Edited by Averous, M. and Balkanski, M., Physical Sciences, Plenum New York, **vol. 55**.
3. *Diluted Magnetic Semiconductors*, (1991). Edited by Jain, M., World Scientific Publishing Co, Pte Ltd, Singapour.
4. *II-VI Compounds and Semimagnetic Semiconductors, (1994)*. Edited by Heinrich, H. and Mullin, J.B., (1995) Fourth International Workshop on Semimagnetic (Diluted Magnetic) Semiconductors Proceeding, Mat. Sciences Forum, Trans. Techn. Publications Ltd, **vol.182-184**.
5. *Diluted Magnetic Semiconductors*, (1988). Edited by Furdyna, J.K. and Kossut, J., Semiconductors and Semimetals (Academic Press, San Diego, **vol. 25**.
6. Galaska, R.R. (1979) *Physics of Semiconductors*. Edited by Wilson, B.L.H., Institute of Physics, Bristol, p. 133.
7. Averous, M., Fau, C., Charar, S., El Kholdi, M., Ribes, D., and Golacki, Z. (1993) *Phys. Rev.* **B47**, 10261.
8. Bindilatti, V., Anisimov, A.N., Oliveira Jr, N.F., Shapira, Y., Goiran, M., Yang, F., Isber, S., Averous, M., and Demianuck, M. (1994) *Phys. Rev.* **B50**, 16464.
9. Isber, S., Averous, M., Shapira, Y., Bindilatti, V., Anisimov, A.N., Oliveira Jr, N.F., Orera, V.M., and Demianuck, M. (1995) *Phys. Rev.* **B51**, 15211.
10. Bauer, G. and Pascher, H. (1991) *Semimagnetic Semiconductors and Diluted Magnetic Semiconductors*, Edited by Averous, M. and Balkanski, M., Physical Sciences, Plenum New York, **vol. 55** p. 209.
11. Giriat, W. and Furdyna, J.K. (1988) *Diluted Magnetic Semiconductors*, (1988). Edited by Furdyna, J.K. and Kossut, J., Semiconductors and Semimetals (Academic Press, San Diego, **vol. 25** p. 1.
12. Furdyna, J.K., Giriat, W., Mitchell, D.F., and Sproule, F. (1983) *J. Solid State Chem.*, **46**, 349.

13. Balzorotti, A., Czyzyk, M., Kisiel, A., Motta, N., Podgorny, M., and Zimmal-Starnawska, M. (1984) *Phys. Rev.* **B30**, 2295.

14. Groves, S. and Paul, W. (1964) *Int. Conf. Semiconductors Physics*, Proceeding, Paris, p. 41.

15. Benecke, C. (1989) *Thesis*, TU Berlin D, 83.

16. Benecke, C and Gumlich, H.E. (1991) *Diluted Magnetic Semiconductors*, Edited by Jain, M., World Scientific Publishing Co, Pte Ltd, Singapour, p. 85.

17. Kossut, J. (1976) *Phys. Stat. Sol. (b)* **78**, 537.

18. Bastard, G., Rigaux, C., Guldner, Y., Mycielski, J., and Mycielski, A. (1978) *J. Phys.*, **39**, 87.

19. Coquillat, D. (1991) *Diluted Magnetic Semiconductors*, Edited by Jain, M., World Scientific Publishing Co, Pte Ltd, Singapour, p. 55.

20. Bhattacharjee, A.K., Fishman, G., and Coqblin, B. (1983) *Physica B* **117-118**, 449.

21. Pidgeon, C.R. and Brown, R.N. (1966) *Phys. Rev.* **146**, 575.

22. Shapira, Y. (1991) *Semimagnetic Semiconductors and Diluted Magnetic Semiconductors*, Edited by Averous, M. and Balkanski, M., Physical Sciences, Plenum New York, **vol. 55**, p. 121.

23. Isber, S., Fau, C., Charar, S., Averous, M., and Golacki, Z. (1995) *II-VI Compounds and Semimagnetic Semiconductors, (1994)*. Edited by Heinrich, H. and Mullin, J.B., Fourth International Workshop on Semimagnetic (Diluted Magnetic) Semiconductors Proceeding, Mat. Sciences Forum, Trans. Techn. Publications Ltd, **vol.182-184**, p. 657.

24. Jain, M. and Robins, M. (1991) *Diluted Magnetic Semiconductors*, Edited by Jain, M., World Scientific Publishing Co, Pte Ltd, Singapour, pp. 1-39.

25. Bernick, R.L. and Kleinnan, L. (1970) *Solid Stat. Commun.* **8**, 569.

26. Mittchell, D.L. and Wallis, R.F. (1966) *Phys. Rev.* **151**, 581.

27. Gorska, M. Anderson, J.R., and Golacki, Z. (1987) *Diluted Magnetic (Semimagnetic) Semiconductors*, Edited by Aggarwal, R.L., Furdyna, J.K., and Von Molnar, S. Mat. Research Society Symposia Proceeding, Pittsburgh, **vol. 89**, pp. 119-123.

28. Isber, S., Charar, S., Fau, C., Averous, M., and Golacki, Z. (1995) *II-VI Compounds and Semimagnetic Semiconductors, (1994)*. Edited by Heinrich, H. and Mullin, J.B., Fourth International Workshop on Semimagnetic (Diluted Magnetic) Semiconductors Proceeding, Mat. Sciences Forum, Trans. Techn. Publications Ltd, **vol.182-184**, p. 663.

29. Averous, M., Lombos, B., Bruno, A., Lascaray, J.P., Fau, C., and Lawrence, M.F. (1987) *Diluted Magnetic (Semimagnetic) Semiconductors*, Edited by Aggarwal, R.L., Furdyna, J.K., and Von Molnar, S. Mat. Research Society Symposia Proceeding, Pittsburgh, **vol. 89**, p. 141.

30. Kossut, J. and Furdyna, J.K. (1987) *Diluted Magnetic (Semimagnetic) Semiconductors*, Edited by Aggarwal, R.L., Furdyna, J.K., and Von Molnar, S. Mat. Research Society Symposia Proceeding, Pittsburgh, **vol. 89**, p. 97.

31. Kuhn, B., Ossau, W., Waag, A. Bicknell-Tassins, R.N., Landwehr, G. (1992) *J. of Crystal Growth* **117**, 871.

32. Chang, S.K., Nurmikko, A.V., Wu, J.W., Kolodziezski, L.A., and Gunshor, R.L. (1988) *Phys. Rev.* **B37**, 119.

33. Jackson, S.R., Nicholls, J.E., Hagston, W.E., Harrison, P., Stimer, T., and Hogg, J.H.C. (1994) *Phys. Rev.* **B50**, 5392.

34. Delporte, E., Peter, G., Berroir, J.M., Delalande, C., Hong, J.M., and Chang, L.L. (1992) *Surface Science* **263**, 570.

35. Gaj, J.A., Grieshaber, W., Bodin-Deshayes, C., Cibert, J., Feuillet, G., Merle d'Aubigné, Y., and Wasicla, A. (1994) *Phys. Rev.* **B50**, 5512.

36. Peter, G. Delporte, E., Bastard, G., Berroir, J.M., Delalande, C., Gil, B., Hong, J.M., and Chang, L.L. (1992) *J. of Luminescence* **52**, 147.

37. Delporte, E., (1995) *II-VI Compounds and Semimagnetic Semiconductors, (1994).* Edited by Heinrich, H. and Mullin, J.B., Fourth International Workshop on Semimagnetic (Diluted Magnetic) Semiconductors Proceeding, Mat. Sciences Forum, Trans. Techn. Publications Ltd, **vol.182-184**, pp. 497-502.

38. Wasiela, A., Peyla, P., Merle d'Aubigné, Y., Nicholls, J.E., Ashenford, D.E., and Lunn, B. (1992) *Semicond. Sci. Technol.* **7**, 571.

39. Haase, M., Qiu, J., De Puydt, J., and Cheng, H. (1991) *Appl. Phys. Lett.* **59**, 1272.

40. Dobrowolska, M., Huo, H., and Furdyna, J.K. (1995) *Acta Physica Polonica* **87**, 95.

41. Delporte, E., *J. of Crystal Growth* **138**, 884.

42. Feit, Z., Kostyk, D., Woods, R.J., and Mak, P. (1991) *Appl. Phys. Lett.* **58**, 343.

43. Partin, D.L. (1988) *I.E.E.E. J. Quantum Electron Q E*, 24.

44. Bauer, G., Pascher, H., and Zawadzki, W. (1992) *Semicond. Sci. Technol.* **7**, 703.

45. Springholz, G., Bauer, G., and Ihninger, G. (1993) *J. of Crystal Growth* **127**, 302.

46. Bauer, G. (1993) *Superlattice and Microstructures* **13**, 27.

47. Shu Yan, Springholz, G., Bauer, G. (1994) *Phys. Rev.* **B49**, 5476.

SURFACE, LEAKY AND SINGULAR MAGNETOPLASMONS ALONG THE INTERFACE OF GYROTROPIC SEMICONDUCTOR

S. T. IVANOV, N. I. NIKOLAEV AND P. I. DANKOV
Department of Physics, University of Sofia
5 J. Bourchier Blvd, 1164 Sofia, Bulgaria

The increasing use of anisotropic elements in many applications ranging from microwaves to optical frequencies has prompted the study of electromagnetic wave propagation in anisotropic media. The detail investigation of magnetoplasmons has been encouraged by the successful use of the hydrodynamic model to study electron layer in parabolic quantum well [1].

Let us consider a plane interface of a gyrotropic semiconductor (SC) and a dielectric with permittivity ε_d or a perfect metallic screen. The x-axis of the Cartesian coordinate system is chosen to be perpendicular to the interface coinciding with the y, z-plane. The SC plasma occupies the region $x > 0$. The Faraday configuration has been considered. In anisotropic medium the E_z- and B_z-components satisfy a system of equations, which can be reduced to a simple biquadratic equation (see e.g.[2]), direct implication of this being the existence of two transverse wave components α_1^2 and α_2^2:

$$\alpha_{1,2}^2 = \kappa^2 \frac{\varepsilon_1 + \varepsilon_3}{2\varepsilon_1} + \frac{\omega^2}{c^2}\frac{\varepsilon_2^2}{2\varepsilon_1} \pm \left\{ \left[\kappa^2 \frac{\varepsilon_1 + \varepsilon_3}{2\varepsilon_1} + \frac{\omega^2}{c^2}\frac{\varepsilon_2^2}{2\varepsilon_1} \right]^2 - \frac{\varepsilon_3 \xi}{\varepsilon_1} \right\}^{1/2}, \quad (1)$$

where $\varepsilon_1 = \varepsilon_L(1 - \frac{\omega_p^{*2}}{\omega^2 - \omega_c^2})$, $\varepsilon_2 = \varepsilon_L \frac{\omega_c}{\omega} \frac{\omega_p^{*2}}{\omega^2 - \omega_c^2}$, $\varepsilon_3 = \varepsilon_L(1 - \frac{\omega_p^{*2}}{\omega_c^2})$ and $\omega_p^* = (4\pi n e^2/m^* \varepsilon_L)^{1/2}$ and $\omega_c = (eB_0/m^*c)^{1/2}$ are the effective plasma and the cyclotron frequencies, respectively; m^* and n are the effective mass and the concentration of the light carriers; ε_L stands for the high frequency dielectric constant of the lattice and $\kappa^2 = k_z^2 - \varepsilon_1 \frac{\omega^2}{c^2}, \xi = \kappa^4 - \varepsilon_2^2 \frac{\omega^4}{c^4}$. In

M. Balkanski (ed.), Devices Based on Low-Dimensional Semiconductor Structures, 171–174.

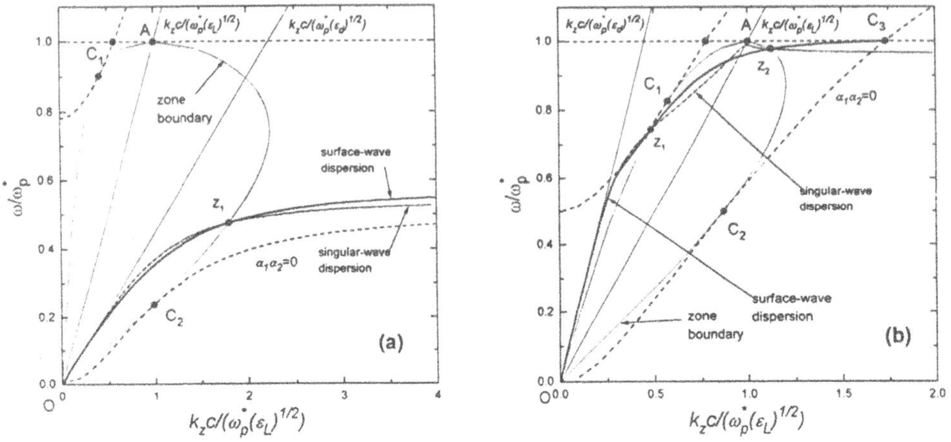

Figure 1. Surface-wave dispersion of the real medium and singular-wave dispersion of the hypothetical medium along the dielectric boundary for: a) weakly magnetized plasma $\omega_c/\omega_p^* = 0.5$ at $\varepsilon_{DL} \equiv \varepsilon_d/\varepsilon_L = 5$; b) strongly magnetized plasma $\omega_c/\omega_p^* = 1.5$ at $\varepsilon_{DL} \equiv \varepsilon_d/\varepsilon_L = 0.15$.

bounded systems any field component is a superposition of two partial waves, which origin from normal modes, but which can not exist independently. In plane geometry the two-plane solution is following:

$$E_z = [A\exp(ikx) + B\exp(-ikx)]\exp[i(k_z z - \omega t)]. \tag{2}$$

Using boundary conditions we can get dispersion equations for plasma dielectric interface:

$$\begin{aligned}(\alpha_1 - \alpha_2)D_D(\omega, k_z) \equiv (\alpha_1 - \alpha_2)\{\alpha_1\alpha_2(\alpha_1 + \alpha_2 + \kappa_d)\varepsilon_1\varepsilon_d + \\ \kappa_d[\alpha_1^2 + \alpha_2^2 + \alpha_1\alpha_2 + \kappa_d(\alpha_1 + \alpha_2)]\varepsilon_1\varepsilon_3 + \kappa_d\kappa^2\varepsilon_3(\varepsilon_d - \varepsilon_3)\} = 0\end{aligned} \tag{3}$$

and for plasma-metal one :

$$(\alpha_1 - \alpha_2)D_M(\omega, k_z) \equiv (\alpha_1 - \alpha_2)(\alpha_1\alpha_2 + \kappa^2\varepsilon_3/\varepsilon_1) = 0, \tag{4}$$

where $\kappa_d^2 = k_z^2 - \varepsilon_d \dfrac{\omega^2}{c^2}$.

Dispersion curves differ for weakly ($\omega_c < \omega_p^*$) and strongly magnetized plasma ($\omega_c > \omega_p^*$) at big k_z. For weakly magnetized plasma the dispersions have an asymptotic behaviour (Fig.1a), while for strongly magnetized plasma they have an end point at the intersection of ω_p^* and $\alpha_1\alpha_2 = 0$ (Fig.1b – the point C_3). Only the plots for dielectric-semiconductor interface are shown. Depending on the ratio $\varepsilon_{DL} = \varepsilon_d/\varepsilon_L$ the cross-points of the

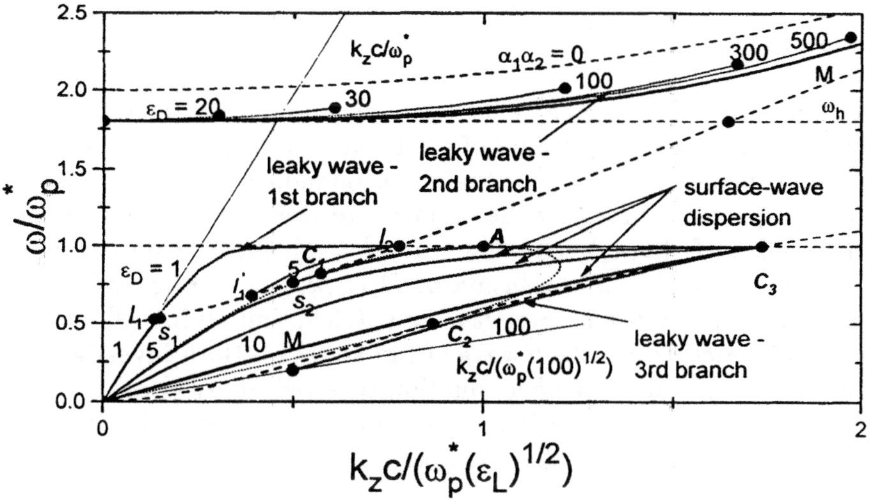

Figure 2. Leaky plasmons for strongly magnetized plasma. at different ε_d. For comparison the surface plasmons are shown.

dispersion with the complex-zone boundary (the loop) may be one , two or three. For metal there is only one crossing point.

The waves in strongly magnetized plasma have been investigated only for metal screen and the mathematical solution for $\omega > \omega_p^*$ was incorrectly interpreted in [3] as physical one. A careful analysis ([4]) shows that the solution for $\omega > \omega_p^*$ has not physical meaning.

The above discussed dispersion are obtained from equations $D_I = 0$ at real k_z. (Here I (interface) stands for D or M). It has been suggested in [5] that a pseudosurface wave propagates in the stop-band, where the dispersion is interrupted (see Fig.2). It was only a suggestion – the dispersion of pseudosurface waves has not been obtained. It, principally, can not be obtained. We solve dispersion equation at complex k_z – $D_I(\omega, k_z = k_z' + ik_z'') = 0$. The imaginary wave number k_z'' describes the radiative attenuation along the z-axes. The numerical results are presented in Fig.2 for strongly magnetized plasma . The stop-band of the structure SC plasma-vacuum is marked with points s_1 and s_2 (Fig.2). Three branches of leaky plasmons have been obtained. And one of them is linked with the gap, but it is not restricted by it (see Fig.2). Moreover they exist even in a system without a gap. One can see that the points l_1 and l_2, which determine the interval of the leaky waves existence, do not coincide with points s_1 and s_2, defining the gap in the dispersion of the surface plasmons. The next region, where the one transverse constant is positive and the other is negative is between the following three lines $\omega_h = \sqrt{\omega_p^{*2} + \omega_c^2}$, upper line $\alpha_{1,2} = 0$ and the light line (Fig.2). These waves in transverse bounded system become pseudosurface waves, which were called anisotropic wave by Ivanov and Alexov [6] .

Finally, the third branch of the leaky plasmons appears at low frequencies for unrealistically big values of ε_d. Their dispersion curves tend to that one of the surface plasmons for the metal boundary.

The dispersion curves for the surface and the leaky wave were received from equations $D_I(\omega, k_z) = 0$, that is at $\alpha_1^2 \neq \alpha_2^2$. On the complex boundary $\alpha_1^2 = \alpha_2^2 = \alpha^2$ and α is a solution of $D_I(\omega, k_z) = 0$. Rao and Uberoi [7] called this solution degenerate wave and Halevi and Quinn [8] proved that it does not describe a surface wave of the kind (2). Both investigations are taking the standard dispersion equation received at $\alpha_1 \neq \alpha_2$. But on the loop $\alpha_1 = \alpha_2 = \alpha$, that is α is 2-tuple root. The two linearity independent solutions in this case should to be taken in the following form:

$$\mathbf{E} = (\mathbf{A_1} + \mathbf{A_2}x)\exp(-\alpha x)\exp[i(k_z z - \omega t)], \tag{5}$$

$$\mathbf{B} = (\mathbf{B_1} + \mathbf{B_2}x)\exp(-\alpha x)\exp[i(k_z z - \omega t)]. \tag{6}$$

For such a field one obtains a new dispersion equations, which formally are the dispersion equations (3,4) at $\alpha_1 = \alpha_2$ and they describe so called singular waves.

We consider a hypothetical medium with $\alpha_1 = \alpha_2$ and proved that its dispersion passes through the intersection points of the real dispersion curve and the loops. This is shown for two cases in Fig.1.

Conclusion

Surface waves in strongly magnetized plasma have an end point. In the region, where one α_1^2 is positive and another α_2^2 is negative (that is in region of pseudosurface waves) three branches of leaky waves exist. The complex-zone boundary is a set of points which at different frequencies in the range of ω $(0 - \omega_p^*)$ describe singular waves.

References

1. Dempsey, J. , Halperin, B. I. (1992) *Phys. Rev. B*, **45**, 1719.
2. Wallis, R. F. (1982) *Electromagnetic Surface Modes* (edited by A. D. Boardman), Wiley, p.575.
3. Halevi, P. (1981) *Phys. Rev.* B, **23**, 2635.
4. Ivanov, S.T., Ivanova, K.M. & Alexov, E.G. (1993) *Physica Scripta* **49**, 228.
5. Wallis, R. F., Brion, J. J., Burstein, E., Harstein, A. (1974), *Phys. Review B*, **9**, 3424.
6. Ivanov, S., T. & Alexov, E., G. (1990) J. *Plasma Phys.* **43**, 51.
7. Rao, U.J. & Uberoi, C. (1978) *Phys.Rev.* B **18**, 294.
8. Halevi, P. & Quinn, J.J. (1980) *Solid State Comm.* **33**, 467.

OPTICAL AND THEORETICAL ASSESSMENT OF GAAS QUANTUM WELLS HAVING SUPERLATTICES AS BARRIER LAYERS

V.DONCHEV, I.IVANOV AND K.GERMANOVA
Faculty of Physics, Sofia University
5 J.Bourchier blvd., Sofia-1126, Bulgaria

The replacement of the AlGaAs alloy barriers of a GaAs quantum well (QW) by short period GaAs/AlGaAs superlattices (SLs) is beneficial because the use of SL structures as prelayers suppress the defect diffusion from the substrate to the QW and relieves the strain caused by the small lattice mismatch between GaAs and AlGaAs. Besides, a better control of the structural system parameters is possible which is important especially in the case of growth on patterned substrates.

The aim of this work is to study the electronic states and the optical properties of GaAs QWs embedded in GaAs/AlAs SLs in order to determine to what extent they are influenced by the presence of a SL in the barrier region.

The samples were planar structures grown by MBE on (100) GaAs substrates. They consist of the following undoped layers: i) 26 periods of $(GaAs)_8/(AlAs)_4$ SL, ii) 4 monolayers (MLs) of AlAs; iii) a 50A thick GaAs layer (QW); iv) 20 periods of $(GaAs)_8/(AlAs)_4$ SL, and v) a 500A thick $Al_xGa_{1-x}As$ cap layer with x=0.30.

Photoluminescence (PL) measurements were performed at 2K with a resolution of 0.2nm using a He-Ne laser with an excitation density of 0.5 W/cm^2. One sample was measured also at 77K by PL and photoluminescence excitation (PLE) spectroscopy using an argon pumped Ti-sapphire laser with an excitation density of $1W/cm^2$.

Theoretical assessment was made using two approaches:

1) The effective mass approximation (EMA) [1] applied to a single GaAs QW of 18 MLs having uniform AlGaAs alloy barriers with the mean Al content x in the SL (x=4/12=0.33). One-band model is used and nonparabolicity of CB and light hole (lh) band is taken into account.

M. Balkanski (ed.), Devices Based on Low-Dimensional Semiconductor Structures, 175–178.

2) The empirical tight binding (ETB) approach with the surface Green function matching method. The algorithm is described in [2,3]. In the present calculations the inhomogeneous region was a finite $(GaAs)_8/(AlAs)_4$ SL (10 or 20 periods) with an embedded centered GaAsQW of 18 MLs. An ideal finite SL and a single QW with AlGaAs barriers (x=0.33) were also studied for comparison.

Table 1 shows that the EMA and ETB approaches give similar energies for e1-hh1 and e1-lh1 transitions of a single QW. When the SL barrier is taken into account these energies increase which can be explained by an increase of the effective potential barrier. The SL barrier is equivalent to AlGaAs barrier with x=0.5. This value is higher than the average Al content x=0.33.

TABLE 1. Calculated and experiment transition energies (in meV) for a GaAs QW of 18ML. The localization of the involved hole states is given, too (QW: in the QW of 18ML; W/SL: in the wells of the SL). The e1 state is localized in the QW.The values in parentheses in the 6th column are the exciton binding energies.

Tran-	Hole loca-	Single QW		QW embedded in a $(GaAs)_8/(AlAs)_4$ SL		
sition	lization	EMA	ETB	ETB	ETB-E_{exc}	experiment
e1-hh1	QW	1625	1624	1639	(-12) 1627	1625 (PL)
e1-lh1	QW	1653	1648	1670	(-15) 1655	1643 (PL)
e1-hh2	W/SL			1701	1690	
e1-hh3	W/SL			1703	1692	
e1-hh4	W/SL			1705	1694	
e1-hh5	W/SL			1708	(-11) 1697	1689 (PL)
e1-hh6	W/SL			1710	1699	
e1-hh7	QW			1714	1703	
e1-lh2	W/SL			1743	1732	
e1-lh3	W/SL			1753	(-11) 1742	1725 (PLE)
e1-lh4	W/SL			1755	1744	

The comparison with experiment is done after subtracting the exciton binding energies E_{exc} from the calculated transition energies (column 6 in Table 1). In the case of e1-hh1 and e1-lh1 transitions which are localized in the QW we use as a first approximation the values of E_{exc} corresponding to a single QW and calculated by an interpolation formula [4]. The PL peaks (see Fig.1) with energies close to these transitions have very small amplitude, which indicates that an other recombination channel is dominant.

The group of transitions e1-hh2 to e1-hh7 which have very close energies do not exist in the spectrum of the ideal SL, nor of the single QW. The

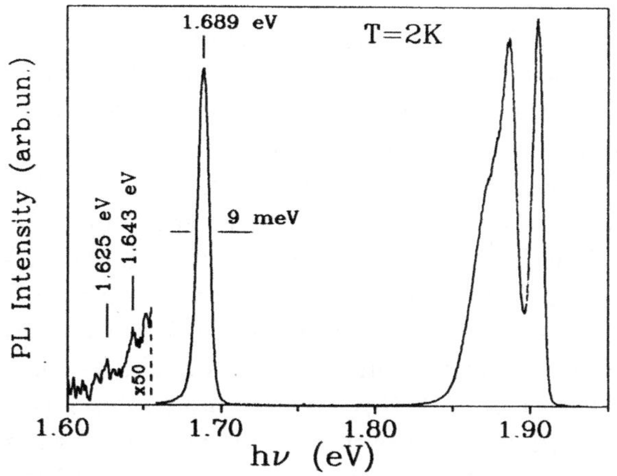

Figure 1. Typical PL spectrum of a 50A GaAs QW embedded in a $(GaAs)_8/(AlAs)_4$ SL.

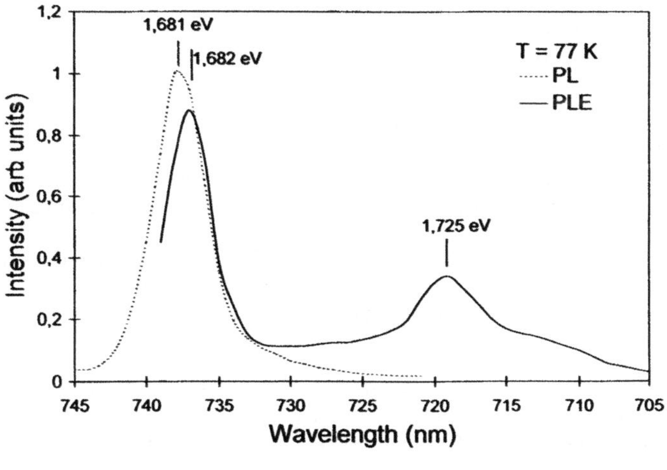

Figure 2. PL and PLE spectra of a 50A GaAs QW embedded in a $(GaAs)_8/(AlAs)_4$ SL. PL is excited with 712 nm. Detection wavelength for PLE is 740 nm.

states hh2 to hh6 correspond to the SL hh-band. However they have non-zero spectral strength in the QW. The state hh7 is localized mainly in the QW. So, we suppose that this group of transitions gives rise to the large PL peak at 1.689eV in Fig.1. The value for E_{exc} in this case is found from temperature dependent PL.

The group of transitions following in energy (e1-lh2 to e1-lh4) is also specific for the system under study. It includes hole states corresponding to the lh-band of the SL. Its separation from the hh-band is about 42 meV. This value corresponds to the energy distance between the two PLE peaks in Fig.2. This result supports our assumption about the origin of the large

PL peak at 1.68-1.69eV.

The sharp doublet around 1.89eV in Fig.1 is explained by the contributions of the lowest SL hh exciton and localized excitons unintentionally introduced trough the well-width fluctuations in the SL.

In conclusion the replacement of the AlGaAs alloy barriers by a GaAs/AlAs SL modifies the electronic structure of the QW and give rise to optical transitions which do not exist in the spectrum of a single QW, nor of an ideal SL. It is found that in the case of a 18 MLs QW some of these new transitions are dominant in the emission spectrum of the system.

The transition energies calculated by ETB method are always a little bit larger than the experimental ones. This could be due to the ETB parameterization used (see [2]). However the model describes well the general physical picture and reveals the main tendencies in the system considered.

We are grateful to Dr. St.Vlaev for performing ETB calculations and helpful discussions, to prof. F.K.Reinhart and D.Martin for supplying the samples and disposing the PLE instrumentation and to D.Batovski and A.Yordanov for technical help. This work was supported by the Bulgarian National Science Fund.

References

1. Herman,M.A., Bimberg,D. and Christen,J. (1991) Heterointerfaces in quantum wells and epitaxial growth processes: Evaluation by luminescence techniques, *J.Appl.Phys.* **70**, R1-R51.
2. Vlaev, S., Velasco, V.R., and Garcia-Moliner,F. (1994) Electronic states in graded-composition heterostructures, *Phys.Rev.B* **49**, 11222-11229; (1994) Tight-binding calculation of electronic states in a triangular symmetrical quantum well, *Phys.Rev.B* **50**, 4577-4580; (1995) Tight-binding calculation of electronic states in an inverse parabolic quantum well, *Phys.Rev.B* **51**, 7321-7324.
3. Vlaev, S. (1995) A tight-binding calculation of transition energies in triangular symmetrical QWs, in M.Balkanski and I.Yanchev (eds.), *Fabrication Properties and Applications of Low-Dimensional Semiconductors*, Kluwer Academic Publishers, Dordrecht, pp.141-142.
4. Oelgart, G. *et al.* (1994) Experimental and theoretical study of excitonic transition energies in GaAs/AlGaAs quantum wells, *Phys.Rev.B* **49**, 10456-10465.

Anisotropy of Optical Characteristics of Low-dimensional and Bulk Many-Valley Semiconductors

A.Svizhenko, V.Vasetskii, V.Poroshin, O.Sarbey
Institute of Physics, Ukrainian National Academy of Sciences, Prospect Nauki 46, Kiev–22, 252650 Ukraine, tel/fax: (380)(44) 265 0987/1589
Zh.Gumenjuk-Sychevskaya, F.Sizov
Institute of Semiconductor Physics, Prospect Nauki 43, Kiev, 252650 Ukraine, tel: (380)(44) 265 6296

Many–valley cubic semiconductors are optically isotropic crystals in near and medium IR range of spectrum. However, many–valley structure of energy bands gives rise to a number of peculiarities of optical characteristics. In this work it is shown theoretically that absorption coefficient by interband optical transitions in IV–VI many–valley semiconductor superlattices depends on a light polarization. We also report about experimental observation of self–induced birefringence of intense IR light (CO_2 laser radiation) in bulk crystals of n–Ge.

1. Superlattices (SL) on the basis of narrow-gap cubic many-valley semiconductors A^4B^6 ($PbTe/Pb_{1-x}Sn_xTe$) .

Due to strong anisotropy of free carrier effective mass there is a difference of carrier spectrum quantization between straight [111] and oblique $<\bar{1}11>$ valleys and valleys become energetically uniquivalent. As a consequence an allotted direction appears along [111] (SL axes) .

For each valley absorption coefficient $\alpha_q(h\omega)$ for interband dipole transitions depends on polarization of incident light. Since the valleys give different contributions to dielectric function the total absorption $\alpha_\Sigma(h\omega)$ is anisotropic.

We calculated the energy structure and absorption coefficient $\alpha_q(h\omega)$ for interband dipole transitions for SL $PbTe\ /\ Pb_{0.8}Sn_{0.2}Te$ ($d \geq 500$ Å).

The total absorption coefficient is

$$\alpha_\Sigma(h\omega) = \frac{\pi^2 e^2 h}{4 m_0^2 \varepsilon_\infty^{\frac{1}{2}} d} \frac{(\overline{e\cdot p_{cv}^1})^2}{p_\perp^2} \sum_n \theta(\omega-\omega_n)(1-\Delta_n) +$$

$$+\frac{\pi^2 e^2 h}{4 m_0^2 \varepsilon_\infty^{\frac{1}{2}} d} \sum_q^4 \frac{(\overline{e\cdot p_{cv}^q})^2}{p_x p_y} \cdot \sum_{n'}^3 \theta(\omega-\omega_{n'})(1-\Delta_{n'}+O(\Delta_{n'}^2)),$$

where $\omega_n = (E_g + E_{cn} + E_{vn})/h$, E_{cn}, E_{vn} – the n–th subband energies, $\Delta_n = \frac{(E_{cn}^2 - E_{vn}^2)^2}{h^4\omega^4} << 1$, $|p_{cv}^q|^2$- matrical element of transition fo q–th valley in dipole approximation, $\bar{e}$ – light polarization vector.

The first term refers to the straight valley [111], the second – to the oblique valleys $<\bar{1}11>$.

It is shown that absorption coefficients in oblique $\alpha_{cvn'n'}^{<\bar{1}11>}$ and straight valleys $\alpha_{cvnn}^{[111]}$ are much different and depend on polarization. For instance:

if $\bar{n}\|[001]$ and $\bar{e}\|[100]$ then

$\alpha_{cvn'n'}^{<\bar{1}11>} = 3.86\cdot 10^3 cm^{-1}$, $\alpha_{cvnn}^{[111]} = 0.57\cdot 10^3 cm^{-1}$,

and

$$K = \alpha_{cvn'n'}^{<\bar{1}11>} / \alpha_{cvnn}^{[111]} \approx 6.7,$$

2. Self-Induced birefringence of IR light in n-Ge.

In bulk many–valley semiconductors strong anisotropy of optical characteristics in the IR range of spectrum is connected with the redistribution of free carriers between degenerate conduction band minima (equivalent valleys). The main reason for this redistribution is the different carrier heating in valleys with the different orientation with respect to the direction of electric field $\bar{E}$ of the light wave. As a consequence of the redistribution the contribution of free carriers to the dielectric function becomes anisotropic, causing self–induced effects of the light beams.

In this work we report the experimental proof of self–induced birefringence of the intense IR light (CO_2 laser radiation) in many–valley semiconductor n–Ge.

The experiments were performed by IR light $\lambda = 10.6\mu$ at 300K and 77K. Samples were cut in $(1\bar{1}0)$ plane from n–Ge doped with $5\cdot 10^{16} cm^{-3}$ Sb. Plates with 3.2 mm thickness had wedge angles of 2°.

We found that a linearly polarized light of a small intensity (I<3 MW/cm^2) passes through the sample with no change of polarization. Contrary to this, increasing

M. Balkanski (ed.), Devices Based on Low-Dimensional Semiconductor Structures, 179–180.

the intensity, we obtained elliptically polarized outgoing light with the major axis of the ellipse along the polarization vector of the incident light if light polarization direction did not coincide with <001>, <110> or <111>. The coefficient of ellipticity $\xi(\Theta)$ strongly depends on the orientation of the electric field of the incident wave relatively to the crystallographic axes of the sample and on the temperature, sharply increasing with decreasing of temperature.

A phenomenological theory developed by us (see Ref.1) excellently describes the experimental results.

Comparison of results of the calculations of ellipticity coefficient $\xi(\Theta)$ with experimental results gives the third order nonlinearity constants: $\varepsilon_2 = 2.68 \cdot 10^{-8} esu^{-2}$ (T=300K) and $\varepsilon_2 = 1.25 \cdot 10^{-7} esu^{-2}$ (T=77K).

The constant ε_2 can be expressed by the redistribution of electrons between the valleys for any polarization of the incident light.

For E II [111] (it is known that for this case the concentration of the electrons in the valley located on this crystallographic axis n_1 is increased, but in the other three valleys the concentration of the electrons n_3 decreased):

$$\frac{n_1}{n_3} = \frac{4}{1 - \varepsilon_2 E^2 \dfrac{3\omega^2}{8\pi e^2 N (m_\perp^{-1} - m_{II}^{-1})}} - 3$$

where E – the amplitude of laser wave.

The heating of electrons by electromagnetic field and redistribution of free hot electrons between equivalent valleys were calculated by us also from the energy balance

- by electron temperature approximation
- with account of scattering on acoustic and optic phonons and electron–electron interaction

Calculated dependence together with experimental dependence of n1/n3 on electric field is presented in Fig 1. It well coincides with data from experiments on birefringence.

The change of refractive index due to the redistribution of hot free carriers between equivalent valleys at 77K was about $4 \cdot 10^{-2}$ at the light intensity $5\,MW\ /\ cm^2$. Such a change of refractive index is rather high one compared to other nonlinearity mechanisms and can be used in some optical applications.

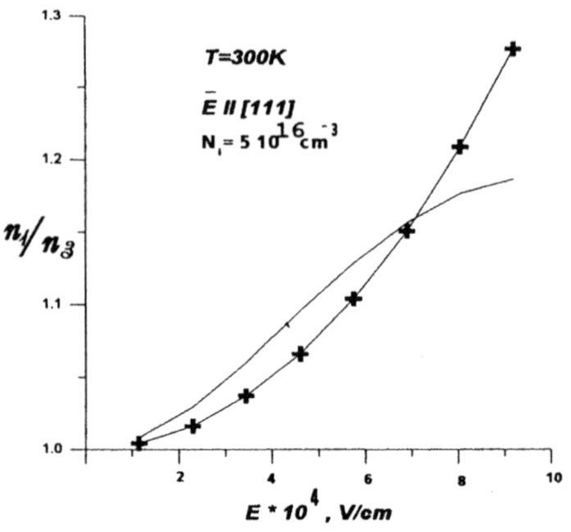

Fig 1. n1/n3 vs field E

-+- experiment

--- calculation

References

1.Proceedings of 22nd International Conference on The Physics of Semiconductors, Vancouver, Canada, 1994, V.1, p.261–265

HOT HOLE EFFECTS IN STRAINED MULTI-QUANTUM-WELL HETEROSTRUCTURES Ge/GeSi

V.Ya.ALESHKIN, N.A.BEKIN, I.V.EROFEEVA, V.I.GAVRILENKO, Z.F.KRASIL'NIK, O.A.KUZNETSOV, M.D.MOLDAVSKAYA and V.V.NIKONOROV
Institute for Physics of Microstructures of Russian Academy of Sciences Nizhny Novgorod, Russia

Last years strained heterostructures on base Ge and Si are of considerable interest due to wide range of properties which can be changed through band-structure engineering. We deal with heterostructures $Ge/Ge_{1-x}Si_x$ ($x \approx 0.1$, $d_{Ge} \leq 200A$) which were grown on moderately doped Ge (111) substrates by CVD techique. In these heterostructures quantum wells in the valence band are realized in germanium layers while the energy structure of conductance band is not quite clear yet. Theoretical calculations of the energy spectra give the negligible value of the conductance band offset (it seems to be the reason of the absence of the electron effects in the our experiments) while in the valence band the depth of the quantum well is about 100 meV. Ge layers are uniformly compressed in the plane of the structure because of the mismatch of lattice periods of Ge and $Ge_{1-x}Si_x$ This deformation may be represented as a result of hydrostatic compression and equivalent uniaxial tension P_{equiv} of Ge layers; the latter is responsible for the valence band splitting. Earlier we have observed the cyclotron resonance (CR) absorption line of high mobility light 2D holes in split valence band in Ge layers [1] and revealed the electric field effects on the hole systems [2]. In this paper we report the further investigation of hot hole CR, that give mush information about the heating and energy spectrum of 2D holes . Also hot hole far infrared emission from selectively doped heterostructures was observed. The results of this experiment are shortly discussed.

Hot hole CR in undoped multi-quantum-well (MQV) $Ge/Ge_{1-x}Si_x$ heterostructures was investigated in d.c. and pulsed electric fields. The electric voltage was applied to the sample via alloyed ohmic contacts deposited on the surface of the heterostructure.

Typical CR spectra of photoexcited holes are presented in Fig.1,a. The position of the CR line at $E = 0$ corresponds to the effective mass $m_c = 0.07m_0$ and mobility calculated from halfwidth of the line is $\mu \approx 10^5$ cm^2/V·s. Application of d.c. fields about few V/cm results in the remarkable shift of CR line to higher magnetic fields. This effect is clearly seen for both the photoexcited holes and the holes produced by

M. Balkanski (ed.), Devices Based on Low-Dimensional Semiconductor Structures, 181–184.

impact ionization of residual shallow acceptors (Fig.1,b). The results of the theoretical calculations of energy momentum law are shown in Fig.2,a.

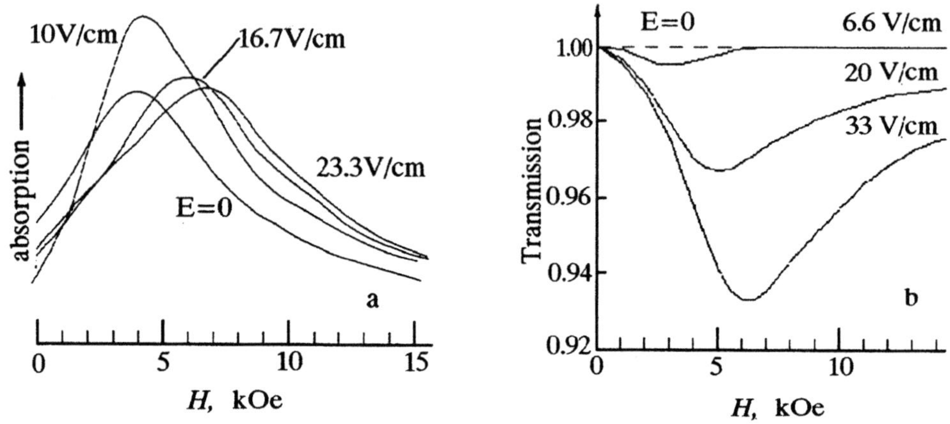

Figure 1. CR spectra in undoped sample #306 in d.c. electic fields (λ=2.1 mm, T=4.2 K); a - CR absorption by photoexcited holes, b - absolute measurements of changes in sample transmission due to the impact ionization of residual shallow acceptors.

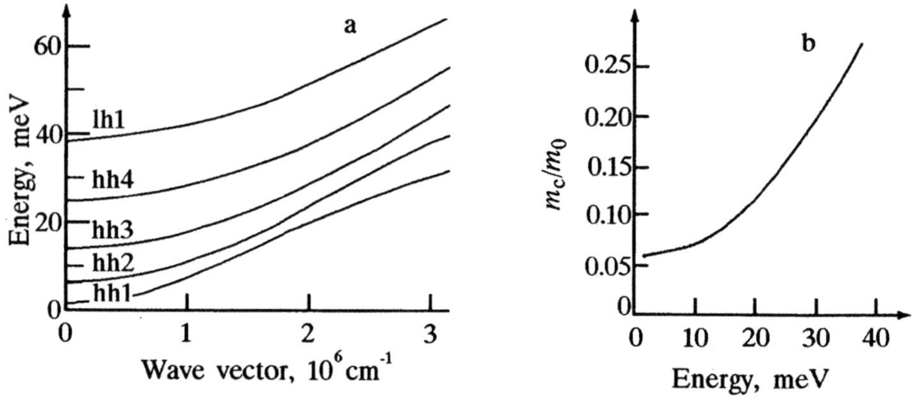

Figure2. Calculated 2D hole energy spectra in Ge/$Ge_{1-x}Si_x$ heterostructure #306 (a) and energy dependence of cyclotron resonance mass in the lowest subbund hh1 (b) ($d_{QW} = 200A$, $x = 0.12$, $\mathbf{k}\|[112]$, $P_{equiv} = 4kbar$).

The calculated effective mass value $m^* \approx 0.06m_0$ at the bottom of lowest subbund (Fig.2, b) is in good agreement with observed cyclotron mass of 2D holes. The shift of the CR line is evidently a result of both the carrier heating and the strong nonparabolicity of the energy-momentum law. In higher electric fields the CR

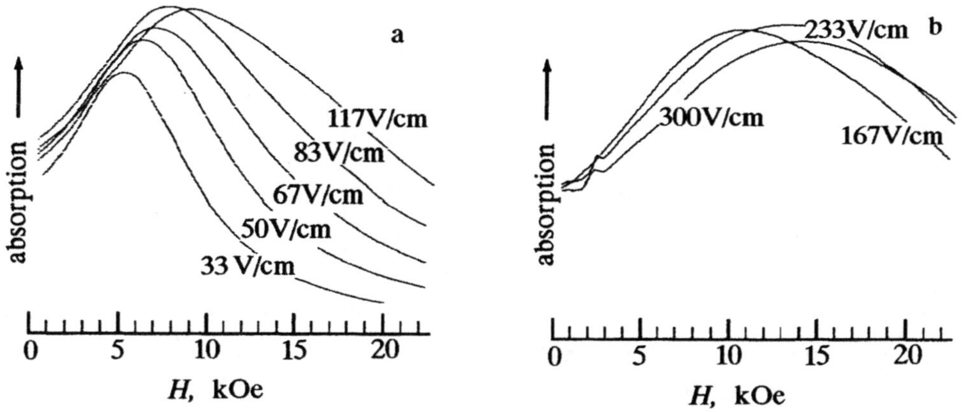

Figure 3. Hole CR absorption spectra in undoped sample #306 in pulsed electric fields ($\lambda = 2.3mm$, T= 4.2K).

investigations were carried out using the pulsed technique (Fig.3). In the latter case the holes were excited at the impact ionization of residual shallow acceptors in the heterostructure thus modulating the microwave absorption in the sample [3]. In strong electric fields the effective mass increases up to an enormous value ($m_c \approx 0.3m_0$ at E » 300 V/cm). (Fig.3 a,b) that corresponds to the carrier heating up to $T_e \geq 200$ K. The latter proves the realization of the streaming motion of the carriers and indicates the possibility of the population inversions of 2D holes in **E**⊥**H** fields just similar to those in bulk p-Ge [4].

The obtained absolute values of CR transmission (Figure 1b) together with the relative increase of the integral intensity of CR line with the rise of pulsed electric fields up to its saturation at E ≥ 200 V/cm (Figure 3a,b) corresponding to the complete ionization of shallow acceptors allow to estimate the total concentration of residual impurities in the heterostructure: $n_s \approx 2 \cdot 10^{11}$ cm^{-2} (per 162 layers of the structure) that roughly corresponds to the volume concentration $n \approx 3 \cdot 10^{14}$ cm^{-3}. Because of adiabatic heating of the sample (together with the substrate) during the strong electric field pulse its temperature increases significantly over 4.2K that results in the thermoionization of shallow acceptors behind the trailing edge of the pulse [3]. Measuring the intensity of CR absorption line just after the electric field pulse as a function of the dissipated Joule energy and taking into account the temperature dependence of the specific heat of the germanium substrate we obtain the temperature dependence of the concentration and estimate E_B» 2-4 meV that is sufficiently smaller than that in bulk Ge because of the strain induced valence band splitting.

Hot hole far IR emission from selectively doped sample #125 ($n_s \approx 2 \cdot 10^{11}$cm^{-2}) was detected from both the plane and the edge of the sample in electric field. The strong electric field up to 500 V/cm was applied to the sample along the quantum layers. In both cases the intensity of the emission detected with Ge:Ga and n-GaAs demonstrates the same dependence on electric fields. The emission intensity increases

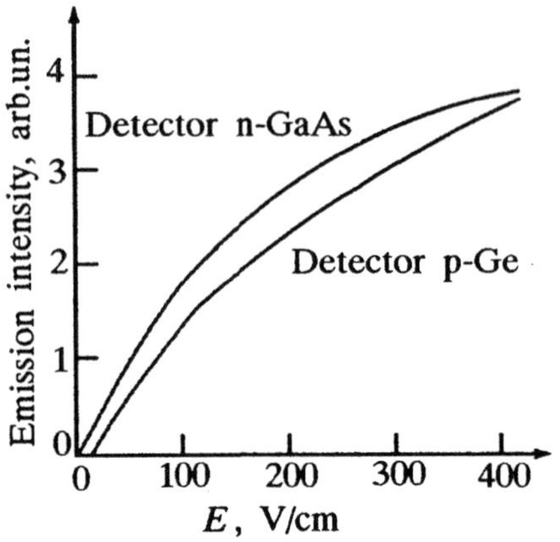

Figure 4. Hot hole far IR emission intensity from selectively doped sample #125 (in the sensitivity ranges of n-GaAs and Ge:Ga detectors) versus the electric field.

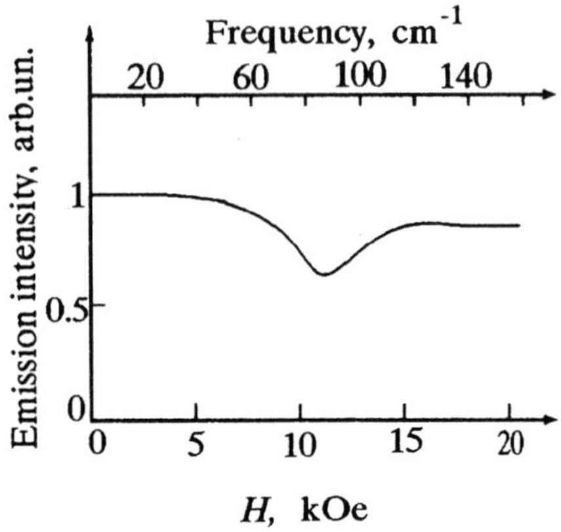

Figure 5. Hot hole far IR emission intensity from the edge of selectively doped sample #125 vs the magnetic field at the rejected filter n-InSb or vs the resonant frequency of the filter: detector Ge:Ga, E=330 V/cm.

monotonously with electric field (Fig.4). The spectral characteristics of the emission from the sample edge were investigated using the cooled tunable by magnetic field filter n-InSb (Fig.5). The spectrum of the signal from the detector is shown to be completely determined by the frequency dependence of the sensitivity of Ge:Ga photoresistor; no features associated with peculiarities of the emission from the sample were observed. The latter indicates that the emission spectrum is broader than that of the detector and that the radiation results predominantly from the indirect intraband transitions of hot carriers rather than from the intersubband transitions.

The research described in this publication was maid possible in part by Grant #1-028 from Russian Scientific Program "Physics of Solid State Nanostructures", Grant #R8H300 from International Science Foundation and Russian Government and Grant #94-842 from INTAS.

References

1. Gavrilenko,V.I. et al. (1994) Cyclotron resonance in Ge layers in $Ge/Ge_{1-x}Si_x$ heterostructures, *JETP Lett.* **.59**, 348-352; (1994) *Jpn.J.Appl.Phys.* **33,** 2386-2387.

2. Gavrilenko,V.I. et al. (1994) Hot carriers in multi-quantum-well heterostructures $Ge/Ge_{1-x}Si_x$, in. Extended Abstracts of *1994 Int. Conf. Solid State Devices and Materials,* Japan Society of Applied Physics, Yokohama, 1994, 503-505.

3. Gavrilenko V.I., Korotkov A.L., Krasil'nik Z.F., and .Nikonorov V.V. (1991) Far IR cyclotron resonance and luminescence of hot holes in p-Ge, *Opt. Quantum Electronics,* **23**, .S163-S176.

4. *Opt. Quantum Electronics* (1991) **23,** N2, Special Issue on Far-infrared Semiconductor Lasers.

ENHANCEMENT OF AVERAGE VELOCITY OF HOT CARRIERS IN SAW-TOOTHED HETEROSTRUCTURES

A.B.KOZYREV and V.A. KOZLOV

Institute for Physics of Microstructures of Russian Academy of Sciences
46, Uljanov St., 603600, Nizhny Novgorod, Russia

The response time of many electronic devices is defined mainly by the average electron velocity and by the device dimensions. The ways to reduce the dimensions of electronic devices are well known. We consider some opportunities of the increase of the average velocity of ensemble of electrons in heterostructures (HS). The average velocity of electrons in semiconductors in strong field is saturated due to the scattering with emission of longitudinal optical phonons (LOP). The motion of an electron consists of two stages: acceleration and optical phonon emission. Acceleration of an electron by the electric field gives the increase of energy. Emission of LOP leads to the loss of energy and, hence, to the loss of velocity of electron. Such cyclic motion at low temperatures and strong fields results in the distribution functions of the streaming type [1] and in the occurrence of the inversion [2]. The usage of the peculiarity of these distributions gives an opportunity to increase the average electron velocity at certain conditions. We consider for this purpose the HS with special layers intended for the stimulation of emission of LOP at assigned places. The spatially localized emission of LOP gives an opportunity to change the differential mobility in strong field and enhance the average velocity of hot carriers in the case of vertical transport. It seems that this effect will be useful both for improving of existing devices and for constructing new devices utilizing the effect of enhancement of velocity.

The main idea is to find such potential profile (bottom of the conduction band) that has a long life time with energy a little bit less than LOP energy and after LOP emission it could gain energy very quickly. For this purpose we propose to use saw-toothed HS which potential profile is shown in Fig.1. HS represents itself the pile of thick and thin layers: in the thick layer there is a smooth rise of potential and in the thin layer there is a strict fall of potential. At appropriate applied electric field the potential profile has the form shown in Fig 1. The motion of the electron in the thick layer with high energy of a little bit less than LOP will be approximately free of scattering until electron reaches the thin layer with strong electric field. In the thin layer electron quickly achieves the energy exceeding the energy of LOP and immediately emits the LOP. After the LOP emission electron looses its energy and its momentum becomes equal to zero. But electron is located now in the thin layer of HS

M. Balkanski (ed.), Devices Based on Low-Dimensional Semiconductor Structures, 185–188.

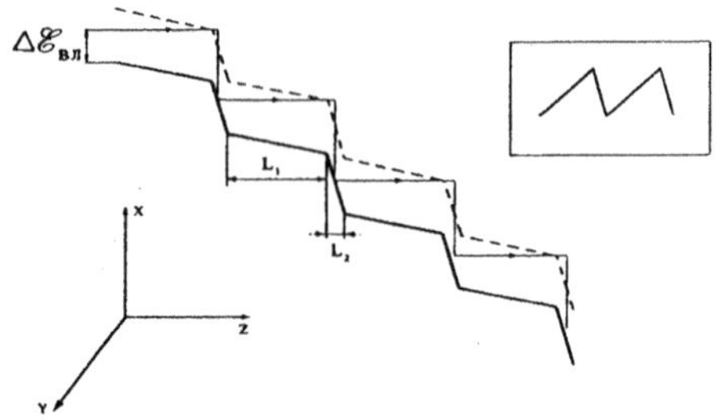

Figure 1. Profile of the conduction band.

with strong field. It means that momentum p_z increases rapidly till electron leaves thin layer and comes to the thick one where electric field is approximately equal to zero so the electron has constant momentum and moves in the vicinity of LOP energy. At the same time the transversal momentum p_t keeps its small value during acceleration so that the momentum p_z in thick layer is of the order of p_0. By that means a specific electron trajectory consistent with the period of HS may arise at certain values of the jump of potential (bottom of the conduction band) and at certain value of applied voltage. In this case the time of flight in the energy range near LOP energy becomes longer in compare with one in the case of the streaming in homogeneous film. Distribution functions of electrons in saw-toothed HS averaged over direction of applied field are shown in Fig.4 and Fig.5. For comparison the distribution functions of electrons in homogeneous semiconductor film are presented in Fig.2 and Fig.3. Simulation was carried out for the following semiconductor parameters: $\theta_D = 430K$ ($\hbar\omega_0 = 5.91\cdot10^{-21}$); $p_0 = 2.68\cdot10^{-26}$ kg·m/sec; $E_0 = 6.51\cdot10^5$ V/m; $m^* = 0.67m_0$; T=77 K. Parameters of HS which were used in simulation are given in Table 1. In the case of homogeneous semiconductor film distribution functions are stretched in the direction of electric field and have usual for streaming conditions [1,2] form. Calculations were carried out for two injectors: solid line - $p_t^2 = p_x^2 + p_y^2 \neq 0$ and dashed line - $p_x^2 + p_y^2 = 0$. One can see from Fig.4 and Fig.5 that distribution functions in the case of saw-toothed HS are located in momentum space near the point $p_z \approx p_0$, $p_t \approx 0$. Such self-consistent motion of electron reduces the acceleration time and gives a possibility to transmit the same distance more quickly than in homogeneous materials. Results of simulation show that the drift velocity can be increased by 1.5 times in comparison with one in homogeneous film at the same value of the average applied field. The average time of flight through HS under consideration has minimum at determined applied voltage. This means that it is possible to achieve NDC at high frequencies in this system due to the bunching effect.

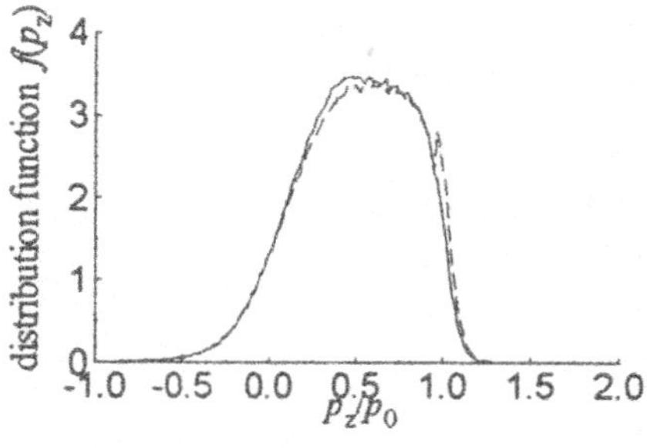

Figure 2. Distribution function $f(p_z) = \int\limits_{p_x, p_y} f(\bar{p}) dp_x dp_y$

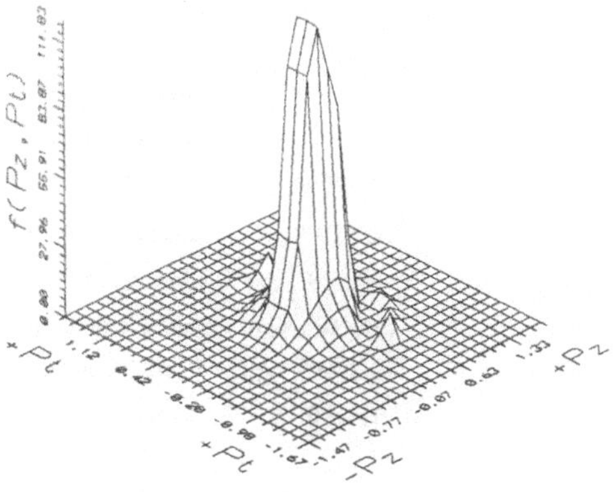

Figure 3. Distribution function

$$f(p_t, p_z) = \int f(\bar{p}) dp_t d\varphi$$

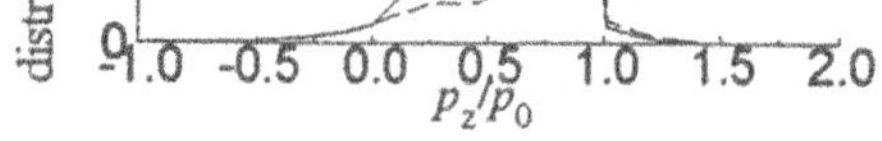

Figure 4. Distribution function $f(p_z) = \int\limits_{p_x, p_y} f(\bar{p}) dp_x dp_y$

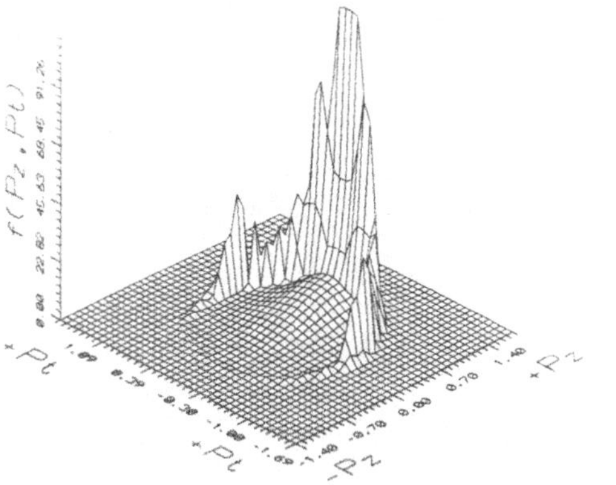

Figure 5. Distribution function

$$f(p_z) = \int\limits_{p_x, p_y} f(\bar{p}) dp_x dp_y$$

TABLE 1. Parameters of HS which were used in simulations

Parameter	Value	Description
L_Σ	$5.66 \cdot 10^{-6}$ m	thickness of homogeneous semiconductor film
E	$2.5 \cdot 10^{5}$ V/m	external electric field applied to homogeneous film
$\Delta\mathcal{E}_{inp}$	$0.9\hbar\omega_0$	energy of electron injected in working space
L_1	$1.1 \cdot 10^{-6}$ m	thickness of thick layer of saw-toothed HS
L_2	$3.4 \cdot 10^{-7}$ m	thickness of thin layer of saw-toothed HS
E_1	$3.3 \cdot 10^{3}$ V/m	electric field in thick layer of saw-toothed HS
E_2	$9.8 \cdot 10^{4}$ V/m	electric field in thin layer of saw-toothed HS
L_Σ	$7.4 \cdot 10^{-6}$ m	total thickness of saw-toothed HS

The research described in this publication was made possible in part by Grant # N0L000 from International Science Foundation and Grant # N0L300 from International Science Foundation and Russian Government.

References

1. Kurosawa T., Maeda H. (1971) Monte Carlo calculation of hot electron phenomena, J.*Phys.Soc.Japan*, **31**, 668 .

2. Andronov A.A., Kozlov V.A. (1973) Low temperature negative differential SHF conductivity at unelastic electron scattering, *ZhETP Lett*. **17**, 124.

PATH-INTEGRAL CALCULATION OF THE ELECTRON DENSITY OF STATES IN MIS-STRUCTURES

G. SLAVCHEVA and I. YANCHEV*
Institute of Biophysics, Bulgarian Academy of Sciences
Acad. G. Bontchev St. bl.23, 1113 Sofia
**Faculty of Physics, Sofia University St. Kliment Ochridski,*
5, James David Bourchier blvd., 1126 Sofia, BULGARIA

1. Introduction

The problem of determining the electron spectrum of disordered structures is currently being pursued with renewed interest owing to its importance from both the technological and fundamental point of view. Semiconductor device characteristics critically depend on the electronic density of states in the band gap tail region. The lack of information near the semiconductor band edges is a well known disadvantage of the experimental methods for obtaining the electronic density of states spectra at the semiconductor-dielectric interface. A number of theoretical approaches have been proposed to solve this problem but the advantage of the path-integral method in real space and time is that it provides an adequate treatment of both the states near the band gap center and the band tail states. This is the reason of choosing the latter for our problem.

Electrons confined to two dimensions at the interface between two materials (e.g. semiconductor-dielectric, which in most of the cases is amorphous) represents an adequate physical model of a quasi 2-D random system. The purpose of the present work is to calculate the density of states of the electrons localized in the 3-D random potential relief generated by charged centers at or near the interface employing the concept of an electron moving in a random system containing dense and weak scatterers [8]. In any treatment of charge fluctuations, their distribution across the interfacial plane plays a role. Further we shall assume the general case of completely random charge distribution at the semiconductor-insulator interface, that is Poisson distribution (the limiting case of dense scatterers would be considered via Gaussian approximation to the Poisson distribution).

M. Balkanski (ed.), Devices Based on Low-Dimensional Semiconductor Structures, 189–198.

2. Theoretical background and formulation of the problem

Our considerations follow the path integral approach of Edwards and Gulyaev [8] to study the disordered systems. The basic model applied to quasi 2-D case considers the electron motion in the field of N scattering centers (ions) confined in area S. The ions are randomly distributed with surface charge density σ = N/S.

We shall be interested in short-range random potential fluctuations near the interface z, ρ « d, where d is dielectric thickness. According to [10] the Green's function or equivalently the potential generated by impurity charged center localized at a point $\mathbf{R}_i$ evaluated at r(τ)=(ρ(τ),z(τ)), at a moment of time τ, has an approximate form:

$$G(\rho(\tau)-\mathbf{R}_i,z(\tau)) \approx \frac{q}{4\pi\varepsilon_0\kappa}\left\{\frac{1}{\sqrt{|\rho(\tau)-\mathbf{R}_i|^2+z^2(\tau)}}-\frac{1}{\sqrt{|\rho(\tau)-\mathbf{R}_i|^2+(2d+z(\tau))^2}}\right\} \tag{2.1}$$

Here $\kappa=\dfrac{k_i+k_{sc}}{2}$ is the mean dielectric constant of the semiconductor and insulator.

The total impurity potential V(**r**(τ)) is given by the superposition principle:

$$V(\mathbf{r}(\tau))=\sum_{i=1}^{N}G(\rho(\tau)-\mathbf{R}_i,z(\tau))=\sum_{i=1}^{N}V(\mathbf{r}(\tau)-\mathbf{R}_i) \tag{2.2}$$

where V(**r**(τ)- $\mathbf{R}_i$) is the potential between the electron and i^{th} ion.

The electron propagator can be written as:

$$\mathrm{G}(\mathbf{r}_2,\mathbf{r}_1,t)=\int \mathcal{D}[\mathbf{r}(t)]e^{iS_l/\hbar} \tag{2.3}$$

where

$$S_l=\int_0^{t}d\tau\left[\frac{m^*}{2}\dot{\rho}^2(\tau)+\frac{m^*}{2}\dot{z}^2(\tau)-\eta V(\mathbf{r}(\tau))\right] \tag{2.4}$$

is the action, V(**r**(τ)) is given by eq. (2.2) and the parameter η measures the strength of the scattering potential. After averaging over all ionic configurations we obtain the average propagator:

$$\overline{G} = \int \mathcal{D}[\mathbf{r}(\tau)] exp\left\{\frac{i}{\hbar}\int_{o}^{t} d\tau \frac{m*}{2}\left[\dot{\rho}^2(\tau) + \dot{z}^2(\tau)\right] + \right.$$
$$\left. + \sigma \int d^2\mathbf{R}\left\{ exp\left\{-\frac{i}{\hbar}\int_{o}^{t} d\tau \eta G(\rho(\tau) - \mathbf{R}, z(\tau))\right\} - 1 \right\}\right\} \tag{2.5}$$

We can expand the exponent in series and define the following higher-order correlators:

$$\sigma\int d^2\mathbf{R} G(\rho(\tau_1) - \mathbf{R}, z(\tau_1)) = \langle V(\mathbf{r}(\tau))\rangle \equiv V_0$$
$$\sigma\int d^2\mathbf{R} G(\rho(\tau_1) - \mathbf{R}, z(\tau_1)) G(\rho(\tau_2) - \mathbf{R}, z(\tau_2)) = W^{(2)}(\rho(\tau_1) - \rho(\tau_2), z(\tau_1), z(\tau_2))$$

$$\sigma\int d^2\mathbf{R} G(\rho(\tau_1) - \mathbf{R}, z(\tau_1)) G(\rho(\tau_2) - \mathbf{R}, z(\tau_2)) \ldots G(\rho(\tau_n) - \mathbf{R}, z(\tau_n)) =$$
$$= W^{(n)}(\rho(\tau_1), \rho(\tau_2), \ldots, \rho(\tau_n), z(\tau_1), z(\tau_2), \ldots, z(\tau_n)) \tag{2.6}$$

where V_o is the mean potential of the system ($E_o = qV_o$ is the mean potential energy of the system).

Hence,

$$\overline{G} = \int \mathcal{D}[\mathbf{r}(\tau)] exp\left\{\frac{i}{\hbar}\int_{0}^{t} d\tau \frac{m*}{2}\left[\dot{\rho}^2(\tau) + \dot{z}^2(\tau)\right] - E_0 t - \right.$$
$$-\frac{\eta^2}{2\hbar^2}\int_0^t d\tau_1 \int_0^t d\tau_2 W^{(2)}(\rho(\tau_1) - \rho(\tau_2), z(\tau_1), z(\tau_2)) + \ldots +$$
$$\left. + \frac{(-1)^n}{n!}\left(\frac{i}{\hbar}\right)^n \eta^n \int_0^t\int_0^t \ldots \int_0^t d\tau_1 \ldots d\tau_n W^{(n)}(\rho(\tau_1), \ldots, \rho(\tau_n), z(\tau_1), \ldots, z(\tau_n)) + \ldots \right\} \tag{2.7}$$

As can be seen the problem of electron motion near the disordered interface could be formulated as path-integration of the non-local action functional. The origin of non-locality is a direct consequence of averaging over all ionic configurations. It is obvious that the form of the correlation functions depends on the electron-ion potential. The electronic density of states is related to the average electron propagator by the relation:

$$\rho(E, z) = \frac{1}{\pi\hbar} \mathrm{Re}\int_0^\infty dt e^{\frac{i}{\hbar}Et} \langle G(\mathbf{r}_2, \mathbf{r}_1, t)\rangle_V = \frac{1}{2\pi\hbar}\int_0^\infty dt exp(iEt/\hbar) Tr\langle G(\mathbf{r}_2, \mathbf{r}_1, t)\rangle \tag{2.8}$$

Here the average over potential realisations propagator is given by eq. (2.7)

3. Statistical considerations and density of states calculation

In the problem we are studying, the nth-order correlator of the random potential is given by:

$$W^{(n)}(\rho(\tau_1),\dots,\rho(\tau_n),z(\tau_1),\dots,z(\tau_n)) = \langle \delta\varphi(\rho(\tau_1),z(\tau_1))\dots\delta\varphi(\rho(\tau_n),z(\tau_n))\rangle \tag{3.1}$$

where $\langle\ \rangle$ stands for averaging over all possible ion configurations and:

$$\delta\varphi(\rho,z) = -\int d^2k_{\|}e^{ik_{\|}\cdot\rho}\int_{-d}^{\infty} dz_o G(k_{\|},z,z_o)\delta\Sigma(k_{\|},z_o) \tag{3.2}$$

is the potential fluctuation-solution of the linearized 3-D Poisson equation [2]. The Fourier transform of the interface charge density fluctuation about its mean value is denoted by $\delta\Sigma(k_{\|},z(\tau))$. The Green's function is given by the approximate expression obtained using the parallel plate capacitor approximation [2]:

$$G(k_{\|},z,z_o) = -\frac{q}{4\pi\varepsilon_o}\frac{1}{k_{\|}}\frac{sh[k_{\|}(z-w)]sh[k_{\|}(z_o+d)]}{D(k_{\|})}, \quad \text{for } z_0 < 0 < z \tag{3.3}$$

with $D(k_{\|})$ given by:

$$D(k_{\|}) = k_i sh(k_{\|}w)ch(k_{\|}d) + k_{sc}ch(k_{\|}w)sh(k_{\|}d) + \frac{(C_{ss}+C_I)}{\varepsilon_o}\frac{1}{k_{\|}}sh(k_{\|}w)ch(k_{\|}d) \tag{3.4}$$

Here $k_{\|}$ is the in-plane 2-D wave vector, k_i and k_{sc} are respectively insulator and semiconductor dielectric constants, C_{ss},C_I are respectively the surface states and inversion layer sheet capacitances. As has been pointed out in [2] the binary correlation function in the case of Poisson distribution in the interfacial plane and density $\bar{\rho}(z_o)$ normal to the interface is given by:

$$\langle\delta\Sigma(\rho_o,z_o)\delta\Sigma(\rho'_o,z'_o)\rangle = \bar{\rho}(z_o)\delta(z_o-z'_o)\delta(\rho_o-\rho'_o) \tag{3.5}$$

Let us consider the surface charge density correlator satisfying the following equation for Poisson distribution in the interfacial plane [1]:

$$\langle\delta\Sigma(\rho_o)\delta\Sigma(\rho'_o)\rangle = \sigma\delta(\rho_o-\rho'_o) \tag{3.5a}$$

The characteristic functional of the Poisson charge distribution $\delta\Sigma(\rho)$ is Gaussian and is given by [3]:

$$\Phi[\delta\Sigma(\rho)] = \exp\left[-\frac{D}{2}\iint\delta\Sigma(\rho)\delta\Sigma(\rho')\delta(\rho-\rho')d\rho d\rho'\right] \tag{3.6}$$

where D=const.=σ and the delta-kernel $\delta(\rho-\rho')$ is the correlation function of the Poisson distribution.

We can obtain all moments of the surface charge distribution calculating subsequent functional derivatives of the Gaussian characteristic functional. Since we can set $\delta\Sigma(\rho)=0$ we obtain that all odd-order moments of the surface charge distribution will be equal to zero and therefore only the even-order moments rest. By mathematical induction the general expression for the 2nth correlator has been derived which reads:

$$\langle \delta\Sigma(\rho_1)\delta\Sigma(\rho_2)\ldots\delta\Sigma(\rho_{2n})\rangle = -\frac{\delta^{2n}\Phi}{\delta\Sigma(\rho_1)\delta\Sigma(\rho_2)\ldots\delta\Sigma(\rho_{2n})} =$$

$$= \frac{D^n}{2^n n!} \sum_{P\{k_1,k_2,\ldots,k_{2n}\}} \prod_{i=1}^{n-1} \delta\left(\rho_{k_{2i+1}} - \rho_{k_{2i+2}}\right) \tag{3.7}$$

where $\sum_{P\{k_1,k_2,\ldots,k_{2n}\}}$ stands for summing over all possible permutations of the numbers $\{k_1, k_2,\ldots,k_{2n}\}=\{1,2,\ldots,2n\}$

In order to obtain the 2nth order potential correlation function we substitute the Fourier transform $\delta\Sigma(k_{||})$ in eq. (3.1) and perform subsequently the integration with δ-functions over ρ_j and over $k_{||}$ - vector. A general conclusion of the above calculations is the statement that the 2n-order correlation function of the random potential can be splitted in a product of pair correlators. Therefore the general problem of 2n-order correlator calculation is reduced to the more simple one of a binary correlator calculation. It is useful to introduce the Fourier-transform of the binary correlation function:

$$\langle V(k)V(k')\rangle = \int_{-\infty}^{\infty} d(z\pm z')e^{-ik_z(z\pm z')} G(k_{|},z,0)G(k_{|},z',0) \tag{3.8}$$

where the Green's function is given by eq. (3.3) and (3.4) and has a symmetric property $G(-k_{||},z)=G(k_{||},z)$. We have assumed that all the interface charge is distributed across the interfacial plane i.e. $\overline{\rho}(z_o) = \sigma\delta(z_o)$ where σ is the surface charge density and thus the z-dependence of the charge distribution can be neglected. The rigorous derivation of the Fourier-transform of the binary correlator is given in [12]. It reads:

$$K^{(2)}(k_{|},k_z) = \left(\frac{q}{4\pi\varepsilon_0}\right)^2 \frac{\sigma}{4\kappa^2}\left(\frac{1}{w}+\frac{1}{d}\right)\frac{1}{\left(k_{|}^2+Q^2\right)\left(k_{|}^2+k_z^2\right)} \tag{3.9}$$

$$\alpha = \left[\frac{(k_i + k_{sc})\varepsilon_0}{(C_{sc}+C_i+C_{ss}+C_I)}\right]^2 \tag{3.10}$$

where we have defined inverse screening radius $Q^2=1/\alpha$ and $C_i = \frac{\varepsilon_o k_i}{d}, C_{sc} = \frac{\varepsilon_o k_{sc}}{w}, C_{ss} \approx qN_{ss}$ are respectively the dielectric capacitance, semiconductor capacitance and the capacitance corresponding to the surface states density N_{ss}.

Let us introduce a non-local harmonic trial action $S_o(\omega_\parallel, \omega_\perp)$, depending on the oscillator frequency in a plane parallel to the interfacial plane $\omega_\parallel$ and on the frequency of oscillations in direction perpendicular to the interfacial plane $\omega_\perp$, thereby accounting for the anisotropy of the problem. We shall be concentrated on the averaging of the propagator $G(\mathbf{r}_2, \mathbf{r}_1 ; t)$ over the trial action. Let us define the non-local harmonic oscillator propagator

$$G_o(\mathbf{r}_2,\mathbf{r}_1;t,\omega_\parallel,\omega_\perp) = \int \mathcal{D}[\mathbf{r}(\tau)] \exp\left[\frac{i}{\hbar} S_o(\omega_\parallel,\omega_\perp)\right] \tag{3.11}$$

and the average propagator:

$$G(\mathbf{r}_2,\mathbf{r}_1;t) = G_o(\mathbf{r}_2,\mathbf{r}_1;t,\omega_\parallel,\omega_\perp)\left\langle \exp\left[\frac{i}{\hbar}(S - S_o(\omega_\parallel,\omega_\perp))\right]\right\rangle_{S_o(\omega_\parallel,\omega_\perp)} \tag{3.12}$$

If we limit ourselves with the first term in a cumulant expansion, we get:

$$G_1(\mathbf{r}_2,\mathbf{r}_1;t,\omega_\parallel,\omega_\perp) = G_o(\mathbf{r}_2,\mathbf{r}_1;t,\omega_\parallel,\omega_\perp)\left\langle \exp\left[\frac{i}{\hbar}(S - S_o(\omega_\parallel,\omega_\perp))\right]\right\rangle_{S_o(\omega_\parallel,\omega_\perp)} \tag{3.13}$$

Therefore we need to calculate averages of the form $\left\langle S - S_o\left(\omega_\parallel,\omega_\perp\right)\right\rangle_{S_o\left(\omega_\parallel,\omega_\perp\right)}$ and the propagator $G_o(\mathbf{r}_2,\mathbf{r}_1;t,\omega_\parallel,\omega_\perp)$.

Consider the task of averaging $\langle S_l\rangle_{S_o}$. Taking into account (eq. (3.1) - (3.7)) we get the following expression for the $2n^{th}$ term of the sum:

$$\left\langle S_l\right\rangle^{(2n)}_{S_o\left(\omega_\parallel,\omega_\perp\right)} = \frac{1}{(2n)!}\left(\frac{i}{\hbar}\right)^{2n-1} \prod_{i=1}^{2n} \frac{(q\sigma)^n}{2^n n!} \int_0^t d\tau_i \sum_{P\{k_1,k_2,\dots,k_{2n}\}} \prod_{j=1}^{2n-1} *$$

$$* \int d^2 k_\parallel^{k_j} \left\langle exp\left\{ i k_\parallel^{k_j}\left[\rho\left(\tau_{k_j}\right) - \rho\left(\tau_{k_{j+1}}\right)\right]\right\}\right\rangle_{S_o\left(\omega_\parallel\right)} *$$

$$* \int dk_z^{k_j} \left\langle exp\left\{ i k_\parallel^{k_j}\left[z\left(\tau_{k_j}\right) - z\left(\tau_{k_{j+1}}\right)\right]\right\}\right\rangle_{S_o\left(\omega_\perp\right)} K^{(2)}\left(k_\parallel^{k_j}, k_z^{k_j}\right) \tag{3.14}$$

Here we have splitted the total action $S_o(\omega_\parallel, \omega_\perp)$ into two actions $S_o(\omega_\parallel)$ and $S_o(\omega_\perp)$. since the following additive expression for the anisotropic trial action has been obtained:

$$S_o\left(\omega_\parallel,\omega_\perp\right) = \int_o^t d\tau \frac{m*}{2}\left[\dot\rho^2(\tau) + \dot z^2(\tau) - \frac{\omega_\parallel^2}{2t}\int_o^t d\sigma\left|\rho(\tau) - \rho(\sigma)\right|^2 - \right.$$

$$\left. - \frac{\omega_\perp^2}{2t}\int_o^t d\sigma\left|z(\tau) - z(\sigma)\right|^2 - \frac{q^2 E^2}{m*^2\,\omega_\perp^2}\right] \tag{3.15}$$

where E is the external uniform electric field. The anisotropic oscillator propagator $G_o(\mathbf{r}_2,\mathbf{r}_1;t,\omega_\parallel,\omega_\perp)$ is obtained using the variational principle imposed over quadratic classical action:

$$S_o'(\omega_\parallel,\omega_\perp)=S_o(\omega_\parallel,\omega_\perp)+\int_o^{t_o} d\tau\mathbf{f}(\tau).\mathbf{r}(\tau) \tag{3.16}$$

where $\mathbf{f}(\tau)=(\mathbf{f}_{xy}(\tau),\mathbf{f}_z(\tau))$ is an external force term and setting $\mathbf{f}(\tau)=0$. Finally we obtain:

$$G_o(\mathbf{r}_2,\mathbf{r}_1;t,\omega_\parallel,\omega_\perp)=\left(\frac{m^*}{2\pi i\hbar t}\right)^{3/2}\left(\frac{\omega_\parallel t}{2sin\left(\frac{\omega_\parallel t}{2}\right)}\right)^2\left(\frac{\omega_\perp t}{2sin\left(\frac{\omega_\perp t}{2}\right)}\right)*$$

$$exp\left[\frac{i}{\hbar}\left(\frac{m^*\omega_\parallel}{4}cotg\frac{\omega_\parallel t}{2}|\rho_2-\rho_1|^2+\frac{m^*\omega_\perp}{4}cotg\frac{\omega_\perp t}{2}|z_2-z_1|^2\right)\right] \tag{3.17}$$

After computation of the averages taking part in eq. (3.14) up to the second cumulant and performing the 3-D k-integration we obtain the average $\langle S_l\rangle_{S_o}$ Thus substituting G_o and $\langle S-S_o\rangle_{S_o}$ in eq.(3.13) we obtain the expression for the first cumulant of the average propagator G_1.

We shall be interested in the asymptotic behaviour of the density of electronic states, which is given by eq. (2.8). For the sake of simplicity we shall consider the ground state contribution to the density of states [4], i.e. $t\to\infty$ (or equivalently, $\omega_\parallel\to 0,\omega_\perp\to 0$). Thus we obtain:

$$\rho_1^j(E)=\frac{1}{2\pi\hbar}\int_{-\infty}^{\infty}dt\left(\frac{m^*}{2\pi i\hbar t}\right)^{3/2}(i\omega_\parallel t)^2(i\omega_\perp t)*$$

$$*exp\left\{-\frac{3}{4}i\omega_\parallel t-\frac{3}{4}i\omega_\perp t-\frac{i}{\hbar}(E_0-E)-\frac{t^2}{2\hbar^2}\frac{\xi}{\sqrt{\pi}}\int_0^\infty\int_0^\infty\frac{dydy'e^{-Q^2y'}}{(y+y'+A)\sqrt{y+C}}\right\} \tag{3.18}$$

Here after the limiting procedure:

$$A=\frac{\hbar}{2m^*\omega_\parallel},C=\frac{\hbar}{2m^*\omega_\perp},\ \xi=\left(\frac{q}{4\pi\varepsilon_0}\right)^2\frac{\sigma}{4\kappa^2}\left(\frac{1}{w}+\frac{1}{d}\right).$$

The integration over t can be performed exactly [5], yielding:

$$\rho_1(E)=\frac{1}{2\pi\hbar}\left(\frac{m^*}{2\pi\hbar}\right)^{3/2}\omega_\parallel^2\omega_\perp 2^{-3/4}\sqrt{\pi}\,\beta^{-5/2}exp\left[-q^2/8\beta^2\right]D_{3/2}\left(q/\beta\sqrt{2}\right) \tag{3.19}$$

Here we have defined:

$$\beta^2 = \frac{1}{2\hbar^2}\frac{\xi}{2\sqrt{\pi}}\int_0^\infty\int_0^\infty dy dy' e^{-Q^2 y'} j^{-1/2}(y,y')$$

$$j(y,y') = (A+y+y')^2(C+y) \tag{3.20}$$

$q = \frac{1}{\hbar}\left[\frac{3}{4}\left(E_{\omega_1}+E_{\omega_\perp}\right)+E_o - E\right]$, $E_\omega = E_{\omega_1}+E_{\omega_\perp}$ is the total oscillator energy.

Let us introduce the screening energy $E_Q = \frac{\hbar^2 Q^2}{2m^*}$ and $z_1 = \frac{E_Q}{E_{\omega_1}}, z_2 = \frac{E_Q}{E_{\omega_\perp}}$ are the normalized energies. Consider now the density of states deep in the band tail. According to Halperin and Lax [7] the low-energy tail can be reached in two equivalent ways, namely, by letting E→+ ∞ (q→+ ∞) or $q/\beta\sqrt{2} \to \infty$, or keeping E=const. and reducing the magnitude of the potential fluctuation (ξ→0).

$$\rho_1(E) = \left[\frac{(E_Q \cdot Q)^3}{\xi_Q^2}\right] a(v,z_1,z_2) exp\left[-\frac{E_Q^2}{2\xi_Q} b(v,z_1,z_2)\right] \tag{3.21}$$

where we have denoted $\xi_Q = \frac{\xi}{Q}$ which has dimension of the energy square, $v = \frac{E_o - E}{E_Q}$.

Let us consider the asymptotic behaviour for the case of strong screening, i.e. Q→∞ (limiting the expansion of the integrant up to second-order term). Then $a(v,z_1,z_2)$ and $b(v,z_1,z_2)$ are given correspondingly by:

$$a(v,z_1,z_2) = \frac{1}{\pi 2^{3/4} 2^9} \frac{\left[\frac{3}{4}\left(\frac{1}{z_1}+\frac{1}{z_2}\right)+v\right]^{3/2}}{z_1 z_2^2\left[\sqrt{z_2} - \frac{(3z_2 - z_1)e^{-(z_2-z_1)}}{3z_2^2}\right]^2} \tag{3.22a}$$

$$b(v,z_1,z_2) = \frac{3\sqrt{\pi}\left[\frac{3}{4}\left(\frac{1}{z_1}+\frac{1}{z_2}\right)+v\right]^2 z_2^2 e^{z_2 - z_1}}{\sqrt{z_2}(3z_2 - 1)} \tag{3.22b}$$

We have introduced parameters $\omega_{\|}$ and $\omega_{\perp}$ in the trial action $S_o(\omega_{\|},\omega_{\perp})$ which have not yet been determined up to this moment. Following Halperin and Lax [7] we should choose z_1, z_2 or z so as to maximize $\rho_1(E)$. We shall limit our considerations to the most simple case of $\omega_{\|}=\omega_{\perp}=\omega$ in order to obtain the asymptotic behaviour of the electron density of states.

$$a(\nu,z)=\frac{1}{\pi 2^{3/4}2^5}\frac{9\left[\frac{3}{4z}+\nu\right]^{3/2}}{(3z-1)^2} \tag{3.23a}$$

$$b(\nu,z)=\frac{3\sqrt{\pi}\left(\frac{3}{4z}+\nu\right)^2 z^{3/2}}{(3z-1)} \tag{3.23b}$$

The best choice of z is found to satisfy the equation:

$$z^2-\left(1+\frac{9}{4\nu}\right)z+\frac{1}{4\nu}=0 \tag{3.24}$$

Keeping the only physical solution $z\approx\frac{9}{4\nu}$ and substituting it in eq. (3.23a) and (3.23b) we obtain:

$$a(\nu)\approx\frac{2^{5/4}}{\pi 3^{11/2}}\nu^{7/2}\approx 1.8.10^{-3}\nu^{7/2}, \qquad b(\nu)\approx\frac{2\sqrt{\pi}}{3}\nu^{3/2}\approx 1.18\nu^{3/2} \tag{3.25}$$

It should be noted that a similar behaviour of the density of states deep in the tail has been obtained for the case of heavily doped strongly compensated semiconductors with correlated impurity distribution potential [11] where the electron screening is strong and large-scale potential fluctuations take place.

4. Conclusions

We have applied the Feynman path integral method to calculate the density of states for an electron moving in a random potential generated by the charged centers localized at or near the semiconductor-dielectric interface. Taking into account the statistical properties of the surface charge distribution we found the density of states deep in the tail in the first cumulant approximation. Our considerations are generalized to account for all higher-order correlators of the random potential. We found that the density of states deep in the tail for the specific case of equal oscillator frequencies and strong screening decays proportionally to $\exp(-\alpha E^{3/2})$ where the electron energy is measured from the mean potential energy.

It should be noted the similarity of the asymptotic behaviour of the density of states deep in the band tail to that obtained for heavily doped strongly compensated semiconductors with correlated impurity distribution potential [11], in the case of equal frequencies of the harmonic oscillator in the interfacial plane and in direction perpendicular to it.

References:

1. Brews, J. R. and Nicollian, E. H. (1982) *MOS Physics and Thechnology*, John Wiley, New York, pp. 847- 850.
2. Brews, J.R. (1972) Surface potential Fluctuations Generated by Interface Charge Inhomogeneities in MOS Devices, *J. Appl. Phys.* **43,** No. 5 .
3. Feynman, R.P. and Hibbs, A. R.(1965) *Quantum Mechanics and Path Integrals*, McGraw-Hill, New York.
4. Samathiyakanit, V., J.(1974) Path-integratheory of a model disordered system, *Phys. C: Solid State Physics* 7, .2873-74
5. Sa-yakanit, V.(1979) Electron density of states in a Gaussian random potential: Path-integral approach, *Physical Review B*. **19,** No.4, .2268.
6. Halperin, B.I. and Lax, M. (1966) *Phys. Rev* **148,** 722.
7. Halperin, B.I. and Lax, M. (1967) *Phys. Rev.* **153,** 802.
8. Edwards, S.F. and Gulyaev, V. B.(1964) *Proc. Phys. Soc.* **83,**.445-6.
9. Khandekar, D.C. and Lawande, S.V.(1986) Feynman Path Integrals: Some Exact Results and Applications, *Physics Reports (Review Section of Physics Letters)* **137**, No.2 & 3,115-229.
10. Gergel', V.A. and Suris, R.A. (1983) *JETP*. **84,** 727.
11. Yanchev, I. Y., Koinov, Z.G. and Petkova, .M.(1981) *Philosophical Magazine B* **44,** No.2, 307-316.
12. Slavcheva, G. and Yanchev, I. (1991) Potential Fluctuations Due to the Randomly Distributed Charges at the Semiconductor-Dielectric Interfaces in MIS-structures, *Solid State Communications*, **.79,** No.5, 439-441.
13. Janke E., Emde F. and Lösch F. (1960) *Tafeln Höherer Funktionen,* B.G. Teubner Verlagsgesellshaft, Stuttgart.
14. Prudnikov, A.P., Brichkov, Ju.A. and Marichev, O.I. (1981) *Integrals and Series*, Nauka, Moscow.
15. Prudnikov, A.P., Brichkov, Ju.A. and Marichev, O.I. (1983) *Integrals and Series (Special functions),* Nauka, Moscow.

PHYSICS AND APPLICATIONS OF EXCITON SATURATION IN MQW STRUCTURES

A. MILLER, A. R. CAMERON and P. RIBLET
Department of Physics and Astronomy
University of St Andrews,
North Haugh, St Andrews,
Fife, KY16 9SS, Scotland

1. Introduction

The passage of laser light through a semiconductor can produce a rich variety of nonlinear optical phenomena [1-5]. The most sensitive effects occur when the photon energy is close to resonance with the band gap energy of the semiconductor. This has provided a fertile area of research for both new physics and device applications over the past 15 years.

In low dimensional semiconductor structures such as multiple quantum wells (MQW), excitonic resonances are especially important. Device applications currently include laser mode-locking elements, waveguide routing switches, optical soliton control and logic arrays for communications, signal processing and computing. Here we are primarily interested in interactions between laser light and low dimensional semiconductors which cause changes of absorption coefficient and refractive index as a function of excess carrier density. Pump-probe and transient grating techniques employing mode-locked lasers can be used to monitor the dynamics of the carriers on ultrashort timescales via these nonlinearities. In this chapter, we restrict our discussion to room temperature nonlinear optical phenomena which are useful for devices. A new method of measuring in-well electron mobilities using transient gratings is described and some device applications are reviewed.

2. Optical nonlinearities in semiconductors

The optical generation of excess carrier populations in semiconductors produces changes in the real and imaginary parts of the susceptibility. Absorption and

M. Balkanski (ed.), Devices Based on Low-Dimensional Semiconductor Structures, 199–225.

refraction are linked by causality via Kramers-Kronig relations. The familiar relations used in linear optics may be modified [4,5] to account for optical nonlinearites when laser excitation of a quasi-equilibrium of electrons and holes results in renormalized material parameters. Thus, refractive index changes, Δn, and absorption changes, $\Delta\alpha$, may be connected by the integral expression,

$$\Delta n(\omega) = \frac{c}{\pi} \mathcal{P} \int_0^\infty \frac{\Delta\alpha(\Omega) d\Omega}{\Omega^2 - \omega^2} \tag{1}$$

where $\mathcal{P}$ denotes the principal part of the integral. Note that *changes* in refractive index at a given frequency, ω, are induced by *changes* in absorption at *all* other frequencies, Ω.

The optical generation of free electrons and holes is essentially instantaneous, but the decay of the excess carrier population depends on the recombination rate, the diffusion coefficient if a concentration gradient exists, or the sweep-out rate in the presence of an electric field. For instance, the excess carrier density as a function of position and time, $N(x,t)$, after excitation with an ultrashort pulse can be determined from the continuity equation (given here for one-dimensional motion in the absence of an electric field),

$$\frac{\delta N(x,t)}{\delta t} = -\frac{N(x,t)}{\tau_R} + D_a \nabla^2 N(x,t) \tag{2}$$

where D_a is the ambipolar diffusion coefficient for electrons and holes and τ_R is the recombination time. A useful parameter is the diffusion length, i.e. the average distance a free carrier moves before recombining, defined by,

$$L_D = \sqrt{D_a \tau_R} \tag{3}$$

The time for recovery to the equilibrium condition ranges from microseconds to femtoseconds depending on the precise conditions.

3. Optical nonlinearities in quantum wells

Strong room temperature excitonic absorption features produce unique nonlinear optical properties in quantum well semiconductors [6,7]. Excitons are more clearly resolved in quantum wells compared to bulk semiconductors because confinement leads to larger binding energies and enhanced oscillator strengths. Exciton saturation provides sensitive absorptive and refractive optical nonlinearities in low dimensional structures. These nonlinearities can be used to time resolve carrier dynamics and transport phenomena in quantum wells.

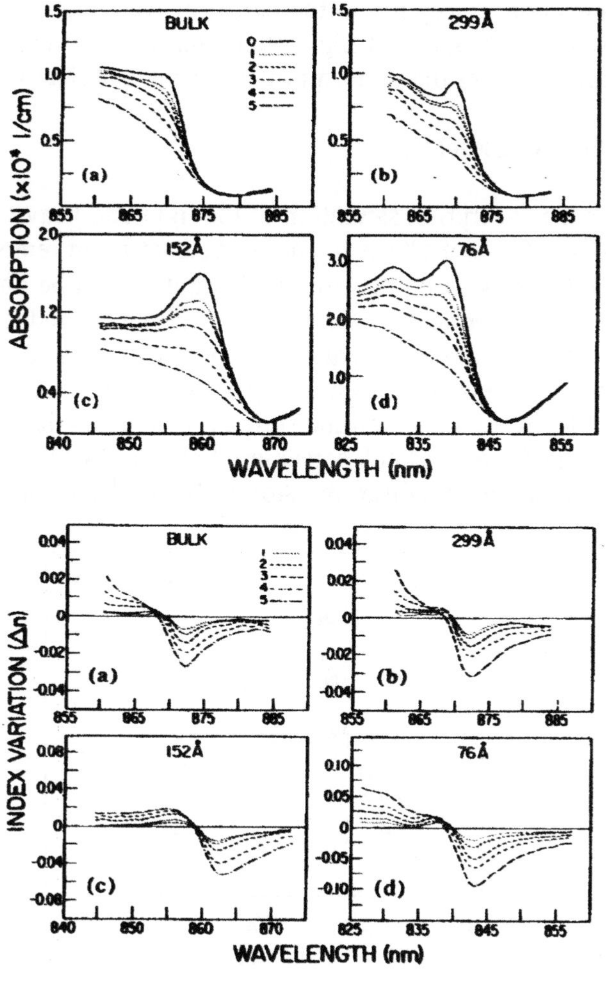

Figure 1. The top set of figures shows absorption changes in bulk GaAs and GaAs/AlGaAs MQWs with three different well widths at increasing optical excitation. The lower set gives the predicted refractive index changes from a Kramers-Kronig calculation. Pump intensities are (0) 0, (1) 670, (2) 1270, (3) 2650, (4) 5400 and (5) 11700 W/cm^2 (Park et al [8]).

Absorption and refractive index changes at the band edges of bulk GaAs and GaAs/AlGaAs MQWs as the carrier density is increased by optical excitation are shown in figure 1. At low carrier densities, the small exciton feature in bulk GaAs becomes more pronounced and splits into a doublet due to the loss of Brillouin zone centre degeneracy of the light and heavy hole valence bands. The enhanced exciton absorption in the narrower wells gives larger absorption and refractive index changes. At high excitation, band filling provides further absorption saturation after the exciton features have completely saturated.

3.1. EXCITON SATURATION

Figure 2 shows the absorption saturation of the light and heavy hole excitons in 6nm wide GaAs/AlGaAs multiple quantum wells at room temperature, using a tunable cw dye laser focussed to ~30μm. Saturation can be observed to set in at well below 1mW of optical power. Accompanying the reduction of the peak height, is an increasing absorption at lower energies. This can result from two causes (i) band gap renormalisation and (ii) lifetime broadening of the heavy hole exciton. The exciton should blue shift as it saturates due to a reduction of the binding energy from Coulomb screening, however this is cancelled by a red shift due to band-gap renormalisation. Above 3mW, the absorption becomes

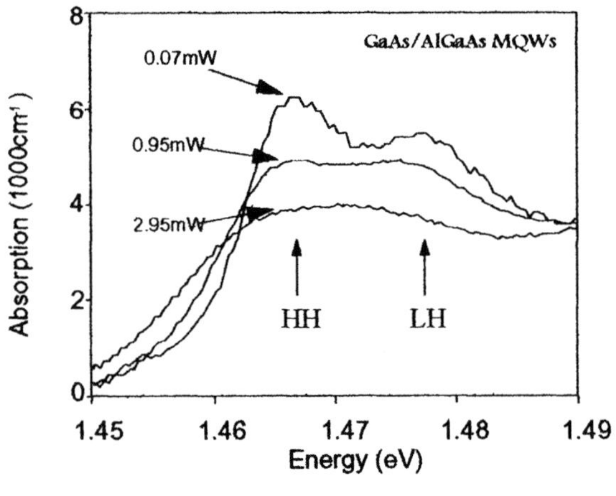

Figure 2. Exciton absorption at low powers in room temperature GaAs/AlGaAs multiple quantum wells.

progressively more difficult to saturate. This is because band-gap renormalisation produces a progressively larger band density of states to saturate. At higher powers and thus higher carrier densities, band filling causes the remaining absorption to saturate.

There are several possible mechanisms which can cause exciton saturation. When the density of excitons approaches that required to fill (real) space (~10^{17} cm^{-3} based on the exciton diameter in GaAs), then the number of additional excitons which can be created is reduced by Pauli exclusion, thus decreasing the level of absorption. However, Knox et al [9] measured a room temperature ionisation time for excitons into free electron-hole pairs of 300fs. Therefore, on picosecond and longer timescales, it can be assumed that the exciton saturation results from free carrier screening.

Saturation of exciton absorption by free carriers, can arise from phase-space filling and Coulomb screening. Lifetime broadening can also reduce the peak height of the exciton absorption feature. It has been shown that although long range Coulomb screening dominates in 3-D semiconductors, this is not the case in lower dimensional semiconductors. Phase-space filling and short range Coulomb effects are more apparent. Phase-space filling arises from Pauli exclusion. At high excess carriers densities, carriers fill the lower band states at low wavevector, k. These are the band states from which the excitons derive, so that band filling of these states will rule out the creation of the excitons associated with them. Exchange screening results from Pauli exclusion forcing the electrons further apart than would be calculated classically.

3.2. TRANSIENT AMPLITUDE GRATINGS

Four-wave mixing provides a useful means of determining the magnitude of the nonlinear coefficients, carrier motion and nonlinear response times. Time resolved experiments can be performed using three input pulses whereby two 'excite' pulses with the same linear polarisation overlapping at the sample in both space and time, create a temporary, spatial modulation of the carriers, figure 3. A time delayed probe pulse can then be diffracted from this grating because of the nonlinearity produced by the carriers.

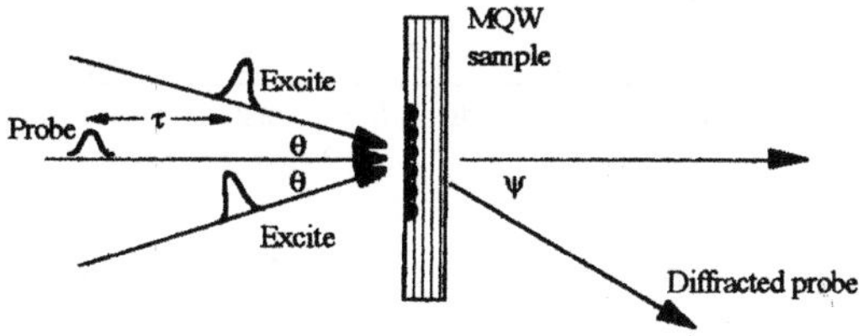

Figure 3. Transient grating configuration

Figure 4 shows the results of transient grating measurements as a function of wavelength at a fixed probe delay for 6.5nm GaAs/AlGaAs MQWs. The maximum diffraction efficiency is noted to occur at a wavelength of 829nm where there is no transmission change, indicating that the grating is primarily refractive rather than absorptive.

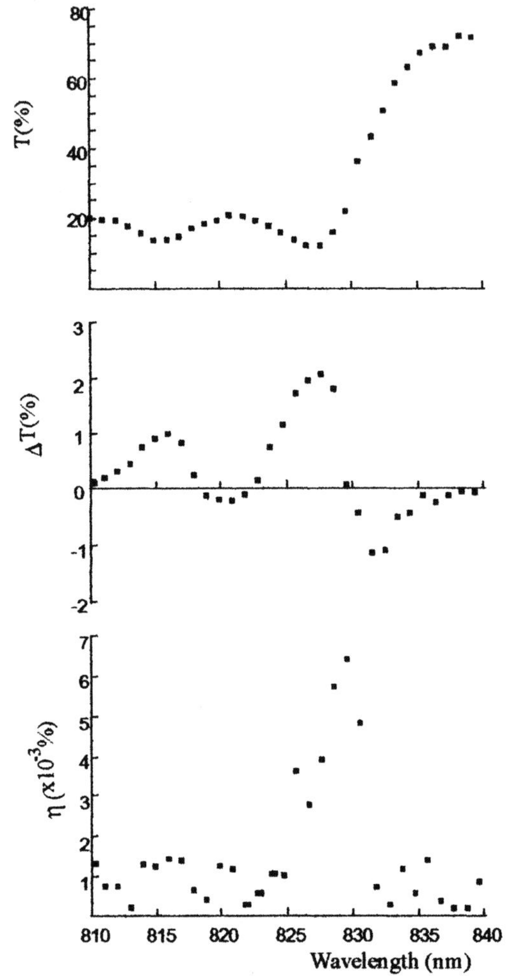

Figure 4. (a) Transmission of GaAs/AlGaAs MQW as a function of wavelength, (b) differential transmission in a pump-probe measurement with fixed delay of 15ps and (c) diffraction efficiency in three-beam transient grating measurement [10].

Figure 5(a) shows the diffraction efficiency as a function of wavelength at an average power of 2mW per excite arm. Diffraction due to the absorption saturation of both light and heavy hole excitons are clearly observed with a maximum diffraction efficiency approaching 10^{-3}. The spot size on the sample was $\omega_0 = 25\mu m$ which implies a refractive index change Δn ~0.05 at 829.5nm. Figure 5(b) shows the power dependence of the diffraction efficiency. As expected from theory, at low pump powers, this is quadratic at the peak (heavy hole) of the spectral diffraction efficiency. Up to about 300μW (~30W/cm^2) the refractive index change per electron-hole pair at 829.5nm can be deduced to be

(a)

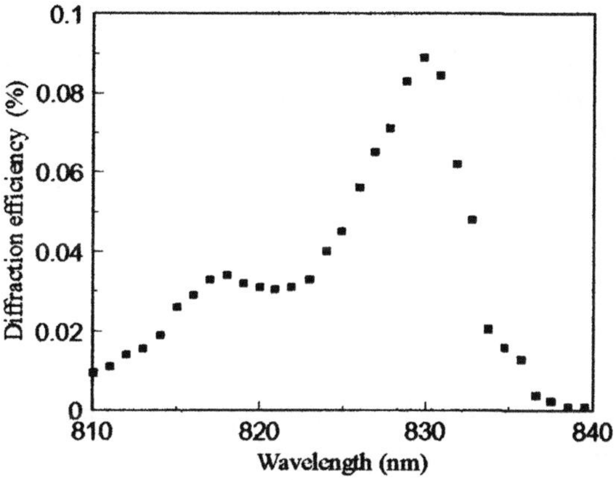

(b)

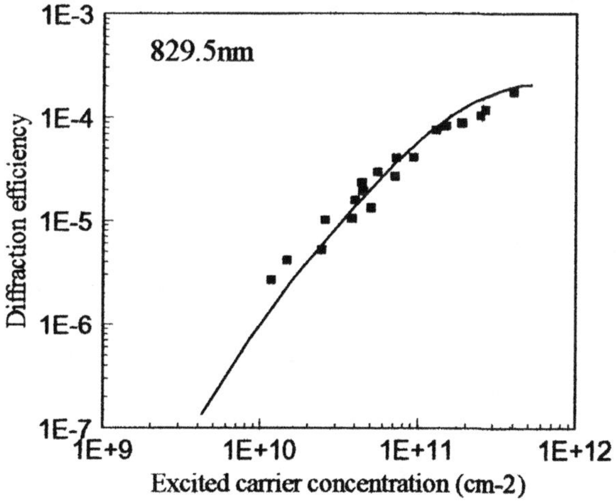

Figure 5. Power dependence of diffracted signal from a transient grating in GaAs/AlGaAs MQWs (a) as a function of wavelength with 2mW average power per pump beam and (b) as a function of carrier density. The solid curve is a theoretical calculation [10].

$n_{eh} \sim 0.8 \times 10^{-19}$ cm^3 from these results. The solid curve in figure 5(b) is a theoretical calculation that takes account of screening and phase space filling which shows a change in slope above 300μW, i.e. a carrier density of ~1.3×10^{11} cm^{-2} (~2×10^{17} cm^{-3}). At this point, the index change Δn ~0.016 after which the exciton saturation limits and the major contribution to the nonlinear refraction comes from band filling [10].

3.3. IN-WELL TRANSPORT

Information about the in-well motion of carriers can be obtained by measuring the decay of the diffracted probe signal as a function of time. The grating decay rate, Γ, is given by,

$$\Gamma = \frac{1}{\tau_D} + \frac{1}{\tau_R} = \frac{4\pi^2 D_a}{\Lambda^2} + \frac{1}{\tau_R} \tag{4}$$

The diffracted signal decays at twice this rate since the diffraction efficiency depends on the square of the refractive index change (or excess carrier density).

Figure 6(a) shows measured decays at different grating spacings produced by changing the angle between the pump beams. By fitting the data as shown in figure 6(b), the in-well ambipolar diffusion coefficient may be deduced from the slope. Values of 13.8 and 16.2 cm^2/s were obtained for the two samples studied. The ambipolar diffusion coefficient is approximately twice the hole diffusion coefficient.

An extension of this technique is to apply an electric field to the MQW sample. Feldmann et al [11] applied a lateral field to cause in-well motion of the photogenerated carriers in a transient grating experiment. The excess electrons and holes move in opposite directions to screen the applied electric field. The diffraction efficiency is then affected by (a) the field-induced redistribution of carriers in k-space due to their acceleration (carrier heating), and (b) screening of the electric field producing strong electro-absorption effects at the exciton frequency. A drift velocity of 5×10^6 cm s^{-1} was deduced from measurements of GaAs/AlAs MQWs at 10K.

3.4. CROSS-WELL TRANSPORT

Cross well motion of carriers can be studied using a four wave mixing configuration which produces a transient grating with a modulation in a direction perpendicular to the wells as shown in figure 7(a). (This is the configuration often used for phase-conjugation by four wave mixing). In this case, the period of the grating is small, Λ ~ 120nm with one period typically

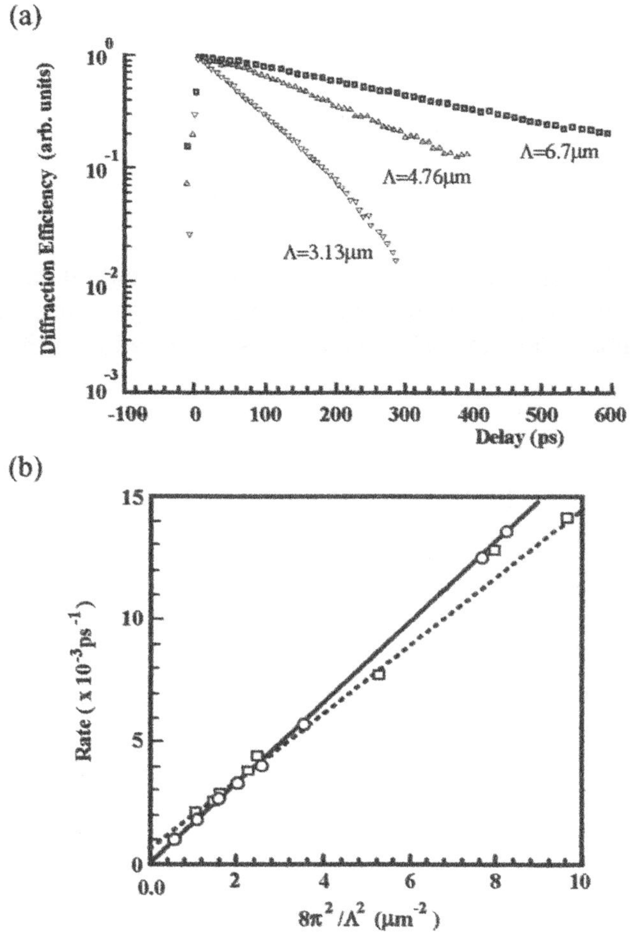

Figure 6. (a) Diffraction of the probe signal as a function of time for grating spacings, Λ = 3.13, 4.76 and 6.7μm. (b) Decay rates plotted to extract ambipolar diffusion coefficients. The results shown are for two different GaAs/AlGaAs MQW samples [10].

covering only a few quantum wells. This grating is not normally observed in bulk semiconductors since diffusion over the small distance washes out the grating very rapidly. In quantum wells, on the other hand, the diffusion is restricted by the wells. Time resolved measurements of the grating decay give information about cross-well transport processes. One geometrical problem occurs in this configuration when there is a small angle, θ, between the grating and the quantum wells, figure 7(b) - the carriers can will move along the wells to wash out the grating due to the concentration gradient in this direction.

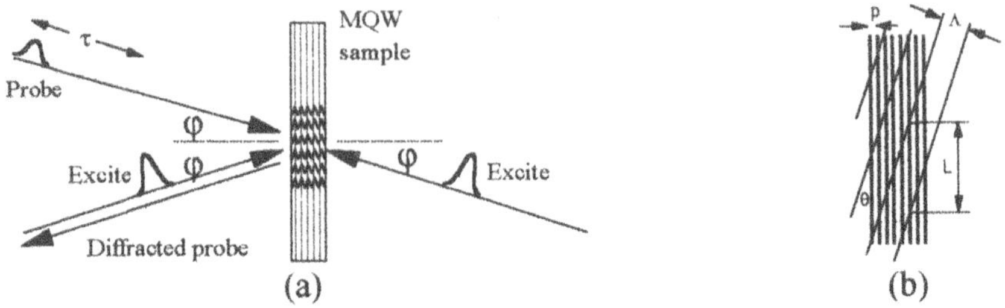

Figure 7. (a) Counterpropagating transient grating configuration, and (b) an illustration of how the quantum wells are positioned relative to the standing-wave intensity maxima at some rotation of the sample.

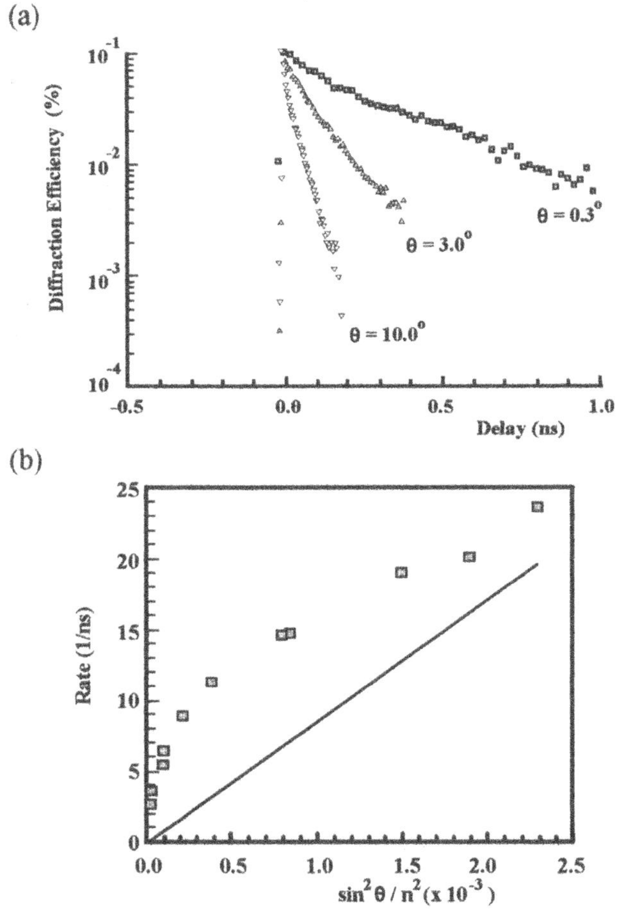

Figure 8. (a) Diffracted probe signal at three angles of the sample, and (b) measured decay rates for the diffracted probe signal [10].

Thus, there will be two components to the grating decay due to in-well and cross-well motion. Thus, taking the sample rotation into account, the grating decay rate will be given by,

$$\Gamma = \frac{4\pi^2 D_a \sin^2\theta}{n^2 \Lambda^2} + \frac{1}{\tau_\perp} + \frac{1}{\tau_R} \tag{5}$$

where $\tau_\perp$ is the grating decay due to cross-well motion.

Figure 8(a) shows the diffracted probe signal as a function of time for several angles of rotation of the sample. Figure 8(b) plots the measured decay rates, 2Γ, versus the geometrical rotation factor given in equation 5. Also plotted is the anticipated relaxation rate due to intrawell diffusion using the diffusion coefficient determined from the previous measurements described in section 3.3. An enhancement in the diffusion rate is explained by the differential emission times of the electrons and holes from the wells (the holes have a lower barrier to cross, leaving an excess of electrons in the wells which are more mobile than the holes).

Norwood et al [12] have used the short period gratings technique to study perpendicular (or cross-well) transport as a function of lattice temperature in GaAs/AlGaAs MQWs with barriers that are sufficiently thick to ensure that tunnelling is minimized. These measurements were carefully carried out under conditions where the grating vector was perpendicular to the wells thus avoiding any contributions from in-well transport. The results were consistent with a model based on thermal activation (or thermionic emission) of carriers from the wells, but from fits to the data, the cross-well transport was found to be diffusion limited by polar-optical-phonon scattering.

3.5. OPTICALLY INDUCED CIRCULAR DICHROISM

Additional insight into the nature of the exciton saturation in MQWs can be obtained by using circularly polarised light. It is possible to create 100% spin polarised electrons because of the interband selection rules close to k=0 in quantum well structures, figure 9. Since phase space filling is sensitive to carrier spin but Coulomb screening is not (within the Boltzmann approximation), this provides a technique for comparing the magnitudes of these two effects [13].

σ_+ and σ_- represent transitions for light with right and left circular polarisations respectively, with σ_+ creating spin-up electrons initially and σ_- creating spin-down electrons. In a pump and probe experiment, the two beams can be chosen to have either the same or opposite senses of polarisation. Figure

10 shows the result of a saturation measurement in GaAs/AlGaAs MQWs at room temperature with the three configurations, (i) OLP - opposite linear polarisation, (ii) SCP - same circular polarisation and (iii) OCP - opposite circular polarisation. This sample was the same as used for the amplitude grating studies described in section 3.2. The experiments were carried out at the peak of the HH exciton absorption, at the same powers densities.

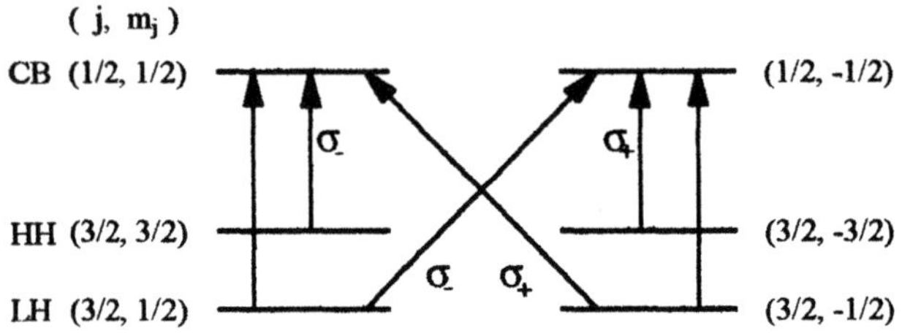

Figure 9. The selection rules for the transitions from the heavy-hole and light-hole valence bands to the conduction band in MQWs. The (j, m_j) refer to the quantum numbers for angular momentum and its component along one direction. The σ_+ and σ_- refer to the transitions excited by each sense of circularly polarised light and correspond to $\Delta m_j = \pm 1$, respectively, where the propagation direction is used to define m_j.

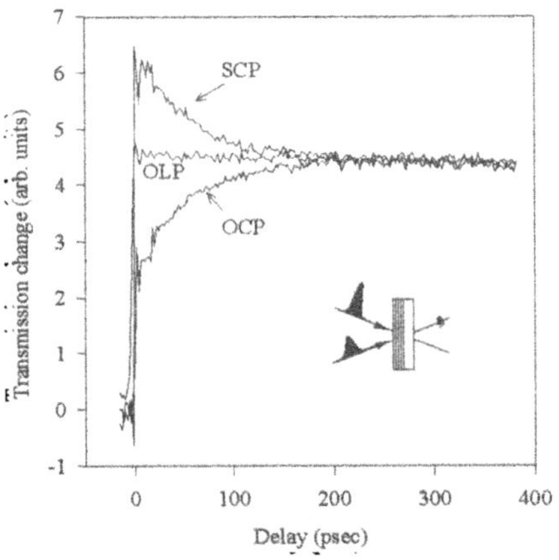

Figure 10. The dependence of probe transmission change on optical time delay in pump-probe measurements for three polarisation configurations, OLP = opposite linear polarisation, SCP = same circular polarisation, OCP = opposite circular polarisation.

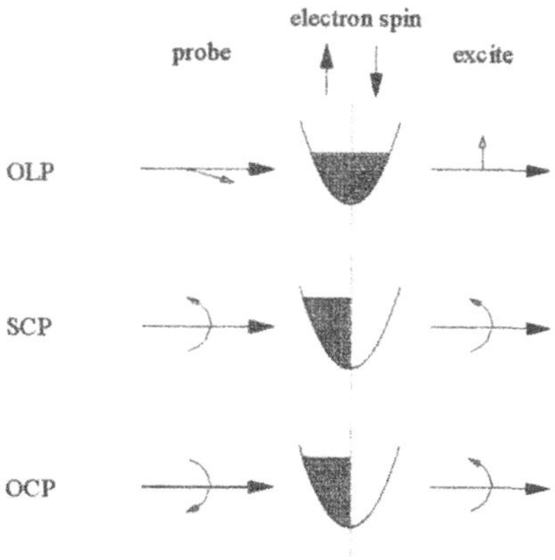

Figure 11. The effect of different polarisations of pump and probe beams.

An illustration of the band-filling conditions for each configuration is shown in figure 11. For the OLP case, equal numbers of spin up and spin down carriers are created and the recovery is set by the carrier recombination time (> 1ns). However, spin relaxation is observed in both SCP and OCP configurations as decreasing and increasing signals respectively. The deduced spin relaxation time for this sample was τ_s = 55ps. Note that the phonon scattering which ionises the excitons into free carriers does not relax the spin polarisation. Phase space filling is enhanced in SCP, since there will be more free carriers created in the spin-up band (say) which therefore causes more saturation of the absorption for the probe beam with the same polarisation state. On the other hand, the probe beam in the OCP configuration will not be sensitive to phase space filling until the spins randomise in direction. Both SCP and OCP will be sensitive to Coulomb screening since this does not depend on carrier spin, which accounts for the onset of the transmission change in the OCP case at zero delay. From a theoretical analysis, it may be concluded that the Coulomb contribution (including lifetime broadening) is of equal magnitude to phase space filling. This refutes a popular misconception that phase space filling is dominant in quantum wells.

3.6. TRANSIENT SPIN GRATINGS

The effect of circularly polarised light on exciton saturation allows the creation of another type of transient grating in a MQW. In this case, the grating is based on the spatial modulation of electron spin orientation within a uniform excess carrier distribution. Diffraction occurs via the spin-dependent free carrier contribution to exciton saturation as described in section 3.5. The decay rate of this grating is dependent on the *electron* diffusion coefficient in contrast to the *ambipolar* motion which is the normal outcome of carrier diffusion after optical generation.

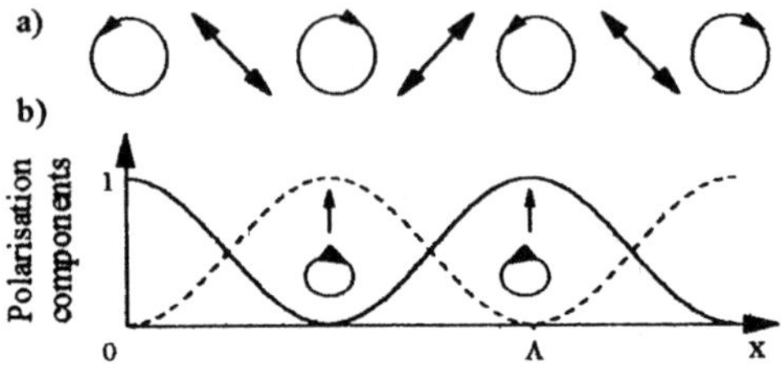

Figure 12. (a) The polarisation modulation produced when two orthogonally polarised light beams interfere. (b) Amplitudes for right and left circular components of the intensity as a function of distance in the plane of the grating.

Polarisation gratings are formed when two beams with crossed linear polarisations intersect. In this case, no intensity modulation exists, and hence no amplitude grating is formed, but a periodic modulation of the electric field polarisation is created across the excitation region. For equal excite intensities, the polarisation changes from linear to circular to orthogonal linear to circular of the opposite sense and back to linear, as shown in figure 12(b). The period of this modulation is identical to that produced for amplitude gratings and is defined by the angle between the two excite beams and the wavelength of light used. The crossed-linear configuration has been used previously to create orientational gratings in semiconductors via the modulation of the linear polarisation through anisotropic state filling [14] (state filling gives preferential excitation of electrons and holes into specific k-directions) or alignment of exciton dipoles [15, 16]. However, Figure 12(a) also shows that crossed-linear polarisations provide a modulation in circular polarisation. An alternative description of this polarisation grating is obtained by separating it into two circularly polarised components with opposite directions of rotation, figure 12(b). The electric field modulation, E(x), along the grating direction, x, can be written in terms of these two circular polarisation components,

$$E(x) = E_0\left\{ sin\left(\frac{\pi x}{\Lambda}\right) e^{-i\frac{\pi}{4}} \frac{\hat{x}+i\hat{y}}{\sqrt{2}} + cos\left(\frac{\pi x}{\Lambda}\right) e^{i\frac{\pi}{4}} \frac{\hat{x}-i\hat{y}}{\sqrt{2}} \right\} + c.c. \qquad (6)$$

where Λ is the grating spacing and $\hat{x}$ and $\hat{y}$ are unit vectors describing the polarisation directions of the incident beams. Therefore, a nonlinearity which is sensitive to circular polarisation will create a grating from two orthogonal linearly polarised beams.

As described in section 3.5, lifting of the degeneracy between the light and heavy hole bands due to confinement in MQW semiconductors, allows access to polarisation sensitive optical selection rules, figure 9. Circularly polarised light, resonant with the heavy hole exciton generates electron-hole pairs with well defined spins. The spin orientation is maintained by the electrons for tens of picoseconds after rapid ionisation of the excitons by longitudinal optical phonons (within 300fs at room temperature [9]). Hole spins may be expected to relax on sub-picosecond timescales at room temperature because of band mixing and the mixed spin character of the valence states[17]. Thus, excitation at the heavy hole exciton in a MQW with crossed linear polarised pulses creates a spatial modulation of the electron spins. At sufficient fluences, electrons saturate the exciton absorption via phase space filling (PSF) and Coulomb screening[7,13]. Since the PSF contribution is spin dependent, a spatial modulation of electron spin orientation will produce a polarisation grating which will diffract a circularly polarised probe beam. Indeed, a linearly polarised probe beam will also be diffracted by this spin-grating since both right and left circular components will be diffracted from the two circular polarisation components of the grating.

The spin-orientational grating decay rate will be determined by a combination of the spin relaxation rate and diffusion of the electrons within the quantum wells. Since hole spin relaxation is rapid, wash-out of the spin-grating can be interpreted in terms of the motion of electrons alone and the decay rate can be written,

$$\Gamma = \frac{4\pi^2 D_e}{\Lambda^2} + \frac{1}{\tau_s} \tag{7}$$

where D_e is the electron diffusion coefficient and τ_s is the electron spin relaxation time. This expression is the same as that for an amplitude grating, where the diffusion coefficient and lifetimes would be replaced by D_a, the ambipolar diffusion coefficient and τ_R the recombination time for electrons and holes.

Experiments were carried out using 1ps pulses from a self-mode-locked Ti:Sapphire laser. Two pump beams and a time delayed probe beam were configured in a standard DFWM forward travelling geometry, figure 3, so that the grating modulation was produced in the plane of the quantum wells. The polarisation of one of the pump beams was rotated by 90^0 to allow the creation of the polarisation grating. The average power in the pump beams was 500μW with ~50μW in the probe beam. The spot sizes on the sample were 30μm (FWHM).

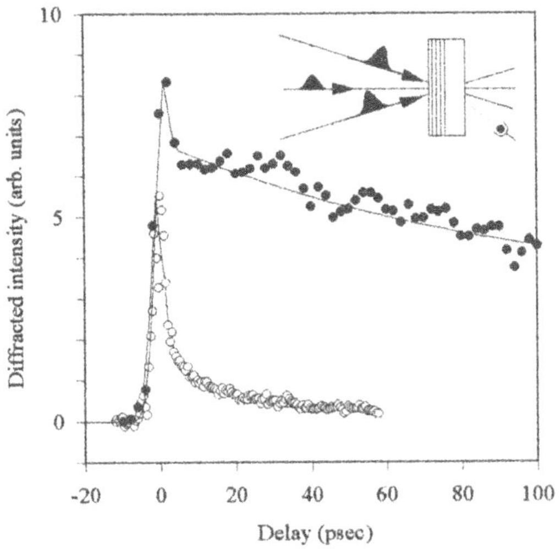

Figure 13. The measured diffracted signal decay rate for a 5μm amplitude grating (●) compared with the corresponding 5μm polarisation grating (○). Exponential decays with time constants of 220 and 13 ps are fitted to the two grating signals.

Figure 13 shows the measured decays for an amplitude grating (parallel linear polarisations) and a polarisation grating (orthogonal linear polarisations) with period Λ=5μm using a linearly polarised probe beam. Both signals are characterised by a single exponential decay. The diffracted signal from the amplitude grating has a decay time of 220ps, whereas the measured decay for the polarisation grating is 13ps. The estimated excess carrier density for the polarisation grating was ~$5x10^{16}cm^{-3}$. Coherence spikes at zero delay in both decay curves are attributed to diffraction of one pump beam from a grating created between the probe and the other pump beam.

The linear polarisation of the diffracted probe beam was observed to be rotated by 90^0, as expected, since there is a π phase shift between the right and left circular polarisation components (spin-up and spin down) of the grating, figure 12.

The decay rates (2Γ) of the diffracted signal obtained for a number of polarisation and amplitude grating periods, Λ, are plotted in figure 14. The spin-relaxation result, τ_s=55ps, is included in figure 14 as the infinite grating spacing limit. The amplitude and polarisation grating results provide the intrawell ambipolar, D_a, and electron, D_e, diffusion coefficients respectively from the gradients, eq. 6. A value of D_a=13.3cm^2/s is in good agreement with previous studies [10]. The deduced hole mobility of μ_h~257cm^2/Vs is in excellent agreement with values for GaAs. The polarisation grating results give a diffusion coefficient of 127 cm^2/s. This clearly shows that the holes no longer limit the diffusion rate within the wells. The electron mobility deduced from this diffusion coefficient is μ_e=4924cm^2/Vs. This value is lower than the typical room temperature value of 8500cm^2/Vs for pure bulk GaAs, however, the difference can be attributed to the high background doping of the sample used here. Ionised impurity scattering significantly affects the electron mobility. Taking into account the background doping, a mobility of ~6000cm^2/Vs may be expected in bulk GaAs. A further decrease may be attributed to interface scattering within the quantum well layers.

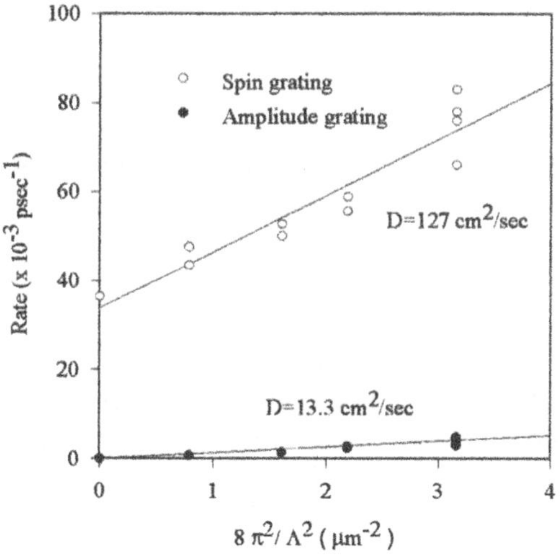

Figure 14. Measured decay rates of the diffracted grating signal against $8\pi^2/\Lambda^2$. Amplitude grating results (●) and polarisation results (○) are plotted together for comparison.

The spin-grating signal has a fast nonlinear response but is still observed at relatively low intensities, ~200kW/cm^2, because of the resonant exciton nonlinearity. This may be compared to >1GW/cm^2 in orientational gratings produced by anisotropic state filling [14].

4. Devices

In this section, we shall consider a number of devices which have employed optical nonlinearities associated with exciton saturation in MQW structures. The main advantage of making use of excitons is the high sensitivity possible compared to most nonlinear optical phenomena. The disadvantages of exciton saturation are the very limited wavelength range overwhich the nonlinearity can be accessed in a given MQW structure, the relatively slow carrier recombination time and the high background absorption. Nevertheless, these disadvantages can be overcome in many cases and successful device demonstrations have been made. Another type of optical nonlineaity associated with excitons occurs in the presence of d.c. electric fields, i.e. the quantum confined Stark effect, which has allowed the successful realisation of all-optical logic elements such as the self-electro-optic effect devices (SEED). We will not consider these devices in this chapter. For a discussion of SEED devices see for example D.A.B. Miller [18].

4.1. BISTABLE ETALONS

Optical bistability may be defined as a condition in which a system has two possible stable output states (transmission or reflection) for a given input power. Optical bistability is achieved by combining an optical nonlinearity with some form of feedback. A Fabry-Perot etalon is a convenient way of achieving optical feedback. In a high finesse cavity the intensity within the etalon, I_{cav}, can be larger than the input intensity, I_i under cavity resonance conditions. This enhances the nonlinear effect, while the inherent feedback from the mirrors can create the conditions for regenerative switching of the etalon transmission (or reflection) and bistability. Bistability in semiconductors is of interest because of the potential for all-optical switches and memory elements for signal routing, image processing, or digital computing. The challenge has been to fabricate devices which combine low switching energies, cascadability and reasonably high speed.

The first demonstration of all-optical bistability in a room temperature GaAs/AlGaAs MQW etalon in 1982 [19] gave considerable stimulation to this field because of the possibilities it seemed to offer for small, practical all-optical logic elements for digital optical computing. The early studies used quantum wells with the substrate removed by selective etching, and high reflectivity dielectric coatings applied to top and bottom surfaces. Although the nonlinearity was originally considered to result primarily from nonlinear refraction associated with exciton saturation in the GaAs system, it is now recognised that the band filling nonlinearity plays an important role in optical bistability at room temperature. However the use of quantum wells provides considerable flexibility since the thin layer of active material can be spaced by low-absorption passive material with a higher band gap. Oudar et al [20] have predicted thresholds for bistability at intensities less than $100 nW/\mu m^2$.

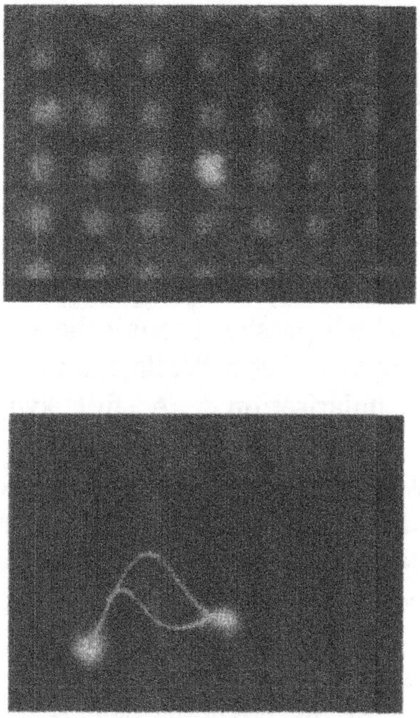

Figure 15. Optical bistability in MQW monolithic microresonators. (a) A 25x25 array of 5x5μm bistable microresonators. The laser spot is located on one resonator. (b) Hysteresis loop obtained on this microresonator [22].

Advances in epitaxial growth technology now allow extremely high quality monolithic etalon structures to be fabricated by MBE or MOCVD growth. This has given significant improvements in performance. Complete etalon structures are produced by creating Bragg reflectors consisting of quarter wave stacks of AlAs/AlGaAs on either side of the GaAs/AlGaAs quantum wells, [21]. Pixellation of the slices to form arrays of bistable elements can then be produced by etching [21] or by quantum well intermixing [22], figure 15.

Recent advances in monolithic bistable etalons by Rivera et al [23] have achieved bistable power thresholds below 100μW, uniform over 15x15 arrays fabricated by reactive ion etching. The low thresholds result from lateral carrier and light confinement in the microresonators.

4.2. A SPIN-POLARISATION SWITCH

A new type of room temperature, all-optical polarisation gate has recently been demonstrated using spin relaxation in MQWs via either absorptive or refractive nonlinearities associated with phase space filling [24,25]. Spin relaxation gives this type of device a fast response (10's of ps for electrons and sub-ps for holes, see section 3.5) however in the configurations employed, the recovery time was further decreased by tunneling in type-II quantum well structures.

Kawazoe et al [24] used the absorption change associated with a 1 ps duration, circularly polarised, pump pulse in resonance with the heavy hole exciton absorption feature in type-II $Al_{0.34}Ga_{0.66}As$ / AlAs QWs. A linearly polarised, probe pulse which passed through the sample at a fixed delay after the pump pulse was analysed after a Wollaston prism which separated the two components of linear polarisation. A full switching was observed by electrically subtracting the signals for the two linear polarisations. A recovery time of less than 1 ps was attributed to (a) scattering of the electrons from the Γ level of the QW to the X level of the barrier, and (b) the ultrafast spin-relaxation time of holes. The switching energy density was $3nJ/cm^2$ and a 80GHz repetition frequency was demonstrated.

Nishikawa et al [25] investigated a GaAs / $Al_{0.51}Ga_{0.49}As$ / AlAs QW device incorporating a monolithic etalon fabricated with two AlAs/AlGaAs Bragg reflectors on either side of the quantum well region. In this case, it is the refractive index change associated with spin-dependent, phase-space filling which produces the switching by a polarisation rotation. This was investigated in a pump-probe configuration similar to Kawazoe et al [24], with a circularly polarised pump beam and linearly polarised probe, but in this case a crossed analyser was used to monitor the linear polarisation rotation. The pump beam energy density for switching was $48fJ/\mu m^2$. A signal decay time of 7ps was achieved by a combination of tunneling in the type-II structure and spin relaxation.

4.3. DIRECTIONAL COUPLERS

Waveguide nonlinear directional couplers and Mach-Zehnder interferomers provide very useful vehicles for the demonstration of all-optical switching [26].

The first report of all-optical switching in waveguides was achieved in a MQW directional coupler switch [27]. The interaction length in a waveguide is typically much longer than in transverse devices, which dictates operating at low values of absorption away from the band gap energy. The switching length can be minimised to around 250μm in GaAs by adopting a zero-gap directional coupler geometry.

Since these devices are single or only a few channels, it is essential that they provide ultrafast responses in order to compete with existing technologies. The use of resonant optical nonlinearities to induce optical switching in these types of device normally has the disadvantage of a long recovery due to carrier recombination. However, the application of an electric field can significantly increase the speed by providing rapid carrier sweep-out [28]. A vertical *pin* diode straddling the waveguide region can provided a large field with the application of only a few volts. Recovery times of less than 10ps can be achieved by this technique.

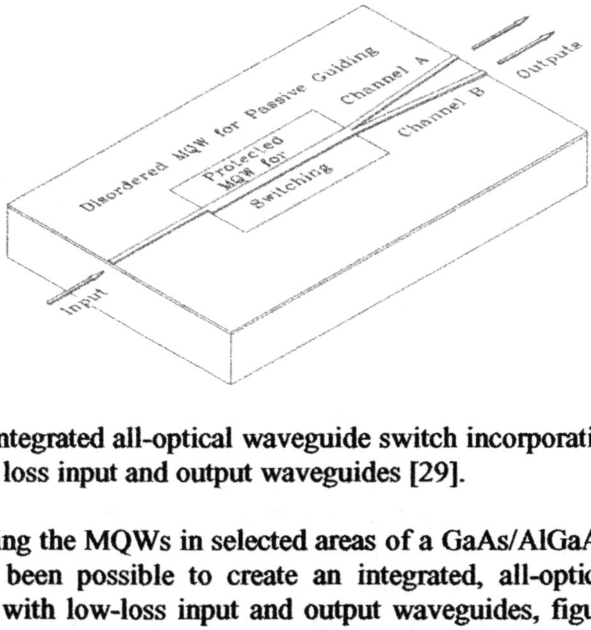

Figure 16. A integrated all-optical waveguide switch incorporating intermixed, low loss input and output waveguides [29].

By disordering the MQWs in selected areas of a GaAs/AlGaAs sample after growth, it has been possible to create an integrated, all-optical, directional coupler switch with low-loss input and output waveguides, figure 16. Partial disordering of the MQWs causes a blue shift of the absorption edge and hence less loss in the non-active regions of the device. The method employed here used the surface oxides of Ga and Al to produce controllable, partial disordering of the quantum wells [29]. By etching these surface oxides, vacancies could be provided which could then travel through the structure under annealing and cause diffusion of the atoms at the interfaces. By protecting the oxide surfaces intermixing is avoided, so that the method can be area selective.

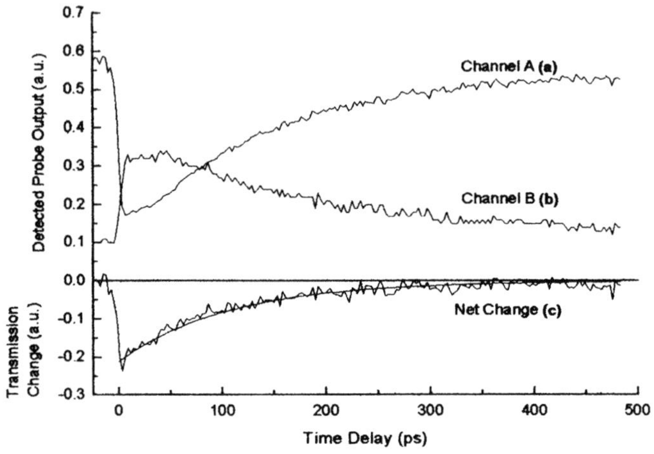

Figure 17. (a) & (b) Switching of the probe transmission through the two output channels A and B, as a function of time delay. (c) Nonlinear transmission component of the pump-probe signal deduced by summing the total probe signal in channels A and B [29].

The recovery of such a device, measured using the pump-probe technique is shown in figure 17. The switch shows a fast 100ps recovery which could be further improved by employing carrier sweep-out.

4.4. SOLITON COMMUNICATIONS

Exciton saturation in MQWs have been suggested as a means of improving the operation of long-haul fibre communications systems employing amplified soliton propagation [30]. Amplifier spacings of 50km are expected, however Gordon-Haus timing jitter and soliton-soliton interactions degrade the system performance. By incorporating MQW saturable absorbers after each amplifier, modelling of soliton propagation predicts a substantial reduction of the jitter after 9000km with 100km amplifier spacings. The requirement on the recovery time of the saturable absorber is around 10ps which can be achieved using material which has been proton bombarded, low temperature grown, or by employing carrier sweep-out in an electric field.

Exciton saturation has been demonstrated as a nonlinear mode-locking mechanism in various laser systems. Ultrashort pulses can be produced from a cw laser by incorporating within the cavity a saturable absorber with relatively slow recovery. Passively mode-locked pulses build up from noise, whereby higher intensities see less loss than lower intensity pulses. The result is a single pulse within the cavity whose duration continues to shorten on each round trip as the saturable absorber sharpens the rising edge on each pass and gain saturation in the laser medium steepens the falling edge. The result is an ultrashort pulse being emitted from the cavity every round trip. The loss should saturate faster than the gain so that the peak of the pulse can be amplified, and the recovery time of the loss should be faster than the gain to avoid build up of spontaneous emission between the pulses. Exciton saturation has the attraction of a low power nonlinearity and a recovery time that can be optimised either during the growth or by post-growth processing.

Silberberg et al [31] first used a GaAs/AlGaAs MQW structure to passively mode-lock a GaAs semiconductor diode laser. The laser cavity was extended by an antireflection coated end facet, through lenses and an anamorphic prism, and focused on to a highly reflecting mirror which had a MQW absorber glued to it. In order to achieve a faster recovery of the excitonic saturation, two methods were compared, (i) a small spot size allowed a 0.4ns optical recovery by diffusion of the carriers along the quantum wells, and (ii) proton bombardment of the sample to produce damage centres as traps provided samples with recoveries down to 150ps which still showed reasonable exciton peaks. A continuous train of 1.6ps pulses at a repetition rate of 2GHz was achieved using the latter method.

More recently, Chen et al [32] have produced pulses with durations down to 0.64ps at up to 350 GHz in monolithic, colliding pulse, mode-locked InGaAsP MQW lasers. A saturable absorber section was located at the symmetric centre of the linear cavity between two cleaved Fabry-Perot facets.

Passive mode-locking of NaCl colour centre lasers was demonstrated by Islam et al [33] with output pulse durations down to 275fs by employing InGaAs/InAlAs quantum wells. The very fast saturable absorption modulation was thought to occur from the generation of a cold nonequilibrium plasma which rapidly thermalises into a Boltzmann distribution.

A further development has been to incorprate Bragg and Fabry-Perot structures into the MQW mode-locking element. Nd:YAG, Nd:YLF and Ti:sapphire cw lasers have been mode-locked by Keller et al [34] employing resonant passive mode-locking whereby a mirror consists of an InGaAs/GaAs MQW structure with a high reflectivity Bragg reflector below it. This is either used as one mirror of the cavity or in an external coupled cavity arrangement. By producing the MQW structure by low temperature MBE, the saturation recovery could be reduced to 25ps. More recently, an anti-resonant Fabry-Perot saturable absorber (A-FPSA) structure with the MQWs incorporated within the etalon has been successful in achieving self-starting, self-mode-locked operation in a number of solid state lasers [35]. This device provides one of the laser high reflectivity mirrors within the free spectral range of the A-FPSA etalon. Absorption reduces the reflectivity in this range, but when saturated at higher incident intensities, the reflectivity increases. A recent achievement was mode-locking of diode-pumped Cr:LiSAF using a MQW A-FPSA as one mirror producing 99fs pulses [36]. Semiconductor lasers have also been successfully mode-locked using A-FPSA structures in an external cavity [37]. This scheme can produce pulses with duration down to 200fs and a commercial system system is now available from Laser Ionics.

5. Conclusions

Excitons in low dimensional structures provide a powerful means of probing ultrafast carrier dynamics. Scattering phenomena, carrier thermalisation, in-well and cross-well transport can be monitored and time resolved via free carrier induced optical nonlinearities. In this chapter, we have given a brief review of these phenomena and illustrated the use of absorption saturation in a number of devices. MQW saturable absorbers for mode-locked lasers have now found commercial application.

Acknowledgements

This work was supported by the Engineering and Physical Sciences Research Council (EPSRC). ARC acknowledges the support of a University of St Andrews Scholarship. We are grateful to Bob Manning, David Hutchings, Michael Snelling, Peggy Perozzo, Patrick LiKamWa and Shawn Shi for valuable discussions and their contributions to the results reviewed in this chapter. We thank Karl Woodbridge for the growth of the MQW samples for the transient grating studies.

References

1. Miller, A., Miller, D.A.B., and Smith, S.D. (1981) Dynamical optical nonlinearities in semiconductors, *Adv. Phys.* **30**, 697-800.
2. Haug, H., ed. (1988) *Optical nonlinearities and instabilities in semiconductors*, Academic Press, San Diego.
3. Miller, A. (1993) Semiconductors, in R.W. Eason and A. Miller (eds.), *Nonlinear optics in signal processing*, Chapman & Hall, London, pp. 66-99.
4. Van Driel, H.M. (1995) Photoinduced refractive index changes in bulk semiconductors, in A. Miller, K.R. Welford and B. Daino (eds.), *Nonlinear optical materials and devices for applications in information technology*, NATO ASI Series, Volume 289, Kluwer Academic Publishers, Dordrecht pp. 141-181.
5. Miller, A., (1995) Nonlinear optical devices, in M. Balkanski and I. Yanchev (eds.), *Fabrication, properties and applications of low-dimensional semiconductors*, Kluwer Academic Publishers, Dordrecht, pp. 383-413.
6. Chemla, D.S. and Miller, D.A.B. (1985) J. Opt. Soc. Am. B **2** 1155-1173.
7. Schmitt-Rink, S., Chemla, D.S., Miller, D.A.B. (1985) Phys. Rev. B **32**, 6601-9.
8. Park, S.H., Morhange, J.F., Jeffery, A.D., Morgan, R.A., Chavez-Pirson, A., Gibbs, H.M., Koch, S.W. and Peyghambarian, N. (1988) Appl. Phys. Lett. **52**, 1201-3.
9. Knox, W.H., Fork, R.L., Downer, M.C., Miller, D.A.B., Chemla, D.S., Shank, C. V., Gossard, A.C. and Weigmann, W. (1985) Phys. Rev. Lett. **54**, 1306-9.
10. Manning, R.J., Crust, D.W., Craig, D.W., Miller, A. and Woodbridge, K. (1988) J. Mod. Opt. **35** 541-551; Miller, A., Manning, R.J., Milsom, P.K., Hutchings, D.C., Crust, D.W. and Woodbridge, K. (1989) J. Opt. Soc. Am. B **6**, 567-578.
11. Feldmann, J., Grossmann, P., Stolz, W., Gobel, E. and Ploog, K. (1992) Semicond. Sci. Technol. **7** B130-2.
12. Norwood, D.P., Swoboda, H.E., Dawson, M.D., Smirl, A.L., Anderson, D.R. and Hasenberg, T.C. (1991) Appl. Phys. Lett. **59**, 219-21; Norwood, D.P., Smirl, A.L. and Swoboda, H.-E. (1995) J. Appl. Phys. **77**, 1113-9.
13. Snelling, M.J., Perozzo, P., Hutchings, D.C., Galbraith, I. and Miller, A. (1994) Phys. Rev. B **49**, 17160-9.

14. Smirl, A.L., Boggess, T.F., Wherrett, B.S., Perryman, G.P. and Miller, A. (1982) Phys. Rev. Lett. **49**, 933-6.
15. Schultheis, L.M., Kuhl, J., Honold, A. and Tu, C.W. (1986) Phys. Rev. Lett. **57**, 1797-1800.
16. Cundiff, S.T., Wang, H. and Steel, D. G., (1992) Phys. Rev. B **46**, 7248-51.
17. Ferreira, R. and Bastard, G. (1994) Solid State Electron. **37**, 851-5.
18. Miller, D.A.B. (1990) Opt. Quantum. Electron. **22** S61-98; Miller, D.A.B. (1995) Quantum well optical switching devices, in A. Miller, K.R. Welford and B. Daino (eds.) *Nonlinear optical materials and devices for applications in information technology*, NATO ASI Series, Volume 289, Kluwer Academic Publishers, Dordrecht pp255-284.
19. Gibbs, H.M., Tarng, S.S., Jewell, J.L., Weinmann, D.A., Tai, K., Gossard, A.C., McCall, S.L., Passner, A. and Wiegmann, W. (1982) Appl. Phys. Lett. **41**, 221-3.
20. Oudar, J.L., Kuszelewicz, R., Sfez, B., Michel, J.C. and Planel, R. (1992) Opt. Quant. Electron. **24**, S193-207.
21. Jewell, J.L., McCall, S.L., Scherer, A., Houh, H.H., Whitaker, N.A., Gossard, A.C. and English, J.H. (1989) Appl. Phys. Lett. **55**, 22-4.
22. Sfez, B.G., Rao, E.V.K., Nissim, Y.I. and Oudar, J.L. (1992) Appl. Phys. Lett. **60**, 607-9.
23. Rivera, T., Ladan, F.R., Izrael, A., Azoulay, R., Kuszelewicz, R. and Oudar, J.L. (1994) Appl. Phys. Lett. **64**, 869-871.
24. Kawazoe, T., Mishina, T. and Masumoto, Y. (1993) Jpn. J. Appl. Phys. **32**, L1756-9.
25. Nishikawa, Y., Tackeuchi, A., Nakamura, S., Shunichi, M. and Yokoyama, N. (1995) Appl. Phys. Lett. **66**, 839-41.
26. Stegeman, G.I. and Miller, A. (1993) Physics of all-optical switching devices, in J.E. Midwinter (ed.) Photonics in Switching, Academic Press, Orlando, pp. 81-145.
27. LiKamWa, P., Stitch, J.E., Mason, N.J., Roberts, J.S. and Robson, P.N. (1985) Electron. Lett. **21**, 26-7.
28. LiKamWa, P., Miller, A., Roberts, J.S. and Robson, P.N. (1991) Appl. Phys. Lett. **58**, 2055-7.
29. Shi, S., LiKamWa, P., Miller, A., Pamulapati, J., Cooke, P. and Dutta, M. (1995) Appl. Phys. Lett. **66**, 79-81.
30. Atkinson, D., Loh, W.H., Afansjev, V.V., Grudinin, A.B., Seeds, S.J. and Payne, D.N. (1994) Opt. Lett. **19**, 1514-6.
31. Smith, P.W., Silberberg, Y. and Miller, D.A.B. (1985) J. Opt. Soc. Am. B **2** 1228-35.

32. Chen, Y.K., Wu, M.C., Tanbun-Ek, T., Logan, R.A. and Chin, M.A. (1991) Appl. Phys. Lett. **58** 1253-5.
33. Islam, M.N., Sunderman, E.R., Soccolich, C.E., Bar-Joseph, I., Sauer, N., Chang, T.Y. and Miller, B.I. (1989) IEEE J. Quant. Electron. **25** 2454-62.
34. Keller, U., Knox, W.H. and 'tHooft, G.W. (1992) IEEE J. Quant. Electron. **28**, 2123-33.
35. Keller, U., Miller, D.A.B., Boyd, G.D., Chui, T.H., Ferguson, J.F. and Asom, M.T. (1992) Opt. Lett. **17**, 505-7.
36. Weingarten, K.J., Kopf, D., Brovelli, L., Kamp, M. and Keller, U. (1994) post-deadline paper, European Conference on Lasers and Electro-Optics (Amsterdam).
37. Delfyett, P.J., Florez, L., Stoffel, N., Gmitter, T., Andreadakis, N., Alphonse, G. and Ceislik, W. (1992) Opt. Lett. **17** 670-2.

INTERFEROMETER WITH NONLINEAR FREQUENCY DOUBLING MIRRORS AS LOSSLESS ALL-OPTICAL SWITCHING DEVICE

S. SALTIEL, K. K0YNOV, I. BUCHVAROV
University of Sofia, Faculty of Physics
Quantum Electronics Department
5 J. Bourchier Blvd., 1164 Sofia, Bulgaria

Nonlinear optical interferometers have proved themselves as useful tools for many applications [1]. For example, the proposed in [2] Twaiman-Green interferometer with two phase conjugate mirrors has been used for measurement of both the real and the imaginary part of cubic susceptibilities [2-5] and for measurement of the fidelity of the phase conjugate process as well [6]. Mach Zehnder interferometer with nonlinear cubic media in both arms [7,8] is suggested for use as an all-optical switching device. Recent investigations of the phase shifting properties of the frequency doubling media [9-12] showed that the process of phase shift is much more efficient if in both arms of the Mach Zehnder interferometer it is used $\chi^{(2)}$ media, tuned for quasi phase matched second harmonic generation (SHG) [12-14].

Phase shifting elements used in all-optical switching devices should fulfill specific requirements: (i) to be able to provide at least $\pi/2$ phase shift; (ii) the fundamental pump power necessary to obtain such phase shift should be as low as possible; and (iii) the losses at fundamental wavelength should be minimized too. It is very difficult to meet all these requirements simultaneously for the known phase shifting elements.

In this report we develop two new ideas. The first idea is to use the nonlinear frequency doubling mirrors as lossless phase shifting elements. Nonlinear frequency doubling mirrors (NFDM's) [15] have been previously used for solid state passive mode locking of solid state lasers [16,17]. As it will be shown later, NFDM satisfies all the requirements listed above. Our second suggestion is for construction of Twaiman-Green interferometer on the base of NFDM's. The proposed interferometer is shown on Fig. 1. The two mirrors are with maximum reflection at fundamental and second harmonic (SH) wavelengths. In front of each mirror there is situated a nonlinear crystal (NC) for SHG. The distances between the NC and the mirrors have to be adjustable in order to control the phase difference $\Delta\phi_0$ between the phase φ_s of the generated SH wave and the phases φ_{1f} and φ_{2f} of the fundamental waves. The two NC are at a quasi phase-matched condition $\Delta k = k_s - k_{1f} - k_{2f} \neq 0$. For our analysis

M. Balkanski (ed.), Devices Based on Low-Dimensional Semiconductor Structures, 227–230.

we chose for one of the arms $\Delta kL = 1$ and for the other $\Delta kL = -1$. In this way, the phase shifts of the reflected fundamental waves from NFDM in both arms have opposite signs.

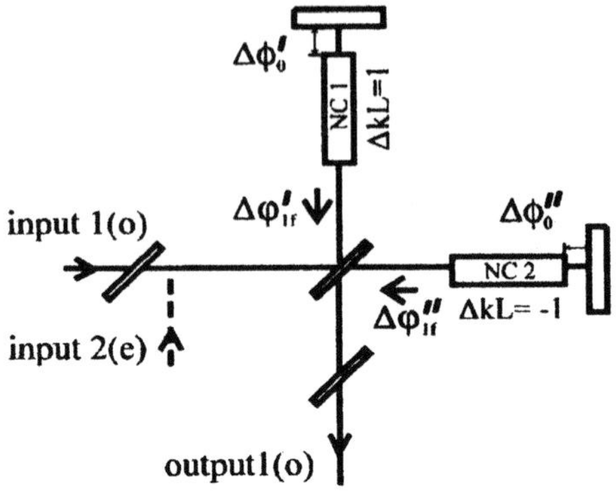

Figure 1.

The proposed interferometer can be constructed in two modifications. In the first one NFDM's employ *ooe* interaction for SHG and in the second one - *oee* interaction for SHG. This two modifications correspond to "on-to-off" and "push-pull" switching devices, respectively. Interfering waves for both modifications of the interferometer are the A'_{1f} and A''_{1f} waves with gained phase shifts $\Delta\varphi'_{1f}$ and $\Delta\varphi''_{1f}$.

The main advantages of the use of NFDM's as phase shifting elements are: a) double pass through the nonlinear media which means an increased phase shift for the same input fundamental power and b) possibility for obtaining 100% (theoretically) reflection at fundamental wavelength by proper adjustment of the phase difference $\Delta\varphi_0$ at the entrance face of the NC on the way back. In this way, by varying the input power or the length of the NC there can be obtained any phase shift and in the same time the losses can be independently optimized by adjustment of $\Delta\phi_0$. For a one pass frequency doubling shifter such independent optimization is not possible.

Theoretical description of the NFDM's as phase shifter was done by numerical solution of reduced amplitude equations in the slowly varying envelope approximation with assumption of zero absorption for all interacting waves:

$$\frac{dA_{1f}}{dz} = -i\sigma_1 A_s A_{2f}^* \exp(-i\Delta k z)$$

$$\frac{dA_{2f}}{dz} = -i\sigma_2 A_s A_{1f}^* \exp(-i\Delta k z) \qquad (1)$$

$$\frac{dA_s}{dz} = -i\sigma_3 A_{1f} A_{2f} \exp(i\Delta k z)\,,$$

where σ_i are nonlinear coupling coefficients. The ratio of the intensities of the two fundamental waves $(A_{2f}/A_{1f})^2$ was taken to be 5.

On Fig. 2. there are shown intensity reflection (dashed line) and phase shift (solid line) of the reflected from the NFDM's fundamental waves. The optimal phase differences for maximum reflection coefficient of the two NFDM were found to be $\Delta\phi_0' = -0.8\pi$ for the arm with NC with $\Delta kL = 1$ and $\Delta\phi_0'' = -0.2\pi$ for the arm with NC with $\Delta kL = -1$.

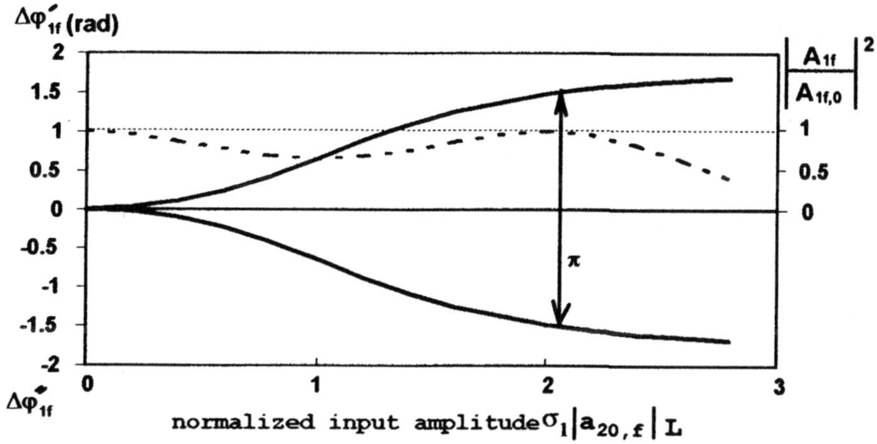

Figure 2

It is seen that for normalized amplitude $\sigma|A_{20,f}|L = 2.2$ for the strong wave at the input of the interferometer the reflection of each NFDM's is ~ 100%. The phase shifts at this level of the fundamental power are $\pi/2$ for one of the arm and $-\pi/2$ for the other and they are enough for a total switching of the interferometer. The switching power level is about 2 times lower than the switching power for the nonlinear Mach Zehnder interferometer [12].

We thank Prof. Boardman for the stimulating discussions and Bulgarian Science Foundation for grant F-405.

References

1. Hariharan, P. (1990) Optical interferometry, *Pep. Prog. Physics* **54**, 339-390.
2. Saltiel, S. M., VanWonterghm, B., Rentzepis, P. M. (1989) Measurement of $\chi^{(3)}$ and phase shift of nonlinear media by means of a phase conjugate interferometer, *Optics Letters* **14**, 183-185.

3. Saltiel, S. M., Goldberg, S., Huppert, D. (1991) Measurement of magnitude and sign of the third order nonlinear susceptibility in lead molibdate crystal, *Optics Communications* **84**, 189-192.
4. Tackx, P., Kauranen, M., Persoons, A. (1994) Determination of resonant and nonresonant third order nonlinearities of organic molecules by phase conjugate interferometry, *Optics Letters* **19**, 1113-1115.
5. Cao, X. F., Jiang, J. P., Hellwarth, R. W., Yu, L. P., Chen, M., Dalton, L. (1990) Bipolaronic Enhanced Third Order Nonlinearity in organic Ladder Polymers, *Proceeding SPIE* **1337**, 114-124.
6. Saltiel, S. M., VanWonterghm, B., Rentzepis, P. M. (1989) Fidelity of Optical Phase Conjugation by Degenerate Four Wave Mixing in Semiconductor Glasses and Ruby, *Applied Physics Letters* **54**, 1842-1844.
7. Al-Hemyari, K., Aitchison, J. S., Ironside, C. N., Kennedy, G. T., Grant, Sibbett, W. (1992) Ultrafast all-optical switching in GaAlAs integrated interferometer in 1,55 mm spectral region, *Electronics Letters* **28**, 1090-1092.
8. Stegeman, G. I., Wright, E. M. (1990) All-optical waveguide switching, *Optics and Quantum Electronics* **22**, 95-122.
9. De Salvo, R., Hagan, D. J., Sheik Bahae, M., Stegeman, G., Van Stryland, E. (1992) Self-focusing and self-defocusing by cascaded second order effects in KTP, *Optics Letters* **17**, 28-30.
10. Stegeman, G., Sheik Bahae, M., Van Stryland, E., Assanto, G. (1993) Large nonlinear phase shifts in second order nonlinear-optical processes, *Optics Letters* **18**, 13-15.
11. Assanto, G., Stegeman, G., Sheik Bahae, M., Van Stryland, E. (1995) Coherent Interactions for All-Optical Signal Processing via Quadratic Nonlinearities, *IEEE Journal of Quantum Electronics* **31**, 673-680.
12. Hutchings, D. C., Aitchison, J. S., Ironside, C. N. (1993) All-optical switching based on nondegenerate phase shifts from a cascaded second order nonlinearity, *Optics Letters* **18**, 793-795.
13. Assanto, G., Stegeman, G., Sheik Bahae, M., Van Stryland, E. (1995) All-optical switching devices based on large nonlinear shifts from second harmonic generation, *Applied Physics Letters* **62**, 1323-1325.
14. Ironside, C. N., Aitchison, J. S., Arnold, J. M. (1993) An All-Optical Switch Employing The Cascaded Second Order Nonlinear Effects, *IEEE Journal of Quantum Electronics* **29**, 2650-2654.
15. Stankov, K. A. (1988) A Mirror with an Intensity Dependent Reflection Coefficient, *Applied Physics B* **45**, 191-194.
16. Stankov, K. A., Jethwa, J. (1988) A new mode-locking technique using nonlinear mirror, *Optics Communications* **66**, 41-44.
17. Buchvarov, I., Saltiel, S., Gagarscii, S. (1995) Nonlinear doubling mode-locking of feedback controlled pulsed Nd:YAG laser, *Optics Communications* **118**, 51-54.

WAVEGUIDE-BASED DEVICES : LINEAR AND NONLINEAR COUPLING

A D BOARDMAN AND K XIE
Photonics and Nonlinear Science Group
Joule Laboratory
Department of Physics
University of Salford
Salford, M5 4WT
United Kingdom

1. Introduction

It is now universally accepted that,in the near future[1], practically all the information received by us, or sent to someone else, will be carried on globally distributed optical fibre networks. In the somewhat longer term (see figure1), there will be no exceptions

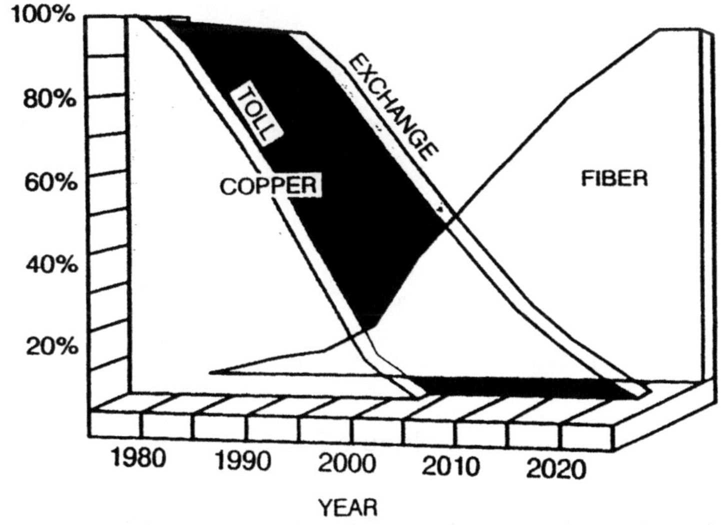

Figure 1. Projections of one major telephone company for the installation of optical fiber. They predict that by early in the next century, the entire local (exchange) and long distance (toll) network will be composed of optical fiber.

to this technology either for long distance communications [*TOLL*] or local communications to and from our homes and businesses [*EXCHANGE*] as information is drawn off the information superhighways. The problem today is that the perceived

M. Balkanski (ed.), Devices Based on Low-Dimensional Semiconductor Structures, 231–292.

requirement for all-optical switching has not yet been met. This form of technology is critically demanded by the type of fibre systems anticipated, which will want to work at terabit data rates. Such a requirement creates a strong pressure for sub-picosecond

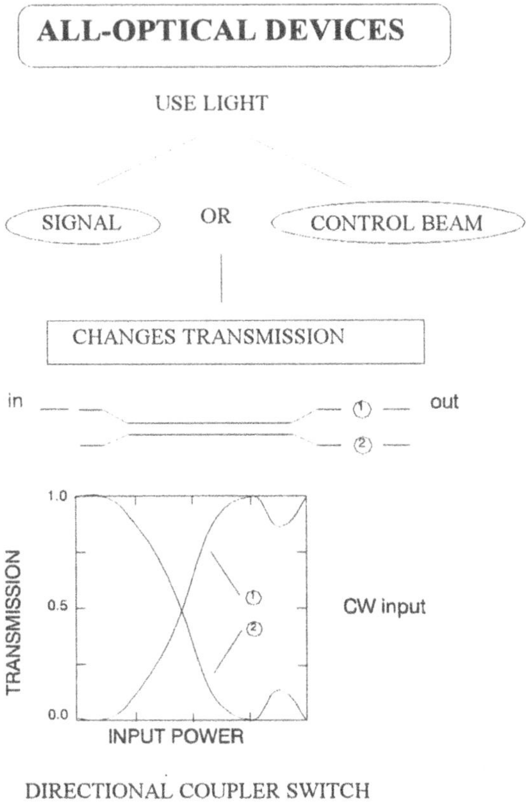

Figure 2

speeds and is going to be beyond the ability of not only today's equipment, but of those currently anticipated for the future too. Hence, the search is on for all-optical (i.e. photonic) devices. If this search fails then an interesting impasse between the ability to deliver information (data), optically, upto a certain location (a computer, for example) and the ability to process it, will develop. The photonic devices required will need the right sort of materials, so the search for them is pulling along with it a huge amount of material assessment to see whether they have desirable optical properties and these must include nonlinearity[2,3]. This will be required in order to permit both signals and controls to be set up, using light beams entirely. In other words, the *power* of a light beam is needed to change a property like transmission of a waveguide to achieve, for example, a switching effect. As shown in figure 2, it is possible to imagine making a simple physical waveguide arrangement[1], called a directional coupler, in which varying the input power leads to a pretty good switch. This is, in theory, of course, but what about the materials that could be used? How good are they? Indeed, what is meant by the term good or desirable?

First, there are some basic points that are rather well-known, but are still worth mentioning. For many years now it has been appreciated that all-optical nonlinearity falls broadly into the categories called *resonant* or *non-resonant.* Basically, resonance, as might be expected, involves a material becoming optically nonlinear because the photons passing through it actually force electrons from the valence band to the conduction band. These *real* or *actual* electron transitions can in principle, give rise to rather large nonlinearities i.e. it is, after all, a resonance phenomenon and should be expected to do so. Unfortunately, the downside is that the process tends to be slow and absorptive and usually works *because* of the latter. Unless there is some *design way*[4] out of this problem, it is not particularly useful, for the kind of fast switching desired, nor, at first sight, is it a good process around which to base waveguiding.

On the other hand, *non-resonant* nonlinearity involves moving bound electrons out of position. In quantum language, these are virtual transitions. The bad news is that it is a relatively *weak* process but the good news is that it is *fast.* The famous non-resonant material is silica glass (used for fibres) but semiconductors can be operated in both non-resonant and resonant modes. Although optical communication fibres are made of silica glass, which, incidentally, has rather a very small nonlinearity, its ability to absorb light at the operational wavelength is low. As a material, because of the way it is used, it is one of the success stories of nonlinear optics. Semiconductors, resonant or non-resonant, are widely used, however, to try out new ideas in guiding and the hope always exists that all-optical switching in semiconductors will be successful. This desire is prompted by the need to integrate compact layers into optical chip structures. Indeed, a measure of success has been reported. Yet it has been argued in the literature that the prospects for both non-resonant and resonant semiconductors for applications is marginal[1]. A summary of these conclusions, by P W Smith, are shown in figure 3. Nevertheless, semiconductors and especially, low-dimensional semiconductor structures that are AlGaAs-based, operating below half the bandgap, are an extremely important *source of nonlinearity.* They have favourable properties for switches and logic devices and, with more complicated geometrical waveguide design, may, eventually, be made to work at the sort of low powers demanded by telecommunications. At this point it is worth recommending the reader to study the pioneering analysis by Stegeman[2] who has defined some very useful figures of merit for *third-order nonlinearity*, through which a judgement on the usefulness of a third-order nonlinear material can be made. For example, if a normalised nonlinear phase shift $W = \Delta\phi/2\pi$ occurs in a directional coupler then $W > 1$, which implies that $\Delta n/(\alpha\lambda) > 1$, where Δn is the nonlinear change in refractive index, α is the absorption coefficient and λ is the operating wavelength. There are other figures of merit relating to two-photon ($T = \dfrac{2\pi}{\Delta\phi}$) and

TYPICAL MATERIALS

- GLASS: nonresonant
- SEMICONDUCTOR

resonant (above [close to] bandgap)

non-resonant (below bandgap)

	glass (non-resonant)	semiconductor (non-resonant)	semiconductor (resonant)
nonlinearity	very small	small	large
response time	~ fsec	~ fsec	~ nsec
absorption	very small	small-moderate	large
thermal effects	small	significant	large
two-photon absorption	small-moderate	small-large	insignificant
primary applications	fibre devices	CW devices below 1/2 gap	bistable devices SEED; VSTEP
CONCLUSION	useful	marginal	marginal

- NOTE THESE CONCLUSIONS BY P W SMITH[REF.1]
- SEARCH IS ON FOR OTHER MATERIALS OR NEW SYSTEM DESIGNS

Figure 3

three-photon ($V = \frac{2\pi}{\Delta\phi}$) absorption. The combination leads to $\Delta\phi < \frac{2\pi W}{1 + WT + WV}$, which generates the appropriate nonlinear phase shift, $\Delta\phi$, for various device options.

Having now given a short introduction to the motivations, and the *usual* material possibilities, for all-optical switching, the sort of configurations that might be considered will be reviewed and the foundations of how to set about designing the kind of all-optical coupling desired, will be laid down. In the final part of this chapter, the interesting ideas surrounding *cascaded nonlinearity* will be examined in some detail and preliminary comment on this phenomenon will be made in the section dealing with examples of switches. The term cascading will be explained later but, basically, it concerns the use of second-order nonlinearity to achieve an effective nonlinear index change through a mixing of a second-harmonic signal, at frequency 2 ω, with the (fundamental) generating wave at frequency ω. Enhanced values of this kind of nonlinearity can be obtained from asymmetric GaAlAs quantum wells. Thus, once again, the low-dimensional structure is the source of nonlinearity in an all-optical switch. Because cascading is now anticipated, and because suitably engineered low-dimensional structures seem to be a real possibility, the promise for all-optical switching seems to be high at the present time.

2. Linear directional coupler

Before entering upon a discussion of the way in which the power of light is going to be useful, consider a very simply analysis that uses two slab [planar films] waveguides that are parallel to each other and sufficiently close to allow the possibility of guided light going from one guide to another. It is this sort of *linear* power transfer that leads to the idea that it may be possible to *control* light with light. The first task is to appreciate the general idea and the second one is to understand how the mathematics is going to work.

Figure 4 shows two, slab, waveguides, which are labelled 1 and 2, respectively and have the refractive index profile also shown in figure 4. At this stage the guides are not identical but, even if they are, the labelling serves to identify the coupling properties.The waveguides are entirely linear and the relevant Maxwell's equations are

$$curl\,\mathbf{E} = -\mu_0 \frac{\partial \mathbf{H}}{\partial t};\; curl\,\mathbf{H} = \varepsilon_0 \varepsilon \frac{\partial \mathbf{E}}{\partial t} \tag{2.1}$$

where μ_0 and ε_0 are the permeability and permittivity of free space, respectively, $\mathbf{E}$ is the electric field, $\mathbf{H}$ is the magnetic field, ε is a dielectric constant, t is time, $\mu_0\varepsilon_0 c^{-2}$, where c is velocity of light in a vacuum and $\varepsilon = n^2$, which defines n, the refractive index. Clearly, the latter can be any kind of spatial distribution that is desired but, for simplicity, and good practical reasons, it is selected here to have the shape drawn in figure 4. During the operation of this waveguide structure no loss of generality is incurred by assuring that the guided wave has TE polarisation i.e.$E\hat{\mathbf{x}}$ where $\hat{\mathbf{x}}$ is a unit vector. Assuming that only a single mode propagates, the total field in the coupler, for propagation along the z-axis, can be written as[5]

$$E = (\psi_1(z)F_1(x,y)e^{-i\beta_1 z} + \psi_2(z)F_2(x,y)e^{-i\beta_2 z})e^{i\omega t} \tag{2.2}$$

where ω is an angular frequency, and $\psi_1(z)$, $\psi_2(z)$ are *slowly varying* functions of z, so that $\frac{d^2\psi_1}{dz^2}$ and $\frac{d^2\psi_2}{dz^2}$ will be neglected. β_1 and β_2 are the wavenumbers of

A FIRST ANALYSIS OF COUPLING

INDEX AND ELECTRIC FIELD DISTRIBUTIONS

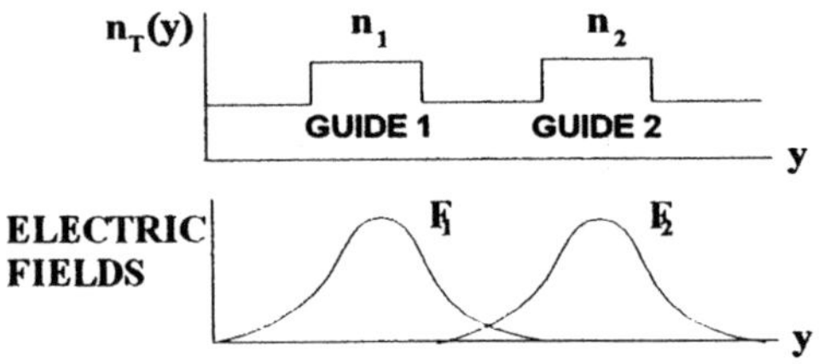

- **$n_T(y)$ is the total refractive index distribution**

Figure 4

the TE single modes that would propagate in each isolated guide. The use of (2.2) in Maxwell's equations gives

$$\left[-2i\beta_1 \frac{d\psi_1}{dz} + \frac{\omega^2}{c^2}\left(n_T^2 - n_1^2\right)\psi_1\right]F_1 + \left[-2i\beta_2 \frac{d\psi_2}{dz} + \frac{\omega^2}{c^2}\left(n_T^2 - n_2^2\right)\psi_2\right]F_2 e^{-i\Delta\beta z} = 0 \tag{2.3}$$

where $\Delta\beta = \beta_2 - \beta_1$ is called the phase mismatch, i.e. it is the difference between the values of the wavenumbers appropriate to each isolated guide, and $\nabla_T^2 F_i + \left(\frac{\omega^2}{c^2} n_i^2 - \beta_i^2\right)F_i = 0$, i=1,2, $\nabla_T^2 = \frac{\partial^2}{\partial x^2} + \frac{\partial^2}{\partial y^2}$. Equation (2.3) can be manipulated into a pair of *generic* coupled equations in the following well-known way. The multiplication of (2.3) by the complex conjugate F_1^*, followed by an integration over x and y gives[5]

$$\left[-2i\beta_1 \frac{d\psi_1}{dz}\langle F_1, F_1\rangle + \frac{\omega^2}{c^2}\left\langle\left(n_T^2 - n_1^2\right)F_1, F_1\right\rangle\psi_1\right] + \left[-2i\beta_2 \frac{d\psi_2}{dz}\langle F_1, F_2\rangle + \frac{\omega^2}{c^2}\left\langle\left(n_T^2 - n_2^2\right)F_2, F_1\right\rangle\psi_2\right]e^{-i\Delta\beta z} \tag{2.4}$$

where the familiar quantum mechanical notation[5]

$$\langle F_1, F_1\rangle = \int\int\left(F_1 F_1^*\right)dxdy \tag{2.5}$$

has been adopted. All the other averages $\langle\ \rangle$ are constructed in the same way and, for slab guides, a unit length along the guide lengh is used. Having performed this task, the natural properties of the coupled waveguide can be exploited[5] i.e. it is obvious from figure 4, that

- there is a small overlap between F_1 and $F_2 \Rightarrow \langle F_1, F_2\rangle \approx 0$
- $\left\langle\left(n_T^2 - n_1^2\right)F_1, F_1\right\rangle \cong 0$: in other words, $n_T^2 - n_1^2 = 0$, where F_1 overlaps itself strongly.

Hence,

$$-2i\beta_1 \frac{d\psi_1}{dz} + \frac{\omega^2}{c^2}\frac{\left\langle\left(n_T^2 - n_2^2\right)F_2, F_1\right\rangle}{\langle F_1, F_1\rangle}\psi_2 e^{-i\Delta\beta z} = 0 \tag{2.6}$$

i.e.

$$i\frac{d\psi_1}{dz} + 2\nu\psi_2 e^{-i\Delta\beta z} = 0 \tag{2.7}$$

where

$$2\nu = \frac{\omega^2}{c^2}\frac{\left\langle\left(n_2^2 - n_T^2\right)F_2, F_1\right\rangle}{\beta_1\langle F_1, F_1\rangle}$$

If these waveguides are identical then synchronism occurs and $\Delta\beta$=0. The generic equations, for this type of design, are therefore [see figure 5]

$$i\frac{d\psi_1}{dz} + 2\nu\psi_2 = 0 \tag{2.8a}$$

$$i\frac{d\psi_2}{dz}+2\nu\psi_1=0 \tag{2.8b}$$

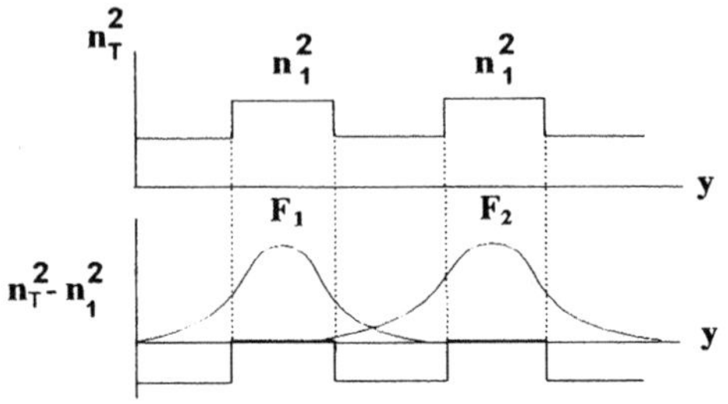

SYNCHRONISM

IDENTICAL WAVEGUIDES

$\Delta\beta = \beta_2 - \beta_1 = 0$

PHASE - MATCHED

- Let $i\kappa \rightarrow 2\nu$

$\frac{d\psi_1}{dz} + 2\nu\psi_2 = 0$
$\frac{d\psi_2}{dz} + 2\nu\psi_1 = 0$
GENERIC FORM

Figure 5

Note that these equations are generic because, although the analysis began with a coupled waveguide problem, they have many uses in physics. For example, they also describe coupled pendula.

Proceeding, for a little while longer with the simple linear case, it is immediately obvious that ψ_2 can be eliminated from equation (2.8a) in favour of ψ_1 by using (2.8b). The result of this manipulation is

$$\frac{d^2\psi_1}{dz^2}+4\nu^2\psi_1=0 \tag{2.9}$$

So, for the boundary conditions, $\psi_1 = 1, \dfrac{d\psi_1}{dz} = 0$ the solutions are

$$\psi_1 = \cos(2\nu z) \tag{2.10a}$$

$$\psi_2 = \mathrm{i}\sin(2\nu z) \tag{2.10b}$$

which demonstrates, rather nicely, the periodic exchange of energy that occurs between guide 1 and 2. If $\Delta\beta \neq 0$ then the generic forms are

$$i\frac{d\psi_1}{dz} + 2\nu\psi_2 e^{-i\Delta\beta z} = 0 \tag{2.11a}$$

$$i\frac{d\psi_2}{dz} + 2\nu\psi_1 e^{i\Delta\beta z} = 0 \tag{2.11b}$$

Although equations (2.11) have a very appealing form, it is a useful move to transform the $e^{\pm i\Delta\beta z}$ factors away with the changes $\psi_1 \rightarrow \psi_1' e^{-i\frac{\Delta\beta}{2}z}$ and $\psi_2 \rightarrow \psi_2' e^{-\frac{\Delta\beta}{2}z}$, where ψ_1' and ψ_2' are *not* the same functions as ψ_1 and ψ_2. For ease of notation, however, the dashes can be suppressed and equations (2.11) can be solved rather easily. In fact, $\psi_1 = A_1\cos(\gamma z) + B_1\sin(\gamma z)$, where $\gamma = \sqrt{(\Delta\beta/2)^2 + 4\nu^2}$ and, if the guide has a length L, then, for an input power distribution $P_1 = 1$, $P_2 = 0$ [at z = 0], the output power from guide 2 is simply[5]

$$P_2 = \sin^2\left[\sqrt{4\nu^2 L^2 + (\Delta\beta/2)^2 L^2}\right] \Big/ \left[1 + [\Delta\beta/(4\nu)]^2\right]$$

From this expression, it can be seen that P_2 is a maximum at $\left(\dfrac{\Delta\beta L}{2}\right) = 0$, but it is rather small elsewhere. In addition, P_1 plunges to zero at $\Delta\beta = 0$. Now $\Delta\beta$ is called the *phase mismatch* and the question in whether $(\Delta\beta L/2)$ can be adjusted *optically*. This would be a very desirable outcome, both in principle, and *practically*. If this can be done then some form of all-optical switching would be achieved. Before such a question can be answered, however, some concept of nonlinearity needs to be introduced, thereby extending the generic equations (2.8) to include nonlinearity. After that, a more formal mathematical machinery will be set up that, hopefully, will draw out an elegant way of looking at nonlinear couplers and will permit the idea that a *soliton coupler* is, potentially, a useful device.

3. Basic nonlinear or soliton switching ideas

All materials polarise under the action of an electric field so, strictly speaking, all materials become, to a greater or lesser extent, nonlinear, when exposed to a laser beam, for example. The way in which a material becomes nonlinear is important, but ,as outlined in the introduction, only the non-resonant route will be selected, for this discussion. When it is said that a material is *linear*, in the context of the discussions here, it really means that this is so for all practical purposes.

The polarisation of a linear material will be defined as

$$P_i = \varepsilon_0 \chi_{ij}^{(1)} E_j \tag{3.1}$$

, where, in general, $\chi_{ij}^{(1)}$ is a second-rank tensor, whose elements are determined by the crystal class of the material under consideration. In fact, it is very easy to locate the forms of $\chi_{ij}^{(1)}$ in the literature, appropriate to each class[6].

Non-resonant nonlinearity can often be accounted for (see [7,8] for easy access to a simplified discussion) by expressing $P_i(\mathbf{r},t)$, the total polarisation of a material, in the following series form

$$\begin{aligned} P_i(\mathbf{r},t) &= \varepsilon_0 \{ \chi_{ij}^{(1)} E_j(\mathbf{r},t) + \chi_{ijk}^{(2)} E_j(\mathbf{r},t) E_k(\mathbf{r},t) \\ &+ \chi_{ijkl}^{(3)} E_j(\mathbf{r},t) E_k(\mathbf{r},t) E_l(\mathbf{r},t) + \ldots \} = P_i^{(L)}(\mathbf{r},t) + P_i^{(NL)}(\mathbf{r},t) \end{aligned} \tag{3.2}$$

where $\mathbf{r} = (x,y,z)$ is a position vector, t is time, $P_i^{(L)}(\mathbf{r},t)$ is the linear contribution and $P_i^{(NL)}(\mathbf{r},t)$ is the nonlinear contribution. Such an expression embodies a number of assumptions. First of all, it assumes that nonlinear optical processes can be accounted for by stopping at the third term. Secondly, it assumes that the nonlinearity is not going to saturate i.e. reach some maximum value, as is often the case with real materials. Thirdly, no account of the possible frequency dependence of $\chi_{ijkl\ldots}^{(n)}$ has been taken. The form (3.2) is adequate, for many theoretical discussions, but, since all our equations work in the frequency domain, care will have to be taken to make sure that the appropriate frequency dependence is included in the nonlinear coefficients, if it becomes necessary.

For an immediate application of (3.2), it can be assumed that the materials involved have a centre of symmetry, in which case $\chi_{ijk}^{(2)} = 0$, identically. This fact not only simplifies (3.2) but, if third-harmonic generation is also neglected, both the linear terms and the *third-order* nonlinear terms operate at the frequency ω : called here the *fundamental* frequency. Maxwell's equations for a scalar field $\mathbf{E}(\mathbf{r},t) = \widetilde{\mathrm{E}}(\mathbf{r},t)\hat{\mathbf{x}}$ reduce to

$$\left(\frac{\partial^2}{\partial x^2}+\frac{\partial^2}{\partial y^2}+\frac{\partial^2}{\partial z^2}\right)E(\omega,\mathbf{r})+n^2(x,y)\frac{\omega^2}{c^2}E(\omega,\mathbf{r})+\omega^2\mu_0 P^{NL}(\omega,\mathbf{r})=0 \tag{3.3}$$

where E(ω,**r**) and P^{NL}(ω,**r**) are the Fourier transforms of $\tilde{E}(\mathbf{r},t)$ and $P^{(NL)}(\mathbf{r},t)$. n^2(x,y) is the refractive index distribution. If, as was done earlier on, E(ω,**r**) is expressed as $E(\omega,\mathbf{r})=\psi(z)\,F(x,y)e^{i\beta z}$, where β is a wavenumber then, introducing ψ_1 and ψ_2, with $\beta=\beta_1$, $\beta=\beta_2$, for the, respective, waveguides the following generic equations, extended to include nonlinearity, exist [for $\beta_1=\beta_2$]

$$i\frac{\partial\psi_1}{\partial z}+\frac{\partial^2\psi_1}{\partial x^2}+2\nu\psi_2+2|\psi_1|^2\psi_1=0 \tag{3.4a}$$

$$i\frac{\partial\psi_2}{\partial z}+\frac{\partial^2\psi_2}{\partial x^2}+2\nu\psi_1+2|\psi_2|^2\psi_2=0 \tag{3.4b}$$

There are quite a few steps on the way to equations (3.4) but they can easily be found from the literature[9,10]. The $2\nu\psi_{1,2}$ terms have already explained explained and the $2|\psi_1|^2\psi_1$, $2|\psi_2|^2\psi_2$ terms show the presence of third-order nonlinearity because P^{NL}(ω,r) reduces to a very simple form, for most materials of interest. In fact, it becomes $\varepsilon_0\,\chi^{(3)}|E|^2E$, where all that remains of the, formidable, fourth-rank tensor $\chi^{(3)}_{ijkl}$ is a simple constant $\chi^{(3)}$. The *cleaning up* of equations (3.4) needed also the transformations $\frac{\omega}{c}x\rightarrow x$, $\frac{\omega}{c}z\rightarrow 2\beta z$, and $\sqrt{\frac{3}{8}\chi^{(3)}}E\rightarrow\psi$.

In equations (3.4) the interpretation of various terms is as follows. Without the nonlinear terms there are three terms of the form $\frac{\partial\psi}{\partial z},\frac{\partial^2\psi}{\partial x^2}$ and $2\nu\psi$. They have the physical interpretations

- $\frac{\partial\psi}{\partial z}$: propagation effects along z
- $\frac{\partial^2\psi}{\partial x^2}$: diffraction or dispersion
- $2\nu\psi$: waveguide coupling

In other words, the $\frac{\partial\psi}{\partial z}$ terms account for any amplitude or phase changes during propagation along z. The $\frac{\partial^2\psi}{\partial x^2}$ can be viewed as representing diffraction if x is actually a spatial coordinate. This term is precisely what is needed to account for the spreading of a laser beam for example i.e. a laser beam in a linear medium will inevitably increase its width, as it propagates, because, at any moment, rays of light arising within it will see the current width of the beam as an aperture through which to diffract. In a linear medium, there is no physical effect to stop this from happening . Of course, the spreading will be slow, but it *will* happen unless nonlinearity is around to combat it. The equations are generic, however, so x is just a symbol. For *beams* that are diffracting x is a space coordinate, but for *pulses* x can be called a time coordinate. In such a case, for clarity, we would write $x \rightarrow \tau$, where τ is time but the meaning is clear, nevertheless.

SOLITON SWITCH

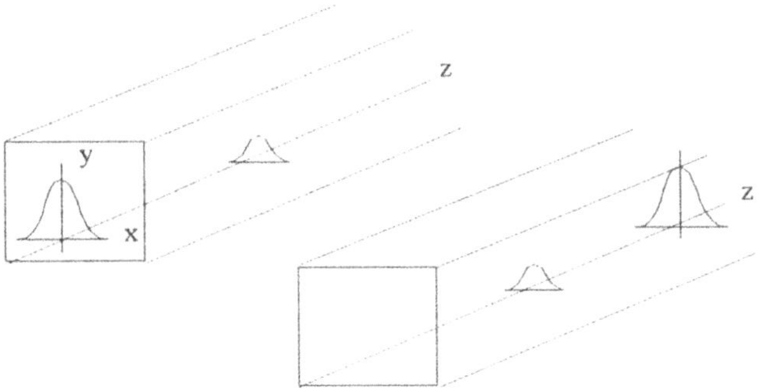

- power transfer
- soliton shapes preserved

Figure6

Equations (3.4) can be used to discuss a nonlinear or a soliton switch. Nonlinear switches in the literature are discussed for CW waves, in which case the $\frac{\partial^2 \psi}{\partial x^2}$ term is weak compared to the $2\nu\psi$ coupling term. In other words, for a light beam, being used in a coupled waveguide structure, the diffraction is weak. This means that points on a beam switch independently and the switch speed depends upon the intensity at the switch point. This model is appropriate to wide beams [or pulses], the shape of the input profile is not preserved and the switching is not sharp i.e. this could be referred to as *dirty switching*. On the other hand, if the $\frac{\partial^2\psi}{\partial x^2}$ terms are strong and comparable to the nonlinear $2|\psi|^2\psi$ terms, and certainly stronger than the $2\nu\psi$ terms, then input shapes are preserved. This is *soliton switching* i.e., as sketched in figure 6, a soliton is launched at the input and energy transfer between guides proceeds to take place, but shapes are always intact. In a soliton switch, therefore, points on a beam cannot switch independently, with the result that clean switching takes place with greater sharpness. It is this quality that makes soliton switching so attractive.

Before leaving this basic, non-mathematical, introduction a number of photonic switching possibilities will be mentioned. They are based upon spatial solitons and can, in principle, be realised in semiconductor structures that are providing reasonably strong nonlinearity through their quantum well properties. A soliton, in these examples, is a light beam that does not spread out due to diffraction, nor does it compress (self-focus) due to nonlinearity. The solitons referred to here are *stationary* states but, of course, an arbitrary input to a waveguide has to evolve to this. It does this through the $\frac{\partial\psi}{\partial z}$ term but it should also be remembered that the $\frac{\partial\psi}{\partial z}$ term also accounts for the phase of the *stationary* amplitude. For a large power input a higher-order soliton will emerge that proceeds to oscillate, or 'breathe', in terms of its peaked structure appearance. Such solitons repeat their shape, at any given propagation distance, in a time or length period called the *soliton period*. Only the lowest-order solitons that have a sech shape are, apparently, of any potential practical value in a device.

Figure 7 shows the operating principle[11] of a potential, all-photonic, switching device, based upon spatial solitons. The nonlinear medium can be a low-dimensional semiconductor structure operating below half the bandgap. The device proposes to use control beams and signal beams. It is well-known that when two spatial soliton beams collide with each other they 'engage in a little dance' and shift in position. This is exploited here. The whole device shown in figure 7 is being viewed from above i.e. what is happening is that all the beams are trapped into guiding tracks (ridges, perhaps) in a thin film planar waveguide. The input and output sections are linear and no attempt is made to incorporate any launching or extraction features. Also reflection at the interfaces is neglected.

If the control beams are absent then the signal beam B reaches F and, even though there is coupling to the two outer signal guides, upto 97% transmission can occur. Suppose now that a control beam equal in magnitude to the signal is switched on in the left-hand channel. Collision of spatial solitons in the nonlinear medium take place, with the consequence that transverse shifts, without change of propagation direction, will occur. The control is coupled out through its allotted channel in the output section but the signal will shift to the left and engage in coupling with its neighbouring guides. The final result is that the signal emerges from the right-hand signal guide. Typically a design could use a wavelength of 1.3μm, and a nonlinear refractive index $1.55 + (5.3 \times 10^{-10})I$, where I is intensity. In the input/output sections, the core refractive index is 1.57, the cladding index is 1.55 and the waveguide thickness is 2.5μm. The power levels involved are P_c = 0.214W/mm [control] and $P_s = 0.4P_c$ to $1.4P_c$ [signal].

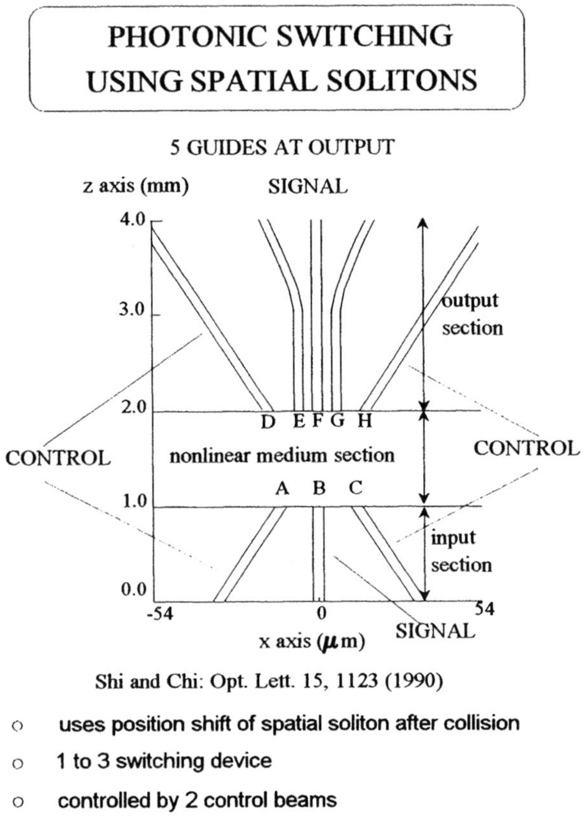

Figure7

Other devices are eminently possible[12,13] with spatial solitons and include soliton logic gates, all-optical switching using spatial vector solitons and asymmetric spatial soliton dragging gates. The latter is particularly appealing because of the possibility of making a NOR gate in a cascaded system. The idea is to drag (literally) a fairly substantial pump beam through at least a beam diameter by using a small signal beam. A spatial aperture [as simple as a hole in a wall] can be used and the game to play is to make the pump, the signal or the signal plus the pump either miss the aperture or go through it, depending upon the application in mind. Finally, it is also possible to imagine a device based upon second-order nonlinearity[14,15] [nonlinear cascading] but a discussion of this fascinating physical process will be left until the end of this chapter.

4. Linear and nonlinear coupler theory

- **LINEAR CW COUPLING**

In this case the generic equations are

$$i\frac{\partial\psi_2}{\partial z}+2\nu\psi_1=0 \tag{4.1a}$$

$$i\frac{\partial\psi_1}{\partial z}+2\nu\psi_2=0 \tag{4.1b}$$

Even though, all the work to be discussed here will use equations in this format, it must be emphasised that they are dimensionless. Both the scales and physical interpretation are a matter of first choosing a material and then, later on, when equations (4) are extended beyond the CW case, to select a beam or pulse interpretation.

Equations (4) are easily solved by using the substitutions

$$\psi_1=\eta_1 e^{i\theta_1},\quad \psi_2=\eta_2 e^{i\theta_2} \tag{4.2}$$

where η_1, η_2 are amplitudes and θ_1, θ_2 are phases. After inserting ψ_1 and ψ_2 into equations (4) it is a very simple matter to set, separately, the real and imaginary parts of the equations equal to zero. Firstly, the substitution into (4.1a) yields

$$i\frac{\partial\eta_1}{\partial z}-\eta_1\frac{\partial\theta_1}{\partial z}+2\nu\eta_2 e^{i(\theta_2-\theta_1)}=0 \tag{4.3}$$

Equating real and imaginary parts equal to zero gives:

$$\text{REAL:}\quad \frac{d\theta_1}{dz}=2\nu\frac{\eta_2}{\eta_1}\cos(\theta_2-\theta_1),\ \frac{d\theta_2}{dz}=2\nu\frac{\eta_1}{\eta_2}\cos(\theta_2-\theta_1) \tag{4.4a}$$

$$\text{IMAGINARY:}\quad \frac{d}{dz}(\eta_1^2)=-4\nu\eta_1\eta_2\sin(\theta_2-\theta_1) \tag{4.4b}$$

$$\frac{d}{dz}(\eta_2^2)=4\nu\eta_1\eta_2\sin(\theta_2-\theta_1) \tag{4.4c}$$

From equations (4.4b) and (4.4c), it is obvious that

$$\frac{d}{dz}\left(\eta_1^2+\eta_2^2\right)=0 \tag{4.5}$$

but this, in fact, is a very nice result. It demonstrates that $\eta_1^2+\eta_2^2 =$ constant. So this constant will be called 2M here i.e.

$$\eta_1^2+\eta_2^2=2M \tag{4.6}$$

where it is recognised that 2M is the total (dimensionless, of course) energy of the coupled modes. The factor of 2 is present because it is our preferred choice of the definition of M i.e. M is the average of η_1^2 and η_2^2. The choice is not fundamental, however. Some other definitions come into view now. These are $\eta_2^2-\eta_1^2=2\mathrm{N}$ [energy difference between two coupled guides, for example] and phase difference $\theta=\theta_2-\theta_1$. Equations (4.4) can now be expressed in the elegant form

$$\frac{dN}{dz}=4\nu\sqrt{M^2-N^2}\,\sin\theta \tag{4.7}$$

$$\frac{d\theta}{dz}=\frac{-4\nu N}{\sqrt{M^2-N^2}}\cos\theta \tag{4.8}$$

Initially, suppose that all the energy is in guide 1 i.e. $\eta_2(0)=0,\ \eta_1^2(0)=-2N=2M$. Hence $(M^2-N^2)_{z=0}=0$. The Hamiltonian of this system is $H=2\nu\{\psi_1^*\psi_2+\psi_2^*\psi_1\}=\nu\sqrt{M^2-N^2}\cos\theta$, however, because this is a conserved quantity. This means that if H = 0, at z = 0, it is zero everywhere i.e. it must be zero, even when $\sqrt{M^2-N^2}>0$, so that $\cos\theta=0$ and $\theta=\pm\pi/2$, $\sin\theta=\pm1$, everywhere. The basic equation for the linear CW coupling described here is, therefore,

$$\frac{dN}{dz}=\pm4\nu\sqrt{M^2-N^2} \tag{4.9}$$

and this equation has a solution

$$N=-M\cos(4\nu z),\ \theta=\pi/2 \tag{4.10}$$

The interpretation of (4.10) is both immediate and interesting. It shows that there is a periodic exchange of energy between the coupled waveguides.

- **NONLINEAR CW COUPLING**

In this case there is no diffraction so the generic equations are

$$i\frac{\partial\psi_2}{\partial z}+2|\psi_2|^2\,\psi_2+2\nu\psi_1=0 \tag{4.11a}$$

$$i\frac{\partial\psi_1}{\partial z}+2|\psi_1|^2\,\psi_1+2\nu\psi_2=0 \tag{4.11b}$$

which have solutions

$$\psi_1=\eta_1 e^{i\theta_1},\quad \psi_2=\eta_2 e^{i\theta_2} \tag{4.12}$$

The substitution of (4.12) into (4.11) very easily gives

$$\frac{d}{dz}(\eta_1^2)=-4\nu\eta_1\eta_2\sin(\theta_2-\theta_1) \tag{4.13a}$$

$$\frac{d\theta_1}{dz}=2\eta_1^2+2\nu\frac{\eta_2}{\eta_1}\cos(\theta_2-\theta_1) \tag{4.13b}$$

$$\frac{d\eta_2^2}{dz}=4\nu\eta_1\eta_2\sin(\theta_2-\theta_1) \tag{4.13c}$$

$$\frac{d\theta_2}{dz}=2\eta_2^2+2\nu\frac{\eta_1}{\eta_2}\cos(\theta_2-\theta_1) \tag{4.13d}$$

The usual definitions $\eta_1^2+\eta_2^2=2M,\ \eta_2^2-\eta_1^2=2N$ and $\theta_2-\theta_1=\theta$ lead to

$$\frac{dN}{dz}=4\nu\sqrt{M^2-N^2}\sin\theta \tag{4.14}$$

$$\frac{d\theta}{dz}=4N-\frac{4\nu N}{\sqrt{M^2-N^2}}\cos\theta \tag{4.15}$$

The Hamiltonian of equations (4.11) is

$$H=2\nu\left(\psi_1^*\psi_2+\psi_2^*\psi_1\right)+|\psi_1|^4+|\psi_2|^4 \tag{4.16}$$

so that

$$H = 2N^2 + 4\nu\sqrt{M^2 - N^2}\cos\theta \tag{4.17}$$

where the *constant* $2M^2$ has been absorbed into the Hamiltonian.

For two coupled waveguides, it is instructive to assume that $N = -\frac{1}{2}\eta_1^2(0) = -M$, $\eta_2^2(0) = 0$, initially. This step implies that H = $2M^2$, initially. It is also very instructive to express the coupling problem in terms of a particle-in-a-potential-well model. It turns out that this is incredibly easy to do and the procedure is as follows. From equations (4.17)

$$\cos\theta = \frac{H - 2N^2}{4\nu\sqrt{M^2 - N^2}} = \frac{\sqrt{M^2 - N^2}}{2\nu} \tag{4.18}$$

since $H = 2M^2$ is *preserved*. Hence,

$$\left(\frac{dN}{dz}\right)^2 = 16\nu^2(M^2 - N^2)\sin^2\theta = -\left\{4(M^2 - N^2)^2 - 16\nu^2(M^2 - N^2)\right\} \tag{4.19}$$

where $\left(\frac{dN}{dz}\right)^2$ is analogous to kinetic energy and the remaining terms add up to the potential energy U(N), where

$$U(N) = 4(M^2 - N^2)^2 - 16\nu^2(M^2 - N^2) \tag{4.20}$$

i.e.

$$\left(\frac{dN}{dz}\right)^2 + U(N) = 0 \tag{4.21}$$

This means that the state of the CW coupler can be, very elegantly, discussed in terms of the position of a *fictitious particle* in the potential distribution U(N) and the nonlinearity can be used to "tune" the device. In the linear limit, equation (4.9) shows that $U(N) = -16\nu^2(M^2 - N^2)$. This is an interesting comparison because it shows that, in equation (4.21), $M \ll 2\nu$ is the linear limit, as expected, intuitively.

Figure 8 shows three examples of U(N). In the first one, figure 8a, $M \ll 2\nu$ which is the linear limit. All that the assumption $M \ll 2\nu$ is saying is that the power-dependent (nonlinear) effects (M) are much less than the linear coupling (ν) effect. In this, linear, limit, the particle begins at N = -M and then "rolls" down the potential hill to $U(N) = -16\nu^2 M$. The "particle" passes through this point and "rolls" up to N = M, at which point the process of "rolling" back to N = -M begins.

In physical terms, N = -M corresponds to the initial excitation of one waveguide $[\eta_1^2(0) = -2M, \eta_2^2(0) = 0]$, the position $U(N) = -16\nu^2 M^2$ corresponds to both guides being equally excited, after coupling of energy from guide 1 to guide 2, and N = +M corresponds to *complete transfer* of energy from guide 1 to guide 2.

The second case, figure 8b, still shows complete, periodic, energy transfer although there is now a "bump" in the potential around N = 0. In this case, $N = -Mcn\left(4\nu z | M^2 / 4\nu^2\right)$, where cn is one of the Jacobian elliptic functions that are well-tabulated in numerical analysis handbooks. The third case is M > 2ν and here the "particle" encounters a positive "bump" at N = 0. This means that "reflection" can occur and, although periodic energy transfer is still possible, it is incomplete. In this case, $N = -M\left[1 - \frac{4\nu^2}{M^2} sn^2\left(2Mz | 4\eta^2 / M^2\right)\right]^{\frac{1}{2}}$, where sn is another member of the Jacobian elliptic function family. This is plotted in figure 8c.

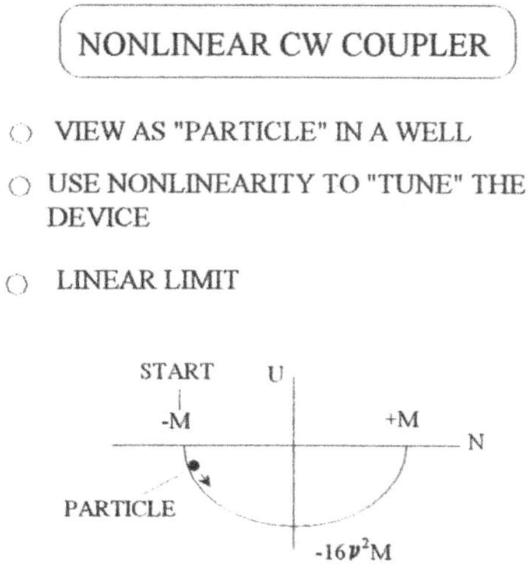

Figure 8a

NONLINEAR CW COUPLER

NONLINEAR (CASE I) : $M<2\nu$

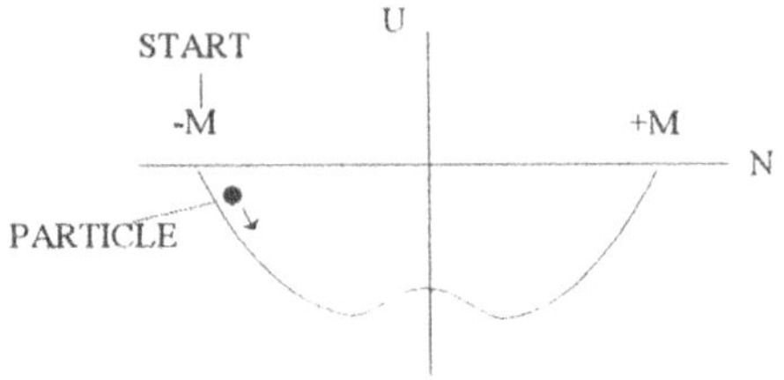

PERIODIC AND COMPLETE ENERGY TRANSFER

Figure 8b

NONLINEAR CW COUPLER

NONLINEAR (CASE II) : $M>2\nu$

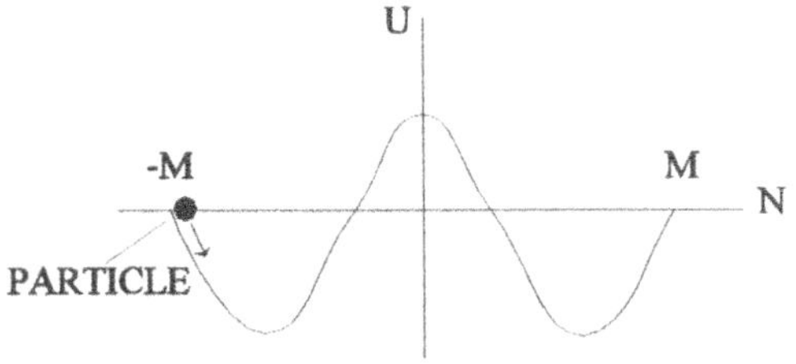

PERIODIC BUT INCOMPLETE ENERGY TRANSFER

Figure 8c

- **THE SOLITON COUPLER**

A description of the soliton coupler can be obtained from the coupled equations

$$i\frac{\partial\psi_1}{\partial z}+\frac{\partial^2\psi_1}{\partial x^2}+2|\psi_1|^2\psi_1+2\nu\psi_2=0 \tag{4.22a}$$

$$i\frac{\partial\psi_2}{\partial z}+\frac{\partial^2\psi_2}{\partial x^2}+2|\psi_2|^2\psi_2+2\nu\psi_1=0 \tag{4.22b}$$

in which the diffractive terms $\frac{\partial^2\psi_{1,2}}{\partial x^2}$ are strong and, in combination with the $|\psi_{1,2}|^2\psi_{1,2}$ terms, permit equations (4.22) to have soliton solutions. It should be remembered that equations (4.22) are *generic* and that the language just used refers to coupled *spatial solitons*. Equations (4.22) also provide a description of *pulse* behaviour with an appropriate re-definition of the variables and the introduction of appropriate length and time scales.

The solution of equations (4.22) is impossible to obtain with exact mathematics, but a numerical simulation will always generate the exact solutions required. A lot can be learned, however, if another approach is adopted. This is the popular variational method[9,16,17] which relies upon adopting some reasonable looking functions and then fixing them, through parameters, to give as a good a description as possible. The adopted functions are trial functions and the mathematical vehicle is the Lagrangian formalism[18,19]. It is quite possible that extremely poor trial functions are selected but their effectiveness can always be checked by comparing the mathematical predictions to the true (numerically-generated) solutions. The idea, to begin with, is to keep the trial functions as simple as possible.

First, it must be recognised that the basic equations (4.22) can be generated, through the Euler-Lagrange equations, from the Lagrangian[9,16,17]

$$L=\sum_i\left[\frac{i}{2}\left[\psi_i^*\frac{\partial\psi_i}{\partial z}-\psi_i\frac{\partial\psi_i^*}{\partial z}\right]-\left|\frac{\partial\psi_i}{\partial z}\right|^2+|\psi_i|^4\right]+2\nu\left[\psi_1^*\psi_2+\psi_1\psi_2^*\right] \tag{4.23}$$

Next, the following simple trial functions can be tried

$$\psi_1=\eta_1(z)\operatorname{sech}(\rho x)e^{i\theta_1(z)},\ \psi_2=\eta_2(z)\operatorname{sech}(\rho x)e^{i\theta_2(z)} \tag{4.24}$$

where, for the moment, fixed-width functions are selected, but note that the amplitudes and phases are now functions of z. It is the evolution of these quantities that will now be sought.

The first step is to *reduce* L by integrating over x, after substituting ψ_1 and ψ_2, i.e. to form

$$\mathcal{L} = \int L dx \tag{4.25}$$

The application of the Euler-Lagrange equations[18,19] then gives

$$\frac{\partial \mathcal{L}}{\partial \eta_1} - \frac{d}{dz}\left(\frac{\partial \mathcal{L}}{\partial\left(\frac{d\eta_1}{dz}\right)}\right) = 0 \tag{4.26a}$$

$$\frac{\partial \mathcal{L}}{\partial \eta_2} - \frac{d}{dz}\left(\frac{\partial \mathcal{L}}{\partial\left(\frac{d\eta_2}{dz}\right)}\right) = 0 \tag{4.26b}$$

$$\frac{\partial \mathcal{L}}{\partial \theta_1} - \frac{d}{dz}\left(\frac{\partial \mathcal{L}}{\partial\left(\frac{d\theta_1}{dz}\right)}\right) = 0 \tag{4.26c}$$

$$\frac{\partial \mathcal{L}}{\partial \theta_2} - \frac{d}{dz}\left(\frac{\partial \mathcal{L}}{\partial\left(\frac{d\theta_2}{dz}\right)}\right) = 0 \tag{4.26d}$$

where

$$\mathcal{L} = -\frac{2\eta_1^2}{\rho}\frac{d\theta_1}{dz} + \frac{4}{3}\frac{\eta_1^4}{\rho} - \frac{2}{3}\eta_1^2\rho - \frac{2\eta_2^2}{\rho}\frac{d\theta_2}{dz} + \frac{4}{3}\frac{\eta_2^4}{\rho} - \frac{2}{3}\eta_2^2\rho + \frac{8\nu}{\rho}\eta_1\eta_2\cos(\theta_2 - \theta_1) \tag{4.27}$$

The soliton coupler problem therefore reduces, quite easily, to the following four, ordinary, nonlinear, coupled, differential equations

$$\frac{d\theta_1}{dz} = \frac{4}{3}\eta_1^2 - \frac{\rho^2}{3} + 2\nu\frac{\eta_2}{\eta_1}\cos(\theta_2 - \theta_1) \tag{4.28a}$$

$$\frac{d\theta_2}{dz} = \frac{4}{3}\eta_2^2 - \frac{\rho^2}{3} + 2\nu\frac{\eta_1}{\eta_2}\cos(\theta_2 - \theta_1) \tag{4.28b}$$

$$\frac{d\eta_1^2}{dz} = -4\nu\eta_1\eta_2\sin(\theta_2 - \theta_1) \tag{4.28c}$$

$$\frac{d\eta_2^2}{dz} = 4\nu\eta_1\eta_2\sin(\theta_2 - \theta_1) \tag{4.28d}$$

In the spirit outlined earlier, the following definitions can be usefully adopted. $\eta_1^2 + \eta_2^2 = 2M$, $\eta_2^2 - \eta_1^2 = 2N$, $\theta_2 - \theta_1 = \theta$. If, initially, the soliton is in guide 1, with guide 2 being empty at $z = 0$, then $N(z = 0) = -M$ and the reduced Hamiltonian is $\mathcal{H} = \frac{4}{3}M^2$. This means that the conclusion for the soliton coupler is that it can also be described as a 'particle in a well' problem, in which

$$\left(\frac{dN}{dz}\right)^2 + U(N) = 0 \tag{4.29a}$$

$$U(N) = \frac{16}{9}(M^2 - N^2)^2 - 16\nu^2(M^2 - N^2) \tag{4.29b}$$

The only difference between the soliton coupler potential and the nonlinear CW potential appears to be the factor $\frac{16}{9}$, as opposed to a factor 4. In other words, there is no essential difference and the coefficient of $(M^2\text{-}N^2)$ comes from an average over the soliton cross-section. The model, i.e. the trial function, said nothing about soliton width, however. It was set to be a constant, but this is not correct. Numerical simulations show that this is unsatisfactory, yet it is used in the literature[17]. A simulation soon shows that soliton width is very important in a switch. It actually makes soliton switching completely different from ordinary nonlinear CW switching. Hence, the width just cannot be ignored in the trial function. Neither can another important parameter of the soliton, which is the chirp, that also, significantly, affects the switching.

The problem above is that a fixed-width trial function was used and, as has just been remarked, this is probably not a good thing to do. Before correcting this, however, it is worth checking the numerical simulations to see what actually happens. Remember that, although the trial functions lead to analytical expressions that can be very useful in discussing the physics of the mechanisms involved, numerical simulations yield *exact* checks on the variational conclusions. This must always be borne in mind when designing devices based on Euler-Lagrange equations.

Figure 9 shows a comparison between the fixed-width variational [trial function plus Euler-Lagrange equations] method and the numerical [exact] conclusions. The simulations were performed by specifying the forms of ψ_1 and ψ_2 on the input plane, at $z = 0$. The results show that difference choices of initial width, ρ, lead to different conclusions. Furthermore, only for *one*, initial, choice of width ρ, does the trial function fixed-width, method (dashed curve) agree with the numerical conclusions (solid curves). This display emphasises just how far apart the theoretical (trial function) and numerical (exact) results are.

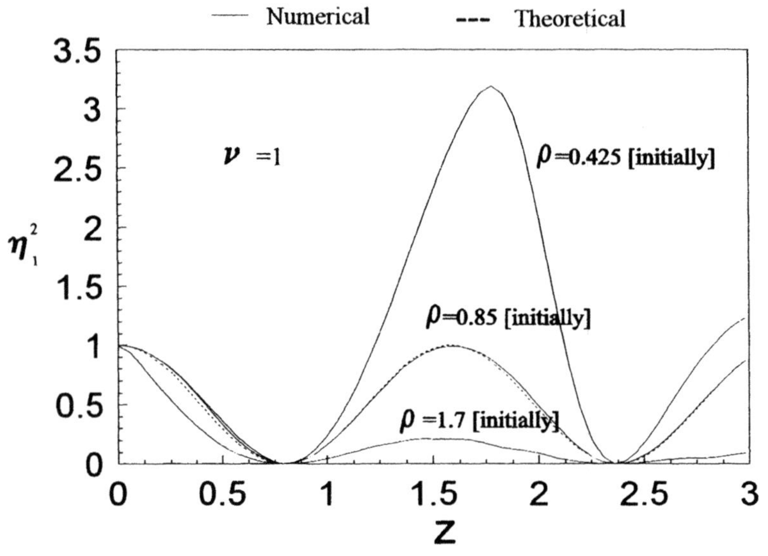

Figure 9

Much better trial functions to use are

$$\psi_{1,2} = \eta_{1,2}(z)\mathrm{sech}[\rho(z)x]e^{i\left[\theta_{1,2}(z)+\frac{C(z)}{2}x^2\right]} \tag{4.30}$$

in which the amplitudes $\eta_{1,2}$, the widths ρ, the phases $\theta_{1,2}$ and the chirp parameters C are all functions of the propagation distance z. As has already been commented, soliton width is very important in switch design and, in fact, makes soliton switching completely different from nonlinear (CW) switching or nonlinear pulse [below soliton threshold] switching.

Figures 10a and b show a numerical comparison between a fixed width and variable width calculations. Figure 10a shows the numerical simulation for the input conditions $|\psi_1(0,x)| = \mathrm{sech}(1.7x), |\psi_2(0,x) = 0$. It demonstrates that $|\psi_1(z,x)|$ falls rapidly and radiates away. On the other hand, figure 10b is the fixed-width result $\eta_1^2 = \rho M[1+\mathrm{cn}(4\nu z|M/4\nu^2)]$, for which $\gamma = 1$, $\rho = 1.7$, $\eta_1(0) = 1$, $\eta_2(0) = 0$. This fixed-width result shows that $|\psi_1|^2 = \eta_1^2(z)\mathrm{sech}^2(\rho\alpha)$ is a periodic function, which is wrong!

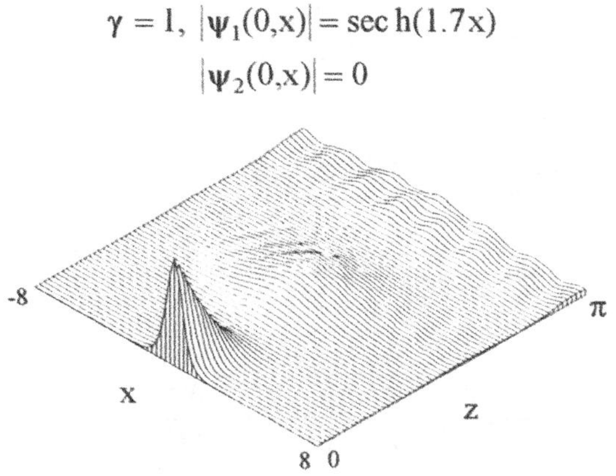

VARIABLE WIDTH CALCULATION

Figure 10a

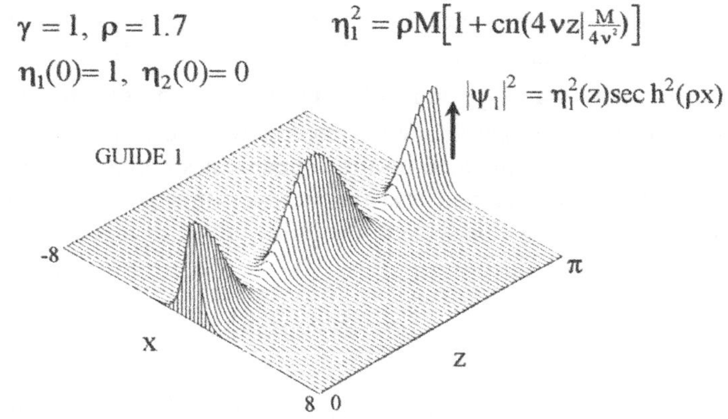

Figure 10b

The Euler-Lagrange equations can now be used to get equations for $\eta_{1,2}(z)$, $\theta_{1,2}(z)$, $\rho(z)$ and $C(z)$. They are

$$\frac{d\theta}{dz}=\frac{8}{3}\rho N-\frac{4\nu N}{\sqrt{M^2-N^2}}\cos\theta \tag{4.31a}$$

$$\frac{dN}{dz}=4\nu\sqrt{M^2-N^2}\sin\theta \tag{4.31b}$$

$$\frac{d\rho}{dz}=-2C\rho \tag{4.31c}$$

$$\frac{d}{dz}\left(\frac{C}{\rho}\right)=\frac{8}{\pi^2}\rho^2\left(\rho-M-\frac{N^2}{M^2}\right) \tag{4.31d}$$

where $\theta=\theta_2-\theta_1, 2N=\eta_2^2-\eta_1^2, 2M=\eta_2^2+\eta_1^2$. These equations can now be used to generate *better* results than were obtained with fixed-width, chirpless, trial functions. Figure 11a shows the fixed-width trial function [variational method] results for η_1^2 as a function of propagation distance z. For these calculations the chirp is kept at its initial value $C_0=0$, ρ is fixed at $\rho_0=1..3$, $\nu=0$ and $N=-M$, at $z=0$. Note that η_1^2 turns out to be a simple, periodic, function of z with η_1^2 always returning to its original value. The new variational theory, which allows for chirp and width to *evolve*, from *initial* values C_0 ρ_0, with z, is shown in figure 11b and demonstrates that $\eta_1^2(z)$ still varies periodically but it does not return always to its initial amplitude. Instead, the amplitude develops a, relatively, slowly varying change with z. This behaviour is confirmed by the numerical simulations shown in figure 11c, thus confirming that improved trial function is rather good at reproducing the exact conditions. To conclude this section, the following observations will be made

- for some value of $\left(\frac{M}{\rho}\right)_{z=0}$ the soliton width is constant
- uncoupled solitons imply that $\left\langle\frac{N}{M}\right\rangle=1$, where $\langle\ \rangle$ denotes an average over z
- for $\left(\frac{M}{\rho}\right)_{z=0}<\frac{1}{1.5}$ the soliton width increases, initially
- if $C=0$, initially, the soliton width keeps on increasing provided that $\left(\frac{M}{\rho}\right)_{z=0}<0.33$ i.e. the soliton radiates during switching
- if $C=0$ and $\left(\frac{M}{\rho}\right)_0>\frac{1}{1.5}$ the soliton width decreases, initially
- if C is large enough, the soliton will radiate.

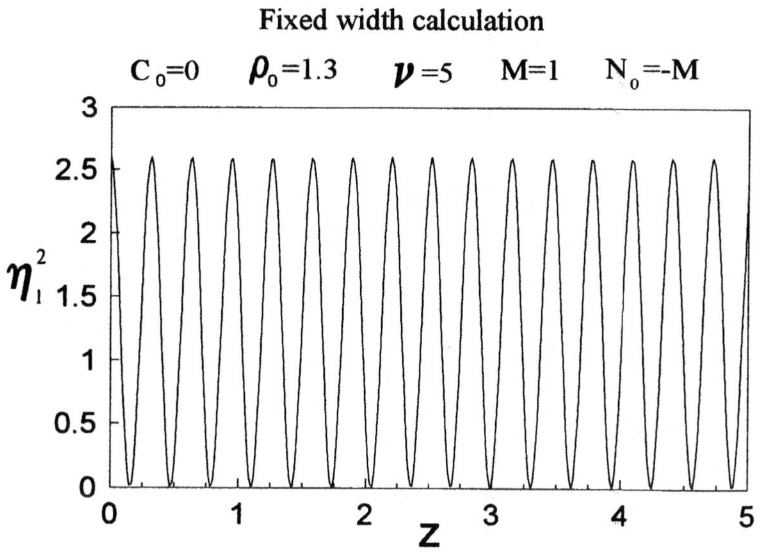

Figure 11a

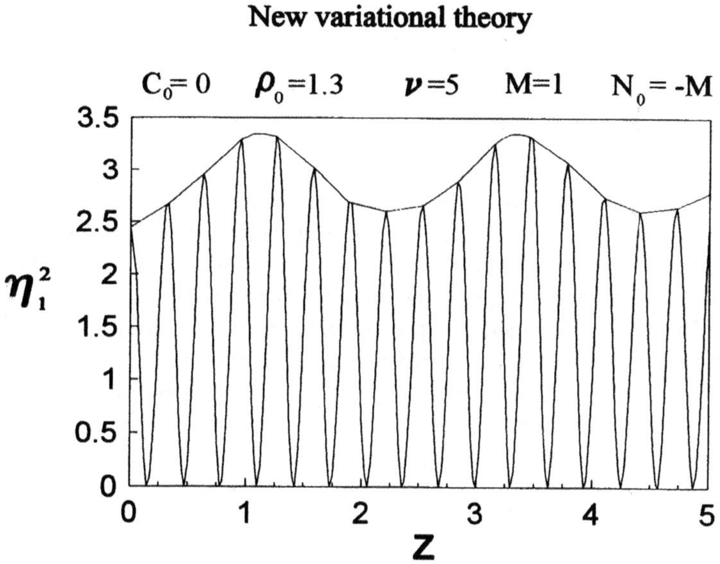

Figure 11b

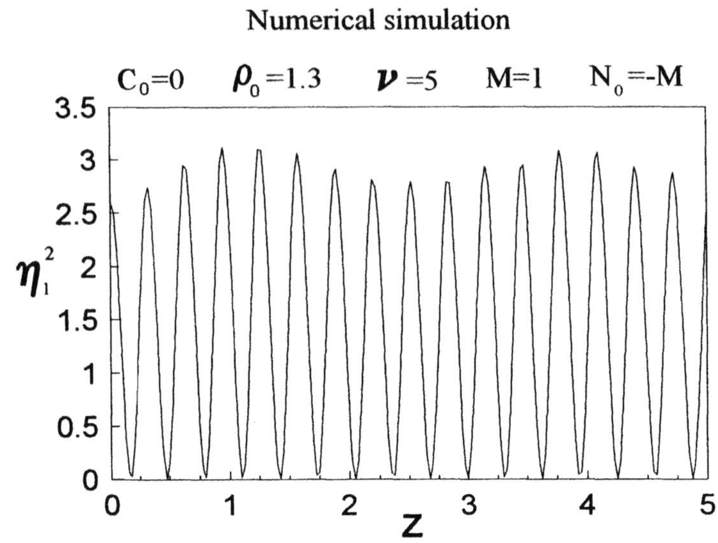

Figure 11c

5. Wave Scattering by a Nonlinear Interface/Film

It is often the case that nonlinear materials, like low-dimensional semiconductors, are accessible through a single surface or in thin film form. It is extremely interesting, therefore, and very important, to consider the behaviour of an interface between a nonlinear material and a linear material or, indeed, another nonlinear material. The ability of such interfaces to impose very desirable switching properties on electromagnetic waves reflected from them is a primary concern. This behaviour can be bistable, or multistable, and may well be associated with the generation of solitons in the neighbourhood of the interface. It seems perfectly possible, in principle, to engineer low-dimensional semiconductor structures to yield such interfaces but it must be borne in mind that 'possible in principle' may never reach the marketplace.

Basically, three, intermediate, possibilities spring to mind when addressing interface problems. These are (a) a linear-linear interface (b) a linear-nonlinear interface (c) a nonlinear-nonlinear interface. All will yield something interesting but the technique used to address (b) will not be the same as the one used for (c).

Case (a), the linear-linear structure, can be set up by looking at a single interface between two, semi-infinite, dielectric media that have dielectric constants ε_1 and ε_2, respectively. If a plane, electromagnetic, wave arrives, through medium 1, at the interface, such that its propagation direction is at an angle θ_i to the normal to the surface, then linear physics demands that $\theta_i = \theta_r$ and $\sqrt{\varepsilon_1}\,\sin\theta_i = \sqrt{\varepsilon_2}\,\sin\theta_t$, where θ_r is the angle of reflection and θ_t is the angle of transmission. These conditions are nothing more than Snell's law. Total reflection will occur, for electromagnetic radiation arriving at the interface, through the medium with dielectric constant ε_1, if $\varepsilon_1 > \varepsilon_2$ and the critical angle, at which this will occur is $\theta_c = \sin^{-1}\left(\sqrt{\frac{\varepsilon_2}{\varepsilon_1}}\right)$ i.e. if $\theta_i > \theta_c$ all of the incident wave will be *reflected* from the interface.

What this means for the second medium, beyond the interface, is that an exponentially decaying field is set up there, very closely confined to the interface region. What then happens if a *beam* of electromagnetic radiation is incident upon the interface, defined above? This is a fascinating problem that has long ago received some attention and has benefited from a very elegant solution. Basically, a beam can be Fourier-analysed into an infinite set of plane waves, travelling at all possible angles to the interface and, therefore, with all possible wavenumbers (momenta). It must be expected, therefore, that the description of the reflection of the *whole beam* will be rather surprising. Geometrically, a ray of electromagnetic energy will reflect from the *point of impact.* It may come as a, relative, surprise that the centre of a beam appears to be reflected from a point displaced, or shifted, from the geometrical point of impact. In other words, if a line coinciding with the centre of a beam is thought of as 'a ray' then the reflection of this 'ray' appears not to emanate from its geometrically expected position. This is called the Goos-Hanchen (G.H.) shift[20,21], after its discoverers, and is illustrated in figure 12, where λ_0 is the vacuum wavelength. At the very least, the question of how nonlinearity will affect this shift should now arise.

The introduction of nonlinearity will begin with a discussion of the nonlinear-nonlinear interface. As a matter of fact, this problem contains a useful possibility that makes its solution rather easy. It is simply this. If two nonlinear materials are joined together at a single interface, then by *matching the materials*, the changes in the parameters across the interface can be turned into mere perturbations. Consider, for example, a linearly polarised field $\mathbf{E} = \overline{\mathrm{E}}\hat{\mathbf{y}}$, where $\hat{\mathbf{y}}$ is a unit vector along the y-direction. Suppose that field is incident upon the interface between two semi-infinite media, such that the interface lies along the x-direction and the normal lies along the z-direction. If the Fourier amplitude of the electric field $\overline{\mathrm{E}}$ is $\mathrm{E}(\omega) \equiv \mathrm{E}$, where ω is the angular frequency, then it satisfies the equation

$$\frac{\partial^2 E}{\partial x^2} + \frac{\partial^2 E}{\partial z^2} + k_0^2 \varepsilon(z, |E|^2) E = 0 \tag{5.1}$$

where $k_0 = \omega/c$, c is the velocity of light in a vacuum and no variation in the y-direction has been anticipated.

The refractive index of each semi-infinite medium is $n_1 + \alpha|E|^2$, $n_2 + \alpha_2|E|^2$, respectively, where $\alpha_{1,2}$ are nonlinear coefficient and the following conditions will be imposed

- $|n_1^2 - n_2^2| << n_1, n_2$
- $|\alpha_1 - \alpha_2| << \alpha_1, \alpha_2$

which guarantee that an *equivalent-particle* theory can be developed. Rather like the idea of an equivalent particle, rolling in a well, was a nice representation of the

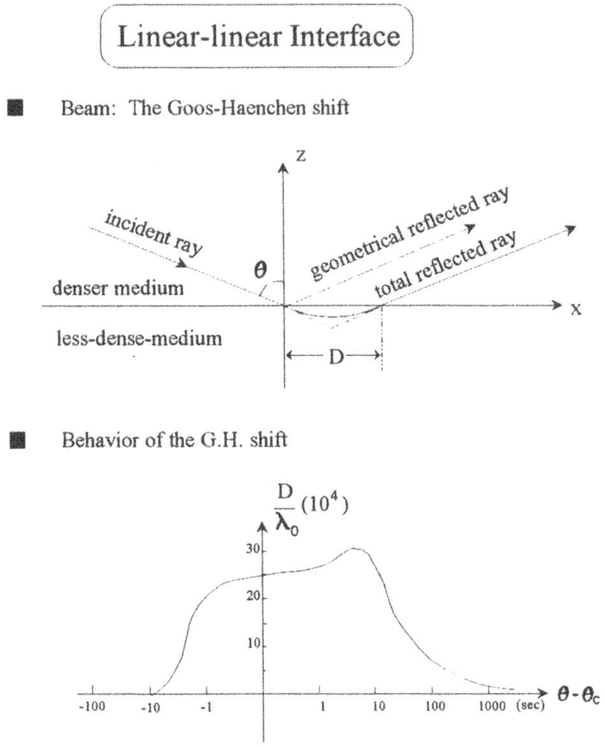

After B.R.Horowitz and T.Tamir, J.O.S.A., vol 61, 586(1971)

Figure 12

couplers in the previous section. Given that the refractive indices are nonlinear, on both sides of the interface, the dielectric function of the nonlinear-nonlinear interface system is

$$\varepsilon(z,|E|^2) = \begin{cases} n_1^2 + 2n_1\alpha_1|E|^2, & z < 0 \\ n_2^2 + 2n_2\alpha_2|E|^2, & z > 0 \end{cases} \tag{5.2}$$

Suppose now that the Fourier amplitude E is

$$E(x,z) = F(x,z)\exp[i\beta k_0 z] \tag{5.3}$$

where F(x,z) is a *slowly varying* function and the *fast* spatial dependence is accounted for through the wavenumber βk_0. Actually, k_0 is factored out here so that a description can be given in terms of an *effective index* β. The transformations involving replacing $\frac{k_0 x}{2\beta}$ by x' and $k_0 z$ by z' can now be performed, but the dashes on x and z will be dropped since we know perfectly well what they really mean. In addition to transforming x and z, F(x,z) can be defined as

$$F(x,z) = \sqrt{\frac{1}{\alpha_1}}\psi(\alpha,z)e^{-[\beta^2-n_1^2]x} \tag{5.4}$$

so that the whole problem is expressible as a *perturbed* nonlinear Schrödinger equation i.e.

$$i\frac{\partial\psi}{\partial x}+\frac{\partial^2\psi}{\partial z}+2|\psi|^2\psi = U(z,|\psi|^2)\psi \tag{5.5}$$

where the perturbation term is[22-25]

$$U(z,|\psi|^2) = \begin{cases} 0 & , z<0 \\ n_1^2 - n_2^2\left[\frac{\alpha_2-\alpha_1}{\alpha_1}\right]|\psi|^2 & , z>0 \end{cases} \tag{5.6}$$

The problem is now defined but the solutions are a little harder to obtain. One approach is to use a technique known as the *inverse-scattering method.* This method has often been referred to in glowing terms and is widely regarded as one of the most important theoretical developments of this century. The main idea is to convert the nonlinear equation into an eigenvalue problem in which the required solution of the original differential equation acts as a scattering potential. It is too much of a deviation to go any further into the method here or to go through the mathematics of applying it. Suffice to say that all readers are exhorted to learn this most elegant of methods[26-30]. In the meantime, however, only the main results will be quoted.

The main question is to ask how the amplitude and position of a beam created near to the surface will evolve[22-25]. In each nonlinear medium the energy incident upon, or moving away from, the surface is carried as nonlinear beam i.e. a spatial soliton. In the absence of an interface, U = 0, such a beam has the mathematical form

$$\psi(\alpha, z) = \eta \operatorname{sech}[\eta(z - z_0)] \exp\left[i \frac{\xi}{2}(z - z_0 + i\theta) \right] \tag{5.7}$$

where η is the peak amplitude of the self-focused beam, z_0 is the location of the beam centre, ξ is the angle the beam makes to the x-axis and $\theta = \theta_0 + \left[\frac{\xi^2}{4} + \eta^2\right]x$ is the phase of the beam. θ_0 is just some arbitrary constant that can be safely set to zero, without destroying any part of the problem. The problem now is to find out what a finite U value (i.e. the presence of the interface) will do.

A spatial soliton incident upon such an interface will refract into the next medium and take off with an amplitude η and a beam centre position z_0. Because it propagates at an angle ξ to the x-axis, then

$$\frac{d\eta}{dx} = 0, \quad \frac{dz_0}{dx} = \xi \tag{5.8}$$

Also the beam angle, obtained from perturbation theory involving the inverse-scattering technique, evolves as[25]

$$\frac{d\xi}{dx} = \left(n_2^2 - n_1^2\right)\eta \operatorname{sech}^2(\eta z_0) + \frac{(\alpha_2 - \alpha_1)}{\alpha_1}\eta^3 \operatorname{sech}^4(\eta z_0) \tag{5.9}$$

and the phase evolves as

$$\begin{aligned} \frac{d\theta}{dx} = \frac{\xi^2}{4} + \eta^2 + \frac{\left(n_1^2 - n_2^2\right)}{2}[1 + \tanh(\eta z_0)] \\ - \frac{3(\alpha_2 - \alpha_1)}{2\alpha_1}\eta^2\left[\frac{2}{3}\tanh(\eta z_0) - \frac{1}{3}\tanh^3(\eta z_0)\right] \end{aligned} \tag{5.10}$$

The combination of equations (5.8) and (5.9) leads to

$$\frac{d^2 z_0}{dx^2} = -\frac{dW}{dz_0} \tag{5.11}$$

which implies that

$$\frac{\xi^2}{2} + W(\eta z_0) = \text{constant} \tag{5.12}$$

where

$$W(\eta z_0) = (n_1^2 - n_2^2)[1 + \tanh(\eta z_0) - \frac{(\alpha_2 - \alpha_1)}{\alpha_1}\eta^2\left[\frac{2}{3} + \tanh(\eta z_0) - \frac{1}{3}\tanh^3(\eta z_0)\right] \quad (5.13)$$

So defining $\gamma = \frac{\alpha_1(n_1^2 - n_2^2)}{\eta^2(\alpha_2 - \alpha_1)}$ means that the potential shapes shown in Fig. 13 can be discussed, as follows. The quantities W_{max} and $W(+\infty)$ are

- $W_{max} = \frac{(n_1^2 - n_2^2)}{3\gamma}\left(1 - \sqrt{1-\gamma}\right)^2\left(2\sqrt{1-\gamma} + 1\right)$
- $W(+\infty) = 2(n_1^2 - n_2^2)\left(1 - \frac{2}{3\gamma}\right)$

Dynamics Of The Beam-Centre

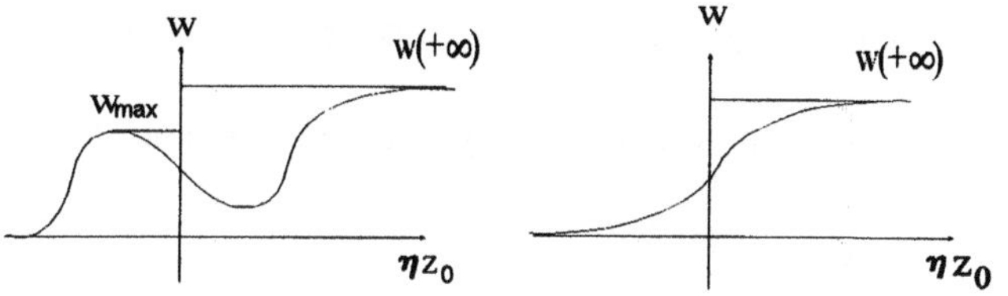

Figure 13

so, for $\frac{2}{3} < \gamma < 1$, the appearance of a minimum for $\eta z_0 > 0$ shows that it is really a well that can *capture* a soliton. In other words, for this condition, a spatial soliton incident upon this nonlinear-nonlinear interface will cause energy to be trapped into the ηz_0 position, corresponding to the minimum in W, forcing it to travel along parallel to the interface. For $\gamma > 1$ total reflection will occur. In fact, for a soliton, incident upon this interface, the following conclusions can be drawn

- $\frac{\xi^2}{2} \geq \max\{W(+\infty), W_{max}\}$: transmission
- $\frac{\xi^2}{2} < \max\{W(+\infty), W_{max}\}$: total reflection

A problem arises now, however, because the formalism just used can not be extended to the linear limit. For example, it can not handle the linear-nonlinear interface or the linear-linear interface. This is partly because the soliton is a purely nonlinear entity, with no linear counterpart, and partly because of the way the solution has been set up. Thus an equivalent particle theory is not applicable to the linear-nonlinear interface problem. Also a plane wave is unstable in a nonlinear medium, thus removing the possibility of another class of problems being investigated. Clearly a new approach[31,32] is needed and this will now be described.

If we return to the basic equations, and consider only electric fields polarised along the y-axis, then the wave equation is

$$\left(\frac{\partial^2}{\partial x^2} + \frac{\partial^2}{\partial z^2}\right)E + k_0^2 \varepsilon_T(z, |E|^2)E = 0 \tag{5.14}$$

where $k_0 = \frac{\omega}{c}$, $\varepsilon_T(z,|E|^2)$ is the total dielectric function describing the whole interface and $\frac{\partial}{\partial y} = 0$, implying uniformity in the y-direction. For a linear-nonlinear interface the refractive index profile is

$$\varepsilon_T(z,|E|^2) = \begin{cases} \varepsilon_1 & , z < 0 \\ \varepsilon(|E|^2) = \varepsilon_2 + \dfrac{|E|^2}{E_c^2} & , z > 0 \end{cases} \tag{5.15}$$

where ε_1 is a linear dielectric constant, ε_2 is the linear part of $\varepsilon(|E|^2)$ and we choose to write the nonlinear coefficient as $\left(1/E_c^2\right)$. There is nothing deep about the latter step. It is only a choice, prompted by the realisation that $\varepsilon(|E|^2)$ is, after all, dimensionless. Equation (5.15) describes a *linear* half-space joining onto a *nonlinear* half-space so the problem about to be analysed really is very different from the previous case and it embodies some difficulties. The important step to take now, however, is to split E into two parts i.e. write

$$E = E_1 + E_2 \tag{5.16}$$

and recognise that E_1 and E_2 satisfy the equations

$$\nabla^2 E_1 + \varepsilon^L E_1 = -[\varepsilon_T - \varepsilon(|E|^2)]E_2 \equiv F_2 \tag{5.17}$$

and

$$\nabla^2 E_2 + \varepsilon(|E|^2)]E_2 = -[\varepsilon_T - \varepsilon^L]E_1 \equiv F_2 \tag{5.18}$$

where ε^L are linear values that are either ε_1 or ε_2, as appropriate. This split can be readily checked by inspection but the idea behind it will now be exposed.

When $F_1 = 0$ and $F_2 = 0$ the media are *disconnected* and we obtain, what will be termed, the zeroth-order solutions $E_1^{(0)}$ and $E_2^{(0)}$. In fact, $E_1^{(0)}$ describes a plane wave, travelling in a linear medium and $E_2^{(0)}$ describes a self-focused beam travelling, of course, in the nonlinear medium. Figure 14a shows clearly what is happening. On the one hand, in the linear medium, a plane wave travels towards the

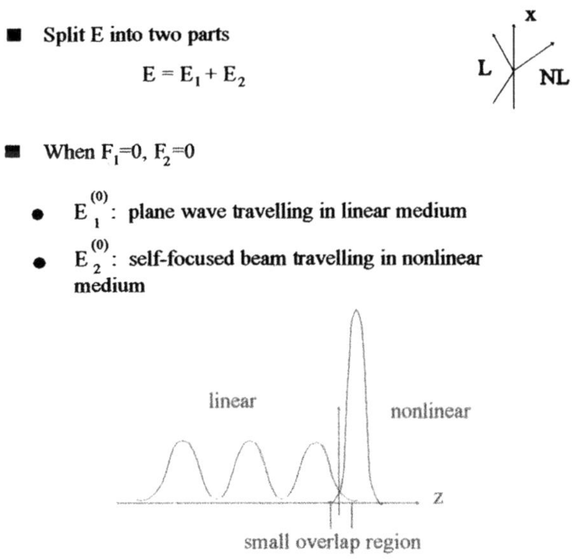

Figure 14a

interface. At the interface, if total internal reflection occurs, $E_1^{(0)}$ has only a *small* exponential tail reaching into the $z > 0$ half-space. On the other hand, a self-focused beam travels close to the interface and so $E_2^{(0)}$ also has only a small exponential tail reaching into the $z < 0$ half-space. Thus F_1 and F_2, which are

$$F_1 = \begin{cases} 0 & , z < 0 \\ -(\varepsilon_1 - \varepsilon_2) & , z < 0 \end{cases} \tag{5.19a}$$

$$F_2 = \begin{cases} -\dfrac{|E|^2}{E_c^2} E_1 & , z > 0 \\ 0 & , z < 0 \end{cases} \tag{5.19b}$$

can be treated as perturbations i.e. the tail of the self-focused wave, in $z < 0$, perturbs $E_1^{(0)}$ to E_1 and the tail of the incoming wave perturbs $E_2^{(0)}$ to E_2, where E_1 and E_2 are now called first-order solutions, as shown in figure 14b.

- The source term

$$F_1 = \begin{cases} 0 & z>0 \\ -(\epsilon_1 - \epsilon_2) E_2 & z<0 \end{cases}$$

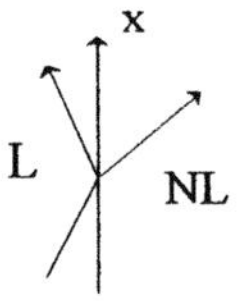

$$F_2 = \begin{cases} -\dfrac{|E|^2}{E_c^2} E_1 & z>0 \\ 0 & z<0 \end{cases}$$

- $E_1^{(0)}$ has only a small exponential tail reaching into the area $z>0$ half-space

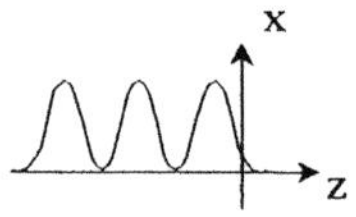

- $E_2^{(0)}$ has only a weak exponential tail stretching into the linear $z<0$ half-space

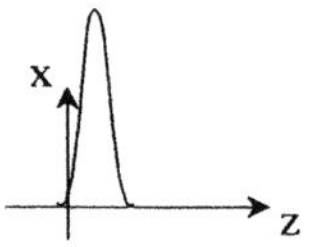

- F_1, F_2 are treated as perturbations
  - The tail of the self-focused wave, in $z<0$, perturbs $E_1^{(0)}$ to E_1
  - The tail of the incoming wave perturbs $E_2^{(0)}$ to E_2

Figure 14b

These zeroth-order solutions are

$$E_1^{(0)} = \begin{cases} C\exp[-\kappa_2 z - i\gamma z] & , z<0 \\ A_0\exp[-\phi_0 - i\kappa_1 z - i\gamma x] + B\exp[i\kappa_1 z - i\gamma x] & , z>0 \end{cases} \tag{5.20}$$

where $\kappa_1 = \sqrt{\varepsilon_1 - \gamma^2}$, $\kappa_2 = \sqrt{\gamma^2 - \varepsilon_2}$ and

$$B = A_0\left(\frac{i\kappa_1 - \kappa_2}{i\kappa_1 + \kappa_2}\right)\exp(i\phi_0),\ C = A_0\frac{2i\kappa_1}{i\kappa_1 + \kappa_2}\exp(i\phi_0) \tag{5.21}$$

Here (κ_1, γ) are wave vector components, A_0 is the incident amplitude, B is the reflected amplitude and ϕ_0 is the phase change experienced on reflection. Hence the result given by (5.20) is quite elementary. The zeroth-order $E_2^{(0)}$ is

$$E_2^{(0)} = \frac{\sqrt{2}\kappa_2^{(so)}E_c}{\cosh\left[\kappa_2^{(so)}(z - z_0)\right]}\exp\left(-i\gamma_s^{(0)}x\right) \tag{5.22}$$

where the s label means self-focusing and $\kappa_2^{(so)} = \sqrt{\gamma_s^{(0)2} - \varepsilon_2}$. The form of equation (5.22) is also well-known. It describes a self-focused beam [i.e. a channel] so that $\left|E_s^{(0)}\right|^2$ has the familiar, soliton-like sech² shape. The disposition, and appearance, of the parameters $\kappa_2^{(so)}$, E_c, z_0 and $\gamma_s^{(0)}$ can easily be checked by substitution. The phase factor $\exp\left(-i\gamma_s^{(0)}x\right)$ merely acknowledges the propagation along the x-axis.

From the information already given it is now possible to obtain the first-order (perturbed) solutions E_1 and E_2. These, after performing a certain amount of work and giving the problem some thought, are

$$E_1 = \begin{cases} Ce^{-\kappa_2 z - i\gamma x} + Qe^{-\kappa_2^{(s)}z - i\gamma_s x} & , z>0 \\ A_0 e^{-i\phi_0 - i\kappa_1 z - i\gamma x} + Be^{-\kappa_1 z - i\gamma x} & \\ +Ge^{i\kappa^{(s)}z - i\gamma_s x} + De^{\kappa_2^{(s)}(z - z_0) - i\gamma_s x} & , z<0 \end{cases} \tag{5.23}$$

where

- $D = -2\sqrt{2}\kappa_2^{(s)}E_c$
- $Q = 2\sqrt{2}\kappa_2^{(s)}\left(\frac{\kappa_2^{(s)} - i\kappa_1^{(s)}}{\kappa_2^{(s)} + i\kappa_1^{(s)}}\right)e^{-\kappa_2^{(s)}z_0}E_c$
- $G = 4\sqrt{2}\frac{\kappa_2^{(s)^2}}{\left(\kappa_2^{(s)} + i\kappa_1^{(s)}\right)}e^{-\kappa_2^{(s)}z_0}E_c$
- $\kappa_2^{(s)} = \sqrt{\gamma_s^2 - \varepsilon_2}$, $\kappa_1^{(s)} = \sqrt{\varepsilon_1 - \gamma_s^2}$

$$E_2 = \frac{\sqrt{2}\kappa_2^{(s)} E_c}{\cosh\left[\kappa_2^{(s)}(z-z_0)\right]} \exp\left[-i\int \gamma_s dx - i\alpha(z-z_0)\right] \tag{5.24}$$

where α, z_0 and γ_s are allowed to change during the propagation.

The evolution equations for the parameters a, z_0 and γ_s are

$$\frac{dz_0}{dx} = \frac{a}{\gamma_s^{(0)}} \tag{5.25a}$$

$$\frac{da}{dx} = -\frac{\sqrt{2}\kappa_2^{(s)2}}{3\gamma_s^{(0)} E_c} \mathrm{Re}\left[Ce^{i\phi} + Q\right] \exp\left(-\kappa_2^{(s)} z_0\right) \tag{5.25b}$$

$$\frac{d\gamma_s^2}{dx} = \frac{2\sqrt{2}\kappa_2^{(s)3}}{\gamma_s^{(0)} E_c} \mathrm{Im}\left[Ce^{i\phi} + Q\right] \exp\left(-\kappa_2^{(s)} z_0\right) \tag{5.25c}$$

where $\phi = \int [\gamma_s - \gamma] dx$. The physical interpretations of a, z_0 and γ_s are as follows.

- a: phase change producing noncollinearity of the channel axis and the interface direction
- z_0: centre of the self-focused beam
- γ_s: perturbed wavenumber, related to the energy flow P_s in the self-focused beam.

This energy flow is

$$P_s = 4\gamma_s\left[\gamma_s^2 - \varepsilon_2\right]^{\frac{1}{2}} E_c^2 \tag{5.26}$$

The dynamics of a self-focused beam near to a linear-nonlinear interface can now be addressed. Attention will be restricted, however, to the simpler case in which the pump beam is absent i.e. $A_0 = 0$. In this case

$$\frac{d^2 z_0}{dx^2} = -\frac{4}{3} \frac{\kappa_2^{(s)3}\left[\kappa_2^{(s)3} - \kappa_1^{(s)2}\right]}{\gamma_s^{(0)2}[\varepsilon_1 - \varepsilon_2]} \exp\left[-2\kappa_2^{(s)} z_0\right] \tag{5.27}$$

$$\frac{d\gamma_s^2}{dx} = -16 \frac{\kappa_1^{(s)}\kappa_2^{(s)5}}{\gamma_s^{(0)}[\varepsilon_1 - \varepsilon_2]} \exp\left[-2\kappa_2^{(s)} z_0\right] \tag{5.28}$$

Equation (5.28) shows, immediately, that $\frac{d^2\gamma_s^2}{dx} < 0$ so that energy is always radiating into the linear medium. The other features that are revealed are as follows. If $\kappa_2^{(s)} > \kappa_1^{(s)}\left(\gamma_s^2 > \frac{\varepsilon_1 + \varepsilon_2}{2}\right)$ then $\frac{d^2 z_0}{dx^2} < 0$ and any high intensity beam is attracted to the interface. If $\kappa_2^{(s)} < \kappa_1^{(s)}\left(\gamma_s^2 \frac{\varepsilon_1 + \varepsilon_2}{2}\right)$ then $\frac{d^2 z_0}{dx^2} > 0$ and any low intensity beam is repulsed by the interface. If $\kappa_2^{(s)} = \kappa_1^{(s)}$, then $\gamma_s^2 = \frac{\varepsilon_1 + \varepsilon_2}{2}$ and P_s becomes the *critical* value $P_{sc} = 2\sqrt{\varepsilon_1^2 - \varepsilon_2^2 E_c^2}$. These interesting conditions are now expressible in terms of P_{sc}. Suppose the initial power is bigger than P_{sc}. A nonlinear beam will then be, initially, attracted to the interface, but after radiating some energy into the linear medium, it will bounce away from the interface. It then takes off into the deep interior of the nonlinear semi-infinite half-space. If the initial power is smaller than P_{sc} a self-focused beam will be repelled by the interface, even if it does radiate some energy into the linear medium. Figure 15 shows this behaviour both as a cross-sectional power contour plot and as a three-dimensional propagation plot.

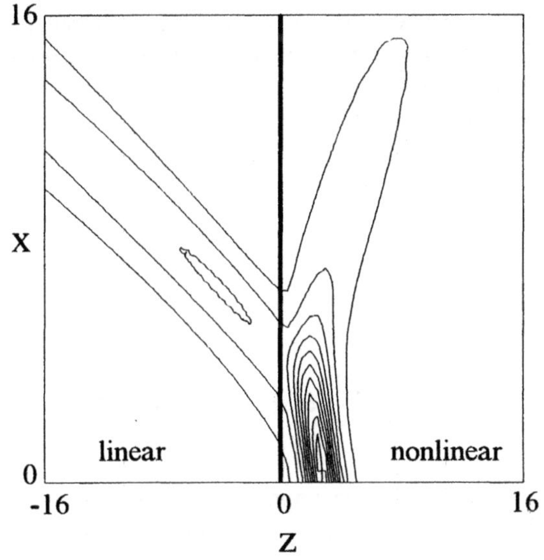

Figure 15a

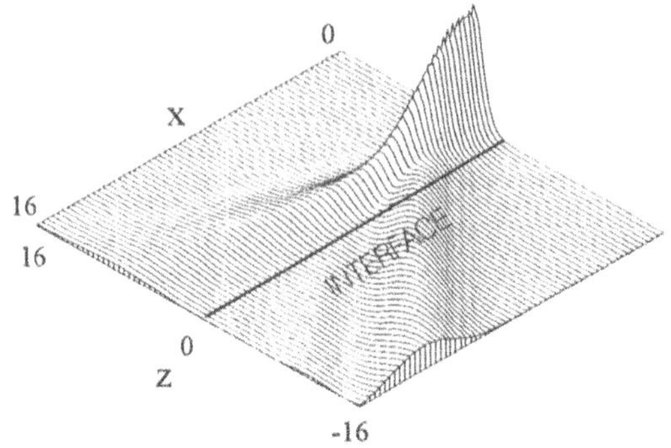

Figure 15b

If, instead of a single interface, there exists a nonlinear thin film, surrounded by linear half-spaces, the system becomes even more interesting. This is quite a large topic so interest here will be restricted to what might be termed a "nonlinear worm". By this we mean a self-focused beam that is attempting to propagate along the film, as shown in figure 16. The beam wriggles about, rather like a worm and radiates energy through the film boundaries, into the linear media, as it progresses. The evolution equation for the centre of the beam is

$$\frac{d^2 z_0}{dx^2} = \frac{8}{3}\frac{\kappa_2^{(s)3}}{\gamma_s^{(0)2}\Delta\varepsilon}\left(\kappa_2^{(s)2} - \kappa_1^{(s)2}\right)\sinh\left(2\kappa_2^{(s)} z_0\right)\exp\left(-2\kappa_2^{(s)} d\right) \tag{5.29}$$

and that for the power, expressed through γ_s^2, is

$$\frac{d\gamma_s^2}{dx} = -32\frac{\kappa_1^{(s)}\kappa_2^{(s)2}}{\gamma_s^{(0)}\Delta\varepsilon}\cosh\left(2\kappa_2^{(s)} z_0\right)\exp\left(-2\kappa_2^{(s)} d\right) \tag{5.30}$$

where $\Delta\varepsilon = \varepsilon_1 - \varepsilon_2$. Clearly what is happening is that the beam is oscillating in position, as it propagates, with a period that is increasing with x. This is sketched in figure 16 and the x-dependent period T(x) is given by

$$\frac{4\pi^2}{T^2(x)} = \frac{16\kappa_2^{(s)9}}{3(\varepsilon_1 - \varepsilon_2)}\left[\kappa_1^{(s)2} - \kappa_2^{(s)2}\right]\exp\left[-2\kappa_2^{(s)} d\right] \tag{5.31}$$

Dynamics of Self-focused Beam in a Nonlinear Film

- The centre of the beam

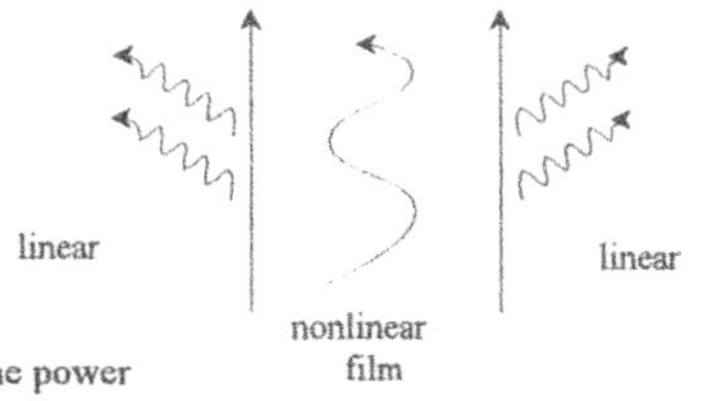

- The power

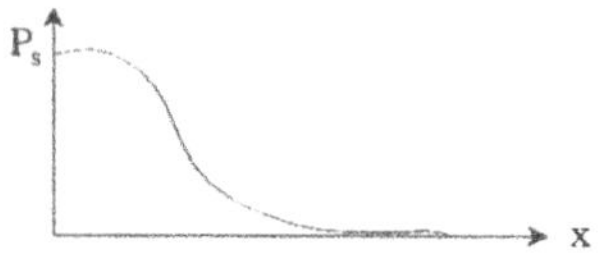

Figure 16a

Dynamics of Self-focused Beam in a Nonlinear Film

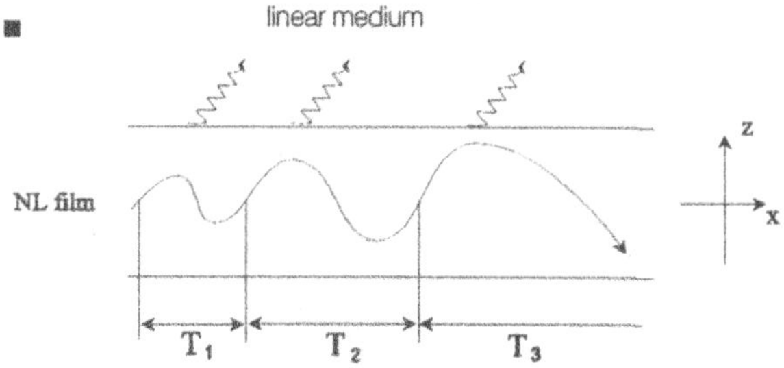

- Frequency of oscillation $\Omega = \frac{2\pi}{T}$
  - Radiation to the linear medium causes γ_s to change
  - $\kappa_1^{(s)}, \kappa_2^{(s)}$ change accordingly
  - $\Omega(x)$: local frequency, function of propagation distance x

Figure 16b

Actually, it is radiation to the linear medium that causes γ_s to change which, in turn, causes changes to $\kappa_1^{(s)}$ and $\kappa_2^{(s)}$. Hence a local frequency $\Omega(x)=2\pi/T(x)$ is set up, which changes with propagation distance. Figure 17 illustrates this behaviour rather nicely.

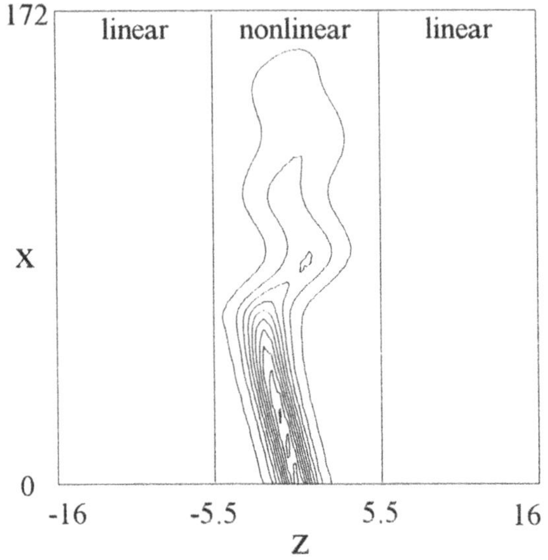

Figure 17a

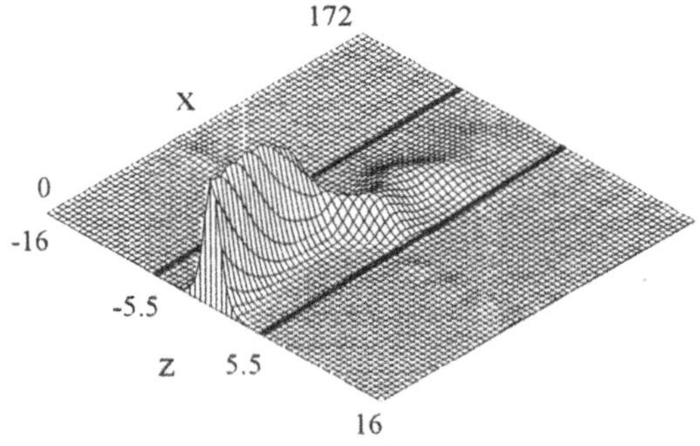

Figure 17b

In the final part of this section, three simple, yet effective, applications can be considered. The first is sketched in figure 18 and can be called an all-optical

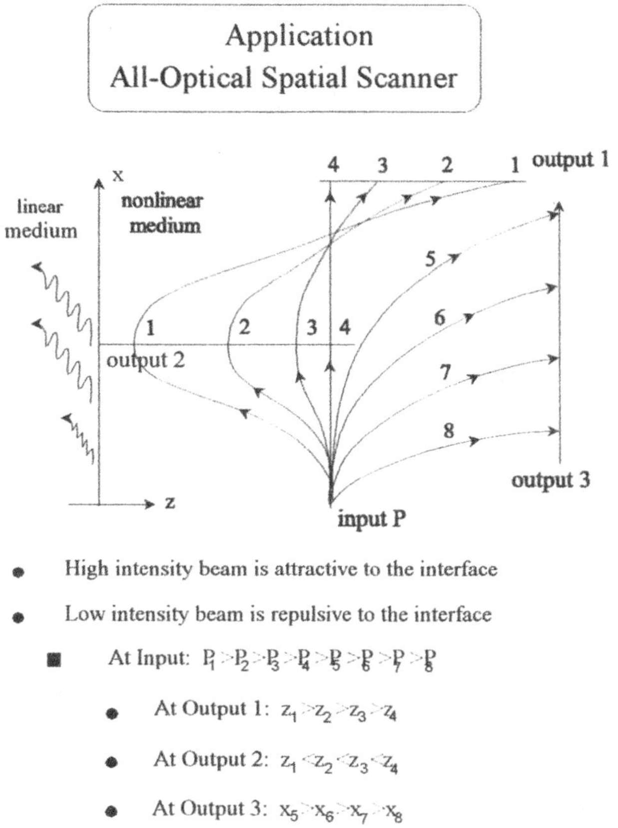

Figure 18

scanner. It is based on a single interface and uses the facts that high intensity beams are attracted to the interface, while low intensity ones are not. If the output ports are imagined to be where they are in figure 18 then at least two outcomes can be achieved. For given input powers $P = P_1 > P_2 > ... > P_8$, the high powers swing towards the interface and the low powers swing away from it, thus giving output 1 or output 3, depending on where you take the measurements. Figure 19 shows an application of nonlinear attenuation. Here the fact that energy in the nonlinear film decays, due to radiation into the nonlinear medium, is being specifically exploited. If the input power is P_{in} and the output power is P_{out} then the attenuation coefficient is $\alpha = \frac{(P_{in} - P_{out})}{P_{in}}$ and this function of P_{in} is shown in figure 19. A low power (P_1) a beam has a large width and the rate of diffraction is small. For such a beam the attenuation is rather low. At the other extreme, we have a high power (P_3) beam that has a much smaller beam width. For this type of beam the tail in the linear medium is small so, even for this case, the attenuation is very low. It is only for intermediate powers, like P_2, that the attenuation tends to rise.

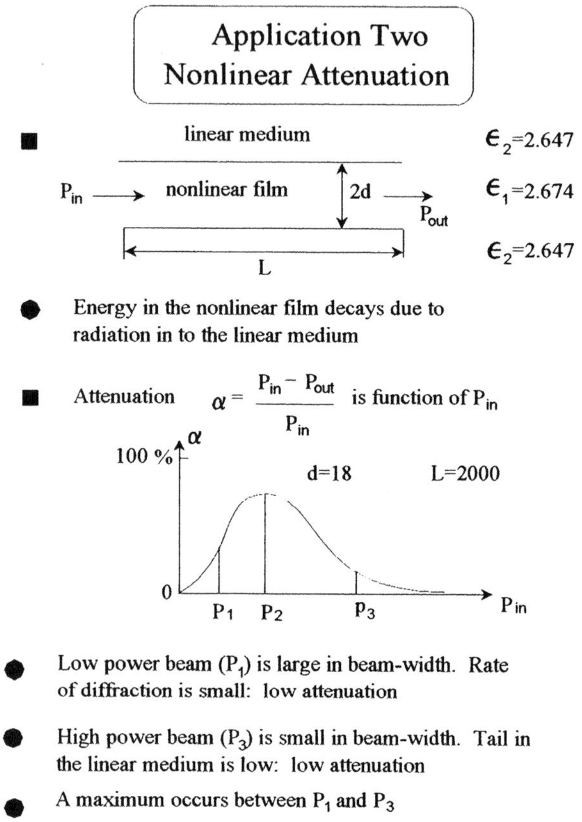

Figure 19

Figure 20 shows a third possibility which is called here the "energy reflector". This name implies that a mirror has been created by the presence of energy contained within a surface wave. As the figure shows, in the absence of surface wave, an incident beam has enough power to be attracted to the interface and there it is trapped. Now, if a surface wave is switched on, an energy barrier is formed and the incident beam is reflected back. In other words a mirror 'made from energy in the surface wave'....an energy reflector...has been created. Figure 21 is a numerical simulation of this phenomenon.

Application Three
Energy Reflector

■

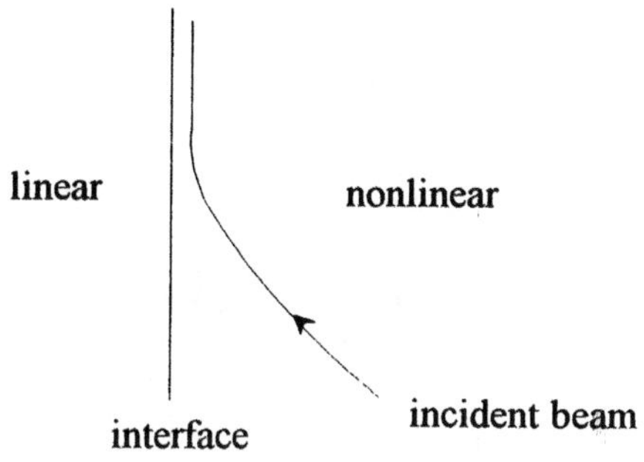

■ Incident beam is trapped by the interface

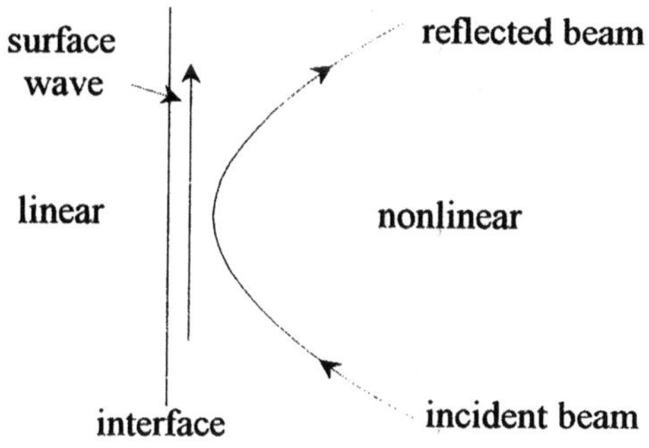

- when the surface wave is switched on, an energy barrier is formed near the interface
- incident beam is reflected back

Figure 20

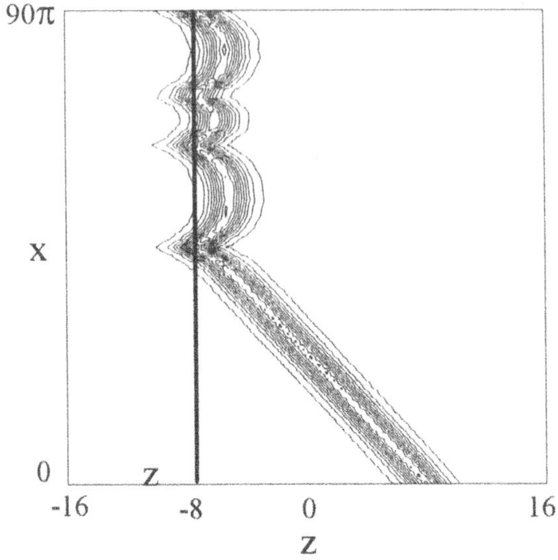

Figure 21a

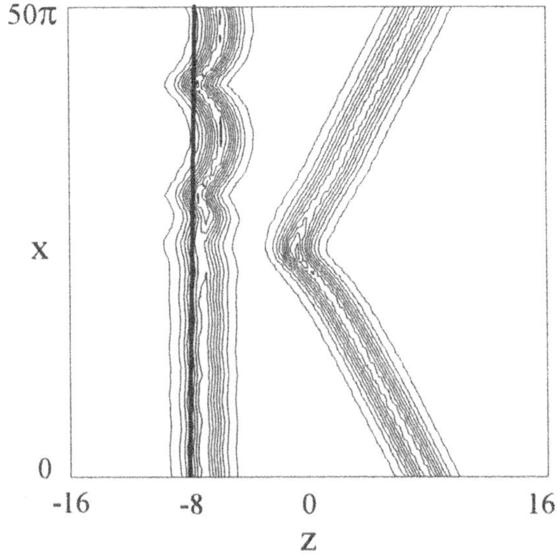

Figure 21b

6. Second-Order Nonlinearity : Cascading

The majority of the literature on envelope or spatial solitons assumes that the nonlinearity is third-order in origin. The solitonic phenomena are also assumed to be driven non-resonantly. In other words, the polarisation **P** induced by an electromagnetic wave, with electric field vector **E**, during its passage through a dielectric material is assumed to be

$$P_i = \varepsilon_0 \left\{ \chi_{ij}^{(1)} E_j + \chi_{ijk}^{(2)} E_j E_k + \chi_{ijkl}^{(3)} E_j E_k E_l + \ldots \right\} \tag{6.1}$$

where i = x,y,z are the, respective, coordinates, ε_0 is the permittivity of free space, $\chi^{(1)}$ is a *linear* susceptibility tensor, $\chi^{(2)}$ is a *second-order* susceptibility tensor and $\chi^{(3)}$ is a *third-order* susceptibility tensor. For materials possessing a centre of symmetry, it is easy to show that $\chi^{(2)}$ has all zero elements. In such materials, any nonlinearity occurs because the third-order susceptibility, $\chi^{(3)}$, makes its appearance as a nonlinear refractive index at the *same frequency as the input* (fundamental) wave. This nonlinear change can be used for soliton propagation, because it can combat dispersion or diffraction.

For $\chi^{(2)}$ to be finite, the material must lack a centre of symmetry. In such a material, *two* field components mix to produce a third i.e. second-harmonic (2ω) generation is possible. Until recently, such materials were not considered to be good candidates for soliton effects. This view has now changed dramatically[14,15,33-36] and some details will be set out below. Linear phase matching is not a requirement and AlGaAs is a suitable material. Indeed, low-dimensional, quantum well, semiconductor structures can be grown in such a way that enhanced second-order effects can be obtained. Furthermore, these quantum well structures can yield *particular* components of the $\chi^{(2)}$ tensor.

Double refraction, shown in figure 22, is a familiar phenomenon and is clearly

Figure 22

related to the crystal symmetry i.e. iceland spar (calcite) has a symmetry lower than the cubic group and the crystal class may reveal the lack of a centre of symmetry and the existence of $\chi^{(2)}$. Important $\chi^{(2)}$ materials have a single optic axis (along the z-axis, for example) and belong to the tetragonal [e.g. $BaTiO_3$], trigonal [e.g. $LiNbO_3$] or hexagonal [e.g. α-quartz] classes. For all of these classes, the linear susceptibility tensor is

$$\chi^{(1)} = \begin{bmatrix} xx & 0 & 0 \\ 0 & xx & 0 \\ 0 & 0 & zz \end{bmatrix} \tag{6.2}$$

where the notation means that $\chi^{(1)}$ is a diagonal tensor with non-zero elements $\chi^{(1)}_{xx}$ and $\chi^{(1)}_{zz}$. In other words, for such materials, the linear tensor has only two independent components. The usually quoted classes for which there is no centre of symmetry are tetragonal [$\bar{4}2m$: KDP, ADP; 3mm : $BaTiO_3$], trigonal [3m : $LiNbO_3$] and hexagonal [622: α-quartz] but, by suitable manipulation, quantum well structures made from semiconductor materials can also be used. It is often not fully appreciated that the well-known semiconductor has a zincblende structure and therefore lacks a centre of symmetry. The finite $\chi^{(2)}$ that exists for GaAs, and other low-dimensional structures based on this material, can be made to have enhanced elements, or different, and more useful, elements by suitable quantum well engineering[15]. The number of finite elements of $\chi^{(2)}_{ijkl}$, for a particular crystal class are already tabulated and are readily available in the literature[6]. Some examples are shown in figure 23.

SECOND-ORDER

$$\chi^{(2)}_{ijk}$$

$\bar{4}2m$ {3 elements}

0	0	0	xyz	xzy	0	0	0	0
0	0	0	0	0	xzy	xyz	0	0
0	0	0	0	0	0	0	zxy	zxy

3m {5 elements}

0	0	0	0	0	xzx	xxz	$\overline{yyy}$	$\overline{yyy}$
$\overline{yyy}$	yyy	0	xxz	xzx	0	0	0	0
zxx	zxx	zzz	0	0	0	0	0	0

Figure 23

$\chi^{(2)}$ materials are of great technical interest because the nonlinearities are very fast i.e. the nonlinearity tends to switch on and off rather rapidly. Such materials are readily available, even in semiconducting form and they enjoy the advantage that devices can be made from them with relative ease. They are attractive, therefore, for integrated/electro-optic applications. As will be seen below, linear phase-matching is not required, since there is a compensating nonlinear phase shift. This is interesting because, classically, envelope and spatial solitons can use such a shift as a power-dependent refractive index, operating at a frequency equal to the input frequency. This kind of index variation usually arises through *third-order nonlinearity* which varies as $|\mathbf{E}|^2\mathbf{E}$, where $\mathbf{E}$ is the electric field vector. Hence if $\mathbf{E}$ varies with time t as $\exp(i\omega t)$, where ω is the angular frequency, then $|\mathbf{E}|^2\mathbf{E}$ also varies as $\exp(i\omega t)$. The conclusion, for third-order material, depends, however, on *rejecting* any third-harmonic (3ω) terms, which is usually justified on the grounds that there is serious linear phase mismatch, producing negligible proportions of 3ω intensity. Consistent, or not, this is the physical reason given for $\chi^{(3)}$ materials but it is now realised that the ω, 2ω production, implied by finite $\chi^{(2)}$, can be treated differently. Basically, a soliton, or an integrated optical device, needs to operate at a single frequency ω, so can the ω and 2ω waves combine to guarantee that this occurs? The answer, broadly speaking, is that *complex conjugate* of the ω-field $[\,E_\omega^*\exp(-i\omega t)\,]$ can combine with the 2ω-field $\left[E_{2\omega}\exp(2i\omega t)\right]$ to produce a contribution varying as $\exp(i\omega t)$. This then looks like an *effective* $\chi^{(3)}$ process and it is called *cascading*. A detailed discussion of this will now be given.

6.1 BASIC FORMALISM

The immediate objectives are to set up basic *scalar equations* for coupled ω and 2ω waves. Vector equations will arise for polarisation interactions but for this discussion only one (TE) polarisation will be assumed throughout. The principal activity is to determine the dependence of stationary states on the various parameters, followed up by a determination of the stability regimes for these states.

The second-order polarisations, associated with the existence of $\chi^{(2)}$ are

- zero frequency (optical rectification)

$$\mathbf{P}^{(2)}(0) = \chi^{(2)}(0;\omega,-\omega): \mathbf{E}_\omega\mathbf{E}_\omega^* + \chi^{(2)}(0;2\omega,-2\omega): \mathbf{E}_{2\omega}\mathbf{E}_{2\omega}^*$$

- fundamental frequency

$$\mathbf{P}^{(2)}(\omega) = \chi^{(2)}(\omega;-\omega,2\omega): \mathbf{E}_\omega^*\mathbf{E}_{2\omega} + \chi^{(2)}(\omega;0,\omega): \mathbf{E}_0\mathbf{E}_\omega$$

- harmonic frequency

$$\mathbf{P}^{(2)}(2\omega) = \chi^{(2)}(2\omega;\omega,\omega): \mathbf{E}_\omega\mathbf{E}_\omega + \chi^{(2)}(2\omega;0,2\omega): \mathbf{E}_0\mathbf{E}_{2\omega}$$

where the components of $\chi^{(2)}$ a have not been identified, the fields are written as vectors and the notation (final frequency: first frequency, second frequency) has been adopted. Since the driving term in Maxwell's equations is $\frac{d^2\mathbf{P}^{(2)}}{dt^2}$, it is clear that $\mathbf{P}^{(2)}(0)$ is not a source term for waves so that $\mathbf{P}^{(2)}(\omega)$ acts as source term for the $\mathbf{E}_\omega$ field and $\mathbf{P}^{(2)}(2\omega)$ acts as a source term for the $\mathbf{E}_{2\omega}$ field. It is the coupling between the ω and 2ω that is the really interesting possibility here.

Maxwell's equations are

$$curl\,\mathbf{H} = \frac{\partial \mathbf{D}}{\partial t} = \frac{\partial}{\partial t}\left\{\varepsilon_0\,\varepsilon\,\mathbf{E} + \mathbf{P}^{NL}\right\} \tag{6.3}$$

$$curl\,\mathbf{E} = -\mu_0 \frac{\partial \mathbf{H}}{\partial t} \tag{6.4}$$

where $\mathbf{H}$ is magnetic field, $\mathbf{D}$ is displacement vector, μ_0 is the permeability of free space, t is time, ε is the linear dielectric constant, $\mathbf{P}^{\mathrm{NL}}$ is the nonlinear polarisation and it will be assumed that $\mathbf{P}^{\mathrm{NL}} = \mathbf{P}^{(2)}$.

Equations (6.3) and (6.4) can easily be manipulated into the coupled equations

$$\frac{\partial^2 E_\omega}{\partial z^2} + \frac{\partial^2 E_\omega}{\partial x^2} + \left(\frac{\omega}{c}\right)^2 \varepsilon_\omega E_\omega + \mu_0\omega^2 P_x^{(2)}(\omega) = 0 \tag{6.5}$$

$$\frac{\partial^2 E_{2\omega}}{\partial z^2} + \frac{\partial^2 E_{2\omega}}{\partial x^2} + \left(\frac{2\omega}{c}\right)^2 \varepsilon_{2\omega} E_{2\omega} + \mu_0(2\omega)^2 P_x^{(2)}(2\omega) = 0 \tag{6.6}$$

where *scalar* fields E_ω, $E_{2\omega}$ are assumed and the linear dielectric constants are evaluated either at ω [ε_ω] or 2ω [$\varepsilon_{2\omega}$]. c is the velocity of light in vacuo and it has been assumed that div$\mathbf{E} = 0$. The latter step may, at first sight, appear to lack self-consistency because div$\mathbf{E} = -(1/\varepsilon_0\varepsilon) d\omega \mathbf{P}^{NL}$ but it is well-established in the literature that, nevertheless, div$\mathbf{E} = 0$ is a good approximation, in the absence of any charge density. Note that $\frac{\partial}{\partial y} = 0$ and that propagation is along the z-axis.

The mathematical treatment of equations (6.5) and (6.6) can be considerably simplified by separating out the *fast* and *slow* dependences. To do this, it is necessary to write

$$\mathrm{E}_\omega(\mathrm{x},\mathrm{z},\mathrm{t}) = \mathrm{E}_\omega(\mathrm{x},\mathrm{z})\exp[\mathrm{i}(\mathrm{k}_\omega \mathrm{z} - \omega \mathrm{t})] \tag{6.7a}$$

$$\mathrm{E}_{2\omega}(\mathrm{x},\mathrm{z},\mathrm{t}) = \mathrm{E}_{2\omega}(\mathrm{x},\mathrm{z})\exp[\mathrm{i}(\mathrm{k}_{2\omega} \mathrm{z} - 2\omega \mathrm{t})] \tag{6.7b}$$

and then to make a slowly varying assumption, through the adoption of the inequalities

$$\frac{\partial^2 E_{\omega,2\omega}}{\partial z^2} << k_{\omega,2\omega} \frac{\partial E_{\omega,2\omega}}{\partial z}$$

where k_ω and $k_{2\omega}$ are the linear wavenumbers $k_\omega^2 = (\omega^2/c^2)\varepsilon_\omega$ and $k_{2\omega}^2 = (2\omega/c)^2\varepsilon_{2\omega}$.

$\chi_{ijk}^{(2)}$ can be considerably reduced through the application of the symmetry operations associated with a given crystal class. This information is available in the literature[6] so that, for example,

- hexagonal : class $\overline{6}$: $\kappa_1 \neq \kappa_2$
  classes $\overline{6}$ m², 3m and 32 : $\kappa_1 = 0$, $\kappa_2 = 0$

and

$$\chi_{xxx}^{(2)}(2\omega;\omega,\omega) = \kappa_1; \chi_{xxx}^{(2)}(\omega;-\omega,2\omega) = 2\kappa_1 : \chi_{yxx}^{(2)}(\omega;-\omega,2\omega) = -2\kappa_2$$

with the polarisations being

$$P_x^{(2)}(\omega) = \frac{\varepsilon_0}{2}\chi_{xxx}^{(2)}(\omega;-\omega,2\omega)E_\omega^* E_\omega = \varepsilon_0\kappa_1 E_\omega^* E_{2\omega} \tag{6.8a}$$

$$P_x^{(2)}(2\omega) = \frac{\varepsilon_0}{2}\chi_{xxx}^{(2)}(2\omega;\omega,\omega)E_\omega^2 = \frac{\varepsilon_0}{2}\kappa_1 E_\omega^2 \tag{6.8b}$$

The coupled equations for E_ω and $E_{2\omega}$ are, therefore

$$2ik_\omega \frac{\partial E_\omega}{\partial z} + \frac{\partial^2 E_\omega}{\partial x^2} + \left(\frac{\omega}{c}\right)^2 \kappa_1 E_\omega^* E_{2\omega} e^{i\Delta kz} = 0 \tag{6.9a}$$

$$2ik_{2\omega} \frac{\partial E_{2\omega}}{\partial z} + \frac{\partial^2 E_{2\omega}}{\partial x^2} + 2\left(\frac{\omega}{c}\right)^2 \kappa_1 E_\omega^2 e^{-i\Delta kz} = 0 \tag{6.9b}$$

where

$$\Delta k = k_{2\omega} - 2k_{2\omega} \equiv \text{linear phase mismatch} \tag{6.10}$$

It is convenient, now, to define the following parameters

$$\Gamma_1 = \frac{\omega}{4cn_\omega}\chi_{xxx}^{(2)}(\omega;2\omega-\omega),\ \Gamma_2 = \frac{\omega}{2cn_{2\omega}}\chi_{xxx}^{(2)}(2\omega;\omega,\omega)$$

where n_ω, $n_{2\omega}$ are the, respective, linear refractive indices at ω and 2ω and the generic form of the coupled equations for CW waves is simply

$$\frac{dE_{2\omega}}{dz} = i\Gamma_2 E_\omega^2 e^{i\Delta kz} \tag{6.11a}$$

$$\frac{dE_\omega}{dz} = i\Gamma_1 E_{2\omega} E_\omega^* e^{-i\Delta kz} \tag{6.11b}$$

Equations (6.11) are much simpler for CW waves because the terms $\frac{\partial^2 E_\omega}{\partial x^2}$ and $\frac{\partial^2 E_{2\omega}}{\partial x^2}$ are omitted for this physical description.

The CW equations can be analysed by introducing the possibility that there will be a nonlinear phase shift $\phi(z)$ i.e. by writing

$$E_\omega = F_\omega(z)e^{i\phi_\omega(z)}, \; E_{2\omega} = F_{2\omega}(z)e^{i\phi_{2\omega}(z)} \tag{6.12}$$

Equations (6.11) can then be solved with two types of boundary conditions which will be called TYPE 1 and TYPE 2 in this discussion. The traditional interest has always been in TYPE 1 boundary conditions which are

$$z = 0: E_\omega = E_0, \; E_{2\omega} = 0 \;\; [\text{no } 2\omega \text{ wave, initially}]$$

which immediately leads to

$$F_\omega^2(z)\frac{d\phi_\omega}{dz} + \frac{\Delta k}{2}\left[F_\omega^2(z) - F_\omega^2(0)\right] = F_\omega^2(0)\left[\frac{d\phi_\omega}{dz}\right]_{z=0} \tag{6.13}$$

Equation (6.13) shows that, when $\Delta k = 0$, i.e. for perfect linear phase matching, $\frac{d\phi_\omega}{dz} = 0$. The conclusion in this case is, therefore

$$\Delta k = 0 \Rightarrow \frac{d\phi_\omega}{dz} = 0$$

and there is no nonlinear phase-shift but a strong 2ω wave is produced. The TYPE 2 boundary condition is that $E_\omega(0) \neq 0$ *and* $E_{2\omega}(0) \neq 0$ i.e. both the ω and the 2ω waves are present at the input plane. If this occurs then a new type of coupled wave becomes possible i.e. stationary states can be generated. The two types of boundary condition are summarised in figure 24

For TYPE 2 boundary conditions the ω and 2ω travel together, from the input, $z = 0$, position and interact with each other. A "balanced propagation" is achieved so that a stationary state is created. A nonlinear phase mismatch ($\Delta\beta$) now exists where

$$\Delta\beta = \frac{d}{dz}\left[\phi_{2\omega} - 2\phi_{\omega}\right] \tag{6.14}$$

and this cancels the linear phase mismatch i.e.

$$\underset{\text{(nonlinear)}}{\Delta\beta} + \underset{\text{(linear)}}{\Delta k} = 0 \tag{6.15}$$

SUSTAINS BALANCED PROPAGATION

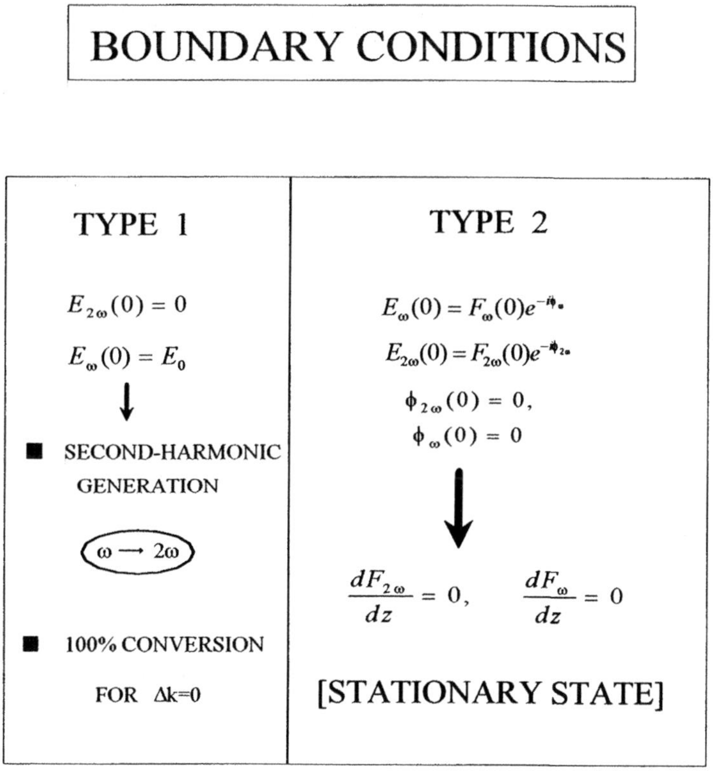

Figure 24

Upto now the discussion has been about scalar, CW, propagation but what will happen if pulses, or finite beams, are used. These possibilities are very important, since devices are most likely to use either a pulse or a beam. The possibilities that can arise are summarised in figure 25. Here it can be seen that stationary soliton-like propagation can occur for both types of boundary condition. For TYPE 1 the power level for $E_{\omega}(0) \neq 0$ must be above a threshold to get soliton-like behaviour, otherwise the input pulse degenerates to radiative noise. For TYPE 2 conditions *exact* soliton-like stationary states occur. They will be called soliton-like because, of course, they are different in form from the solutions of the unperturbed, and uncoupled, nonlinear Schrödinger equation.

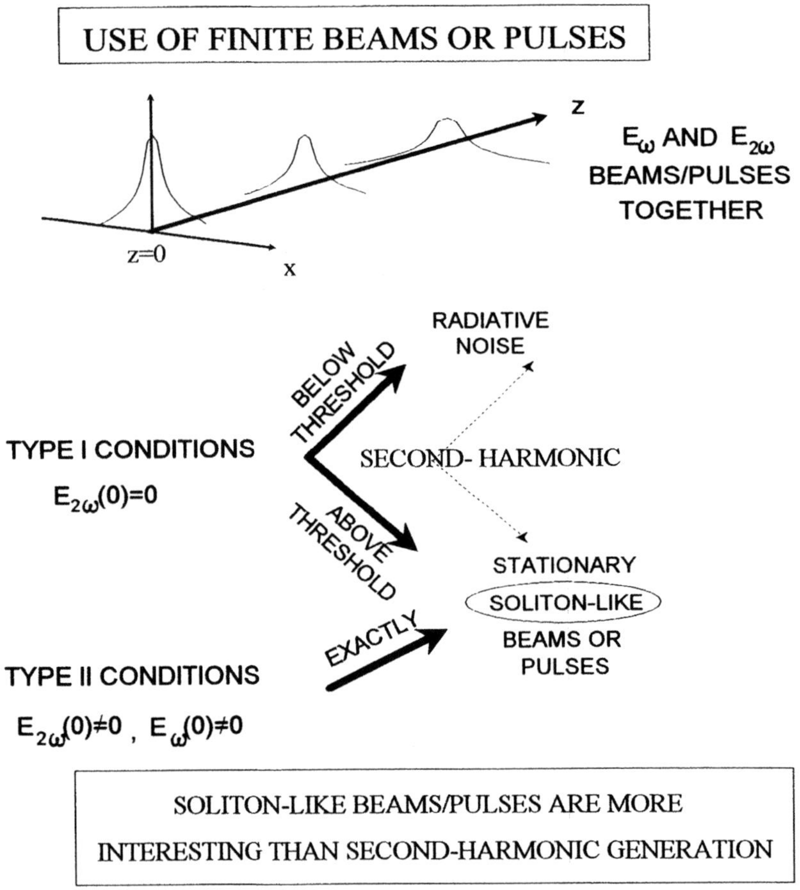

Figure 25

The basic equations that describe pulse or beam propagation can be obtained from the equations given earlier but it will be more revealing, first, to make the substitutions

$$E_\omega = w\exp(i\beta_1 z), E_{2\omega} = v\exp(i\beta_2 z) \tag{6.16}$$

and then follow this step with the transformations $\frac{\omega}{c}\sqrt{2k_\omega\beta_1}\,x \rightarrow x$ and $\frac{\omega}{c}\beta_1 z \rightarrow z$ and the definition $\beta = 2\beta_2/\beta_1$. After all of this is done the basic coupled equations for beam, or pulse, propagation become[33,36]

$$i\frac{\partial w}{\partial z} + \frac{\partial^2 w}{\partial x^2} - w + w^* v = 0 \tag{6.17a}$$

$$i\alpha\frac{\partial v}{\partial z} + \frac{1}{2}\frac{\partial^2 v}{\partial x^2} - \frac{\alpha\beta}{2}v + \frac{w^2}{2} = 0 \tag{6.17b}$$

in which $\alpha = k_{2\omega}/2k_\omega$. The assumption that $\alpha\beta >> 1$ permits equation (6.17b) to be approximated by

$$\alpha\beta >> 1 \; : \; v = \frac{w^2}{\alpha\beta} \tag{6.18}$$

If this is the case then v can be eliminated from (6.17a), giving an equation only for w. The *uncoupled* equation, for $\alpha\beta >> 1$, is simply the nonlinear Schrödinger equation and shows that the coupled problem can be expressed in terms of an *effective* third-order nonlinearity. This is the so-called *cascaded limit*.

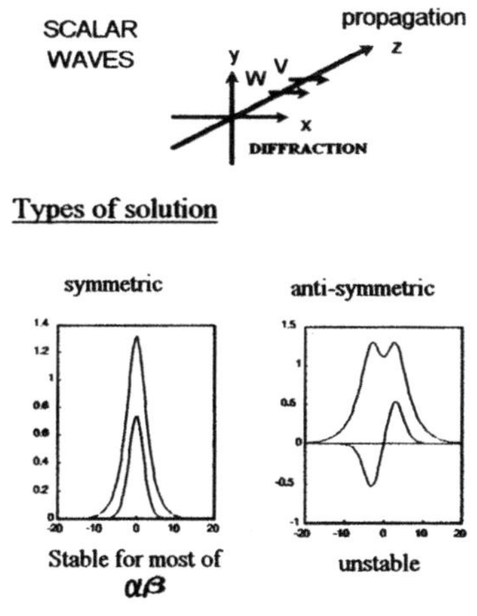

Figure 26

The typical solutions of equations (6.17) are shown in figure 26 and are symmetric or anti-symmetric. It can be shown by computer experiment that the anti-symmetric solutions are always unstable, while the symmetric solutions are stable for most values of $\alpha\beta$. Basing stability conclusions on numerical experiments is unsatisfactory, to say the least, so some stability analysis now needs to be done. The method will use a Lagrangian density L, reduced by integration over α to $\mathcal{L}$ and will also use trial functions as was done for the coupler problems earlier in this chapter. The trial functions have the form

$$w = \eta_1 \exp\left[-\rho_1^2 (x - x_1)^2 + i\frac{\xi_1}{2}(x - x_1)^2 + i\frac{\theta_1}{2}\right] \tag{6.19a}$$

$$v = \eta_2 \exp\left[-\rho_2^2 (x - x_2)^2 + i\xi_2 (x - x_2)^2 + i\theta_2\right] \tag{6.19b}$$

Note that Gaussian functions are going to be tried, $x_{1,2}$ are beam/pulse positions along x, $\xi_{1,2}$ are velocities, $\rho_{1,2}$ are a measure of the half-widths, $\eta_{1,2}$ are amplitudes and $\theta_{1,2}$ are phases. The reduced Lagrangian, together with the Euler-Lagrange equations lead to ten coupled equations[33] for $\eta_i, \rho_i, x_i, \theta_i, \xi_i$ $[i = 1,2]$ but the details will not be given here. The stationary solutions have the property

$$\frac{dx_1}{dz} = \frac{dx_2}{dz} = 0, \frac{d\xi_1}{dz} = \frac{d\xi_2}{dz} = 0, \frac{d\theta_1}{dz} = \frac{d\theta_2}{dz} = 0$$

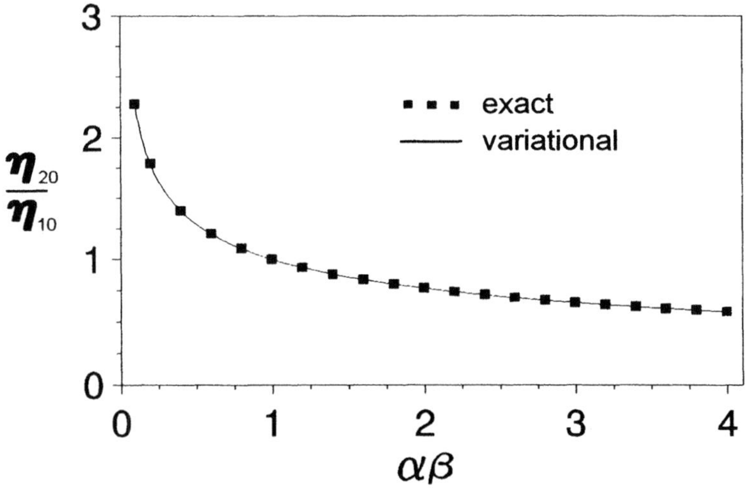

Figure 27

which imply that $x_1 = x_2 = 0$, $\xi_1 = \xi_2 = 0$, $\theta_1 = \theta_2 = 0$. If the stationary solutions are labelled with the subscript 0 then the stationary state is $\eta_1 = \eta_{10}$, $\eta_2 = \eta_{20}$, $\rho_1 = \rho_{10}$, $\rho_2 = \rho_{20}$ and it is interesting to calculate the η_{20}/η_{10} ratio needed to sustain a, stationary, $[2\omega,\omega]$ solitary wave. This is shown in figure 27, where it is very encouraging that the exact [numerical simulation] and the variational results coincide to a remarkable degree of accuracy.

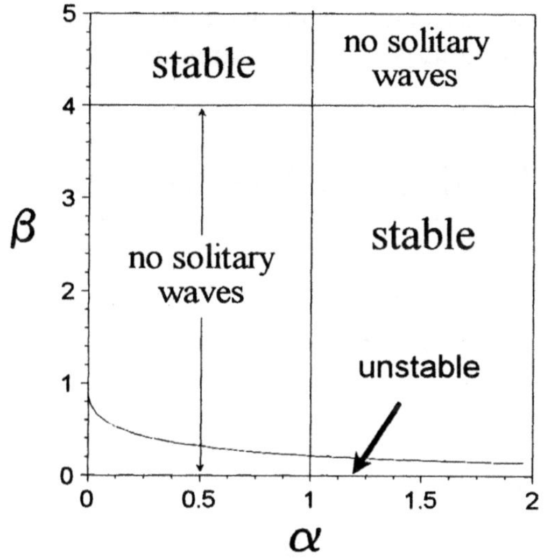

Figure 28

A stability analysis proceeds from the Euler-Lagrange equations by making small changes to phase, beam position or direction or amplitudes or widths. In fact, the perturbations can be classified into symmetric [e.g. direction perturbations] or asymmetric [e.g. amplitude perturbations]. A stationary state exists for $x_1 = 0$, $x_2 = 0$ to $x_1 = \delta x_1$, $x_2 = \delta x_2$. In addition, the velocities become perturbed from 0 to $\xi_1 = \delta\xi$, $\xi_2 = \delta\xi$. The Euler-Lagrange equations then yield

$$\xi = \frac{1}{2M}\left[\frac{\eta_{20}^2}{\rho_{20}}\xi_2 - \frac{\eta_{10}^2}{2\rho_{10}}\xi_1\right] \tag{6.20}$$

$$M = \frac{1}{2}\left[\alpha\frac{\eta_{20}^2}{\rho_{20}} + \frac{\eta_{10}^2}{2\rho_{10}}\right], \Delta = x_2 - x_1 \tag{6.21}$$

$$\frac{1}{2}\left(\frac{d\Delta}{dz}\right)^2 + \frac{ab}{2}\Delta^2 = 0 \tag{6.22}$$

$$\frac{1}{2}\left(\frac{d\xi}{dz}\right)^2 + \frac{ab}{2}\xi^2 = 0 \tag{6.23}$$

in which a and b are rather complicated combinations of all of the parameters involved. It will not serve the argument, at this stage, to derive a or b but they can be found in the literature[33]. It is sufficient to draw the following conclusion. Since $a > 0$ and $b > 0$ then the solutions are always stable to this type of perturbation.

The stability of these new stationary states that arise from coupled ω, 2ω equations can be best expressed in the form of a (β,α) plot in which domains of stability can be easily seen. This is shown in figure 28. Here it can be appreciated that there are some domains for which no solutions can be found. It is interesting that $\alpha = 1$ [perfect *linear* phase-matching condition] separates existence from non-existence regions. Also $\beta = 4$ is the perfect nonlinear phase-matching condition.

STABILITY AND POWER

- variational stability analysis has established the following stability regimes.

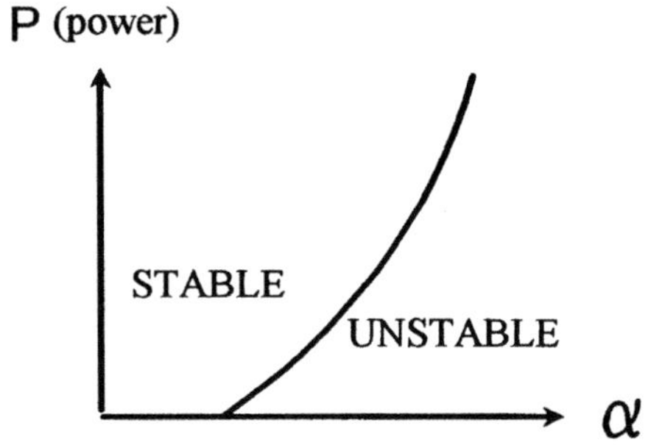

- numerical simulations support the mathematical conclusions
- future work will exploit

  ·crystal symmetry ·vector waves
  ·electro-optics ·magneto-optics

Figure 29

The stability conditions are summed up in figure 29 which shows the variation of power with α and the clear division between stable and unstable regimes. For any given $\alpha < 1$ all solutions are stable but for $\alpha > 1$ the solutions are unstable until a certain threshold power is achieved. Figure 30 [(a) α=1.5, β=2; (b) α=1.5, β=0.1] shows some tests, through numerical simulation, of the stability conclusions. The simulations entirely validate the variational, trial function-based, analysis. Further work in this area will now emerge with an emphasis on more explicit exploitation of crystal symmetry, electro-optic phenomena, magneto-optic effects and vector wave formulations.

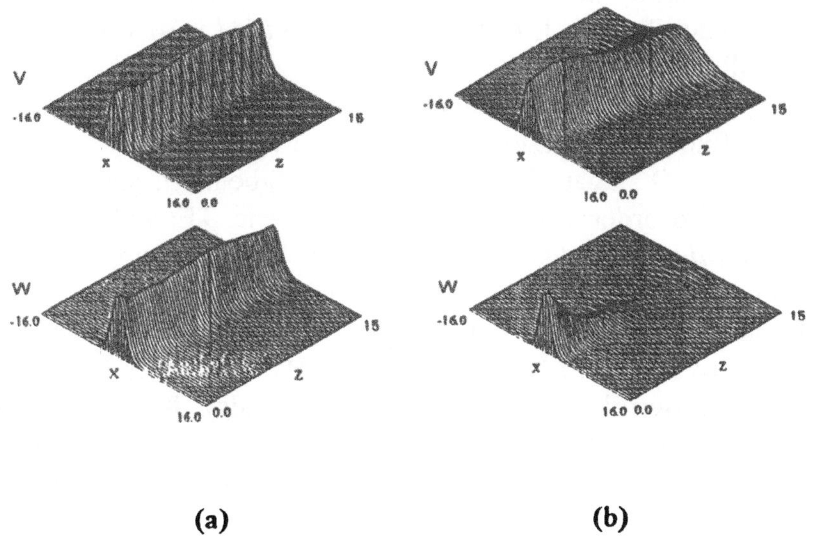

Figure 30

References

1. Smith, P.W. (1993) All-optical devices: materials requirements, *SPIE* **1852**, 2-9.
2. Stegeman, G.I. (1993) Material figures of merit and implications to all-optical waveguide switching, *SPIE* **1852**, 75-89.
3. Eaton, D.F. (1991) Nonlinear optical materials, *Science* **253**, 281-287.
4. Jonusauskas, G., Gadonas, R. and Rulliere, C. (1994) "Fast" optical Kerr gate with "slow" nonlinearity. *Optics Communications* **112**, 80-84.
5. Syms, R. and Cozens, J. (1992) *Optical Guided waves and Devices*, McGraw-Hill, London.
6. Butcher, P.N. (1965) *Nonlinear Optical Phenomena*, **Bulletin 200**, Engineering Experiment station, Ohio state University.
7. Boardman, A.D., Egan, P., Lederer, F., Langbein, U. and Mihalache, D. (1991) Third-order nonlinear electromagnetic TE and TM guided waves in *Nonlinear Surface Electromagnetic Phenomena*, H.E.Ponath and G.I.Stegeman (Eds), North-Holland, 73-287.
8. Boardman, A.D., Egan, P., Twardowski, T. and Wilkins, M. (1990) *Nonlinear Waves in Solid State Physics*, NATO ASI Series B: **247**, A.D. Boardman, M. Bertolotti and T.Twardowski (Eds), Kluwer Academic Publishers, New York and London, 1-51.
9. Boardman, A.D. and Xie,K. (1993) Theory of spatial solitons, *Radio Science* **28**, 891-899.
10. Boardman, A.D. and Xie, K. (1994) Bright spatial soliton dynamics in a symmetric optical planar waveguide structure, *Physical Review A* **50**, 1851-1866.
11. Shi, T. and Chi, S. (1990) Nonlinear photonic switching by using the spatial soliton collision, *Optics Letters* **15**, 1123-1125.
12. Cao, X.D. and Meyorhofer, D.D. (1994) All-optical switching by means of collisions of spatial vector solitons, *Optics Letters* **19**, 1711-1713.
13. Blair, S., Wagner, K. and Mcleod, R. (1994) Asymmetric spatial soliton dragging, *Optics Letters* **19**, 1943-1945.
14. Ironside, C.N., Aitchison, J.S. and Arnold, J.M. (1993) An all-optical switch employing the cascaded second-order nonlinear effect, *IEEE Journal of Quantum electronics* **29**, 2650-2654.

15. Kelaidis, C., Hutchings, D.C. and Arnold,J.M. (1994) Asymmetric two-step GaAlAs quantum well for cascaded second-order processes, *IEEE Transactions on Quantum Electrnics* **30**, 2998-3005.
16. Boardman, A.D., Xie, K. and Zharov, A.A. (1995) Polarization interaction of spatial solitons in optical planar waveguides, *Physical Review A* **51**, 692-705.
17. Pare, C. and Florjanczyk, M. (1990) Approximate model of soliton dynamics in all-optical couplers, *Physical Review A* **41**, 6287-6295.
18. Whitham, G.B. (1973) *Linear and Nonlinear Waves*, Interscience, New York.
19. Morse, P.M. and Feshbach, H. (1953) *Methods of Theoretical Physics*, McGraw-Hill, New York.
20. Lee, D.L. (1986) *Electromagnetic Principles of Integrated Optics*, John Wiley & Sons.
21. Horowitz, B.R. and Tamir, T. (1971) Lateral displacement of a light beam at a dielectric surface, *Journal of the Optical Society of America* **61**, 586-594.
22. Aceves, A.B., Varatharajah, P., Newell, A.C., Wright, E.M., Stegeman, G.I., Heatley, D.R., Moloney, J.V. and Adachihara, H. (1990) Particle aspects of collimated light channel propagation at nonlinear interfaces and in waveguides, *Journal of the Optical Society of America* 7, 963-974.
23. Aceves, A.B. Moloney, J.V. and Newell, A.C. (1989) Theory of light-beam propagation at nonlinear interfaces. I. Equivalent-particle theory for a single interface, *Physical Review A* **39**, 1809-1827.
24. Moloney, J.V., Newell, A.C. and Aceves, A.B. (1992) Spatial soliton optical switches: a soliton-based equivalent particle approach. *Optical and Quantum Electronics* **24**, S1269-S1293.
25. Kivshar, Yu.S., Kosevich, A.M. and Chubykalo, O.A. (1990) Radiative effects in the theory of beam propagation at nonlinear interfaces, *Physical Review A* **41**, 1677-1688.
26. Lax, P. (1968) Integrals of nonlinear equations of evolution and solitary waves, *Communications in Pure and Applied Mathematics* **21**, 467-490.
27. Zakharov, V.E. and Shabat, A.B. (1972) Exact theory of two-dimensional self-focusing and one-dimensional self-modulation of waves in nonlinear media, *Soviet Physics JETP* **34**, 62-69.

28. Lamb, G.L. (1980) *Elements of Soliton Theory*, John Wiley & Sons, New York.
29. Scott, A.C., Chu, F.Y.F. and McLaughlin, D.W. (1973) The soliton: a new concept in applied science, *Proceedings of the IEEE* **61**, 1443-1483.
30. Ablowitz, M.J. and Clarkson, P.A. (1991) *Solitons, Nonlinear Evolution Equations and Inverse Scattering*, Cambridge University Press, United Kingdom.
31. Aliev, Y.M., Boardman, A.D., Xie, K. and Zharov, A.A. (1994) Conserved energy approximation to wave scattering by a nonlinear interface, *Physical Review E* **49**, 1624-1633.
32. Aliev, Y., Boardman, A.D., Smirnov, A.J., Xie, K. and Zharov, A.A., Spatial dynamics of soliton-like channels near to interfaces between optically linear and nonlinear media, *Physical Review E*, [in the press].
33. Boardman, A.D., Xie, K. and Sangarpaul, A. (1996) Stability of scalar spatial solitons in cascadable nonlinear media, *Physical Review A*, [in the press].
34. Desalvo, R., Hagan, D.J., Sheik-Bahae, M. and Stegeman, G.I. (1992) Self-focusing and self-defocusing by cascaded second-order effects in KTP, *Optics Letters* **17**, 28-30.
35. Menyuk, C.R., Schiek, R. and Torner,L. (1994) Solitary waves due to $\chi^{(2)}:\chi^{(2)}$ cascading, *Journal of the Optical Society of America* **11**, 2434-2443.
36. Buryak, A.V. and Kivshar, Y.S. (1995) Solitons due to second-harmonic generation, *Physics Letters A* **197**, 407-412.

Waveguide Mach-Zehnder Intensity Modulator produced via Proton Exchange Technology in $LiNbO_3$

S. TONCHEV, B. YORDANOV, M. KUNEVA, I. SAVATINOVA, M. ARMENISE* and V. PASSARO*
Institute of Solid State Physics, Bulgarian Academy of Sciences, Sofia 1784, Bulgaria
** Politecnico di Bari, Dipartamento di Electrotecnica ed Elettronica, 70125 Bari, Italy*

1. Introduction

Guided-wave Mach-Zehnder intensity modulators are key elements for a variety of applications in optical communications. They usually present a broad operation band and low drive voltages. Two general ways for fabrication of such modulators on $LiNbO_3$ are known - by standard Ti-indiffusion and by proton exchange with additional annealing.

The Ti in-diffusion technique is the most frequently used because it allows the fabrication of waveguides supporting both TE and TM modes by the relatively large increase of both refractive indices and exhibit low propagation loss.

In contrast with conventional technologies mentioned above, the proton exchange (PE) takes place at relatively low temperatures (160 - 250 ^{0}C). This process seems to be very perspective technology for waveguide fabrication in $LiNbO_3$ as the problems connected with the high temperature out-diffusion from the surface of the substrate are avoided. Its main advantage is the fast and simple waveguide formation procedure. The method is relatively non-toxic and inexpensive and makes possible a thermal tuning of the refractive index profile in the protonated layer.

The PE reaction proceeds according to the scheme:

$$LiNbO_3 + xH \rightarrow Li_{1-x}H_xNbO_3 + xLi \quad (1)$$

which takes place in the surface layer of the crystal (several μm depth). The depth of the exchanged region is determined by the process parameters - temperature, duration and proton source. Suitable protonating media are some low-acidity hydrate melts and weak acids having high boiling point and thermal stability like benzoic, pyrophosphoric, palmitic, stearic, oleic, toluic, cinnamic, octanoic, adipic acid. The most commonly used is benzoic acid, which has been characterised in detail as a proton source, but recently some interest is directed also to the other acids producing some different waveguide parameters.

PE leads to a strong extraordinary refractive index increase Δn_e and to a step index profile. In the case of $LiNbO_3$ the index change Δn_e is about 0.11 at λ=0.633 μm which is almost one order of magnitude higher than that obtained by using Ti in-diffusion. The ordinary refractive index slightly decreases: Δn_o = -0.04. This is the reason

M. Balkanski (ed.), Devices Based on Low-Dimensional Semiconductor Structures, 293–296.

for the polarising effect of protonated regions which reveal waveguiding action only for TE or TM modes depending on crystal orientation. It is important to note that the protonated region has increased photorefractive resistance compared to virgin $LiNbO_3$. Another advantage of PE is the possibility to combine it with other technologies, like Ti in-diffusion for example (TIPE technology) for fabrication of segments with different Δn on a common substrate, used for birefringence tuning of waveguides.

A serious disadvantage of PE-technology is the electrooptical properties deterioration in protonated $LiNbO_3$ at high concentration of hydrogen. After strong PE the optical properties of waveguides show some instability and losses which can be avoided or eliminated by melt buffering or/and by additional thermal treatment. Significant restoration takes place at annealing temperatures over 340 ^{0}C.

2. Experimental results

Here we report results for the fabrication and the properties of a Mach-Zehnder modulator using proton exchange technique with post-exchange annealing (APE waveguides). A comparison between this type of modulator and a previously one fabricated by us via Ti-indiffusion [1], but using the same waveguide and electrode parameters is made.

The Mach-Zehnder intensity modulator was fabricated on X-cut $LiNbO_3$ with bulk refractive index n_b=2.1984 for operation at λ=0.633 μm. The principle scheme of the device with the electrode configuration is shown in Fig.1. Specially designed quasi-smooth Y-branches [2] have been used (the dashed area in this figure). Their first branching segment is 2 mm long with a half branching angle of 0.4 ^{0}. Suitable series of segments increase the angle till a waveguide separation of 140 μm is obtained over a length of 5.3 mm. Three-electrodes-two-gaps configuration with a common electrode between the waveguides was used. The electrode gap d=6 μm was so chosen that the alignment of the electrodes to the waveguides was not particularly critical. The length of the electrodes L was 5.2 mm.

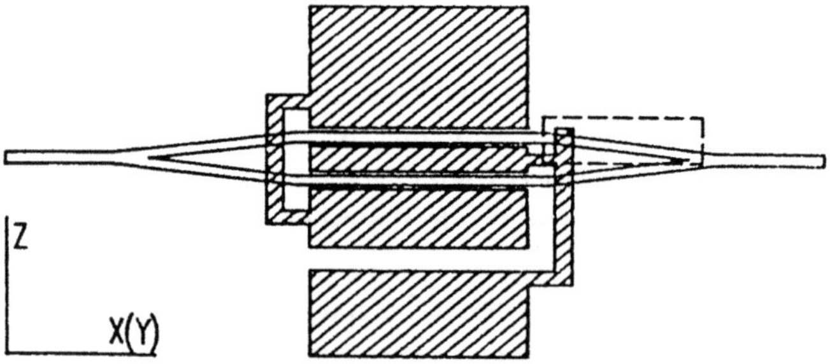

Figure 1. Schematic set-up of the Mach-Zehnder modulator.

The proton exchange reaction was performed in pure benzoic bath at 180 ^{0}C for 20 minutes in a sealed container and a programmable oven to exclude the active evaporation of the acid as well as to form a region with uniform and precisely controllable temperature. Standard photolitographic technique with a mask with channel width of 5.1 μm was used. The depth of as-exchanged planar waveguide is evaluated to be 0.3 μ m using a proton diffusion coefficient $D_x(180\ ^0C)=0.1 \mu m/hr$. The extraordinary index change was measured to be $\Delta n_e=0.0468$. A post-exchange annealing at 350 ^0C for 1 hour was applied to restore the electrooptical properties of the hydrogenated regions. After this procedure the index change was reduced to $\Delta n_e=0.018$. The channel waveguides supported two TE transverse modes at $\lambda=0.633$ μm.

Figure 2 shows a typical low frequency response of the device for a symmetric triangular drive waveform. It is seen that the peak-to-minimum voltage U_π is 14 V. The depth of modulation was 88%. The high frequency response of the modulator measured with 50 ohm load resistance is shown in Fig. 3.

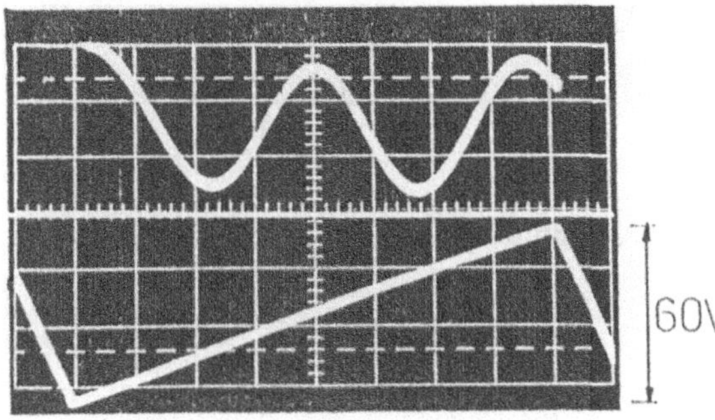

Figure 2. Intensity modulation of the modulator: upper trace - modulated signal, lower trace - applied drive voltage.

The half-wave voltage U_π of an interferometric modulator is given by

$$U_\pi = \frac{\pi d \lambda}{2 L n_e^3 r_{33} G} \tag{1}$$

where G is the overlap of the optical field of the excited mode to the external electric field. The value of U_π measured in the present experiment is approximately six times the value of the Ti-indiffused variant of the same modulator reported in [1]. This result can be explained by an insufficient reconstruction of the electrooptical activity through annealing. According to [3], the post-exchange annealing could restore nearly 70% of the degraded electrooptical r_{33} coefficient. Probably in this experiment we have not recovered the electrooptical properties at such a degree due to the insufficient time of annealing. Also, the high drive voltage could be related by the fact that a smaller electrooptical overlap G has been achieved in the proton-exchanged version of modulator. Anyway, it is clear that by an appropriate selection of the processing parameters for the proton exchange (time, temperature and acid concentration), the post-exchange annealing (time, temperature and environment) and the electrode configuration (electrode gap to channel width ratio), one could build a Mach-Zehnder intensity modulator quite comparable with the Ti-indiffusion one, using far more easier and simpler technology.

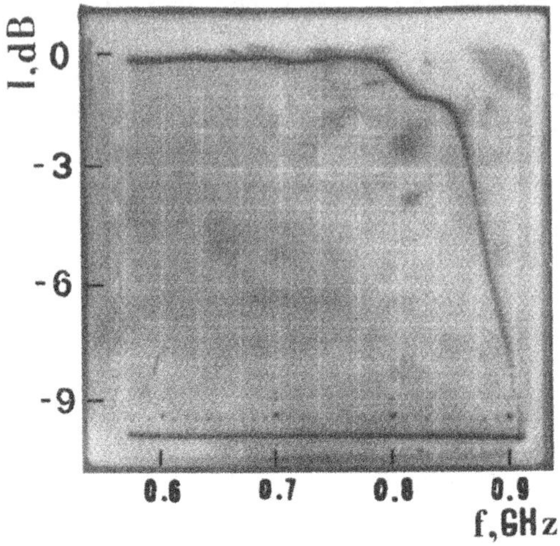

Figure 3. High frequency response of the modulator.

3. Conclusions

Broadband high efficient Mach-Zehnder modulator fabricated in $LiNbO_3$ via PE reaction is presented. Propagation and curvature losses are minimised using specially designed quasi smooth Y-branches. The measured drive voltage of the device is 6 times higher than this of a Ti in-diffused variant of the modulator. Further investigation of the electrooptical properties of APE waveguides is necessary.

4. Acknowledgements

This work was financially supported by the Bulgarian Ministry of Education, Science and Technology under Contract F-419. The financial contribution of the European Commission under Copernicus Project COP-959 is also gratefully acknowledged.

5. References

1. Tonchev, S., Todorov, R., Zilling, K. and Savatinova, I. (1990) Mach-Zehnder type modulator for integrated optics, *J.Opt.Commun.* **11**, 147-150
2. Johnson, L.M. (1985) Integrated optical components for fiber sensors, *SPIE* **578**, 233-236
3. Loni, A. (1987) An experimental study of proton-exchanged Lithium Niobate optical waveguides, PhD Dissertation, Glasgow University

EVANESCENT FIELD COUPLING BETWEEN A SINGLE MODE OPTICAL FIBER AND A PLANAR WAVEGUIDE

A. ANDREEV, K. PANAJOTOV, E. KARAKOLEVA, AND B. ZAFIROV
Institute of Solid State Physics
72 Tsarigradsko Chaussee blvd, 1784 Sofia, Bulgaria

The interest in the properties of a waveguiding structure consisting of a coupled single-mode optical fiber and planar waveguide (PWG)- fig.1, has grown recently because of the various applications it could have [1], [2], [3].

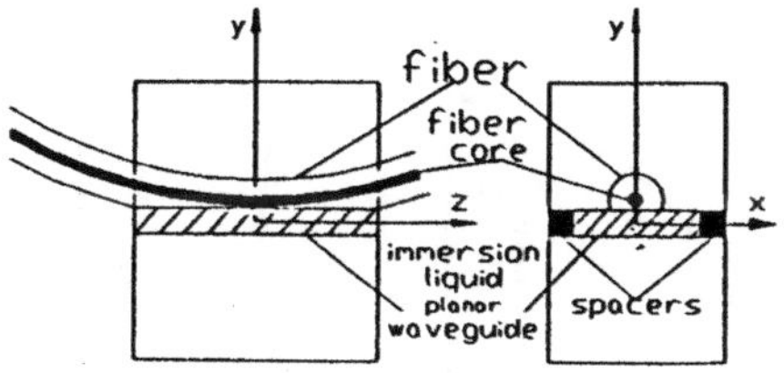

Figure 1. **Longitudinal and cross-sectional view of the distributed fiber-to-planar waveguide coupler.**

Theoretically, fibers have been treated as equivalent PWGs [4] and the coupled mode formalism has been applied [5], but this approach has not lead to a good qualitative description of the system [1]. The reason for this inconsistence with the theory is in the restriction of the theoretical treatment of the field components to one dimension. A more precise theory of the light transfer between a fiber and a waveguide slab has been developed by Marcuse [6]. To treat the evanescent field coupling between a slab and a fiber Marcuse canceled the infinite set of PWG's modes by placing two parallel hypothetical perfectly conducting surfaces at distances $\pm D_x$ in the transverse direction x, and then using the CMT presented the total field $\Psi(x,y,z)$ of the coupled waveguides in a scalar approximation as a superposition of the fiber- and the PWG- modes. The coupled mode approach is generalized for the case when the coupling coefficients become functions of

M. Balkanski (ed.), Devices Based on Low-Dimensional Semiconductor Structures, 297–300.

the propagation distance z [5]. The fiber output power distribution in dependence on the structure parameters as a PWG refractive index n_s, PWG thickness d_s, polishing depth s_0 (i.e., the minimum distance between the fiber core and polished coupler half-block surface), are experimentally and numerically investigated. There is very good agreement between the experiment and Marcuse's model- fig.2, [7]. The exact position of the minimum

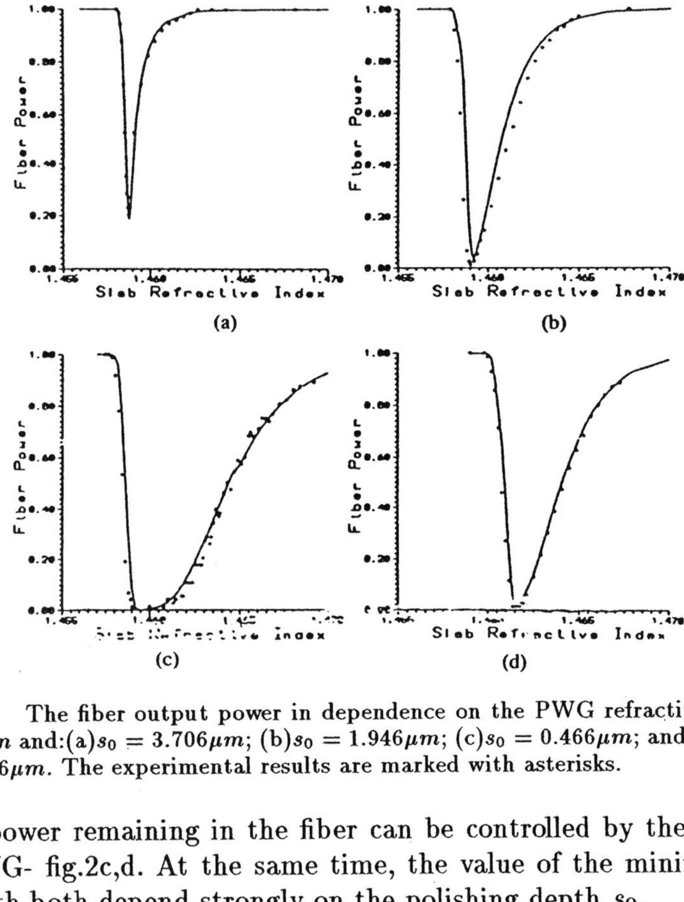

Figure 2. The fiber output power in dependence on the PWG refractive index n_s for $d = 1.7\mu m$ and:(a)$s_0 = 3.706\mu m$; (b)$s_0 = 1.946\mu m$; (c)$s_0 = 0.466\mu m$; and (d)$d = 1.1\mu m$, $s_0 = 1.946\mu m$. The experimental results are marked with asterisks.

of the power remaining in the fiber can be controlled by the thickness of the PWG- fig.2c,d. At the same time, the value of the minimum and its halfwidth both depend strongly on the polishing depth s_0.

Different theoretical approaches of Marcuse's model are employed to show that the light spreading into the PWG plays an important role in the power transfer. The light transfer between the fiber and the PWG depend strongly on the relationship between their propagation constants β_{f0}- for the opical fiber, and β_s- for the PWG, [7]. For the values of n_s at which $\beta_s < \beta_{f0}$, the light enters the PWG only in the vicinity of the fiber and then is coupled back into the fiber, fig.3a. For the values of n_s at which $\beta_s \approx \beta_{f0}$- fig.3b, there are three well pronounced peaks for the minimum distance between the fiber and the PWG ($z = 0$) and the same three peaks, with changed

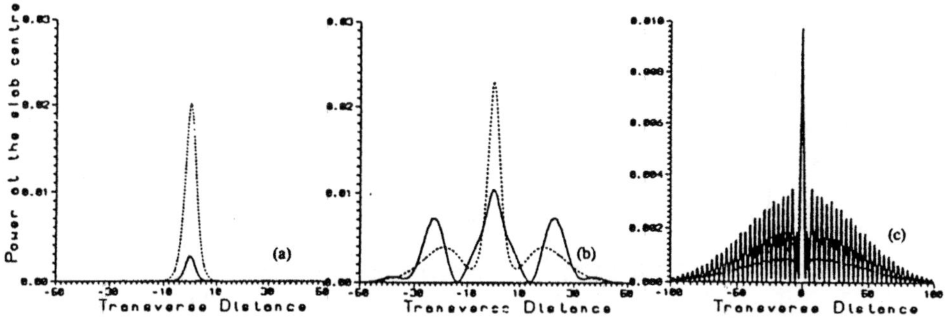

Figure 3. **The total optical power | $\Psi(x,y,z)$ $|^2$ at the centre of the PWG ($y=0$) as a function of the transverse distance $x(\mu m)$, for $s_0 = 0.466\mu m$ at $z = 0\mu m$ (dotted line) and $z = 3000\mu m$ (solid line) for different n_s: (a)1.4578; (b)1.4588; (c)1.4598.**

amplitudes, remain at $z = 3000\mu m$, where there is no coupling. When β_s is considerably larger than β_{f0}, the PWG acts as a sink for the optical power in the fiber, so that the total power launched initially into the fiber is spread out laterally in the PWG- fig.3c. There is only one peak at $z = 0$ and a great number of peaks at $z = 3000\mu m$. The same results have been demonstrated using BPM in order to trace the light spreading in the x-direction of the slab [8].

The Marcuse's model has been expanded to include a second (odd-symmetry) mode of the symmetric and asymmetric PWG in [7], and [8]. For the case of asymmetric PWG the two possible linear states of the light polarization have been taken into account [9]. It has been shown that this distributed coupler can function as a polarizer and modulator by changing the refractive indices of the PWG (n_s) and the cover medium (n_c) as well as the PWG thickness (d_s) [9]. The calculated fiber output power P_f as a function of the PWG refractive index n_s is presented for two values of the cover medium refractive index and for two values of the polishing depth, fig.4a-d. The decrease of the polishing depth s_0 leads to a deepening and widening in the minima of the optical power output. For a symmetric PWG -fig.4(a,c), the coupler is polarization insensitive. When the cover medium is air ($n_c = 1.0$)- fig.4(b,d), the coupler acts as a polarizer for given values of the PWG refractive index n_s. In this way we show that the drain of the optical power from the fiber to the PWG can be driven not only by changing the PWG's parameters but also by changing the refractive index of the cover medium.

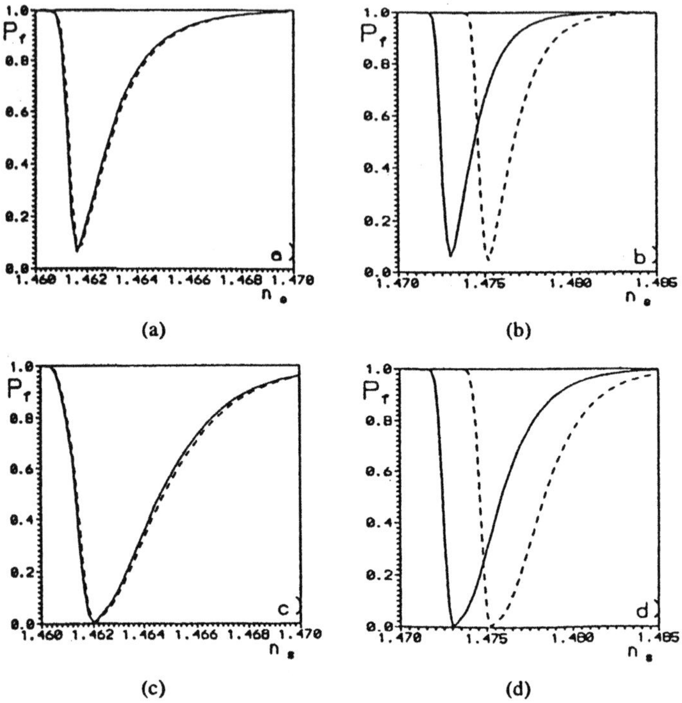

Figure 4. The fiber output power P_f for the TE- mode (solid line) and TM-mode (dashed line) in dependence on n_s for $s_0 = 3\mu m$ and: (a)$n_c = 1.453$; (b)$n_c = 1.0$, and for $s_0 = 2\mu m$: (c)$n_c = 1.453$; (d)$n_c = 1.0$.

References

1. Andreev,A.T., Panajotov, K.P. and Zafirova, B.S. (1993) Single-mode fiber to planar waveguide coupling, *SPIE*, **1973**.
2. Millar, C., Brierley, M. and Mallinson, S. (1987) Exposed- Core Single - Mode - Fiber Channel- Dropping Filter Using a High- Index Overlay Waveguide, *Optics Lett.*, **12**, pp.284-286.
3. Johnstone, W., Thursby, G., Moodie, D., Varshey, R. and Culshaw, B. (1992) Fiber optic wavelength channel selector with high resolution, *Electron. Lett.*, **28**, pp.1364-1365.
4. Andreev, A., Panajotov, K. and Zafirova, B., (1992) Polished fiber-optic coupler with intermediate planar waveguiding layer - part I, *J. Lightwave Tech.*, **10**, pp.882-887.
5. Panajotov, K., Andreev, A. and Zafirov, B. (1992) Polished Fiber-Optic Coupler with Intermediate Planar Waveguiding Layer-II Part, *J. Lightwave Technol.*, **10**, pp.1363-1366.
6. Marcuse, D. (1989) Investigation of coupling between a fiber and an infinite slab, *J. Lightwave Technol.*, **7**, pp.122-130
7. Andreev, A. and Panajotov, K. (1993) Distributed Single - mode Fiber - to - Single - mode Planar Waveguide Coupler, *J. Lightwave Technol.*, **11**, pp.1985-1989.
8. Panajotov K. and Andreev, A. (1994) Distributed Coupling Between a Single - mode Fiber and a Planar Waveguide, *JOSA B*, **11**, pp.826-834.
9. K. Panajotov (1994) Polarization Properties of a Fiber - to - Asymmetric Planar Waveguide Coupler, *J. Lightwave Tech.*, **12**, pp.983-988.

PSEUDOMORPHIC HEMTS: DEVICE PHYSICS AND MATERIALS LAYER DESIGN

THOMAS GRAVE
Siemens AG
Corporate Research and Development
D-81730 Munich
Germany

ABSTRACT: In this paper, pseudomorphic HEMTs on GaAs substrate and both lattice-matched and pseudomorphic HEMTs on InP substrate are treated. First, the basic requirements are considered that must be met by any HEMT epitaxial layer sequence. They are compared to the fundamental possibilities for layer growth that exist within the system of III-V compound semiconductors, and the five most common HEMT layer structures are discussed. An elementary introduction into HEMT device physics is given with special emphasis on the influence of specific layer schemes on charge control. The small-signal equivalent network is used to point out differences between intrinsic and extrinsic device due to the presence of parasitics. Guidelines for HEMT design with respect to analog low-noise and power applications are given. Most optimization considerations are illustrated by means of GaAs-based HEMT examples. Properties of InP-based devices are treated more concisely. Overviews over the state-of-the-art concerning high-frequency, low-noise and power performance are presented for both types of HEMTs, and a selection of future concepts for further improvements is outlined.

1. Introduction

Present Situation. Today, high electron mobility transistors (HEMTs) are the most important analog electronic devices for the frequency range between 10 and 100 GHz. They achieved this position by their enormous versatility: HEMTs are used in low-noise as well as power circuits, offering outstanding performance in both applications.

Particularly as low-noise devices, HEMTs are rivaled by no other technology. For power purposes, two additional types of III-V compound transistors might be considered, but they are potential competitors for the HEMT only in the lower part of its frequency range. The first alternative is the well established GaAs MESFET, usable up to approximately 30 GHz, but with inferior power gain. Recently, heterobipolar transistor (HBT) technology has established a second possibility. It is

M. Balkanski (ed.), Devices Based on Low-Dimensional Semiconductor Structures, 301–348.

presently maturing rapidly, with prospective applications up to, say, 50 GHz. For higher frequencies, however, HEMTs will stay the only available solid-state power device for the foreseeable future.

Consequently, HEMTs are the only choice if a *universal* technology for microwave and millimeter wave ICs is required. This situation is reflected by the widespread use of HEMTs - particularly GaAs-based pseudomorphic HEMTs - in analog communication and information systems by the major companies in this field worldwide. How did this prominent role in high-frequency electronics evolve historically?

HEMT History. The first papers on the pseudomorphic HEMT's predecessor, the lattice-matched AlGaAs/GaAs HEMT, were published in 1980 by Mimura *et al.* [1] and Delagebeaudeuf *et al.*[2]. The AlGaAs/GaAs HEMT makes use of the fact that a two-dimensional electron gas (2DEG) of extremely high carrier mobility forms at the heterointerface between AlGaAs and GaAs if the AlGaAs is appropriately *n*-doped (Dingle *et al.* [3], Störmer *et al.* [4]). Due to the small discontinuity of the conduction band energy at the heterointerface and the missing barrier between 2DEG and substrate, the current that can be achieved with these devices is relatively low, and the confinement of the carriers in the channel is poor by today's standards. With sufficiently short gates, however, low noise figures and high current-gain cutoff frequencies can be obtained so that these HEMTs became rapidly acknowledged to be well suited for analog receivers and digital ICs.

It was soon recognized that an energy barrier between channel and substrate would significantly improve the HEMT concept. Such a barrier can be achieved by the introduction of a pseudomorphic InGaAs channel between the GaAs buffer and the supply layer, transforming the HEMT into a PHEMT. First PHEMT results were published in 1985 by Rosenberg *et al.* [5] who used GaAs also for the supply layer, and Ketterson *et al.* [6] who introduced the standard layer sequence of today, i.e. GaAs buffer, InGaAs channel and AlGaAs supply. The new device was soon characterized in more detail (Ketterson *et al.* [7]), and its high potential for microwave and millimeter wave applications was demonstrated (Henderson *et al.* [8]).

However, the PHEMT's performance not only as a low-noise but also as a power transistor could be fully exploited only after the so-called double-heterojunction (DH) layer sequence had been developed. In this case, there is an *n*-AlGaAs supply layer not only on top of the channel but also below it. Power DH-HEMTs with a lattice-matched GaAs channel were reported by Hikosaka *et al.* [9] in 1985. The presently most common power HEMT structure was created in 1988 and 1989 when Smith *et al.* [10,11] sandwiched a pseudomorphic InGaAs channel between the two *n*-AlGaAs supplies.

All HEMT variants discussed so far are based on layer sequences grown on GaAs substrate. The development of the other important family of HEMTs, i.e. HEMTs grown on InP, began about 1982. In these transistors, InAlAs barrier layers lattice-matched to the substrate are grown below and above the InGaAs channel. Their advantage over GaAs-based devices is twofold: the conduction band discontinuity at the channel/barrier interfaces is significantly larger, as well as the indium content in the channel. This means increased electron concentration in the channel, together with improved carrier transport properties. An InGaAs channel lattice-matched to InP contains 53% indium. For pseudomorphic channels, even higher In mole fractions are possible. For comparison: in GaAs-based HEMTs, lattice-matched channels do not contain any indium at all, and in the pseudomorphic case, the content is usually limited to about 25%.

The first report on the successful fabrication of InP-based lattice-matched HEMTs (LMHEMTs) was made by Pearsall *et al.* early in 1983 [12], but due to the difficulties with the growth of the material, it took some more years until the devices showed the expected performance (Hirose *et al.* 1985 [13], Peng *et al.* 1987 [14]).

First results with pseudomorphic channels on InP were reported in 1986 (Kuo *et al.* [15]), and in 1988, InP PHEMTs became the first devices to exhibit a current-gain cutoff frequency f_T in excess of 200 GHz (Mishra *et al.* [16]). The result was achieved with an indium content of 62% in the channel. However, on InP substrate, the advantages of PHEMTs over LMHEMTs are not as pronounced as in the GaAs system.

The domain of HEMTs grown on InP are ultra-high frequency low-noise applications. At high drain voltages, gate leakage currents and impact ionization in the channel are hard to suppress so that these devices have for a long time not been regarded to be suitable for power application. Recently, attempts have been made to overcome these problems. They will be discussed at the end of this paper when advanced future concepts are presented.

Scope of this Article. In the present paper, we will treat HEMTs only with analog applications in mind. Within the family of HEMTs grown on GaAs, LMHEMTs and PHEMTs differ with respect to layer sequences and device physics. Since the AlGaAs/GaAs LMHEMT has become outdated for analog applications, we will not discuss it here. For HEMTs grown on InP, the situation is substantially different: LMHEMTs and PHEMTs share the same layer sequences and are often very similar in performance. There is only a difference in indium content in the channel, in many cases very small. For these reasons, *we shall discuss both LMHEMTs and PHEMTs on InP*. Thus, the subject of this review could be more precisely described by *HEMTs with InGaAs quantum well channels.*

In principle, we intend to avoid lengthy theoretical discussions, but rather want to look at HEMT device physics and epitaxial layer design from a practical point of view. This means: the present article tries to clarify the problems usually faced by a device designer when a HEMT layout for analog low-noise or power applications has to be developed. Therefore, a basic understanding of

- the possibilities and limitations of epitaxial layer design,
- the basic current-voltage characteristics of HEMTs,
- the physics of charge carrier control in HEMT channels,
- and the performance requirements for a specific application

is tried to be given in the following text.

2. PHEMT Epitaxial Layer Structure

2.1. EPITAXIAL GROWTH

The binary III-V compound semiconductors are all of the same zinc-blende crystallographic structure. Solid solutions can be formed among them, leading to a large number of ternary and quaternary alloys. Aside from technical problems, it is generally possible to grow these alloys on each other (by means of various epitaxial methods), provided

- the lattice constants are equal (*lattice-matched growth*),
- or the lattice constants do not differ too much, and the thickness of the layer mismatched to the growth substrate is kept below a critical limit. In this case, the strain within this layer will not relax, and the formation of crystalline dislocations is avoided (*pseudomorphic growth*).

If two semiconductors with different energy gaps are grown on each other, discontinuities of the conduction and valence band edges arise at the heterointerface. In most cases, the conduction band offset between adjacent layers ΔE_C is close to 2/3 of the total band gap difference ΔE_G.

There are basically two techniques that can be used for the growth of HEMT layers: molecular beam epitaxy (MBE) and metal-organic vapor phase epitaxy (MOVPE, also called MOCVD, metal-organic chemical vapor deposition).

Today, the majority of HEMT wafers is grown by MBE. This is caused by a number of reasons: the interfaces between layers are very smooth, changes in layer composition and doping can be made very abruptly, and the control of background impurities is comparatively easy. A disadvantage is the small wafer throughput of most MBE systems.

The high production capacity that can potentially be achieved has for a long time been judged the major advantage of MOVPE. Moreover, alloys containing phosphorus can be easily grown, contrary to MBE. On the other hand, MOVPE has a number of severe drawbacks. The control of background impurities is painstaking (especially if the layers contain aluminum) and only practicable if a large number of identical wafers has to be fabricated. Additionally, it is nearly impossible to "switch off" high doping concentrations abruptly enough if an undoped layer must be grown over a heavily doped one (which is the case in DH-PHEMTs, for instance). And finally, the MOVPE principle relies on extremely toxic gases like arsine and phosphine. Their handling necessitates extensive safety precautions and, in most countries, is prohibited close to residential areas.

2.2. BASIC LAYER SEQUENCE

PHEMTs belong to the family of field effect transistors (FETs): the current flow between two ohmic contacts - source and drain - is controlled by a third electrode, the gate. In most cases, the gate is a metal/semiconductor contact (Schottky contact).

The main difference to conventional FETs is the quantum well (QW) channel. This is a thin layer of *undoped* material where the energy gap E_G is lowest compared to the rest of the layers (see Figure 1). This means that electrons from the doped high-bandgap supply (or barrier) layer, located between the gate and the QW, are transferred into the QW because this is energetically favorable for them. Here, they are locally separated from their parent donor ions. Since the Coulomb potentials connected to the donors possess a finite range, electrons in the channel would be scattered by them if the doping would extent too close to the supply/channel interface. Therefore, to prevent mobility degradation from ionized impurity scattering, a few nanometers of the high-bandgap material directly adjacent to the channel are usually left undoped, forming the so-called spacer layer.

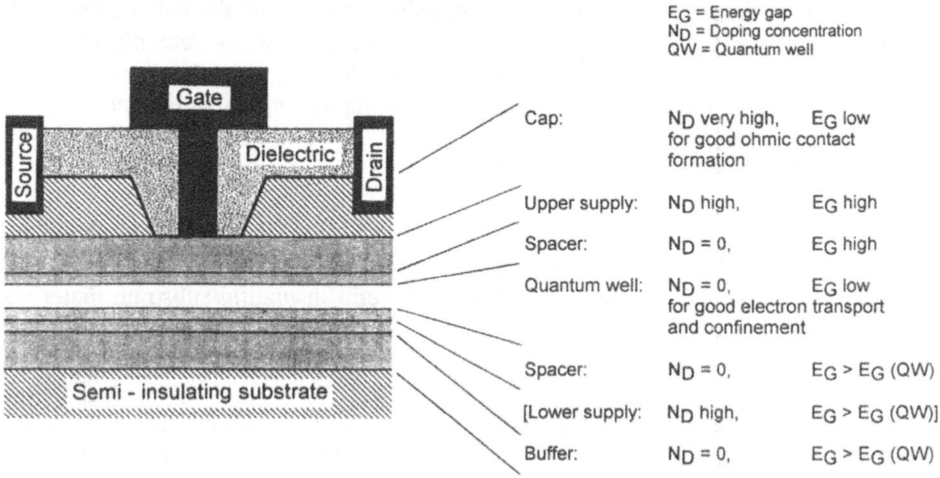

Figure 1. PHEMT layer schematic. N_D is the doping concentration.

Below the channel, there is also material with higher E_G. This creates an energy barrier for the electrons that prevents them from penetrating into the buffer where the control by the gate voltage is ineffective and transport properties are poor. In some types of PHEMTs (among them the most popular layer sequence on GaAs substrate), this lower barrier is not as high as the upper one. There are also cases in which there is additional doping in the region below the QW in order to get high electron concentration in the channel, for instance in power PHEMTs. This additional doping is usually only applied when the energy barrier below the channel is as high as the upper one (DH-PHEMTs).

Between the lower barrier layer (or the channel, if there is no separate barrier employed) and the semi-insulating substrate, the buffer layer extends, usually about one micrometer thick. This large thickness has several reasons: initial growth defects that originate from particles on the substrate surface, for instance, are reduced during the growth process, and any impurities from the growth chamber atmosphere or diffusing out of the substrate are gettered in the lower part of the buffer layer far away from the channel. Impurities in the buffer influence the device behavior even if they are several hundred nanometers away from the channel, especially under high-voltage bias conditions, since the energy barrier between channel and buffer only reduces electron injection into the buffer but does not completely prevent it. Moreover, the number and energetical position of states in the buffer bandgap determines where the Fermi level is pinned in the gap. This influences the position of the quantum well relative to the Fermi level and, hence, its population with electrons.

The uppermost layer in the PHEMT epi structure is the cap layer. It is necessary since one of the crucial requirements for the upper supply or barrier layer is that it has a large energy gap. It is hardly possible to make good ohmic contacts to large-bandgap materials, and therefore an extra layer optimized for the formation of source and drain contacts must be introduced. The cap should have a low bandgap and a high donor density. In the region of the intrinsic transistor, i.e. below the gate electrode, it must be etched away, since here we want the current flow to take place in the quantum well. Outside the intrinsic device, the cap can contribute to a highly conductive connection to the metal contacts by forming a shunt to the channel.

2.3. MATERIALS FOR PHEMT LAYERS

We now want to discuss which of the III-V compounds can be possibly used for the individual layers of the basic PHEMT structure. Figure 2 shows the bandgap energy *vs.* the lattice constant for these compounds. Two semi-insulating substrate materials are available for epitaxial growth: GaAs and InP.

2.3.1. *PHEMTs on GaAs Substrate*

To GaAs, there exists no III-V compound of the same lattice constant but with smaller bandgap. Thus, for small-gap channels on this substrate, there is only the choice of *pseudomorphically* grown InGaAs. If the lattice constant is not allowed to be too different from that of GaAs, the InGaAs bandgap is not very much smaller, and the barrier between buffer and channel is low. In practice, indium contents between 20 and 25% are used.

Fortunately, there are at least a number of compounds lattice-matched to GaAs but with higher E_G. These are AlAs, $In_{0.48}Al_{0.52}P$ and $In_{0.48}Ga_{0.52}P$. The growth of materials containing phosphorus is difficult with MBE so that their use has only very recently started together with efforts to fabricate PHEMTs based on MOVPE.

The conventional choice for a high-bandgap semiconductor lattice-matched to GaAs is AlGaAs. Two basic types of PHEMTs on GaAs exist: in the first case, the InGaAs channel is grown directly on a GaAs buffer, and high-bandgap AlGaAs is used only for the upper supply layer. Until recently, this has been the standard PHEMT design on GaAs, frequently called single-heterojunction PHEMT (SH-PHEMT). In the second case, AlGaAs is also grown between GaAs substrate and InGaAs channel to create a higher energy barrier to the substrate. This design is usually called double-heterojunction PHEMT (DH-PHEMT). It is considered to be more difficult to grow because the growth front of AlGaAs tends to be rough creating an imperfect interface to the channel.

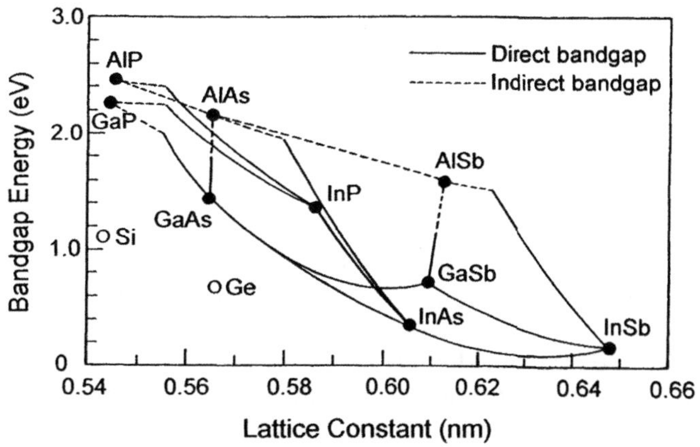

Figure 2. Bandgap energy *vs.* lattice constant for major III-V compounds.

The aluminum content x in n-$Al_xGa_{1-x}As$ supply layers is limited by two effects. For $x > 0.45$, the bandgap changes from direct to indirect. The conduction band minimum is not located at the center of the Brillouin zone any more but at the three-fold degenerate X point close to the zone edge [17]. This means an enormous increase in the density of states in the lowest part of the conduction band and leads to insufficient transfer of electrons into the InGaAs channel. The second effect limiting the aluminum content in $Al_xGa_{1-x}As$ is the deep donor level DX which inevitably appears together with a shallow donor level when an n-type silicon doping is present (see discussion in section 5.1.). This causes ineffective donor activation and undesirable transient charge redistribution effects between these levels for x larger than about 0.23. Thus, for PHEMTs grown on GaAs substrate, the conduction band discontinuity between InGaAs quantum well and AlGaAs supply is limited to approximately 0.3 eV in most practical cases.

The uppermost layer in a PHEMT structure on GaAs substrate is usually a highly n-doped cap, also made of GaAs.

2.3.2. *LMHEMTs and PHEMTs on InP Substrate*

On InP, the conditions for the formation of deep quantum well channels are more favorable. Due to the aforementioned difficulties with the MBE growth of phosphides, the buffer grown on this substrate is usually not InP but lattice-matched $In_{0.52}Al_{0.48}As$ with about the same bandgap (Figure 2). On top of this buffer or lower barrier layer, a lattice-matched $In_{0.53}Ga_{0.47}As$ channel is grown, followed by the upper $In_{0.52}Al_{0.48}As$ barrier. The conduction band discontinuity between channel and barriers is about 0.52 eV in this LMHEMT which is considerably larger than in the PHEMT on GaAs. As a low-bandgap cap, the same material is used as for the channel ($In_{0.53}Ga_{0.47}As$).

The indium content in the channel can easily be increased to over 70%, creating a PHEMT. Compared to the case of GaAs, the performance advantage of a PHEMT over an LMHEMT on InP is smaller. Sometimes, it may even be obscured by different qualities of the employed manufacturing processes.

2.4. FIVE PROTOTYPE HEMTS

The layer structures of the most popular HEMT types are compiled in Table 1. For comparison, the very first HEMT, i.e. the lattice-matched AlGaAs/GaAs HEMT, is also included. As already mentioned, this HEMT lacks an energy barrier between channel and buffer. Its triangular quantum well at the buffer/supply interface supports only relatively small drain currents and does not effectively suppress electron injection into the buffer at high drain voltages.

These deficiencies are partly removed with the second layer sequence in Table 1, the SH-PHEMT. Presently, this is the "standard" HEMT device for low-noise applications in the microwave and millimeter wave frequency range. However, the energy barrier to the buffer is quite small so that it cannot effectively confine the electrons to the channel anymore if the drain voltage approaches the order of 5 V.

TABLE 1. The five most popular HEMT layer structures

Cap	GaAs	GaAs	GaAs	$In_yGa_{1-y}As$ $y = 0.53$	$In_yGa_{1-y}As$ $y = 0.53$
Upper supply	$Al_xGa_{1-x}As$ $x \leq 0.23$	$Al_xGa_{1-x}As$ $x \leq 0.23$	$Al_xGa_{1-x}As$ $x \leq 0.23$	$In_xAl_{1-x}As$ $x = 0.52$	$In_xAl_{1-x}As$ $x = 0.52$
Cond. band discont. ΔE_C	≈ 0.20 eV	≈ 0.30 eV	≈ 0.30 eV	≈ 0.52 eV	> 0.52 eV
Channel	GaAs lattice matched	$In_yGa_{1-y}As$ $y \approx 0.20$ pseudomorphic	$In_yGa_{1-y}As$ $y \approx 0.20$ pseudomorphic	$In_yGa_{1-y}As$ $y = 0.53$ lattice matched	$In_yGa_{1-y}As$ $y \geq 0.53$ pseudomorphic
Cond. band discont. ΔE_C	0	≈ 0.12 eV	≈ 0.30 eV	≈ 0.52 eV	> 0.52 eV
Lower supply or buffer	GaAs	GaAs	$Al_xGa_{1-x}As$ $x \leq 0.23$	$In_xAl_{1-x}As$ $x = 0.52$	$In_xAl_{1-x}As$ $x = 0.52$
Substrate	GaAs	GaAs	GaAs	InP	InP
Remarks	historically first HEMT	single heterojunction PHEMT on GaAs	double heterojunction PHEMT on GaAs	standard HEMT on InP	advanced HEMT on InP

In the DH-PHEMT on GaAs (third column of Table 1), the confinement is improved. Additionally, a higher current than in the standard PHEMT can be realized. These properties enable the DH-PHEMT to be used as a power transistor. Its development meant a breakthrough for PHEMTs which thereby became a universal microwave technology applicable for receivers *and* transmitters.

The last two types of transistors in Table 1, the LMHEMT and PHEMT on InP, differ with respect to indium content of the channel and, hence, quantum well depth. InP-based devices are mostly used for purposes where lowest noise figures or extreme high-frequency performance are required. Though they are capable of even higher drain currents than GaAs DH-PHEMTs, they are difficult to optimize for power applications (see section 9.2.).

2.5. LIMITATIONS FOR PSEUDOMORPHIC CHANNEL GROWTH

Up to now, we have only discussed the composition of the layers in a PHEMT but not their individal thicknesses. For pseudomorphic layers, these are limited depending on the amount of mismatch to the growth substrate.

2.5.1. *Pseudomorphic Growth on GaAs Substrate*

This situation was for the first time investigated by Matthews and Blakeslee [18] who used experiments with alternating layers of GaAs and GaAsP to derive a theoretical expression for the critical layer thickness L_C which cannot be exceeded if misfit dislocations must be avoided. The validity of the model for the system InGaAs/GaAs was demonstrated by Andersson *et al.* [19] by means of photoluminescence measurements. The thickness L_C up to which the mismatch of the $In_yGa_{1-y}As$ layer can be accomodated by elastic strain is given by

$$0.07y = \frac{a\,(1 - \sigma / 4)\left[\ln\,(L_C\sqrt{2} / a) + 1\right]}{2\sqrt{2}\,\pi\, L_C\,(1 + \sigma)} \tag{1}$$

where y is the indium content, a is the GaAs lattice constant (a = 0.565 nm), and σ is Poisson's ratio (a dimensionless combination of stiffness constants which is 0.23 for GaAs). The curve defined by Equation 1 is shown in Figure 3.

For $y = 0$, the growth is lattice-matched, and the magnitude of L_C is infinite (this is nearly the case in the AlGaAs/GaAs HEMT). On the other hand, for large y, the critical thickness is only a few nanometers. Quantum wells that are too narrow contain only a small number of charge carriers, and the transport properties of these carriers are degraded due to excessive interface scattering. The best compromise with respect to transistor performance (number of carriers in the channel and carrier mobility) is found around $y = 0.2$ and a channel thickness of 12 nm. Two other commonly cited combinations are also indicated in Figure 3.

2.5.2. *Pseudomorphic Growth on InP Substrate*

Compared to the situation on GaAs, even lattice-matched quantum wells have a large ΔE_C to the barrier layers so that there is less need to drive pseudomorphic growth towards its crystallographic limits. For lattice-matched QWs, the thickness is mainly determined by device physical considerations and not by fundamental limitations. In case of pseudomorphic growth, one should expect the calculations of Matthews and Blakeslee [18] to hold also for InP.

In 1985, however, People and Bean [20] published a model for critical layer thicknesses that is not based on the mechanical equilibrium considerations of Ref. [18] but on energy balance. Recently, this new model has been experimentally verified for InGaAs/InAlAs heterostructures by Taguchi *et al.* [21]. It leads to significantly larger critical thicknesses than the older model. For $In_yGa_{1-y}As$ with y = 0.8, an L_C of 20 to 25 nm is given in Ref. [21], opposed to 3.5 nm predicted by the mechanical equilibrium model. For y = 0.7, the values are 50 to 100 nm opposed to about 8 nm. This means that the channel dimensions of InP PHEMTs up to quite high indium percentages in the QW could be still determined by device physical considerations rather than by growth limitations. In most published InP PHEMTs, the QW thicknesses are below the maxima given by the energy balance model but larger than allowed by the theory based on mechanical equilibrium.

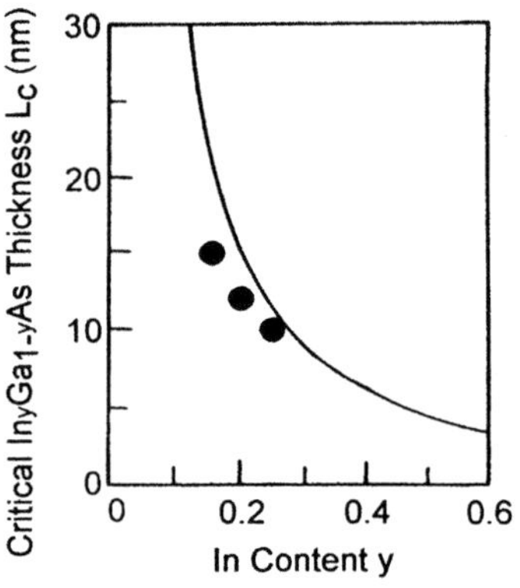

Figure 3. Critical InGaAs thickness as a function of In content after Matthews and Blakeslee [18].

Up to now, most published PHEMTs in the system InGaAs/AlGaAs on GaAs do not exceed the limitations given by the model of Matthews and Blakeslee. The performance of these devices, however, could considerably profit from deeper or wider quantum wells.

2.6. HEMT CROSS-SECTIONAL DIMENSIONS

The conduction band structures of HEMTs on GaAs and InP substrate are compared in Figure 4. The profiles are approximately to the same scale to give an impression of the larger QW cross section of the HEMTs on InP.

Apart from the mere QW sizes, several other differences can be seen. The built-in voltage $q\phi_B$ at the gate/semiconductor interface is considerably smaller for InAlAs than for AlGaAs (in practice, 400 compared to over 700 mV). This is one - but not the only - reason for the smaller thickness d of the supply layer of InP-based HEMTs since it means a smaller depletion depth under the gate. In the standard PHEMT on GaAs and the HEMTs on InP, the barrier layers below the channel are normally undoped (here, the conduction band energy E_C increases with respect to the Fermi level E_F as a function of depth because E_F is pinned in the buffer at or below mid-gap). In the DH-PHEMT, generally both barriers are doped causing a more symmetrical shape of the QW. Here, without the doping layer below the channel, the bottom of the QW would be too high relative to the Fermi level due to the position of E_F in the AlGaAs buffer.

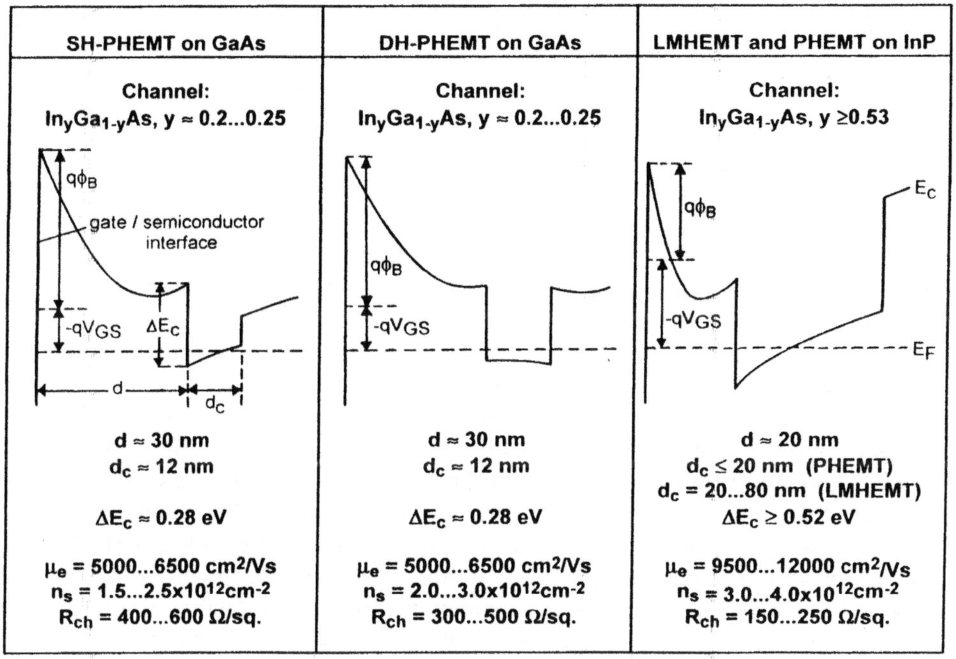

Figure 4. Conduction band structures and typical quantum well transport data for HEMTs on GaAs and InP substrate. μ_e: electron mobility, n_s: electron concentration in channel, R_{ch}: channel resistance.

3. Transport Properties of InGaAs Channels

3.1. VELOCITY-FIELD CHARACTERISTICS IN THICK LAYERS

The different indium concentrations in the QWs of GaAs- or InP-based HEMTs are one major reason for their different performances. This can be seen in Figure 5 which shows stationary electron drift velocities as a function of electric field for the compounds GaAs, $In_{0.53}Ga_{0.47}As$ and InAs. The experimental data were obtained with a constant electric field in bulk crystals or thick lattice-matched epitaxial layers. In a narrow HEMT channel, the transport is not stationary, and the situation is additionally complicated by scattering of the electrons at the adjacent interfaces, by scattering at the Coulomb potentials of the dopants behind these interfaces, and by the presence of strain. Thus, Figure 5 represents an idealized situation.

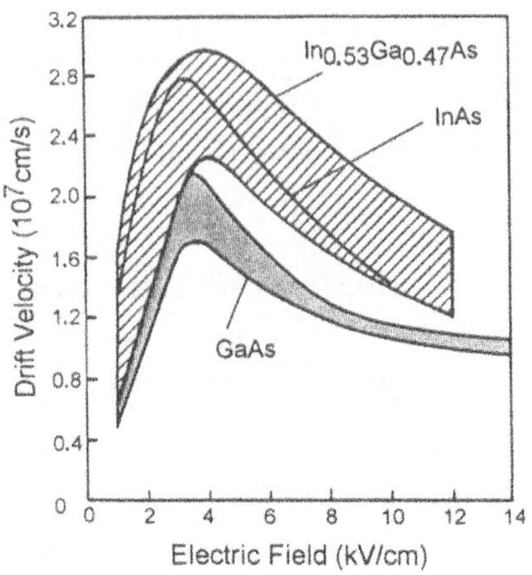

Figure 5. Electron drift velocities in GaAs, InGaAs and InAs *vs.* electric field.

The shaded area for GaAs comprehends the data by Ruch and Kino [22], Braslau and Hauge [23], and Houston and Evans [24]. It is composed of several distinct regimes. Up to fields E of about 3 kV/cm, the electron velocity v is determined by a constant low-field mobility μ_e (ohmic transport), where $v = \mu_e E$. Beyond the peak velocity v_p measured at fields close to 4 kV/cm, the stationary drift velocity decreases with the field until a high-field value v_{sat} is reached. The hatched area for $In_{0.53}Ga_{0.47}As$ is a compilation of data by different authors done by Littlejohn *et al.* [25]. For unstrained $In_{0.53}Ga_{0.47}As$, μ_e, v_p and v_{sat} are higher than for GaAs. The scattering of the data is considerable, probably due to different quality of the epitaxial growth, but the general superiority of the $In_{0.53}Ga_{0.47}As$ transport properties compared to GaAs is clearly visible.

The case of pure InAs is the most complicated in Figure 5. The low-field mobility is the highest of all three materials discussed here. For high fields, reliable experimental data are difficult to achieve. Thus, a theoretical curve by Brennan and Hess [26] is shown in the figure. The idealized calculation does not take impact ionization into account. However, the low bandgap of InAs (see Figure 2) causes the impact ionization rate to be considerable for the electric fields of interest here, and its neglect has a significant influence on the result of the calculation. Absence of impact ionization allows the electrons to pick up high energies from the electric field without being scattered back into states near the conduction band edge. This means that there is a high transfer rate of electrons from the central band minimum to the higher minima where the transport properties are extremely degraded. This is reflected by the curve in Figure 5 which does not indicate any superiority of the carrier peak velocity in InAs over the velocity in $In_{0.53}Ga_{0.47}As$.

In reality, however, this is not quite true because impact ionization actually plays an important role in high-field transport in InAs. It acts as a permanent cooling mechanism which prevents the electrons from attaining enough kinetic energy for transitions to the satellite valleys, thus confining them to the central band minimum.

According to Brennan and Hess [18], this eliminates the decrease of the electron velocity beyond the peak value at approximately 3 kV/cm in Figure 5 but leads to a continuous rise of v up to fields of 100 kV/cm where a new maximum of v_{sat} = 8 x 10^7 cm/s is reached.

Due to the small dimensions of high-field domains in HEMTs, transport is not stationary, transient effects occur, and the relevance of the results discussed above is difficult to estimate. Nevertheless, indications remain that an increase of the indium content in PHEMT channels to values close to 100% might lead to some drawbacks. Improvement of the transport properties can be coupled to an increase of the impact ionization rate in the channel causing breakdown of the device already at low bias voltages (besides other undesirable effects like large output conductance and the creation of holes).

3.2. LOW-FIELD TRANSPORT IN THIN LAYERS

Compared to bulk material, the situation in thin layers is further complicated by the influence of the adjacent material layers, the interfaces and, often, by the presence of strain. Thus, information of general validity on thin layer transport is even more restricted because carrier velocities and mobilities are not only governed by "intrinsic" properties of the layer material but also by conditions imposed by the layer surroundings and the quality of the epitaxial growth. Additionally, the situation is obscured by the problems with high-field transport. For these reasons, mainly Hall mobility data are available for quantum well structures, and we will make only very few qualitative statements on low-field transport.

Generally, with increasing In content, the effective electron mass in InGaAs decreases, and the 300 K mobility is expected to go up. Strain, on the other hand, increases the effective mass (Liu *et al.* [27]). When the In content y is raised from 0 (unstrained case on GaAs substrate) to 0.3 (strained case), increasing strain and alloy scattering cause a decrease of the mobility at 77 K by about a factor of two (Nguyen et al. [28]). At 300 K, the decreasing effective mass by far dominates the counteracting effects, and the mobility improves with In concentration for InGaAs grown on both InP and GaAs substrates (Bhattacharya [29]).

3.3. TRANSPORT IN PRACTICAL HEMT CHANNELS

In this complex situation, we prefer to restrict ourselves to the statement of typical low-field mobilities for the three types of HEMT quantum well channels of Figure 4. In practice, the physical influences discussed above are mostly obscured by effects of the individual epitaxial growth quality and scattering at the Coulomb potentials of the dopant atoms in the supply layers. The proper optimization of the spacers that separate these dopants from the quantum well is important if high mobilities μ_e shall be combined with maximum electron sheet densities n_s. Both n_s and μ_e determine the channel sheet recistance R_{ch}. The quantities given in Figure 4 are typical numbers that can be obtained by Hall effect measurements at HEMT quantum well structures. They do not represent record data.

The HEMTs based on InP feature electron mobilities far superior to those on GaAs. Additionally, their larger conduction band discontinuity at the channel-barrier interface leads to substantially higher n_s. In combination, this means that InP based HEMTs possess significantly smaller sheet resistances than GaAs based devices. Before we discuss how these 2DEG properties translate into HEMT performance, we briefly sum up some basic current-voltage characteristics that the HEMT shares with all field-effect transistors.

4. Current-Voltage (I-V) Characteristics of PHEMTs

4.1. OUTPUT CHARACTERISTICS

For a PHEMT, the dependence of the drain current I_D on the drain-source voltage V_{DS} (output characteristics) is similar to that for a conventional III-V MESFET (MEtal-Semiconductor Field-Effect Transistor, i.e. FET with Schottky gate electrode). As an example, Figure 6 shows an output characteristics of a "standard" SH-PHEMT on GaAs substrate.

In the saturated current regime, I_D is not really constant for a fixed gate-source voltage V_{GS} but still increasing with V_{DS} which means that there is a finite output conductance g_{ds} given by

$$g_{ds} = (\partial I_D / \partial V_{DS}) \Big|_{V_{GS} = \text{const.}} \tag{2}$$

Additionally, the curves I_D (V_{DS}) are not equidistant for the different gate-source voltages. This indicates that I_D is not a linear function of V_{GS}, i.e. that the transconductance g_m is not constant with V_{GS}. The transconductance is defined by

$$g_m = (\partial I_D / \partial V_{GS}) \Big|_{V_{DS} = \text{const.}} \tag{3}$$

PHEMTs are distinguished from other FETs by their high g_m. In the example of Figure 6, the peak value is approximately $g_m / W_G = 525$ mS/mm (W_G : gate width). In combination with $g_{ds} / W_G = 25$ mS/mm, this results in a voltage gain $g_m / g_{ds} = 21$.

4.2. TRANSFER CHARACTERISTICS

The nonlinearity of the dependence of I_D on V_{GS} can be better seen in the transfer characteristics and the transconductance characteristics (Figure 7). The transconductance g_m (V_{GS}) exhibits a pronounced maximum for V_{GS} close to 0.25 V.

It is one of the aims of this paper to give some insight into the underlying mechanisms so that measures to influence the shape of this curve in a purposeful way can be identified.

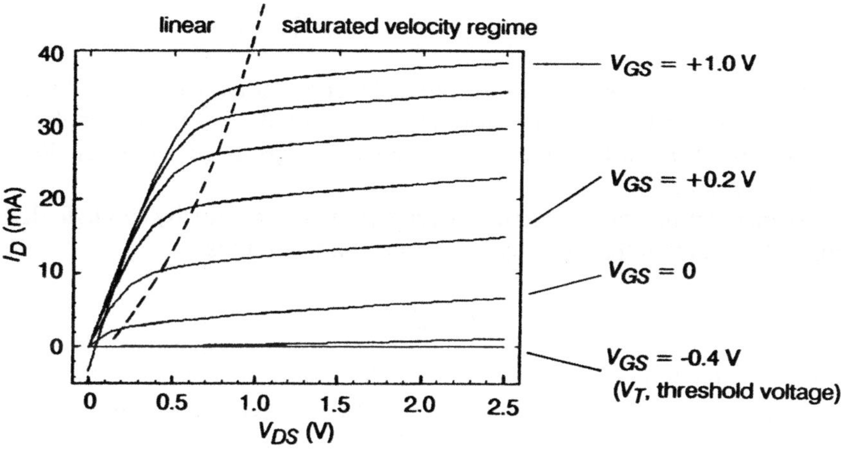

Figure 6. Output characteristics I_D (V_{DS}) of an SH-PHEMT on GaAs substrate with W_G = 80 µm and L_G = 0.25 µm.

4.3. TRANSCONDUCTANCE OF AN IDEALIZED PHEMT

In an idealized PHEMT, all charge carriers are confined in the QW in a sheet of negligible thickness. This 2DEG is assumed to be located at $z = d + \Delta d$ (see Figure 8) where d is the thickness of the barrier layer between gate and channel, and Δd takes

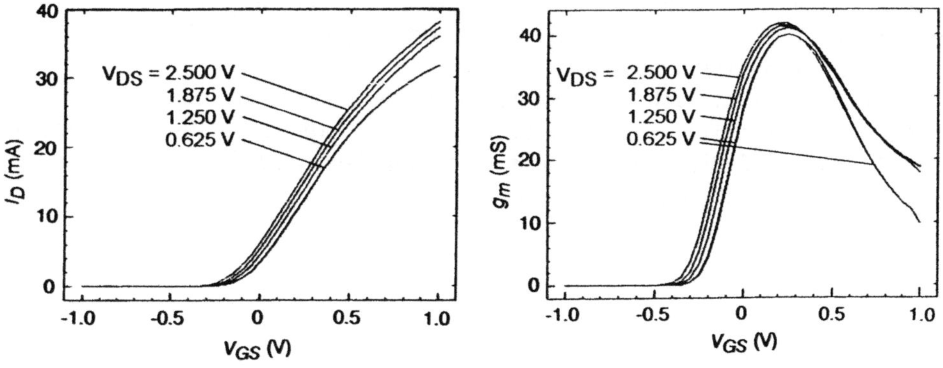

Figure 7. Transfer characteristics I_D (V_{GS}) and transconductance characteristics g_m (V_{GS}) of an SH-PHEMT on GaAs substrate with W_G = 80 µm and L_G = 0.25 µm.

into account that the effective center of the electron distribution is not directly at the barrier-channel interface but somewhere in the QW.

The sheet charge of the 2DEG is $Q_s = q\, n_s\, L_G\, W_G$ (L_G : gate length), and the sheet capacitance under the gate C_s is given by

$$C_s = \partial Q_s / \partial V_{GSi} = [\varepsilon / (d + \Delta d)]\, L_G\, W_G \tag{4}$$

where ε is the dielectric constant of the barrier layer material. The gate voltage V_{GSi} of the *intrinsic* transistor must be distinguished from the externally applied V_{GS} because both quantities differ due to the presence of parasitic elements. We will discuss this in chapter 6.

If we assume that the charge carriers move with uniform velocity through all parts of the transistor, and that this velocity is v_{sat}, the drain current is

$$I_D = Q_s\, v_{sat} / L_G \tag{5}$$

which leads to the following expression for the transconductance of the intrinsic device, g_{mi} (which must also be distinguished from the externally measured g_m):

$$g_{mi} = \partial I_D / \partial V_{GSi} = [\varepsilon\, v_{sat} /(d + \Delta d)]\, W_G = C_s\, v_{sat} / L_G\ . \tag{6}$$

An important consequence of Equation 6 is: for maximum transconductance, the distance d between gate and QW must be minimized. Unfortunately, this also results in a maximum sheet capacitance which is directly proportional to g_{mi} and can be reduced only by a shortening of the gate length.

Additionally, a second problem arises with the reduction of d. At the interface between gate and barrier layer, the built-in voltage $q\phi_B$ causes the conduction band edge E_C to bend upwards with respect to the Fermi energy E_F thus establishing a charge depletion layer. If d is too small, the QW will be lifted relative to E_F, and n_s will be reduced. In extreme cases, it might be possible that the QW cannot be adequately filled with electrons even if maximum positive gate bias is applied to counteract the surface depletion.

Therefore, it is favorable if the doping concentration N_D in the barrier layer is high. High doping has two consequences: it leads to a high n_s (presumed, the charge transfer into the QW is effective), and it causes a more rapid decrease of the surface potential since the depth d_d of the depletion zone is given by

$$d_d \approx (2\, \varepsilon\, \phi_B / q\, N_D)^{1/2}\ . \tag{7}$$

The common barrier material for GaAs-based PHEMTs, AlGaAs, can be practically doped up to levels of about N_D = 3...5 x 10^{18} cm^{-3}. Higher concentrations are useless because they do not translate into higher carrier densities due to decreasing donor activation and increasing electron trapping in deep levels (DX centers).

InP-based HEMTs usually possess InAlAs barrier layers. Better activation compared to AlGaAs allows silicon dopings up to $N_D = 1.5 \times 10^{19}$ cm^{-3} so that very small barrier thicknesses d can be employed. Together with the large v_{sat} of $In_{0.53}Ga_{0.47}As$, this a key contribution to the realization of ultra-high transconductances with InP-based HEMTs.

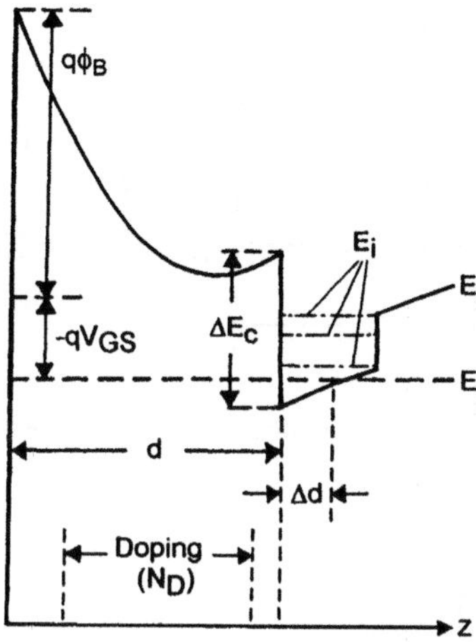

Figure 8. Conduction band diagram of a PHEMT with homogeneously doped supply layer.

However, there are two major reasons for which a high doping cannot be applied throughout the whole thickness d of the barrier layer. The first one was already mentioned in section 2.2.: since the Coulomb potentials of the donor ions have a finite range, the doping must be separated from the QW interface by a spacer typically of the order of 2 nm.

The second reason is the bad quality of Schottky contacts to highly doped semiconductors. Such contacts are characterized by low barriers, high leakage currents under reverse bias conditions, and low reverse breakdown voltages. Therefore, an undoped layer is also appropriate at the gate/barrier interface.

Extremely small barrier thicknesses d can be achieved by the concept of *delta doping* as discussed by Schubert [30] (also called *planar doping* or *pulse doping*). Here, the supply donors are concentrated to a few atomic monolayers. This can be realized by MBE growth interruption of the host crystal during deposition of the impurities. In AlGaAs, planar donor concentrations up to $n_{delta} = 7 \times 10^{12}$ cm^{-2} are reasonable. Analogous to the case of bulk material doping, even higher concentrations can be used in InAlAs due to the better activation efficiency. Figure 9 shows a conduction band diagram for a delta-doped PHEMT which can be identified by the linear potential drop in undoped material.

The preceding discussion provided some insight how the peak value of g_m can be optimized but did not explain the particular shape of the curve g_m (V_{GS}). To understand this, charge control in a PHEMT channel must be discussed in more detail.

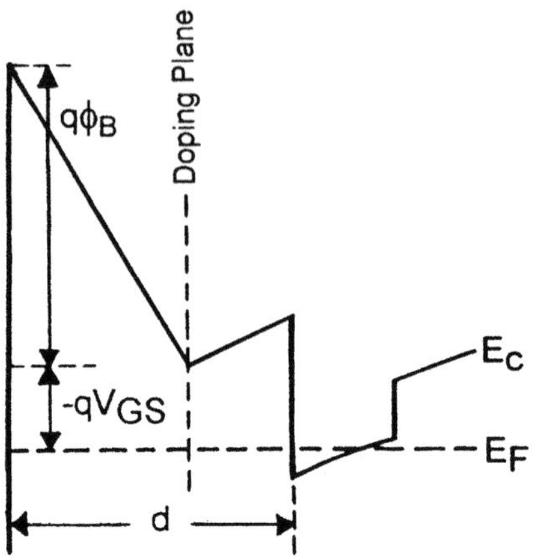

Figure 9. Conduction band diagram of a delta-doped PHEMT.

5. Charge Control in PHEMTs

5.1. SCHRÖDINGER'S AND POISSON'S EQUATIONS

The local distribution of the charge carriers in the direction z perpendicular to the semiconductor-gate interface is described by two equations which have to be solved simultaneously: Schrödinger's and Poisson's equations. A discussion of these equations can be found in the review by Nguyen *et al.* [28] and in the article by Ando and Itoh [31].

Schrödinger's equation is given by

$$[- (\hbar^2 / 2m^*) (\partial^2 / \partial z^2) - qV(z) + \Delta E_C(z)] \, \phi_i(z) = E_i(z) \, \phi_i(z) \tag{8}$$

where $\hbar$ denotes Planck's constant, m^* is the effective mass of the electrons in the conduction band, $V(z) = E_C(z) / (- q)$ is the electrostatic potential, $\Delta E_C(z)$ is a function that describes the conduction band discontinuities at the heterojunctions that enclose the quantum well, ϕ_i (z) is the envelope electron wavefunction, and E_i are the Eigenvalues of the electron energy in the quantum well. The envelope wavefunction is connected to the complete wavefunction $F_i\,(z, \boldsymbol{r})$ by

$$F_i\,(z, \boldsymbol{r}) = \phi_i(z) \exp\,(iq\boldsymbol{r}) \tag{9}$$

where the vectors $\boldsymbol{q}$ and $\boldsymbol{r}$ are in a plane parallel to the heterojunction.

Poisson's equation is

$$\partial^2 V(z) / \partial z^2 = [q / \varepsilon(z)] [N_D^+(z) - N_A^-(z) + N_{DX}^+(z) - n(z)] \tag{10}$$

where the dielectric constant ε is different for the particular layers and therefore a function of z. $N_D^+(z)$ is the conzentration of ionized shallow donors, and $N_A^-(z)$ the concentration of ionized shallow acceptors. $N_{DX}^+(z)$ is a term which is important for GaAs-based PHEMTs with AlGaAs barrier layers but not for InP-based HEMTs with InAlAs barriers. It denotes the concentration of occupied "DX" type deep donor levels in doped AlGaAs. $n(z)$ is the three-dimensional electron concentration.

The quantity $N_{DX}^+(z)$ depends on the Al content of the AlGaAs layers and on their (silicon) doping concentration. Two different models that describe this dependency are discussed in the papers by Chand *et al.* [32] and Schubert and Ploog [33]. Both papers agree in the observation that for Al contents exceeding $x \approx 0.2$ the introduction of silicon does not only lead to the formation of shallow donor levels but simultaneously to a concentration of deep donors that has to be taken into account.

Schubert and Ploog postulate that the energetical position of the deep levels is independent of the Al content. However, an increase of the Al concentration causes the deep-state density to rise relative to the desired shallow-state density. This results in ineffective doping activation since an increasing part of the donor electrons will not be transfered into the quantum well or the supply layer conduction band but will be trapped in the DX levels. Here, it does not contribute to current transport but only may participate in slow trap charge redistribution processes which interfere with the regular high-frequency transport mechanisms.

In the model by Chand *et al.*, it is not the density of the DX centers that depends on the Al concentration, but their energetical position. It is assumed that the DX energy is separated from the energy of the X conduction band minimum by a constant difference which means that the DX energy rises with Al content, together with the X valley. This means that the DX occupation probability rises with Al content.

Both DX models lead to the conclusion that Al concentrations substantially exceeding, say, 23% are problematic if trap charge redistribution effects shall be avoided (especially at temperatures below 300 K), i.e. effects having to do with voltage and time dependend variations of the term $N_{DX}^+(z)$ in Equation 10.

The local electron concentration $n(z)$ is the sum of the local electron concentrations in the subbands connected to the individual energies E_i , with the sum over all envelope wavefunctions normalized to one:

$$n(z) = \sum_i |\phi_i(z)|^2 n_i. \tag{11}$$

The electron concentrations n_i in subbands i are given by

$$n_i = (kTm^* / \pi\hbar^2) \ln [1 + \exp ((E_F - E_i) / kT)] \qquad (12)$$

where k is Boltzmann's constant and T the temperature. Please note that not only $n(z)$ but also all other terms in the bracket of the right-hand side of Equation 10 include occupation probabilities which depend on the position of E_F.

A simultaneous solution to Equations 8 and 10 is subject to two major boundary conditions. The first one is the potential at the semiconductor-gate interface which is given by $V(z = 0) = V_{GS} - \phi_B$, and the second one is the continuity of the electric displacement $\varepsilon (z)\, \partial V(z) / \partial z$ at all heterojunction interfaces. When looking for a first guess for $V(z)$ to start with the iterative solution process, one might assume that $V(z)$ is a straight line in regions where no doping is introduced, for instance in the layer between delta-doping and gate, but has to keep in mind that mobile charge $n(z)$ might be transfered to parts of such regions so that $\partial^2 V(z) / \partial z^2$ is not equal to zero any more.

We shall now discuss the numerical solution of Schrödinger's and Poisson's equations for two specific cases.

5.2. CHARGE CONTROL IN GaAs-BASED SH-PHEMTS

The first example is an SH-PHEMT on GaAs substrate with a gate-to-channel separation of about 35 nm and a homogeneously doped $Al_{0.2}Ga_{0.8}As$ supply layer (2 nm undoped spacer to the channel). The channel is 12 nm $In_{0.2}Ga_{0.8}As$. Electrostatic potential and electron wavefunction are shown in Figures 10 and 11 for two different

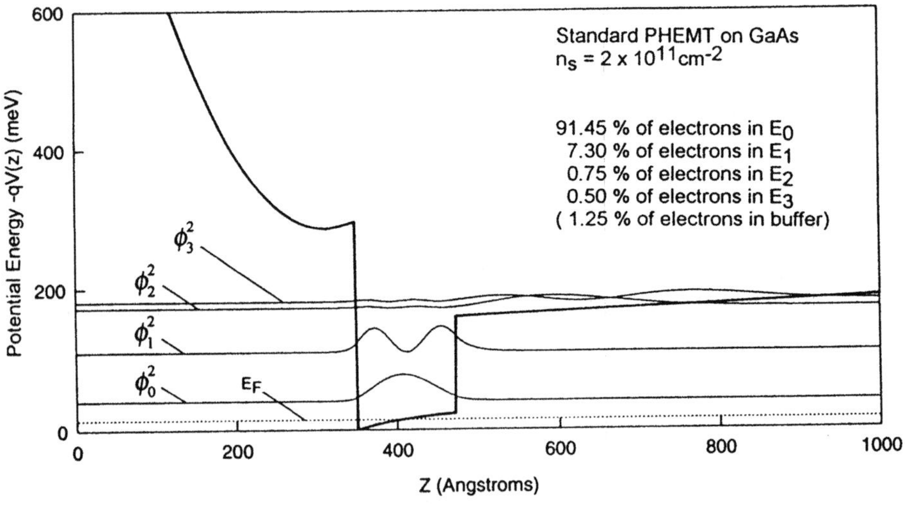

Figure 10. **Conduction band diagram and electron wavefunction in an SH-PHEMT on GaAs for V_{GS} = - 0.39 V. Program PSIGaAs by A. Zrenner. Results courtesy H. Riechert.**

gate voltages. In Figure 10, V_{GS} is negative and the number of electrons in the quantum well is small, i.e. the device is close to pinch-off (under the assumption v_{sat} = 1.5 x 10^7 cm/s, n_s = 2 x 10^{11} cm^{-2} roughly corresponds to a drain current of 50 mA/mm). The squares of the envelope wavefunctions show that close to 99% of the electrons reside in the ground state or the first excited state in the quantum well.

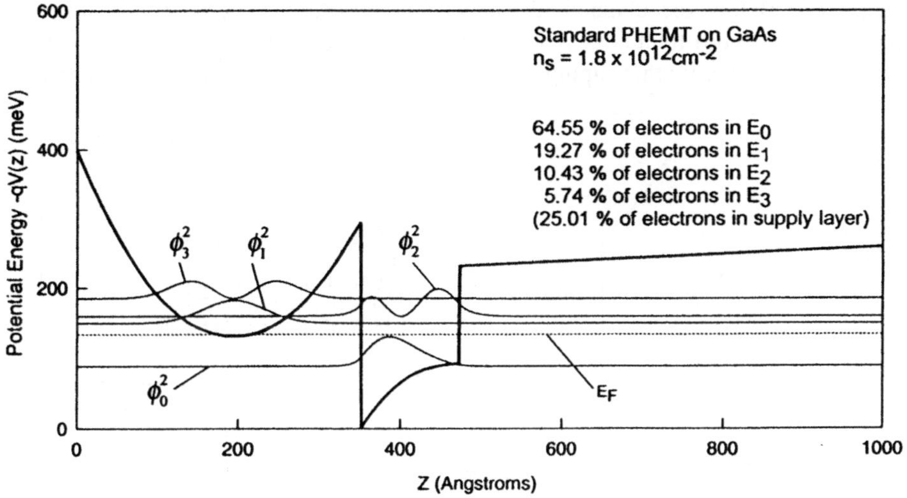

Figure 11. Conduction band diagram and electron wavefunction in an SH-PHEMT on GaAs for V_{GS} = + 0.43 V. Program PSIGaAs by A. Zrenner. Results courtesy H. Riechert.

However, there is a small number of electrons (1.25%) populating the excited levels E_2 and E_3. The wavefunctions belonging to these levels are not centered in the quantum well but extend far into the GaAs buffer below it. This is a consequence of the very small energy barrier between channel and buffer.

In Figure 11, the gate voltage is positive, and the number of carriers in the channel is high (n_s = 1.8 x 10^{12} cm^{-2} corresponds to about 430 mA/mm if the same v_{sat} is used as above). Only the carriers belonging to energy states E_0 and E_2 are confined to the quantum well. The remaining 25% (states E_1 and E_3) sit in a "parasitic" channel which has developed in the AlGaAs supply. In contrast to the case of Figure 10 where the ground state electrons are located precisely in the center of the channel, the ground state electrons (i.e. the majority of carriers) now form a large sheet charge concentrated near the gate-supply heterojunction. According to Poisson's equation, this gives rise to a local electric field. Due to the boundary condition of field continuity at all heterojunctions, this field is also present in the supply region adjacent to the channel bending the conduction band down to the Fermi level. Of course, the electrons in this parasitic channel suffer from degraded transport properties.

5.3. CHARGE CONTROL IN GaAs-BASED DH-PHEMTS

The second example is a DH-PHEMT on GaAs substrate with a gate-to-channel separation of about 30 nm and a homogeneously doped $Al_{0.2}Ga_{0.8}As$ upper supply layer, also with 2 nm undoped spacer. As previously, the channel is 12 nm $In_{0.2}Ga_{0.8}As$. Below the channel, there is a second $Al_{0.2}Ga_{0.8}As$ supply layer: the doped region of 6 nm thickness is separated from the channel by a 4 nm spacer. The undoped material below this supply is again $Al_{0.2}Ga_{0.8}As$. In Figure 12, the carrier

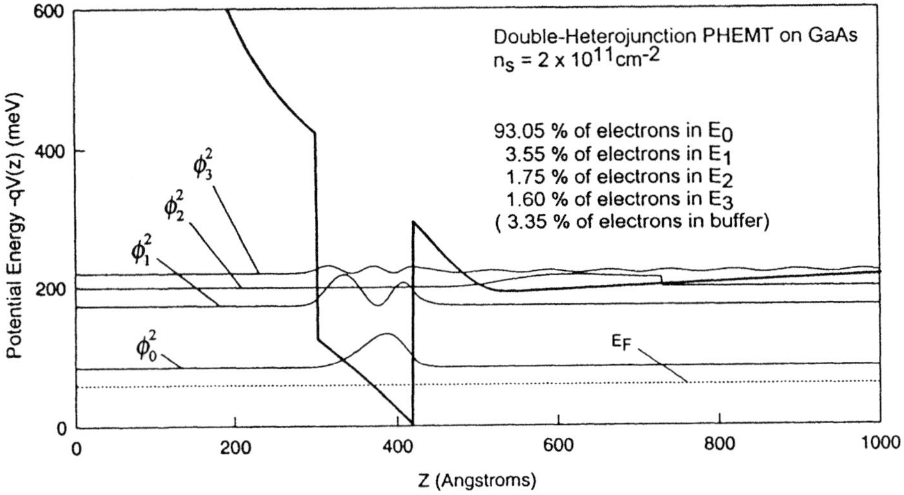

Figure 12. Conduction band diagram and electron wavefunction in a DH-PHEMT on GaAs for V_{GS} = - 0.64 V. Program PSIGaAs by A. Zrenner. Results courtesy H. Riechert.

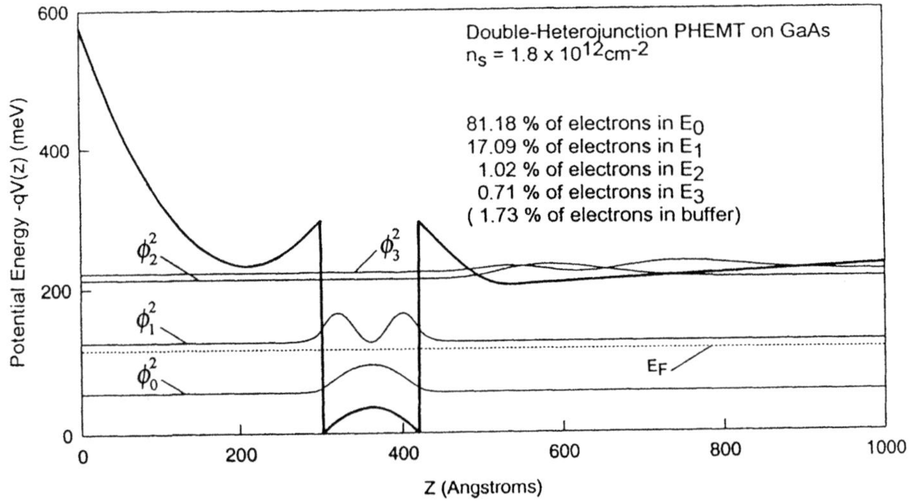

Figure 13. Conduction band diagram and electron wavefunction in a DH-PHEMT on GaAs for V_{GS} = + 0.26 V. Program PSIGaAs by A. Zrenner. Results courtesy H. Riechert.

density is the same as in Figure 10, i.e. 2×10^{11} cm^{-2}. As a consequence of the doping below the QW, the gate voltage necessary to deplete the channel this far is more negative than in Figure 10 though the gate-to-channel separation is smaller.

In the DH-PHEMT, the percentage of electrons in states E_2 and E_3 (3.35%) is higher than in the SH-PHEMT by a factor of 2.7. Although not all of these electrons are located in the lower supply (Figure 12 shows a finite magnitude of $\phi_3{}^2$ inside the quantum well), it is quite clear that parasitic conduction below the channel is larger in the DH-PHEMT. This is caused by the doping below the channel which bends the conduction band down. The barrier below the channel is higher than in the SH-PHEMT but its upper part is thin so that electrons in the excited states can tunnel through it. The majority of the electrons, however, is located inside the quantum well, close to the lower heterojunction.

The conduction band diagram of a DH-PHEMT with positive gate bias is shown in Figure 13. The carrier density of 1.8×10^{12} cm^{-2} is achieved with a smaller gate voltage compared to the SH-PHEMT. But more important, there are no parasitic electrons in the upper supply. This means that all electrons are "fast", and the same number of carriers supports a larger drain current than in the SH-PHEMT. The ground state wavefunction is centered in the channel (contrary to the SH-PHEMT where it is close to the upper heterojunction). Surprisingly, there are still a few electrons found below the quantum well.

5.4. CHARGE CONTROL AND TRANSFER CHARACTERISTICS

Transfer and transconductance characteristics of an SH-PHEMT have already been shown in Figure 7. Figures 14 presents the corresponding characteristics of a DH-PHEMT.

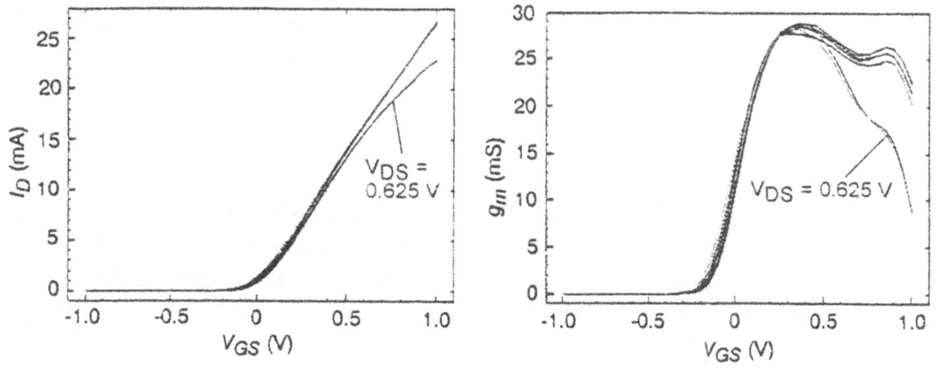

Figure 14. Transfer characteristics I_D (V_{GS}) and transconductance characteristics g_m (V_{GS}) of a DH-PHEMT on GaAs substrate with $W_G = 60$ μm and $L_G = 0.25$ μm.

In Figure 14, it can be seen that the dependence of the drain current I_D on the gate voltage V_{GS} is more linear compared to Figure 7. This is reflected in the transconductance $g_m(V_{GS})$ which features a distinct plateau for V_{GS} between + 0.2 and + 0.9 V. Such a plateau cannot be found in the corresponding plot for the SH-PHEMT (Figure 7).

These differences in the transconductance characteristics of SH-PHEMT and DH-PHEMT can be explained by their different charge control behavior. In an SH-PHEMT, when V_{GS} increases from the pinch-off voltage towards more positive values, the 2DEG is generated in the middle of the quantum well and moves (while n_s grows) towards the upper heterojunction. The maximum g_m is obtained when V_{GS} is such that the 2DEG is close to the heterojunction (i.e. the distance to the gate is minimized) but the conduction band in the supply is still well above E_F. For even more positive V_{GS}, the 2DEG density n_s will not grow anymore, but the supply will be increasingly populated with (slow) carriers, and g_m drops.

In the DH-PHEMT at threshold, the 2DEG does not develop in the center of the channel but shifted towards the lower heterojunction. With increasing V_{GS}, the peak of the electron distribution also moves upwards. In contrast to the SH-PHEMT, the maximum n_s is already reached when the wavefunction sits approximately symmetrically in the quantum well. This means that high carrier concentration in the channel can be achieved without creating an extremely high electric field at the upper heterojunction that forces the conduction band in the supply down to E_F. This happens only for even higher V_{GS} so that the onset of parasitic conduction in the supply is delayed. Consequently, in a DH-PHEMT, carriers confined to the quantum well are modulated without simultaneous modulation of slower carriers during a fraction of the total gate voltage swing that is significantly larger than in an SH-PHEMT. However, a transfer characteristics as linear as shown in Figure 14 can be achieved only with small threshold voltage, i.e. with comparatively thin supply layer.

Unfortunally, the favorable behavior of the DH-PHEMT is combined with a drawback: for a given supply thickness, the effective separation of the 2DEG from the gate is some nanometers larger in a DH-PHEMT compared to an SH-PHEMT. Therefore, the SH-PHEMT can be better optimized for high *peak* transconductances when this separation has to be minimized.

5. 5. MODULATION EFFICIENCY

The notion of "Modulation Efficiency" was introduced by Foisy *et al.* [34]. It is an attempt to extend the validity of the simple formulae that describe the *ideal* HEMT (in which all carriers are completely confined to an infinitely thin channel and move with uniform velocity v_{sat}) in a phenomenological way so that the more complex transport that takes place in a *real* HEMT is included.

Because of the absence of slow carriers inside or outside the channel, the ideal HEMT is the device in which the modulation efficiency η is at its theoretical limit. This means that *maximum* current modulation is achieved by *minimum* charge modulation, and $\eta = 1$.

In the *real* HEMT, only a part of the charge carriers travel in the quantum well with a velocity v_{sat}. Other parts of the electrons

- are also confined to the channel but have not yet reached high speed ("gradual channel" part),
- move in the supply layer at even lower speed,
- may be trapped in the supply so that changes in V_{GS} only redistribute trapped charge,
- or may have penetrated deep into the buffer.

This has the consequence that far more charge has to be modulated for a given current change than in a device that has only fast carriers in the channel, and $\eta < 1$.

The expressions that yield the intrinsic transconductance of the idealized device (Equations 4 to 6) may now be appropriately modified. For the real device, the channel charge Q_s in Equation 5 must now be replaced by the total charge Q_G under the gate, moving with the smaller effective velocity $v_{sat}\ \eta$, and Equation 5 now reads

$$I_D = Q_G\, v_{sat}\ \eta \,/\, L_G\,. \tag{13}$$

The intrinsic transconductance is thus given by

$$g_{mi} = (\partial Q_G \,/\, \partial V_{GSi})\,(v_{sat}\,\eta \,/\, L_G) = C_{GSi}\, v_{sat}\,\eta \,/\, L_G \tag{14}$$

where C_{GSi} is the gate-source capacitance of the intrinsic HEMT.

Finally, we want to derive an expression for a major figure of merit for HEMTs, the current-gain cutoff frequency or transit frequency f_T. It is often represented as the inverse of the effective transit time of the carriers through the intrinsic region: $f_T = (2\pi\tau)^{-1}$. For the real device,

$$f_T = v_{sat}\,\eta \,/\, (2\pi\, L_G)\,. \tag{15}$$

This equation shows that good modulation efficiency results not only in high transconductance but also in high cutoff frequency. For $\eta = 1$, f_T is at its theoretical limit.

Comparison of Equations (14) and (15) yields

$$f_T = g_{mi} \,/\, (2\pi\, C_{GSi})\,. \tag{16}$$

In Equation 16, the time-of-flight across the intrinsic transistor is considered to be the only time constant determining f_T. However, a more detailed discussion in section 7.1. will show that this is a simplification and additional contributions to f_T must be taken into account.

6. Small-Signal Equivalent Network

6.1. INTRINSIC AND EXTRINSIC NETWORK ELEMENTS

Up to now, we have only dealt with the *intrinsic* HEMT which is essentially the physical part of the device directly below the gate contact. In practice, this intrinsic HEMT cannot be connected to its electrical environment without any losses. These losses determine which percentage of the device's intrinsic performance can be actually measured *extrinsically*. The elements which determine the amount of losses are called parasitics.

Figure 15 shows a small-signal equivalent network which includes the elements of the intrinsic HEMT and the parasitics that connect the inner device to the three terminals that are accessible to the measurement (source, gate and drain pads). In the figure, it was tried to arrange the network elements in a way that their physical origin is reflected. For a complex device like the HEMT, the choice of a specific equivalent network never can be unique. Therefore, approaches slightly different from that in Figure 15 are also possible. These ambiguities can result in doubts if a specific element is of intrinsic or extrinsic character. A good discussion of HEMT network modeling is given in the article by Das [35].

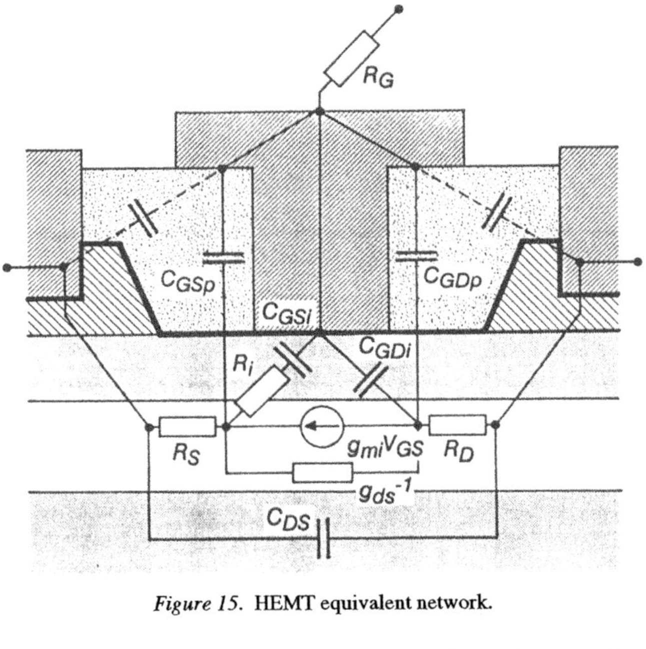

Figure 15. HEMT equivalent network.

In the network in Figure 15, elements of the intrinsic device are the current source g_{mi}, the gate-source capacitance C_{GSi}, an intrinsic series resistance to this capacitance, R_i, sometimes referred to as the "channel" resistance, the gate-drain capacitance C_{GDi}, and the output conductance g_{ds}. Extrinsic elements are the source, drain and gate resistances R_S, R_D and R_G, the gate-source and gate-drain parasitic capacitances C_{GSp} and C_{GDp}, and the drain-source capacitance C_{DS}.

One of the most important consequences of the existence of a finite R_S is that the extrinsically measured transconductance g_m is reduced compared to the intrinsic value g_{mi}:

$$g_m = g_{mi} / (1 + R_S g_{mi}) . \tag{17}$$

The presence of parasitic capacitances cannot be seen in the dc properties of the HEMT but has a very important effect on its high-frequency performance. The total gate-source and gate-drain capacitances are given by

$$C_{GS} = C_{GSi} + C_{GSp} , \tag{18}$$

$$C_{GD} = C_{GDi} + C_{GDp} , \tag{19}$$

and Equation 16 has to be modified to

$$f_T = g_{mi} / [2\pi (C_{GS} + C_{GD})] . \tag{20}$$

There is yet another cutoff frequency which depends on a larger number of equivalent network elements. As a figure of merit, it may therefore be valued higher than f_T. It is the maximum frequency of oscillation or power-gain cutoff frequency f_{max} given by [36, 37]

$$f_{max} = f_T [4 g_{ds} (R_S + R_i + R_G) + 2 (C_{GD} / C_{GS}) ((C_{GD} / C_{GS}) + g_{mi} (R_S + R_i))]^{-1/2} . \tag{21}$$

An optimization of f_T and f_{max} can be achieved by minimizing all network elements, with the exception of g_{mi} which must be as large as possible.

6.2. OPTIMIZATION OF NETWORK ELEMENTS

A high g_{mi} implies a small gate-to-channel separation and, hence, is coupled to a high C_{GSi}, according to Equation 14. This conflict can only be solved by realizing a short gate length. This is a major reason why L_G is such a key issue for a fast device.

R_S can be minimized by observing a couple of rules: the ohmic source and drain contacts should be placed close to the inner device, and the remaining distance should be bridged by a highly conductive cap as the topmost layer in the epi sequence. Highly doped caps with small band gaps also assist in the formation of ohmic contacts with low resistivities. The cap is intersected by the recess etch trough. Below this trough, current flow is intended in the quantum well only. Under the sole aspect of R_S reduction the recess length should be as short as possible, but we will see later that a tradeoff is necessary for the sake of other device properties, i.e. that the recess length must be larger than L_G and some finite length of "ungated channel" at both sides of the gate must be allowed. The resistance of the ungated channel depends on the sheet resistance of the 2DEG, i.e. on the electron concentration n_s and mobility μ_e in the quantum well, and on the threshold voltage V_T. If V_T is zero or even positive, the surface potential in the ungated channel will deplete the quantum well from carriers so that R_S can achieve undesirably high values if the ungated channel is too long.

The second resistance that has to be minimized is R_G. The requirement to realize short gates leads to high end-to-end resistances of the gate stripes especially if the gate widths are large as it is the case in analog devices (compared to transistors for digital IC use). This situation results in the concept to attach a gate overlayer with a large cross section and, hence, low resistance to a short-length gate stem forming a "T-gate" or "mushroom gate" as sketched in Figure 15.

The magnitude of the output conductance g_{ds} reflects the extent to which the drain current is controlled not only by V_{GS} but also by V_{DS}, a generally unwanted effect. It can be suppressed by choosing a high "aspect ratio" $L_G\ /\ (d + \Delta d)$ which means placing the gate close to the channel so that the control of the gate potential over the charge carriers is good. Usually, aspect ratios between 5 and 10 are judged to be reasonable. Additionally, a large ΔE_C between the quantum well and the lower barrier helps to prevent the electrons from penetrating deep into the buffer where they can escape from the control by V_{GS}. This is the reason that - among the GaAs-based PHEMTs - the DH-PHEMT is superior to the SH-PHEMT with respect to g_{ds}.

A high g_{ds} cannot only be caused by electrons in the buffer below the channel but also by real-space transfer of carriers into the upper supply or the highly conductive cap between gate and drain. This can be reduced by a long ungated channel at the drain side of the gate. For reasons of process technology, the recess trough is usually symmetrical with respect to the gate center so that the ungated channel regions at the source and drain sides of the gate are of equal length. Consequently, the decision for a particular recess length for a given L_G always represents a compromise: short recess is in favor of small R_S, long recess is in favor of small g_{ds}.

The minimization of the parasitic capacitances C_{GSp} and C_{GDp} has an impact on both f_T and f_{max}. Since C_{GDi} is generally much smaller than C_{GSi}, the addition of parasitics is, of course, unfavorable with respect to C_{GS}, but even worse for C_{GD}. In other words, large parasitics degrade f_T, but they degrade f_{max} even more. The most common reason for the existence of large parasitics is the presence of a thick dielectric layer covering the HEMT surface for the purpose of device passivation. Often, the gate overlayers are supported by these passivating layers (as shown in Figure 15) causing a capacitive coupling between overlayer and substrate (or even between overlayer and ohmic contacts, as also indicated in the figure). This undesirable effect can be avoided by choosing a thin dielectric - preferably with a small dielectric constant - which does not completely fill the space between overlayer and substrate. Even if a passivation is completely omitted, a small part of the parasitic capacitances never can be avoided. This part is the so-called fringe capacitance which arises at the edges of any metal strip on top of a semiconductor simply as a consequence of Poisson's law [38, 39].

7. PHEMT High-Frequency Properties

7.1. BIAS DEPENDENCE

In section 5.5., f_T was treated as the inverse of the time-of-flight of the carriers across the region under the gate. We now want to identify additional components which contribute to the total delay τ because this enables us to understand the dependence of the high-frequency properties on V_{DS} and I_D. The discussion is mainly based on the treatments by Moll *et al.* [40] and Greiling and Nguyen [41].

The components of the total delay divide into *intrinsic* and *extrinsic* delays, analogous to the intrinsic and extrinsic network components. The most important part of the *intrinsic* contribution is the time-of-flight across the inner HEMT (in section 5.5., this was considered the only existing time constant):

$$\tau_i = L_G / v_{sat}\, \eta = C_{GSi} / g_{mi}\,. \tag{22}$$

At the drain end of the gate, there exists a dipole zone or drain depletion region the extension of which depends on device geometry (length of ungated channel at drain side) but also on drain voltage. Therefore, the transit time τ_d through this domain, the drain delay, is proportional to V_{DS}.

The third intrinsic part of the total delay is the channel charging time τ_c. It is the inverse RC constant for changing the charge state of the channel and is inversely proportional to the number of carriers in the channel, n_s, or to the drain current, I_D.

The *extrinsic* contributions to the total delay are less dependent on bias conditions. They are essentially the fringe and parasitic capacitance charging time τ_{par} and the pad capacitance charging time τ_{pad}. Since pads must have a minimum size, the relative contribution of pad capacitances is large for devices with small gate periphery, and the best current-gain cutoff frequencies can be measured at HEMTs with wide gates.

In summary, f_T is given by

$$f_T = (2\pi\, \tau)^{-1} = [2\pi\, (\tau_i + \tau_d + \tau_c + \tau_{par} + \tau_{pad})]^{-1}\,. \tag{23}$$

7.2. PROPERTIES OF DIFFERENT EPI LAYER DESIGNS

Current-gain cutoff frequency and maximum frequency of oscillation for three different types of GaAs-based PHEMTs are shown as a function of drain current in Figure 16. The devices in the figure are all of the same gate width and have comparable gate lengths. Pad configuration and technology process are also the same so that the influence of parasitics is identical for the three devices. However, the similarity of the f_T curves does not mean that g_m and C_{GS} are identical, too. The standard SH-PHEMT without delta-doping has the largest gate-to-channel separation and, hence, the smallest C_{GS}, but also the smallest g_m. This results in f_T values

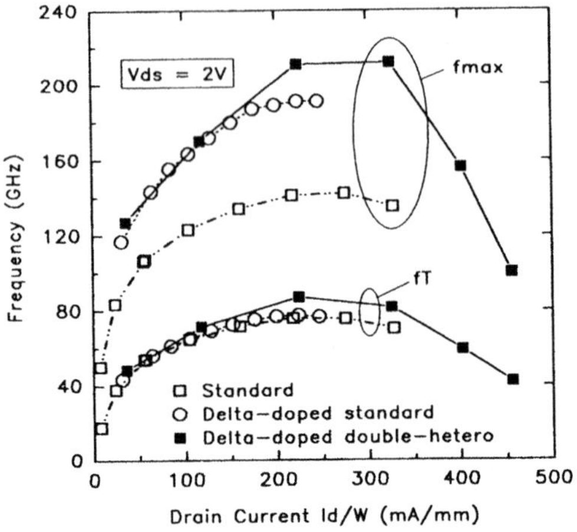

Figure 16. Cutoff frequencies f_T and f_{max} of PHEMTs on GaAs substrate. All devices passivated, $L_G = 0.18...0.20$ µm, $W_G = 180$ µm.

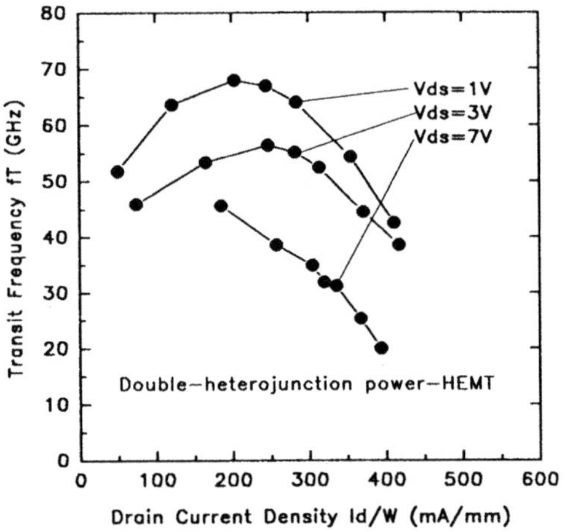

Figure 17. Current-gain cutoff or transit frequency f_T of a power DH-PHEMT on GaAs substrate. Device passivated, $L_G = 0.23$ µm.

comparable to those of the two delta-doped PHEMTs with smaller d and higher g_m. The dependence of f_T on I_D reflects the fact that for high I_D, g_m stays highest for the DH-PHEMT, as already demonstrated in Figure 14, and that the the smallest channel charging times τ_c are also realized in the DH-PHEMT.

The f_{max} data reveal additional information. The large d in the standard SH-

PHEMT leads to a large g_{ds} being the main reason for the low f_{max}. In the delta-doped SH-PHEMT, the gate is close to the channel, the aspect ratio is high, g_{ds} low, and f_{max} is significantly improved. The delta-doped DH-PHEMT has a comparable d, but contrary to the SH-PHEMTs, the electrons are all confined to the quantum well even for high I_D (see Figure 13) so that modulation efficiency and f_{max} are high for a very large current interval.

Figure 17 illustrates the decrease of f_T with V_{DS} caused by the increase of drain delay. The data are taken from a DH-PHEMT optimized for K-band power applications (see also section 9.3.). The wide extension of the recess trough at the drain side of the gate of this device causes a long drain depletion region which allows to apply V_{DS} up to 10 V. On the other hand, this is the reason for the strong increase of the drain delay with V_{DS} which can be seen in the figure.

7.3. CUTOFF FREQUENCIES: STATE OF THE ART

An overview over published f_T and f_{max} record results is given in Table 2.

TABLE 2. HEMT cutoff frequencies: state of the art

Device	L_G (nm)	g_m/W_G (mS/mm)	f_T (GHz)	f_{max} (GHz)	Author (Organization)
PHEMT on GaAs	150	640	100	350	Lester et al. (Cornell/GE 1988) [42]
	100	650	150	250	Nguyen et al. (Cornell 1989) [43]
	80	740...920	-	270	Chao et al. (GE 1989) [44]
	100	700	120	290	Tan et al. (TRW 1990) [45]
LMHEMT on InP	80	1150	250	220...300	Nguyen et al. (Hughes 1990) [46]
	150	1300	165	405	Chao et al. (GE 1990) [47]
	150	900	-	455	Ho et al. (GE 1991) [48]
PHEMT on InP	100	1160	210	300	Mishra et al. (Hughes 1989) [49]
	50...65	1700...1740	300...340	250...280	Nguyen et al.(Hughes 1992) [50,51]
	100	1550	305	340	Wojtowicz et al. (TRW 1994) [52]

Most of the results in Table 2 were achieved with unpassivated devices which are not handicapped by the presence of large parasitic capacitances. The collection of data clearly shows the improvement of transport properties with increasing indium content of the channel or quantum well depth, consequently leading to the highest g_m and f_T for the PHEMTs on InP with up to 80% In. Surprisingly, the superiority of the InP-based HEMTs over the GaAs-based devices with respect to f_{max} is less pronounced. This has essentially two reasons. The large energetical separation of the central conduction band valley from the satellite valleys in InGaAs with high In content allows the electrons to pick up very high energies from the electric field before they are scattered into the satellites with their degraded transport properties. These high energies facilitate real-space transfer to the barrier layers even though the quantum well is deeper compared to the GaAs-based HEMTs. The second effect is the early onset of impact ionization which leads to the presence of holes and a reduction of modulation efficiency.

8. Optimization of Low-Noise HEMTs

8.1. FUKUI FORMULA

The noise figure *NF* of a transistor is defined as the signal-to-noise ratio *S/N* of the input signal devided by the *S/N* of the output signal. For frequencies $f < 20$ GHz, *NF* can be well approximated by the semi-empirical formula given by Fukui [53]:

$$NF\,(\text{dB}) = 10 \log \left[1 + 2\pi\, K f\, C_{GS}\, ((R_S + R_G) / g_{mi})^{1/2}\right], \tag{24}$$

$$NF\,(\text{dB}) = 10 \log \left[1 + K\, (f / f_T)\, (g_{mi} / (R_S + R_G))^{1/2}\right]. \tag{25}$$

The quantity K is a fitting factor which takes the properties of the channel transport into account in a phenomenological way. All physical effects which make a HEMT superior to a MESFET with respect to noise performance are essentially contained in this factor. Since Fukui's equation is a simplification, it does not very well describe the noise figures at higher frequencies f especially because it does not consider the drain region as an additional noise source. Anyhow, the formula works well as a basis to identify the major criteria for a good low-noise HEMT.

8.2. LOW-NOISE BIAS CONDITIONS

First of all, the parasitics R_S, R_G and C_{GSp} must be kept as small as possible. However, the minimization of the total input capacitance C_{GS} is in conflict with the maximization of g_m which is both required by Equation 24. This makes the proper choice of the bias V_{GS} important. Figure 18 shows a one-dimensional calculation of C_{GS} (V_{GS}) by Deutschmann *et al.* [54] in comparison to measured data. In this example, the device is pinched off for $V_{GS} < -0.5$ V. The measurements were performed at a device of extremely large gate area (L_G x W_G = 100 x 100 μm^2), because in this case, the influence of parasitics is small compared to the intrinsic capacitance so that the behavior of the charge carriers can be studied without disturbing influences. This is the reason why C_{GS} is negligible for $V_{GS} < V_T$ in Figure 18. When V_{GS} is increased, electrons begin to populate the channel, and C_{GS} grows rapidly. For a finite range of V_{GS} around 0 V, the increase of C_{GS} with V_{GS} reduces. This is the regime of high modulation efficiency where the 2DEG in the channel is already established, and any additionally generated free electrons will also be confined to the quantum well. With V_{GS} further increasing to values significantly above 0 V, electrons begin to accumulate in the upper supply, and the increase of C_{GS} with V_{GS} becomes more rapid again.

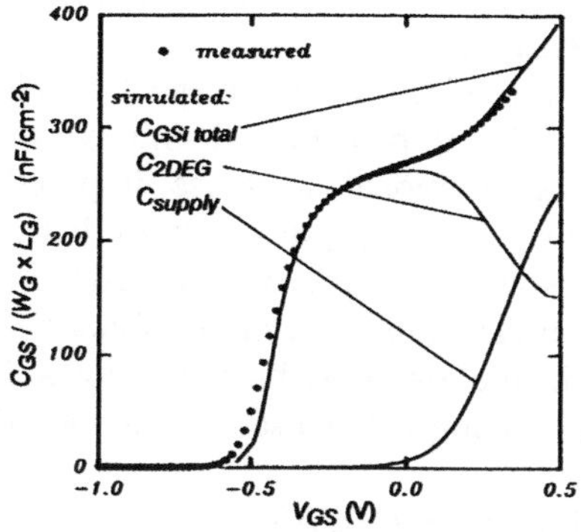

Figure 18. Gate-source capacitance C_{GS} as a function of gate voltage V_{GS}: comparison of simulation and measurement for a GaAs-based SH-PHEMT with $L_G \times W_G = 100 \times 100\ \mu m^2$.

It is evident from Figure 18 that a good bias regime for low-noise operation is expected below the shoulder in the curve C_{GS} (V_{GS}) because C_{GS} is still quite small while g_m increases drastically with V_{GS}. In fact, for most PHEMTs, the lowest noise figures are achieved for relatively small currents I_D / W_G = 40...100 mA/mm, i.e. for V_{GS} not very far above pinchoff. Clearly, it must be avoided that electrons also populate the region below the channel with its inferior transport properties under this bias condition. Therefore, a device like the one shown in Figure 10 will exhibit a lower *NF* than the one in Figure 12.

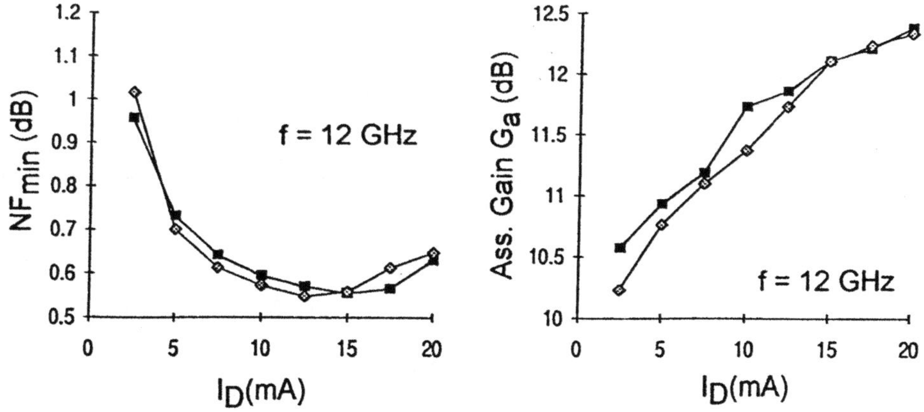

Figure 19. Noise figure *NF* and associated gain G_a of a standard SH-PHEMT on GaAs for f = 12 GHz. Mean values from two different wafers, devices passivated and mounted in ceramic package, L_G = 0.2 μm, W_G = 180 μm. MBE wafer growth by Walter Schottky Institute, Garching, Germany.

Figure 19 shows an example how the measured *NF* depends on I_D in a GaAs-based PHEMT. The relatively broad minimum followed by a slow increase with I_D is a typical feature of PHEMTs. Lattice-matched AlGaAs/GaAs devices show a more pronounced increase at the right-hand side of the minimum due to the early onset of parasitic conduction in the supply.

Optimization of bias conditions with respect to *NF* only makes sense if a reasonable associated gain G_a can be achieved simultaneously. Therefore, the device has to meet requirements additional to those imposed by the Fukui formula. Essentially, it has to achieve a high f_{max} at low-noise bias. As we have demonstrated in Figure 16, f_{max} increases with I_D in the current regime of interest here, and so does G_a (see Figure 19). The broad minimum of the function *NF* (I_D) exhibited by the PHEMT thus allows to shift the low-noise bias point towards higher current density in the interest of better amplification without sacrificing too much *NF* performance.

8.3. NOISE FIGURE: STATE OF THE ART

Published outstanding *NF* data are compiled in Table 3.

TABLE 3. HEMT noise figures and associated gain: state of the art

Device	L_G (nm)	g_m/W_G (mS/mm)	f (GHz)	NF (dB)	G_a(dB)	Author (Organization)
PHEMT on GaAs	150	800	94 18	2.4 0.5	5.4 15.1	Duh et al. (GE 1990) [55]
	100	700	93.5	2.1	6.3	Tan et al. (TRW 1990) [45]
	150	680	60	1.6	6.5	Katoh et al.(Mitsubishi 1993) [56]
PHEMT on GaAs (InGaP-Supply, MOVPE)	150	> 420	12 50	0.41 1.2	13.0 5.8	Takikawa et al. (Fujitsu 1993) [57]
LMHEMT on InP	200	-	63.5	0.8	8.7	Mishra et al. (Hughes 1989) [49]
	150	900	93 18	1.4 0.3	6.6 17.2	Chao et al. (GE 1990) [47]
LMHEMT on InP (MOVPE)	130	700	12	< 0.30	15	Fujita et al. (Toshiba 1993) [58]
PHEMT on InP	100	1000...1200	95	1.3	8.2	Tan et al. (TRW 1991) [59]
	150	800	12	0.23	16	Hwang et al. (Avantek/HP 1993) [60]

The table shows a clear advantage of the InP-based HEMTs over the PHEMTs on GaAs. Analogous to the situation with f_T and f_{max}, the InP HEMT's superiority with respect to *NF* is more pronounced than with respect to G_a. Most of the data were measured on-wafer, i.e. not with packaged devices. Therefore, many of the cited *NF* values are subject to considerable uncertainty. The transconductances are mostly maximum values not measured under low-noise bias.

9. Optimization of Power HEMTs

9.1. BASIC REQUIREMENTS

During low-noise transistor operation, the high-frequency component modulating the dc gate voltage has only a very small amplitude. Thus, in such small-signal applications, the device properties are of interest only close to the particular bias point. This is completely different for power transistors: large-signal operation means that the gate voltage may sweep over the full range between maximum gate forward bias and pinchoff (or even beyond). Figure 20 schematically shows the case of class "A" power operation with an idealized, purely resistive load. In class A, the large-signal bias point is such that I_D is about half the maximum drain current of the device, I_{Dmax}, and that V_{DS} is also about half the maximum value that can be applied, V_{DSmax}.

In reality, the load of a power transistor probably is an impedance. This means that the modulation of the gate voltage does not sweep the bias point along a straight line any more but along some closed loop enclosing large areas of the output characteristics. However, the idealized class A case is sufficient for a basic discussion of the properties that are desirable for a power HEMT.

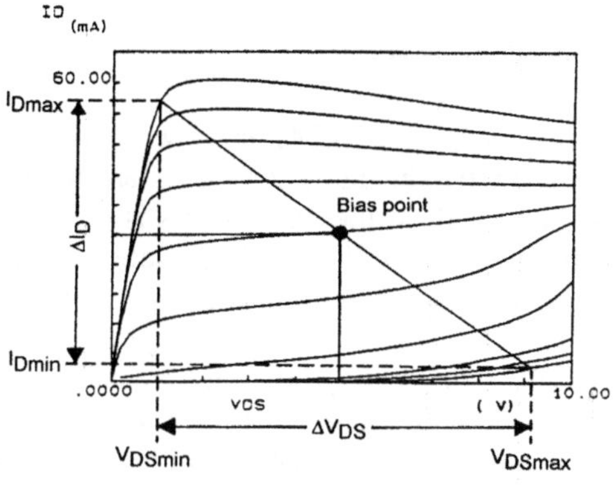

Figure 20. Power HEMT output characteristics with large-signal bias point and idealized load line.

Under these conditions, the maximum output power that can be achieved is given by

$$Pout\ max = \frac{1}{\sqrt{2}}\frac{\Delta V_{DS}}{2}\frac{1}{\sqrt{2}}\frac{\Delta I_D}{2} = \frac{1}{8}\Delta V_{DS}\,\Delta I_D. \tag{26}$$

According to this relation, the usable drain voltage and drain current swings have to be optimized for maximum output power. To the very first order, this means that a good power device should realize high values of V_{DSmax} and I_{Dmax} simultaneously. As can be seen in Figure 20, the minimum voltage V_{DSmin} along the load line is not zero due to the magnitude of R_S and R_D which determines the extent of the linear regime in the output characteristics. Additionally, I_{Dmin} is of finite size if there are leakage currents present preventing a perfect pinchoff at V_{DSmax}. Complete pinchoff at V_{DSmax} can only be achieved if the gate-drain breakdown voltage fulfills the condition $V_{GDbd} > V_{DSmax} - V_T$. Unfortunately, in the presence of high I_{Dmax}, only a low V_{GDbd} can be realized, and vice versa. This necessitates a tradeoff which is additionally influenced by considerations of frequency performance. We will return to this issue in the next section.

Another important property of a power transistor, the power added efficiency *PAE*, is defined by

$$PAE = (P_{out} - P_{in}) / P_{dc} = P_{out} (1 - G_a^{-1}) / P_{dc} \tag{27}$$

where P_{out} and P_{in} represent the high-frequency output and input power, $G_a = P_{out} / P_{in}$ is the power gain, and P_{dc} is the power of the dc bias. Efficiency is an important figure of merit because high *PAE* means small power losses and, thus, small self-heating effects and long battery operation. The theoretical limit in class A operation, $PAE = 1/2$, is reached for infinite gain G_a. The gain at a particular signal frequency is bias dependent. This imposes an additional boundary condition for the proper choice of the large signal bias.

9.2. POWER HEMT DEVICE DESIGN

A schematic cross-section of a power HEMT is shown in Figure 21. It differs from the HEMTs in Figure 1 and 15 by the introduction of an additional recess, the "wide recess" of length L_{WR}. The depth of the wide recess is usually adjusted in a way that the highly doped supply layer below the cap is depleted by the surface potential, but not the channel. Choice of the proper L_{WR} allows to find a compromise which combines reasonable R_S and R_D with sufficiently large V_{GDbd}. As discussed by Huang *et al.* [61], a long L_{WR} will improve the gate-drain breakdown but reduce f_T at a given bias V_{DS} (because the drain depletion zone is long, increasing the drain delay), whereas a short L_{WR} will cause an earlier gate-drain breakdown but a higher f_T at the

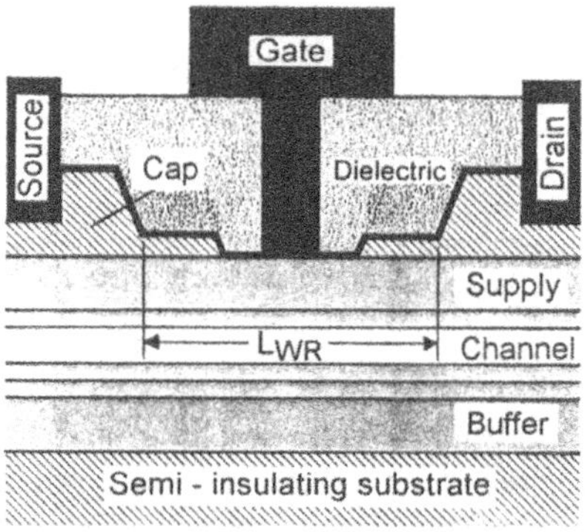

Figure 21. Power HEMT schematic cross-section.

same V_{DS} (because the drain delay is short).

These physical relations imply that the lateral dimensions of a power HEMT are a function of the intended frequency range: L_{WR} must decrease with increasing f, and the large signal bias V_{DS} must decrease accordingly to the reduced breakdown voltage. To realize reasonable output power again, I_{Dmax} must be sufficiently enhanced.

High I_{Dmax} can be realized by the use of epitaxial layer sequences that feature high n_s. In this respect, among the GaAs-based HEMTs, the DH-PHEMT is clearly superior to the SH-PHEMT. Moreover, the DH-PHEMT's enhanced energy barrier at the heterointerface below the channel is extremely welcome: it is the reason for a significantly improved confinement of the charge carriers to the quantum well at high V_{DS}. Thus, the DH-PHEMT not only provides the amount of current needed for power at ultra-high frequency, but also offers the best transport properties at high bias, in other words: it is the best epitaxial concept for both low and high frequency.

In principle, a compromise with respect to f_T, R_S and V_{GDbd} can also be tried by use of a single recess, i.e. by extension of the inner recess to a length of the order of L_{WR}. In most HEMT epi sequences this would expose a large area of a supply layer that contains aluminum. Huang *et al.* [62] have demonstrated that this can result in problematic high-frequency behavior due to slow surface effects. A double recess according to Figure 21 helps to avoid this. It is evident that an unsymmetrical shape of the double recess is desirable (shorter recess at the source side) because it does not degrade R_S if V_{GDbd} is enhanced. In practice, this cannot be easily realized with the small lateral dimensions typical for HEMTs since lithographical alignment of the gate to the wide recess is difficult.

Up to now, we have not mentioned power HEMTs on InP substrate. This is because InP-based layer sequences of the type introduced so far (see Table 1) are not particularly well suited for power application. Problems arise due to the small bandgap of InGaAs with In contents of 53% and above (early onset of impact ionization), due to locally high fields (early breakdown of the gate-drain diode), and due to deconfinement of the carriers by real space transfer into the supply or buffer layer. Recently, concepts have been developed to tackle these problems. One of these will be discussed in the last section about advanced HEMT concepts.

9.3. EXAMPLE: A POWER PHEMT FOR K-BAND

In this section, we use a PHEMT designed for application in 15 to 30 GHz medium-power MMICs as an example to which result the numerous optimization tradeoffs can lead. Figure 22 shows the output characteristics of this device.

The epitaxial layer structure used is essentially the one discussed in Figures 12 and 13. Hall measurements at a control sample resulted in $n_s = 2.6 \times 10^{12}$ cm^{-2}. In the MMICs, the large signal bias point is $V_{DS} = 5$ V, $I_D = 48$ mA where f_T is about 35 GHz. In most parts of the prospective load line, f_T is well above the intended application frequency.

Figure 23 presents the result of an on-wafer load-pull measurement of the output

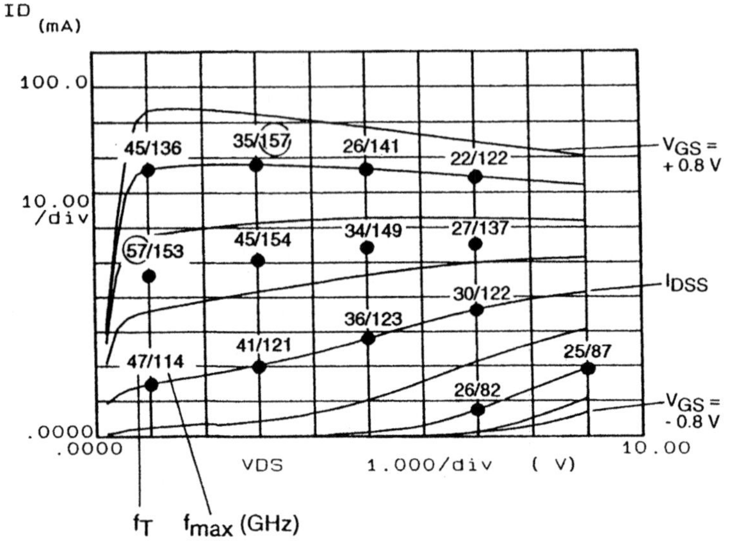

Figure 22. Output characteristics I_D (V_{DS}) and cutoff frequencies f_T / f_{max} of a power DH-PHEMT on GaAs substrate. L_G = 0.25 µm, W_G = 240 µm, L_{WR} = 1.30 µm.

power of such a device at f = 10 GHz. The power gain G_a is 17.5 dB, and the output power at 1 dB gain compression, P_{-1dB}, is 24.5 dBm (corresponding to 780 mW/mm). The figure demonstrates that G_a is constant for a large range of P_{in}. This good linearity is a consequence of the fact that g_m varies only slightly over a large fraction of the load line. The small *PAE* at low P_{in} is caused by P_{dc} being large compared to P_{out} under these conditions. According to Equation 27, the peak *PAE* of about 55% cannot be achieved in class A operation. Therefore, the measurement result is an indication that the PHEMT is already driven into so-called class AB conditions for the

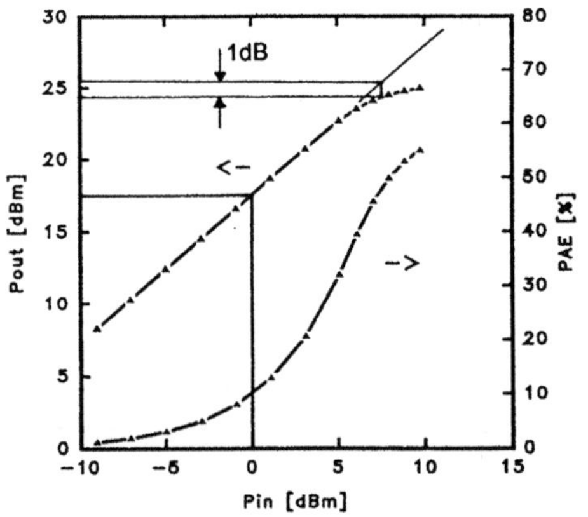

Figure 23. P_{out} *vs.* P_{in} at f = 10 GHz for a DH-PHEMT like the one in Figure 22, but with W_G = 360 µm. V_{DS} = 7 V. MBE growth by H. Riechert, Siemens.

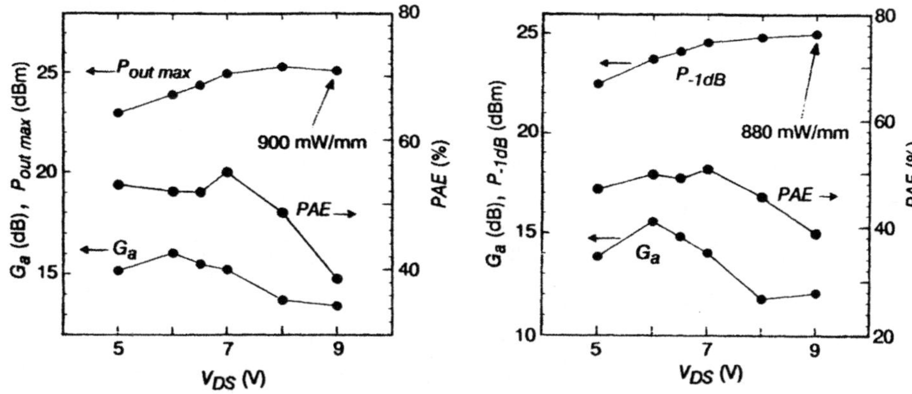

Figure 24. $P_{out\ max}$ and P_{-1dB} *vs.* V_{DS} at f = 10 GHz for the DH-PHEMT of Figure 23. MBE growth by H. Riechert, Siemens.

higher values of P_{in}, i.e. that the gate bias reaches values more negative than V_T.

Figure 24 shows P_{out}, G_a and *PAE* as a function of drain bias. The power saturation of the device is very abrupt (as could already be seen in Figure 23) which results in P_{-1dB} being only marginally smaller than $P_{out\ max}$. Beyond about V_{DS} = 7 V, *PAE* and G_a decrease due to reduced modulation efficiency. Additional results achieved with this device are contained in Table 4.

9.4. OUTPUT POWER: STATE OF THE ART

A compilation of outstanding power HEMT results is given in Table 4.

For a proper understanding of the data, some additional remarks are necessary. If *PAE* exceeds 50%, this is an indication that the mode of operation was not class A but class AB or B. It is technically very difficult to maintain a specific output power P_{out} / W_G measured with small W_G also for very large W_G (i.e. for a large number of gate fingers in parallel) and to achieve high *absolute* P_{out}. Thus, values of P_{out} / W_G realized with large W_G deserve the most regard in Table 4. Additionally, it is more demanding to develop a device with high linear P_{-1dB} than with high P_{out} at unspecified gain compression.

As expected from the preceding discussion, a clear tendency can be seen in Table 4 that the large signal bias voltage V_{DS} goes down with increasing frequency. This implies that power is achieved by a large current swing. To increase drain current, some of the transitors listed in Table 4 are "doped-channel" HEMTs. In these devices, not all but part of the total doping sheet concentration is applied directly in the quantum well and not in the supply layers. Strictly speaking, this is a violation of the original HEMT principle which is based on a local separation of the impurities from the current path. Ironically, this violation is made in HEMTs intended for the highest frequencies where best transport properties are demanded.

TABLE 4. HEMT output power, gain and efficiency: state of the art

Device	f (GHz)	W_G (μm)	V_{DS}(V)	P_{out}/W_G (mW/mm)	P_{-1dB}/W_G (mW/mm)	G_a(dB)	PAE(%)	Author (Organization)
DH-PHEMT on GaAs	4.5	400	8	830	-	17.2	63	Kao et al.(GE 1992) [63]
			14	1260				
	10	1200	≤ 8.5	935	-	8.9	56	Huang et al. (Raytheon 1991) [64]
	18	400		870		7.4	47	
	10	360	7		790	14	51	Grave, Schleicher and Riechert (Siemens 1994)
			9		880	12	39	
			10	1120		13.6	50	
			10		860	16.1	42	
doped channel DH-PHEMT on GaAs	18	75	-	1000	-	6.8	50	Kim et al. (TI 1989) [65]
"inverted" PHEMT on GaAs	25	120	5	810		10.9	59	Zhou et al. (HP 1989) [66]
					680	12.9	45	
DH-PHEMT on GaAs	32	400	7	922	-	6.7	37	Huang et al. (Raytheon 1993) [67]
		1200		619		4.1	22	
DH-PHEMT on GaAs	44	900	5.4	766	-	3.0	24	Smith et al. (GE 1991) [68]
PHEMT on GaAs, 2 doped channels	60	50	3.95	1000	~900	2.7	25.5	Saunier, Tserng (TI 1989) [69]
doped channel PHEMT on GaAs	60	75	[illegible]	1000		4.4	36	Smith et al. (GE 1990) [70]
DH-PHEMT on GaAs	60	400	5.0	550		4.5	25.4	Lai et al. (TRW 1993) [71]
doped channel DH-PHEMT on GaAs	94	160	3.4	392		4.0	13.2	Streit et al. (TRW 1991) [72]
LMHEMT on InP	12	300	4.0 ?	960		-	40	Matloubian et al. (Hughes 1991) [73]
	20	800	4.0	700		≈7	47.1	Matloubian et al. (Hughes 1993) [74]
doped channel LMHEMT on InP	57	450	3.5	444		3.6	17	Matloubian et al. (Hughes 1993) [75]
			3.5		333	2.6	20	

With the doped-channel modification, it is possible to maintain a P_{out} / W_G of about 1 W/mm up to a frequency of 60 GHz. Due to the aforementioned difficulties with high fields in InP-based HEMTs, these have been inferior to GaAs-based power HEMTs up to now. Recently, ideas have been developed to overcome this deficiency. An example will be discussed in the last section.

10. Outlook: Advanced HEMT Concepts

10.1. MOTIVATION FOR ADVANCED HEMTS

In this paper, it has been discussed that InP-based HEMTs profit from their deep quantum wells with high n_s mostly with respect to g_m and f_T but that their superiority over GaAs-based HEMTs is weaker when it comes to f_{max} and is completely gone for power applications. This situation suggests two main directions for improved concepts.

The first one is the search for ideas how to improve the GaAs-based PHEMT with respect to f_T and g_m. The manufacturability of devices on GaAs substrate is so much better that efforts to delay the introduction of InP devices to regular production lines as long as possible are considered worthwhile.

The second direction is the aim to remove the InP-based HEMT's deficiencies as a power HEMT. This would create a transistor family of universally superior performance.

These two goals are certainly of practical importance. Independent from the consideration if there is an immediate technical demand, the ambition exists to drive transitor speed to its ultimate limits in the some-hundred gigahertz range. This defines a third direction for advanced concepts.

Necessarily, the following outlook must remain incomplete since the amount of new ideas is vast and constantly growing.

10.2. ADVANCED HEMTS ON GaAs SUBSTRATE

10.2.1. *Metamorphic Growth*

The basic idea is to grow the typical InP HEMT $In_{0.52}Al_{0.48}As/In_{0.53}Ga_{0.47}As$ barrier/quantum well system also on GaAs substrate (see, for instance, Wang *et al.* [76], Masato *et al.* [77]). Since this system is extremely mismatched to GaAs, it is necessary to grow a thick graded $In_xAl_{1-x}As$ buffer on the substrate where x slowly increases from 0 to 0.52. This buffer contains dislocations and microcracks. These are intended to be buried deep below the active layers of the HEMT but actually are difficult to keep restricted to these regions. Superlattice buffers made of InAlAs/InGaAs layer sequences can also be used as dislocation filters.

10.2.2. *MOVPE Growth Instead of MBE*

As already mentioned in section 2.1., MOVPE is a growth technique with the advantage that layers with phosphorus can be grown with relative ease. This property can be used to realize supply layers that do not contain aluminum, for instance $In_{0.48}Ga_{0.52}P$ (Takikawa and Joshin [57]). In this material, deep DX centers do not exist which results in better donor activation than in AlGaAs. Additionally, the surface depletion region is shallower due to a lower Schottky barrier. The advantage of both effects is the small thickness of the supply layer which can be realized. This leads to excellent suppression of short-channel effects and, hence, excellent gain.

Among all GaAs-based devices, these HEMTs came closest to InP HEMT performance with respect to noise figure (see Table 3. The noise figure *NF* at 12 GHz was measured in package!).

A natural extension of the concept is to use quaternary AlGaInP as a supply instead of InGaP (Bachem et al. [78]). On the one hand, the advantage of the supply without Al is lost, but on the other hand, the conduction band offset is noticeably larger.

The general problems with MOVPE growth, for instance the difficult control of background impurites, have already been mentioned in section 2.1. For HEMTs with phosphide supplies, the additional complication of proper growth of arsenide/phosphide interfaces arises.

10.3. ADVANCED HEMTS ON InP SUBSTRATE

10.3.1. *InAs Channel HEMTs*

These HEMTs are the natural completion to all attemps to grow PHEMTs on InP with increasing indium content in the quantum well, having ever higher f_T and g_m records in mind. The idea is to create a device with maximum conduction band offset between channel and supply. The InAs channel is, of course, extremely mismatched to InP (see Figure 2) and can therefore only be very thin. In the case of Yang *et al.* [79], a 5 nm InAs channel was combined with a lattice-matched InGaAs pre-channel and a conventional InAlAs supply. The high-frequency performance did not yet exceed the results achieved with InP PHEMTs with less In in the channel possibly due to growth difficulties with the InAs layer.

Bolognesi *et al.* [80] have grown a 10 nm InAs channel in combination with AlSb barrier layers. This combination offers a conduction band offset of 1.3 eV. In view of L_G = 0.5 μm, the measured f_T of 93 GHz was promising, but the device suffered considerably from problems due to holes created by impact ionization, for instance parasitic bipolar action.

10.3.2. *InP Channel HEMTs*

The purpose of this development is to find a HEMT with better power performance than the GaAs-based power PHEMT. InP is considered a suitable channel material because it has a high v_{sat}, but the velocity peak is only weak and delayed to higher fields compared to InGaAs. This results in high breakdown fields. The bandgap of InP is similar to lattice-matched InAlAs (see Figure 2). However, there exists a conduction band offset of about 0.39 eV between both materials. This is possible because the valence band offset is in the same direction. Aina *et al.* [81] realized HEMTs with InP channel and InAlAs barrier layers by MOVPE and achieved a very high P_{out} of 1.45 W/mm at V_{DS} = 7 V (f = 30 GHz). A major advantage of InP is the very slow decrease of f_T with V_{DS}, a consequence of the moderate dependence of the drain delay time on the field. But there is a drawback: the low μ_e renders the realization of small parasitic resistances very difficult. R_S and R_D, however, determine the extension of the linear regime in the output characteristics. They are therefore parameters with an important influence on the usable voltage swing and, hence, *PAE*.

An attempt to overcome the R_S handicap is the growth of double-layer channels composed of InP and lattice-matched InGaAs (Jelloian *et al.* [82]).

11. Conclusion

We have discussed basic properties of the two major groups of HEMTs: devices based on GaAs substrate, and devices based on InP substrate. Both have the InGaAs channel in common, but in case of the InP-based HEMTs, the indium content in the quantum well is significantly higher, and the conduction band offset between the barrier layers and the channel is larger yielding a deeper quantum well.

The two families of devices possess specific advantages and drawbacks which can mostly be explained in terms of these physical differences.

The GaAs-based PHEMT is the more mature device. Its substrate material is available in 4" size and less subject to wafer breakage than InP. This makes GaAs PHEMT processing less expensive, and manufacturing is possible on existing MESFET technology lines.

GaAs PHEMTs are inferior to InP HEMTs with respect to f_T, g_m and NF. The main reasons for this are the lower electron velocity v_{sat} (due to the smaller indium mole fraction in the quantum well) and the smaller 2DEG concentration n_s (due to the smaller conduction band offset). However, the performance which can be realized with GaAs PHEMTs is in many cases sufficient for circuits with frequencies close to 100 GHz. Additionally, at any frequency, good large-signal performance is - at the moment - easier to achieve with GaAs PHEMTs. The reason for this is the very same responsible for the smaller v_{sat}: in high fields, the electrons cannot attain such high conduction band energies as they do in channels with higher indium percentage. In combination with the larger bandgap, this shifts the onset of avalanche breakdown towards higher drain voltages.

Though PHEMTs are now a production technology (at least when grown on GaAs) there are still dynamic developments going on. They follow several directions in parallel: it is tried to expand the range of GaAs-based PHEMTs to higher frequencies by the growth of channels more similar to InP HEMTs (concept of "metamorphic" HEMTs), remove the deficiencies of InP HEMTs with respect to power applications to create a HEMT family of universal superiority, and finally, explore how far the frequency range of HEMTs can be extended into the multi-hundred gigahertz range by use of small-bandgap quantum wells like InAs and barriers like AlSbAs.

Acknowledgments

This paper became only possible due to the support of the following colleagues: H. Riechert (MBE wafer growth and one-dimensional device simulation), L. Schleicher (HEMT process development), G. Jäger and F. Raisch (process control), A. Morav and P.W. von Basse (dc characterization), H.J. Siweris, M. Kärner and H. Tischer (HF characterization), J.E. Müller (on-wafer load-pull power measurements), and R. Deutschmann (one- and two-dimensional device simulation). The author is indebted to A. Mesquida Küsters, H. Kniepkamp and W. Kellner for fruitful discussions and critical reading of the manuscript. He further wants to express his gratitude to all other people involved in HEMT processing at both the Siemens Corporate Research and Development and the Siemens Semiconductor Production Group who are not explicitly named here. Part of the low-noise HEMTs discussed in this paper were MBE-grown by the Walter Schottky Institute (Technical University of Munich).

References

1. Mimura, T., Hiyamizu, S., Fujii, T., and Nanbu, K. (1980) A new field-effect transistor with selectively doped GaAs/n-$Al_xGa_{1-x}As$ heterojunctions, *Jap. J. Appl. Phys.* **19**, L225-L227
2. Delagebeaudeuf, D., Delescluse, P., Etienne, P., Laviron, M., Chaplart, J., and Linh, N.T. (1980) Two-dimensional electron gas MESFET structure, *Electron. Lett.* **16**, 667-668
3. Dingle, R., Störmer, H.L., Gossard, A.C., and Wiegmann, W. (1978) Electron mobilities in modulation-doped semiconductor heterojunction superlattices, *Appl. Phys. Lett.* **33**, 665-667
4. Störmer, H.L., Dingle, R., Gossard, A.C., Wiegmann, W., and Sturge, M.D. (1979) Two-dimensional electron gas at a semiconductor-semiconductor interface, *Solid-State Comm.* **29**, 705-709
5. Rosenberg, J.J., Benlamri, M., Kirchner, P.D., Woodall, J.M., and Pettit, G.D. (1985) An $In_{0.15}Ga_{0.85}As$/GaAs pseudomorphic single quantum well HEMT, *IEEE Electron Device Lett.* **EDL-6**, 491-493
6. Ketterson, A., Moloney, M., Masselink, W.T., Peng, C.K., Klem, J., Fischer, R., Kopp, W., and Morkoc, H. (1985) High transconductance InGaAs/AlGaAs pseudomorphic modulation-doped field-effect transistors, *IEEE Electron Device Lett*. **EDL-6**, 628-630
7. Ketterson, A.A., Masselink, W.T., Gedymin, J.S., Klem, J., Peng, C.-K., Kopp, W.F., Morkoc, H., and Gleason, K.R. (1986) Characterization of InGaAs/AlGaAs pseudomorphic modulation-doped field-effect transistors, *IEEE Trans. Electron Devices* **ED-33**, 564-571
8. Henderson, T., Aksun, M.I., Peng, C.K., Morkoc, H.,Chao, P.C., Smith, P.M., Duh, K,-H.G., and Lester, L.F. (1986) Microwave performance of a quarter-micrometer gate low-noise pseudomorphic InGaAs/AlGaAs modulation-doped field effect transistor, *IEEE Electron Device Lett.* **EDL-7**, 649-651
9. Hikosaka, K., Hirachi, Y., Mimura, T., and Abe, M. (1985) A microwave power double-heterojunction high electron mobility transitor, *IEEE Electron Device Lett.* **EDL-6**, 341-343
10. Smith, P.M., Chao, P.C., Lester, L.F., Smith, R.P., Lee, B.R., Ferguson, D.W., Jabra, A.A., Ballingall, J.M., and Duh, K.H.G. (1988) InGaAs pseudomorphic HEMTs for millimeter wave power applications, *IEEE MTT-S Tech. Digest*, 927-930
11. Smith, P.M., Kao, M.Y., Ho, P., Chao, P.C., Duh, K.H.G., Jabra, A.A., Smith, R.P., and Ballingall, J.M. (1989) A 0.15 μm gate-length pseudomorphic HEMT , *IEEE MTT-S Tech. Digest*, 983-986
12. Pearsall, T.P., Hendel, R., O'Connor, P., Alavi, K., and Cho, A.Y. (1983) Selectively-doped $Al_{0.48}In_{0.52}As/Ga_{0.47}In_{0.53}As$ heterostructure field effect transistor, *IEEE Electron Device Lett.* **EDL-4**, 5-8
13. Hirose, K., Ohata, K., Mizutani, T., Itoh, T., and Ogawa, M. (1985) 700 mS/mm 2DEGFETs fabricated from high mobility MBE-grown n-AlInAs/GaInAs heterostructures, *Gallium Arsenide and Related Compounds 1985, Inst. Phys. Conf. Ser. No.* **79**, 529-534
14. Peng, C.K., Aksun, M.I., Ketterson, A.A., Morkoc, H., and Gleason, K.R. (1987) Microwave performance of InAlAs/InGaAs/InP MODFET's, *IEEE Electron Device Lett.* **EDL-8**, 24-26
15. Kuo, J.M., Lalvic, B., and Chang, T.Y. (1986) New pseudomorphic MODFETs utilizing $Ga_{.47-u}In_{.53+u}As/Al_{.48+u}In_{.52-u}As$ heterostructures, *IEDM Tech. Digest*, 460-463

16. Mishra, U.K., Brown, A.S., and Rosenbaum, S.E. (1988) DC and RF performance of 0.1 μm gate length $Al_{.48}In_{.52}As$-$Ga_{.38}In_{.62}As$ pseudomorphic HEMTs, *IEDM Tech. Digest*, 180-183
17. Pollak, F.H. (1993) Energy gaps of AlGaAs, in S. Adachi (ed), *Properties of Aluminium Gallium Arsenide*, INSPEC, the Institution of Electrical Engineers, London, p. 54
18. Matthews, J.W., and Blakeslee, A.E. (1974) Defects in epitaxial multilayers, *J. Crystal Growth* **27**, 118-125
19. Andersson, T.G., Chen, Z.G., Kulakovskii, V.D., Uddin, A., and Vallin, J.T. (1987) Variation of the critical layer thickness with In content in strained $In_xGa_{1-x}As$-GaAs quantum wells grown by molecular beam epitaxy, *Appl. Phys. Lett.* **51**, 752-754
20. People, R., and Bean, J.C. (1985) Calculation of critical layer thickness versus lattice mismatch for Ge_xSi_{1-x}/Si strained-layer heterostructures, *Appl. Phys. Lett.* **47**, 322-324
21. Taguchi, T., Takeuchi, Y., Matugatani, K., Ueno, Y., Hattori, T., Sugiyama, Y., and Tacano, M. (1993) Critical layer thickness of $In_{0.80}Ga_{0.20}As/In_{0.52}Al_{0.48}As$ heterostructures, *J. Crystal Growth* **134**, 147-150
22. Ruch, J.G., and Kino, G.S. (1968) Transport properties of GaAs, *Phys. Rev.* **174**, 921-931
23. Braslau, N., and Hauge, P.S. (1970) Microwave measurement of the velocity-field characteristic of GaAs, *IEEE Trans. Electron Devices* **ED-17**, 616-622
24. Houston, P.A., and Evans, A.G.R. (1977) Electron drift velocity in *n*-GaAs at high electric fields, *Solid-State Electr.* **20**, 197-204
25. Littlejohn, M.A., Kim, K.W., and Tian, H. (1993) High-field transport in InGaAs and related heterostructures, in P. Bhattacharya (ed), *Properties of Lattice-Matched and Strained Indium Gallium Arsenide*, INSPEC, the Institution of Electrical Engineers, London, p. 107-116
26. Brennan, K., and Hess, K. (1984) High field transport in GaAs, InP and InAs, *Solid-State Electr.* **27**, 347-357
27. Liu, C.T., Lin, S.Y., Tsui, D.C., Lee, H., and Ackley, D. (1988) Cyclotron resonance measurements of the electron effective mass in strained AlGaAs/InGaAs/GaAs pseudomorphic structures, *Appl. Phys. Lett.***53**, 2510-2512
28. Nguyen, L.D., Larson, L.E., and Mishra, U.K. (1992) Ultra-high-speed modulation-doped field-effect transistors: a tutorial review, *Proc. of the IEEE* **80**, 494-518
29. Bhattacharya, P.K. (1993) Low- and high-field transport in pseudomorphic InGaAs based heterostructures, in P. Bhattacharya (ed), *Properties of Lattice-Matched and Strained Indium Gallium Arsenide*, INSPEC, the Institution of Electrical Engineers, London, p. 117-126
30. Schubert, E.F. (1990) Delta doping of III-V semiconductors: fundamentals and device applications, *J. Vac. Sci. Technol.* **A8(3)**, 2980-2996
31. Ando, Y., and Itoh, T. (1988) Analysis of charge control in pseudomorphic two-dimensional electron gas field-effect transistors, *IEEE Trans. Electron Devices* **35**, 2295-2301
32. Chand, N., Henderson, T., Klem, J., Masselink, W.T., Fischer, R., Chang, Y.-C., and Morkoc, H. (1984) comprehensive analysis of Si-doped $Al_xGa_{1-x}As$ ($x = 0$ to 1): theory and experiments, *Phys. Rev. B* **30**, 4481-4492
33. Schubert, E.F., and Ploog, K. (1984) Shallow and deep donors in direct-gap *n*-type $Al_xGa_{1-x}As$:Si grown by molecular-beam epitaxy, *Phys. Rev. B* **30**, 7021-7029
34. Foisy, M.C., Tasker, P.J., Hughes, B., and Eastman, L.F. (1988) The role of inefficient charge modulation in limiting the current-gain cutoff frequency of the MODFET, *IEEE Trans. Electron Devices* **35**, 871-878
35. Das, M.B. (1991) HEMT device physics and models, in F. Ali and A. Gupta (eds), *HEMTs and HBTs: Devices, Fabrication, and Circuits*, Artech House, Boston, London, pp. 11-75
36. Wolf, P. (1970) Microwave properties of Schottky-barrier field effect transistors, *IBM J. Res. Dev.* **14**, 125-141
37. Das, M.B. (1985) A high aspect ratio design approach to millimeter-wave HEMT structures, *IEEE Trans. Electron Devices* **ED-32**, 11-17
38. Wasserstrom, E., and McKenna, J. (1970) The potential due to a charged metallic strip on a semiconductor surface, *Bell Syst. Tech. J.* **49**, 853-877

39. Anholt, R. (1991) Dependence of GaAs MESFET fringe capacitances on fabrication technologies, *Solid-State Electr.* **34**, 515-520
40. Moll, N., Hueschen, M.R., and Fischer-Colbrie, A. (1988) Pulse-doped AlGaAs/InGaAs pseudomorphic MODFETs, *IEEE Trans. Electron Devices* **35**, 879-886
41. Greiling, P.T., and Nguyen, L. (1991) Ultrahigh-frequency InP-based HEMTs for millimeter wave applications, *SPIE* Vol. **1475**, 34-41
42. Lester, L.F., Smith, P.M., Ho, P., Chao, P.C., Tiberio, R.C., Duh, K.H.G., and Wolf, E.D. (1988) 0.15 μm gate-length double recess pseudomorphic HEMT with f_{max} of 350 GHz, *IEDM Tech. Digest*, 172-175
43. Nguyen, L.D., Tasker, P.J., Radulescu, D.C., and Eastman, L.F. (1989) Characterization of ultra-high-speed pseudomorphic AlGaAs/InGaAs (on GaAs) MODFETs, *IEEE Trans. Electron Devices* **36**, 2243-2248
44. Chao, P.C., Shur, M.S., Tiberio, R.C., Duh, K.H.G., Smith, P.M., Ballingall, J.M., Ho, P., and Jabra, A.A. (1989) DC and microwave characteristics of sub-0.1-μm gate-length planar-doped pseudomorphic HEMTs, *IEEE Trans. Electron Devices* **36**, 461-473
45. Tan, K.L., Dia, R.M., Streit, D.C., Lin, T., Trinh, T.Q., Han, A.C., Liu, P.H., Chow, P.-M.D., and Yen, H.C. (1990) 94-GHz 0.1-μm T-gate low-noise pseudomorphic InGaAs HEMTs, *IEEE Electron Device Lett.* **11**, 585-587
46. Nguyen, L.D., Jelloian, L.M., Thompson, M., and Lui, M. (1990) Fabrication of a 80 nm self-aligned T-gate AlInAs/GaInAs HEMT, *IEDM Tech. Digest*, 499-502
47. Chao, P.C., Tessmer, A.J., Duh, K.-H.G., Ho, P., Kao, M.-Y., Smith, P.M., Ballingall, J.M., Liu, S.-M.J., and Jabra, A.A. (1990) W-band low-noise InAlAs/InGaAs lattice-matched HEMTs, *IEEE Electron Device Lett.* **11**, 59-62
48. Ho, P., Kao, M.Y., Chao, P.C., Duh, K.H.G., Ballingall, J.M., Allen, S.T., Tessmer, A.J., and Smith, P.M. (1991) Extremely high gain 0.15 μm gate-length InAlAs/InGaAs/InP HEMTs, *Electron. Lett.* **27**, 325-326
49. Mishra, U.K., Brown, A.S., Delaney, M.J., Greiling, P.T., and Krumm, C.F. (1989) The AlInAs-GaInAs HEMT for microwave and millimeter-wave applications, *IEEE Trans. Microw. Theo. Tech.* **37**, 1279-1285
50. Nguyen, L.D., Brown, A.S., Thompson, M.A., Jelloian, L.M., Larson, L.E., and Matloubian, M. (1992) 650-Angstrom self-aligned-gate pseudomorphic $Al_{0.48}In_{0.52}As/Ga_{0.20}In_{0.80}As$ high electron mobility transistors, *IEEE Electron Device Lett.* **13**, 143-145
51. Nguyen, L.D., Brown, A.S., Thompson, M.A., and Jelloian, L.M. (1992) 50-nm self-aligned-gate pseudomorphic AlInAs/GaInAs high electron mobility transistors, *IEEE Trans. Electron Devices* **39**, 2007-2014
52. Wojtowicz, M., Lai, R., Streit, D.C., Ng, G.I., Block, T.R., Tan, K.L., Liu, P.H., Freudenthal, A.K., and Dia, M.R. (1994) 0.10 μm graded InGaAs channel InP HEMT with 305 GHz f_T and 340 GHz f_{max}, *IEEE Electron Device Lett.* **15**, 477-479
53. Fukui, H. (1979) Optimal noise figure of microwave GaAs MESFETs, *IEEE Trans. Electron Devices* **ED-26**, 1032-1037
54. Deutschmann, R., Fischer, C., Sala, S., and Selberherr, S. (1993) Evaluation of effective device parameters by comparison of measured and simulated *C-V* characteristics for conventional and pseudomorphic HEMTs, in S. Selberherr, H. Stippel and E. Strasser (eds), *Simulation of Semiconductor Devices and Processes Vol. 5*, Springer-Verlag, Wien, pp. 461-464
55. Duh, K.H.G., Chao, P.C., Ho, P., Tessmer, A., Liu, S.M.J., Kao, M.Y., Smith, P.M., and Ballingall, J.M. (1990) W-band InGaAs HEMT low noise Amplifiers, *IEEE MTT-S Tech. Digest*, 595-598
56. Katoh, T., Yoshida, N., Minami, H., Kashiwa, T., and Orisaka, S. (1993) A 60 GHz-band ultra low noise planar-doped HEMT, *IEEE MTT-S Tech. Digest*, 337-340
57. Takikawa, M., and Joshin, K. (1993) Pseudomorphic *n*-InGaP/InGaAs/GaAs high electron mobility transistors for low-noise amplifiers, *IEEE Electron Device Lett.* **14**, 406-408
58. Fujita, S., Noda, T., Wagai, A., Ashizawa, Y., and Hosoi, S. (1993) Extremely low noise InGaAs/InAlAs HEMT grown by MOCVD, *Electron. Lett.* **29**, 1557-1558

59. Tan, K.L., Streit, D.C., Chow, P.D., Dia, R.M., Han, A.C., Liu, P.H., Garske, D., and Lai, R. (1991) 140 GHz 0.1 μm gate-length pseudomorphic $In_{0.52}Al_{0.48}As/In_{0.60}Ga_{0.40}As$/InP HEMT, *IEDM Tech. Digest*, 239-242
60. Hwang, T., Chye, P., and Gregory, P. (1993) Super low noise pseudomorphic InGaAs channel InP HEMTs, *Electron. Lett.* **29**, 10-11
61. Huang, J.C., Boulais, W., Platzker, A., Kazior, T., Aucoin, L., Shanfield, S., Bertrand, A., Vafiades, M., and Niedzwiecki, M. (1993) The effect of channel dimensions on the millimeter-wave power performance of a pseudomorphic HEMT, *GaAs IC Symp. Tech. Digest*, 177-180
62. Huang, J.C., Jackson, G.S., Shanfield, S., Platzker, A., Saledas, P.K., and Weichert, C. (1993) An AlGaAs/InGaAs pseudomorphic high electron mobility transistor with improved breakdown voltage for X- and Ku-band power applications, *IEEE Trans. Microw. Theo. Tech.* **41**, 752-759
63. Kao, M.-Y., Fu, S.-T., Ho, P., Smith, P.M., Chao, P.C., Nordheden, K.J., and Wang, S. (1992) Very high voltage AlGaAs/InGaAs pseudomorphic power HEMTs, *IEDM Tech. Digest*, 319-321
64. Huang, J.C., Jackson, G., Shanfield, S., Hoke, W., Lyman, P., Atwood, D., Saledas, P., Schindler, M., Tajima, Y., Platzker, A., Masse, D., and Statz, H. (1991) An AlGaAs/InGaAs pseudomorphic high electron mobility transistor (PHEMT) for X- and Ku-band power applications, *IEEE MTT-S Tech. Digest*, 713-716
65. Kim, B., Matyi, R.J., Wurtele, M., Bradshaw, K., Khatibzadeh, M.A., and Tserng, H.Q. (1989) Millimeter-wave power operation of an AlGaAs/InGaAs/GaAs quantum well MISFET, *IEEE Trans. Electron Devices* **36**, 2236-2242
66. Zhou, G.-G., Chan, K.T., Hughes, B., Mierzwinski, M., and Kondo, H. (1989) A pseudomorphic MODFET structure with excellent linear power performance at mm-wave range, *IEDM Tech. Digest,* 109-112
67. Huang, J.C., Saledas, P., Wendler, J., Platzker, A., Boulais, W., Shanfield, S., Hoke, W., Lyman, P., Aucoin, L., Miquelarena, A., Bedard, C., and Atwood, D. (1993) A double-recessed $Al_{0.24}GaAs/In_{0.16}GaAs$ pseudomorphic HEMT for Ka- and Q-band power applications, *IEEE Electron Device Lett.* **14**, 456-458
68. Smith, P.M., Ferguson, D.W., Kopp, W.F., Chao, P.C., Hu, W., Ho, P., and J.M. Ballingall (1991) A high power, high efficiency millimeter-wave pseudomorphic HEMT, *IEEE MTT-S Tech. Digest*, 717-720
69. Saunier, P., and Tserng, H.Q. (1989) AlGaAs/InGaAs heterostructures with doped channels for discrete devices and monolithic amplifiers, *IEEE Trans. Electron Devices* **36**, 2231-2235
70. Smith, P.M., Chao, P.C., Ballingall, J.M., and Swanson, A.W. (1990) Microwave and mm-wave power amplification using pseudomorphic HEMTs, *Microwave J.* May 1990, 71-86
71. Lai, R., Wojtowicz, M., Chen, C.H., Biedenbender, M., Yen, H.C., Streit, D.C., Tan, K.L., and Liu, P.H. (1993) High-power 0.15-μm V-band pseudomorphic InGaAs-AlGaAs-GaAs HEMT, *IEEE Microw. Guided Wave Lett.* **3**, 363-365
72. Streit, D.C., Tan, K.L., Dia, R.M., Liu, J.K., Han, A.C., Velebir, J.R., Wang, S.K., Trinh, T.Q., Chow, P.-M.D., Liu, P.H., and Yen, H.C. (1991) High-gain W-band pseudomorphic InGaAs power HEMTs, *IEEE Electron Device Lett.* **12**, 149-150
73. Matloubian, M., Nguyen, L.D., Brown, A.S., Larson, L.E., Melendes, M.A., and Thompson, M.A. (1991) High power and high efficiency AlInAs/GaInAs on InP HEMTs, *IEEE MTT-S Tech. Digest*, 721-724
74. Matloubian, M., Brown, A.S., Nguyen, L.D., Melendes, M.A., Larson, L.E., Delaney, M.J., Thompson, M.A., Rhodes, R.A., and Pence, J.E. (1993) 20-GHz high-efficiency AlInAs-GaInAs on InP power HEMT, *IEEE Microw. Guided Wave Lett.* **3**, 142-144
75. Matloubian, M., Brown, A.S., Nguyen, L.D., Melendes, M.A., Larson, L.E., Delaney, M.J., Pence, J.E., Rhodes, R.A., Thompson, M.A., and Henige, J.A. (1993) High-power V-band AlInAs/GaInAs on InP HEMTs, *IEEE Electron Device Lett.* **14**, 188-189
76. Wang, G.-W., Chen, Y.-K., Schaff, W.J., and Eastman, L.F. (1988) A 0.1-μm gate $Al_{0.5}In_{0.5}As/Ga_{0.5}In_{0.5}As$ MODFET fabricated on GaAs substrates, *IEEE Trans. Electron Devices* **35**, 818-23

77. Masato, H., Matsuno, T., and Inoue, K. (1991) $In_{0.5}Ga_{0.5}As$/InAlAs modulation-doped field effect transistors on GaAs substrates grown by low-temperature molecular beam epitaxy, *Jap. J. Appl. Phys.* **30**, 3850-3852

78. Bachem, K.H., Pletschen, W., Winkler, K., Fleissner, J., Hoffmann, C., and Tasker, P.J. (1993) AlGaInP/GaInAs/GaAs-MODFETs with carbon doped p^+-GaAs gate structure, a novel device concept, its implementation and device properties, *Proc. Int. Symp. GaAs and Related Compounds, Inst. Phys. Conf. Ser. No.* **136**, 35-40

79. Yang, D., Chen, Y.C., Brock, T., and Bhattacharya, K. (1992) DC and microwave performance of a 0.1-μm gate InAs/$In_{0.52}Al_{0.48}As$ MODFET, *IEEE Electron Device Lett.* **13**, 350-352

80. Bolognesi, C.R., Caine, E.J., and Kroemer, H. (1994) Improved charge control and frequency performance in InAs/AlSb-based heterostructure field-effect transistors, *IEEE Electron Device Lett.* **15**, 16-18

81. Aina, O., Burgess, M., Mattingly, M., Meerschaert, A., O'Connor, J.M., Tong, M., Ketterson, A., and Adesida, I. (1992) A 1.45-W/mm, 30-GHz InP-channel power HEMT, *IEEE Electron Device Lett.* **13**, 300-302

82. Jelloian, L.M., Matloubian, M., Liu, T., Lui, M., and Thompson, M.A. (1994) InP-based HEMTs with $Al_{0.48}In_{0.52}As_xP_{1-x}$ Schottky layers, *IEEE Electron Device Lett.* **15**, 172-174

BASICS OF PSEUDOMORPHIC HEMTS TECHNOLOGY AND NUMERICAL SIMULATION

A. ASENOV, S. BABIKER, N. CAMERON, S. MURAD,
M. HOLLAND, and S.P. BEAUMONT
Nanoelectronics Research Centre
Department of Electronics and Electrical Engineering
Glasgow University, Glasgow G12 8QQ
Scotland, UK

1. Introduction

This chapter consist of two parts. In the first part we describe the technology used in the Nanoelectronics Research Centre of Glasgow University for fabrication of high performance 200 - 100 nm gate length Pseudomorphic HEMTs (PsHEMTs). Most of the fabrication sequence is fairly standard and gives a good overall impression of the basic technological steps involved in the fabrication of modern PsHEMTs. There are however some specific solutions which distinguish our PsHEMT technology from the recipes adopted by other university and industry based groups. Among those 'secrets' are the use of an As_2 source with a low temperature As_4 cracker in the MBE growth of the vertical layer structure, and a new selective and low damage recess reactive ion dry etching process based on $SiCl_4/SiF_4$ chemistry.

In the second part we introduce the different numerical techniques which are suitable for simulation of Ps HEMTs. We start with one dimensional (1D) simulation of the vertical layer structure, based on numerical solution of the Poisson's equation. The carrier transport and the geometry effects are introduced latter in two dimensional (2D) simulation of the whole device, based on self consistent solution of the Poisson's equation and the Boltzmann transport equation in drift-diffusion (DD), hydrodynamic (HD) or Monte Carlo (MC) approximations. The underlying physics and the limitations of the different approximations are discussed in relation to the PsHEMT operation. The techniques used for discretization and solution of the governing sets of equations are briefly described. The different simulation approaches are illustrated with PsHEMT simulation results from the Glasgow compound FET simulator H2F and the commercial

M. Balkanski (ed.), Devices Based on Low-Dimensional Semiconductor Structures, 349–397.

simulator MEDICI. At the end of this chapter we discuss the need for 3D simulation in relation to the fringing effects, charge discreteness and technology fluctuations in very small deices.

2. PsHEMT Technology

The schematic cross-section of a typical pseudomorphic HEMT is shown in Figure 1. The band diagram in the same figure gives an idea about the bandgap of the different layers and the potential profile in the channel region of the device. The conducting channel is formed in a strained InGaAs layer grown on GaAs or AlGaAs buffer. The current flows in the 2D electron gas (2DEG) at the interface between the channel layer and the AlGaAs layer above it. A uniform or pulsed (δ-) doped layer supplies the electrons in the channel. A thin AlGaAs spacer separates those electrons from the ionised donors in order to reduce the impurity scattering. A heavily doped cap layer protects the 2DEG form the charge trapped on surface states. The low resistance mushroom shape (T-) gate is placed in a recess region on the top of the Schottky layer, close to the 2DEG, to allow better control on the 2DEG carrier concentration and hence on the PsHEMT's current. The source and the drain Ohmic contacts lead the current out of the device.

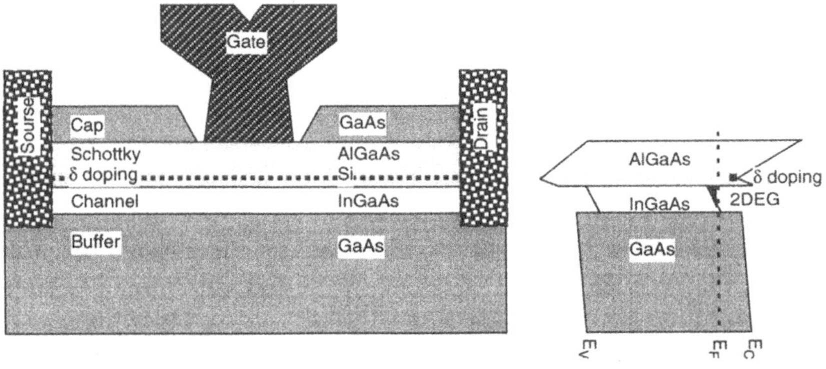

Figure 1. Schematic cross-sectional view of a typical pseudomorphic HEMT and the band diagram in the channel region.

These are four critical technological steps in the fabrication of a PsHEMT: (i) the growth of a high quality vertical layer structure using molecular beam epitaxy (MBE), (ii) the fabrication of a short (200 - 100 nm) T-gate using electron beam (e-beam) lithography, (iii) the gate recessing using wet chemical or selective reactive ion etching (SRIE), and (iv) the formation of low resistance Ohmic contacts.

2.1. VERTICAL LAYER STRUCTURE: MBE

We use an Intervac (Varian) Gen II MBE system to grow the vertical layer structure of our PsHEMTs and to introduce the necessary doping. The system is equipped with a specially designed As_2 source employing a low temperature As_4 cracker. This allows the growth of high quality InGaAs layers at relatively low substrate temperatures. The growth rate is tightly controlled using the standard RHEED oscillations technique. The substrate temperature is monitored using an Ircon series V optical pyrometer. Typical growth rates as a function of the substrate temperature are given in Figure 2 for layers with different compositions [1].

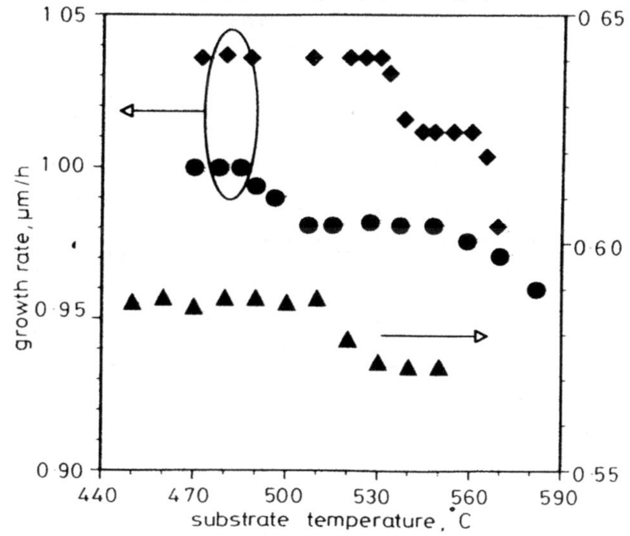

Figure 2. Growth rate as a function of the substrate temperature.

The layer sequence in the latest generation Glasgow PsHEMTs is shown in Figure 3 and detailed in Table 1 [2]. The growth starts with a 100 periods superlatice (impurity trap) to prevent the migration of impurities and defects from the semi-insulating GaAs substrate into the active device area. A 600 nm GaAs buffer separates the superlatice from the PsHEMT's channel and reduces the influence of the Fermi level pinning in the superlatice on the free carrier concentration in the channel. The 10 nm strained $In_xGa_{1-x}As$ channel is grown at 500 - 520° C. The low growth temperature allows the indium fraction x to be increased to 30% (far above the 20 - 25% indium in the conventional PsHEMTs [3]) without noticeable strain relaxation. This increases the spacer/channel conduction band offset and improves the high field transport in the channel. The relatively thin (2.5 nm) $Al_{0.3}Ga_{0.7}As$ spacer improves the carrier transfer efficiency at the cost of of a slight mobility reduction as a result of the increased remote impurity scattering.

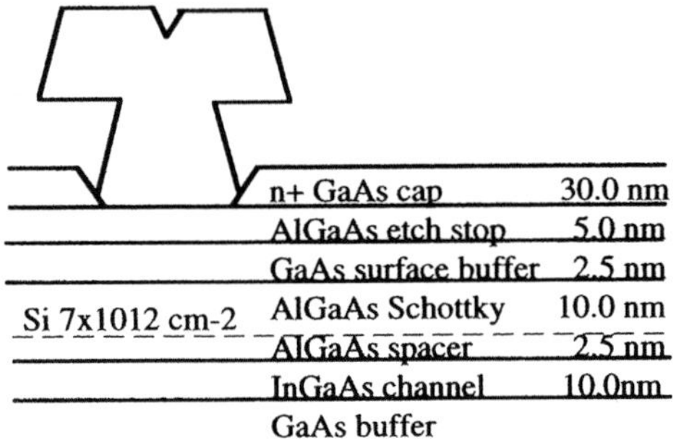

Figure 3. Vertical layer structure of the latest generation Glasgow PsHEMTs

TABLE 1 Detailed information about the layer structure of the Glasgow PsHEMTs.

Layer	A 867 / 8
Cap	30nm $4x10^{18}$ cm^{-3} Si-GaAs
Etch Stop	5nm AlGaAs
Surface Buffer	2.5nm GaAs
Schottky	10nm AlGaAs
Supply	2 ml GaAs Si $7x10^{12}$ cm^{-2} 3 ml GaAs
Spacer	2.5 nm AlGaAs
Channel	10 nm InGaAs
Buffer	600 nm GaAs
Impurity trap	100 x 25.5 Å GaAs/ 25.5 Å $Al_{0.3}Ga_{0.7}As$ 1.1 µm GaAs
Substrate	SI (100) GaAs

The Si δ-doping is encapsulated between 3 and 2 monolayers GaAs. The first 3 monolayers GaAs grown on top of the spacer prevent the incorporation of impurities during the deposition of Si. Together with the second 2 monolayers of GaAs they reduce the diffusion of Si and the spreading of the δ-doping. The doping efficiency at the doping level $7x10^{12}$ cm^{-3} is approximately 70 % The 10 nm $Al_{0.3}Ga_{0.7}As$ Schottky layer is separated from the 5 nm $Al_{0.3}Ga_{0.7}As$ recess etch stopper by a 2.5 nm GaAs buffer. The role of this buffer is to prevent deep oxidation of the Schottky layer in the recess region which leads to a degradation of the device. The overall separation between the channel and the gate is 21.5 nm which results in more than 5 aspect ratio for 100 nm gate length devices, providing the necessary short channel and fringing effects immunity. The heavily doped 30 nm GaAs cap layer screens the 2DEG in the region between the gate and the contacts from the negative charge of the electrons trapped on deep, acceptor type surface states.

2.2. T-GATE: E-BEAM LITHOGRAPHY

The base of the mushroom shape T-gate defines the channel length which is 100 nm or below in modern PsHEMTs. The extended metal on the top reduces the gate series resistance - one of the crucial factors limiting the high frequency device performance. Both the small dimension of the gate and the necessity to carve the gate shape require electron beam (e-beam) lithography.

In the fabrication of the T-gates of our PsHEMTs we use a Leica Cambridge EBPG5-HR Beamwriter. The Ti/Au lift-off process which is described bellow is based on a three layer resist system (Figure 4). The botom 100 nm PMMA layer, which has a high molecular weight and a low sensitivity, defines the base of the gate. The middle 400 nm P(MAA/MMA) layer with high sensitivity defines the mushroom umbrella. The thin 50 nm low molecular weight PMMA layer on the top has a medium sensitivity and forms a small overhang which separates the metal of the gate from the metal on the top of the photoresist facilitating the lift-off process.

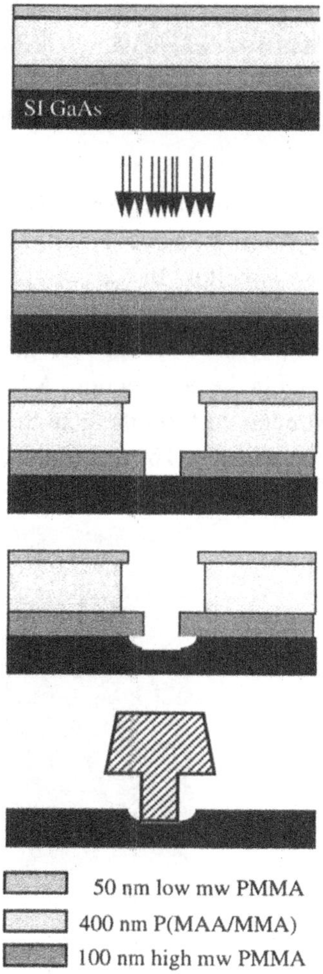

Figure 4. The T-gate process.

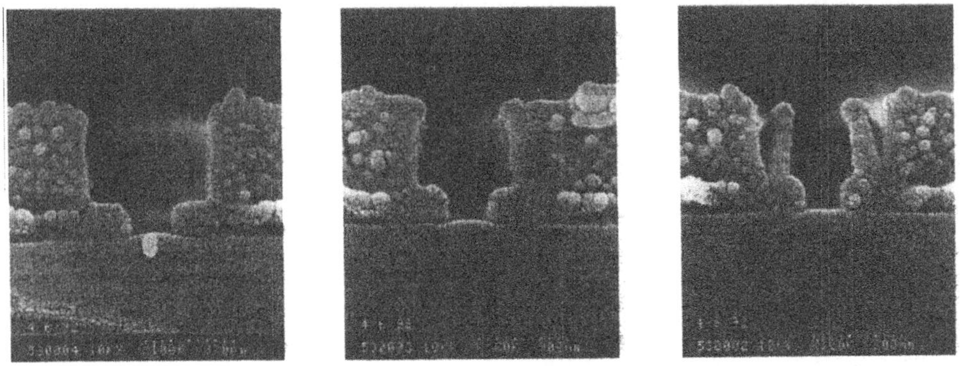

Figure 5. SEMs of T-gate profiles written with the same 80 nm centre line size but different edge, side and centre exposures.

During e-beam exposure the dose is varied along the gate length. In the middle the dose is above the sensitivity threshold of the bottom resist layer. In the rest of the gate the dose is bellow the sensitivity of the bottom layer but increases towards the gate umbrella edges to enhance the mushroom shape. After the resist development the cap layer is recessed trough the gate base window using the SRIE process which is described in the next section. A layer of Ti is deposited and penetrates through the gate-base window to form the gate Schottky barrier followed by gold which fills the whole mushroom cavity. The rest of the Ti/Au layer is lifted-off by stripping the photoresist.

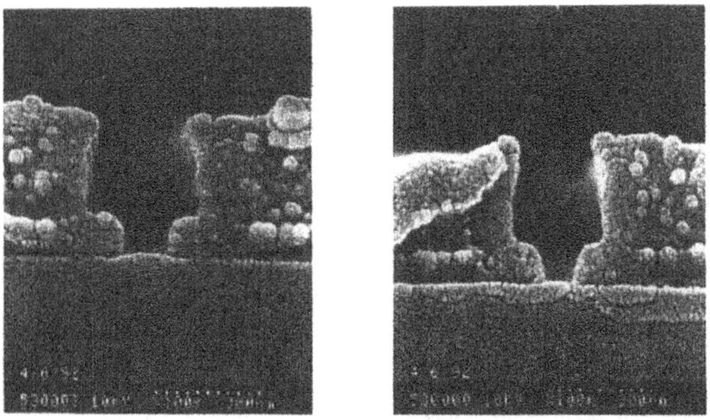

Figure 6. SEMs of T-gate profiles written with the same exposure ratios but different centre line sizes of 40 and 80 nm

The selection of the exposure doses in the different regions is an empirical process which requires a lot of experiments. Figure 5 illustrates the results of such experiments in which the edge, side and centre exposures are varied in order to obtain an optimal cavity shape. If once the optimal exposure ratios are established gates with different base lengths can be fabricated by only varying the size of the central line (Figure. 6).

2.3. GATE RECESS: DRY ETCHING

The etching of the gate recess is critical to the PsHEMT performance. The distance between the gate and the channel determines the transconductance. The offset and the shape of the recess region control the device parasitics such as coupling capacitances and access resistances [4] (see also Section 3.2). The etching process should allow precise control of the recess depth and the recess offset shape.

In the fabrication of our PsHEMTs we use a damage free Selective Reactive Ion Etching (SRIE) process based on $SiCl_4/SiF_4$ chemistry in a Plasma Technology µP80 reactor. Extremely high selectivity in etching GaAs/AlGaAs was achieved [5] for an appropriate choice of the process parameters. The fraction of $SiCl_4$ in the chamber varies between 10 and 30 % at relatively high pressure of 150 mtorr. The RF power is in the range of 10 to 20 W and for DC biases is between 20 and 40 V.

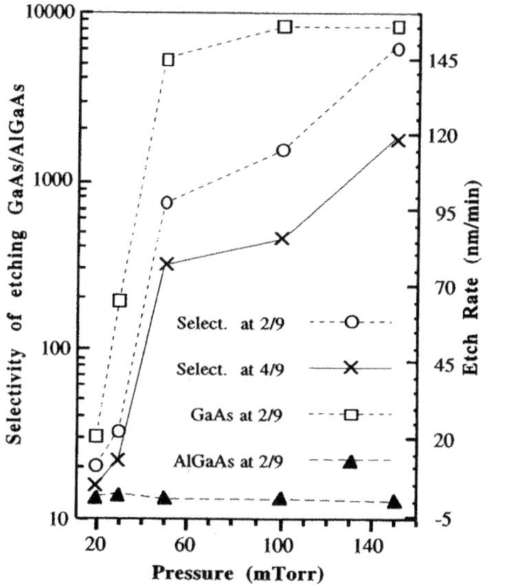

Figure 7. The etch rates of GaAs, AlGaAs and the selectivity in SiCl4/SiF4 plasma as a function of the pressure for two flow rate ratios: 2/9 and 4/9 sccm.

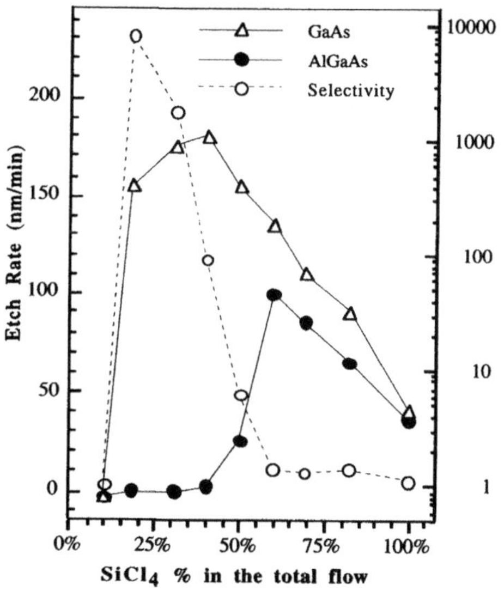

Figure 8. The etch rates of GaAs and AlGaAs and the selectivity in SiCl4/SiF4 plasma as a function of SiCl percentage in the total flow at 150 mtorr.

Due to the high selectivity the recess etching stops at the interface between the GaAs cap and the AlGaAs etch stopper. The gate offset and the shape of the recess is controlled by overetching. However, if the cap layer is relatively thin the lateral cap etching slows down or even stops, because of a back deposition of some of the reaction products. In this case the addition of traces of oxygen enormously increases the lateral etching rate and allows large gate offsets to be achieved. The shapes of the recess when the etching reaches the GaAs/AlGaAs interface and after significant overetching are shown in Figure 9.

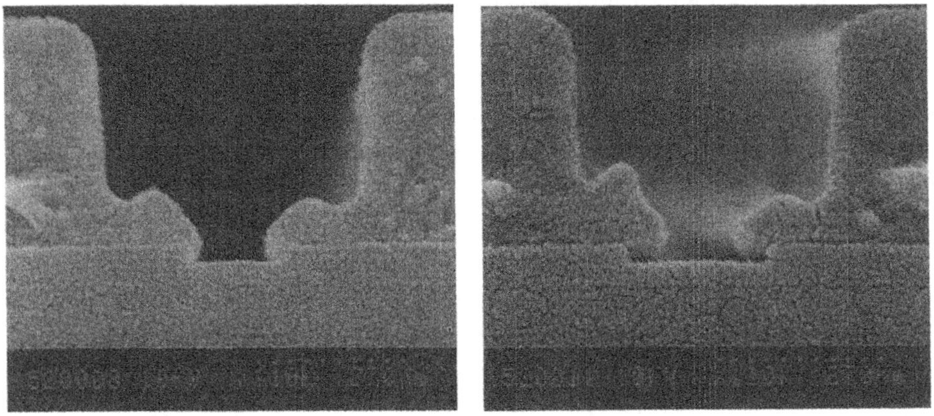

Figure 9. Cross-sections of recess shapes (a) after 75 s etching (b) after 115 s etcing.

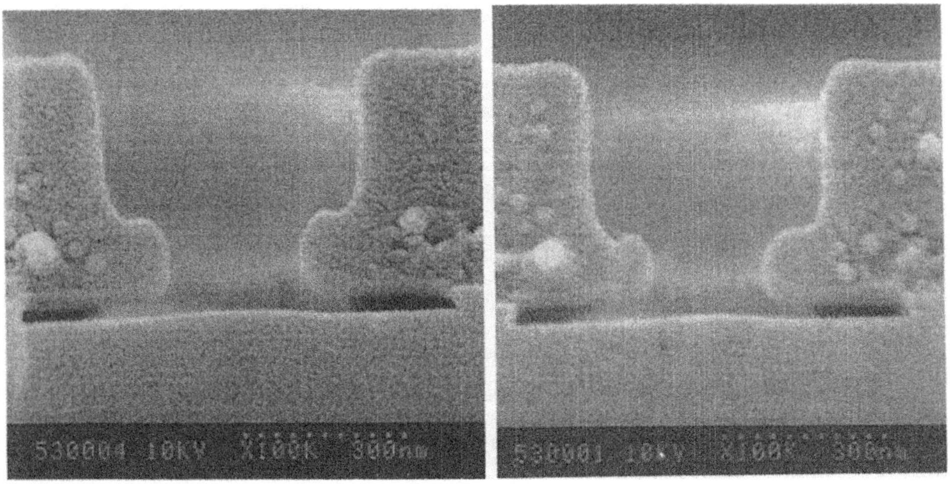

Fgure 10. Cross-sections of recess shapes for two gate lengths.

The recess shape and the gate offset does not depend on the actual gate length (Figure 10). This allows the fabrication of PsHEMTs with gate lengths bellow 100 nm without any further changes in the etching conditions.

The recess region is above the PsHEMTs channel and any damage due to the SRIE will affect the device performance, reducing the 2DEG carrier concentration and degrading the channel mobility. Due to the low DC bias our SRIE gate recess process is virtually damage free [6]. This was proved in a number of experiments involving control structures and real devices. Figure 11 presents etching and transport measurement results obtained on a control structure with a fully depleted cap. Simulations of such layer structure predict that the carrier concentration in the channel will not be affected by the removal of the cap, provided that there is no damage upon etching. No detectable modification of the carrier concentration and mobility was observed after 10 min etching which is 50 times larger than the conventional overetching time. Very similar are the results obtained by etching and measuring the current in a real device without a gate (Figure 12). The drain saturation current initially falls down during the etching of the heavily doped cap layer which protects the channel from surface charges and provides a parallel current path. However, when the cap is fully removed both the carrier concentration and the current remain constant up to 300 sec overetching, indicating that no noticeable damage is generated.

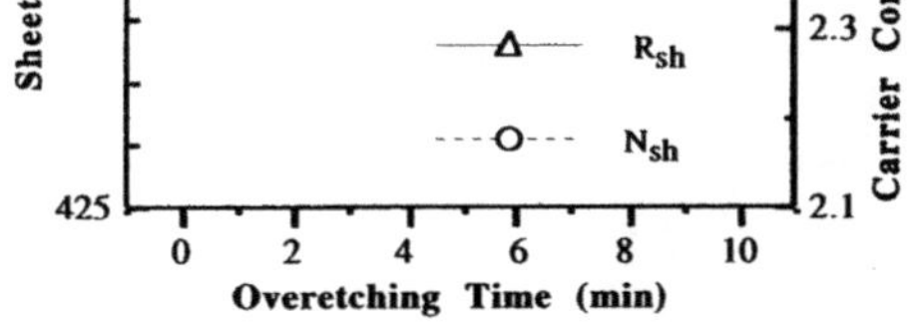

Figure 11. The sheet resistance and carrier concentration as a function of etching time in $SiCl_4/SiF_4$ plasma.

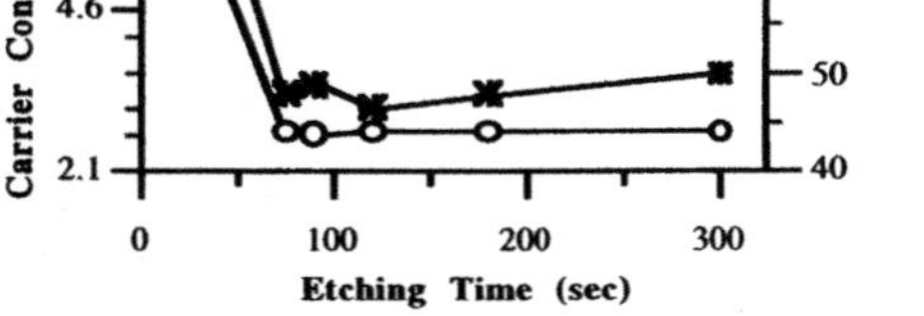

Figure 12 The sheet resistance and drain saturation current as a function of etching time in $SiCl_4/SiF_4$ plasma.

2.4. OHMIC CONTACTS

The fabrication of ohmic contacts with low contact resistance is a very important technological step. Any additional series resistance introduced by the contacts, especially at the source side of the device reduces the effective transconductance and degrades the RF device performance. This is a crucial issue in devices with 100 nm or below gate length, where the contact resistance becomes comparable with the resistance of the channel region [7].

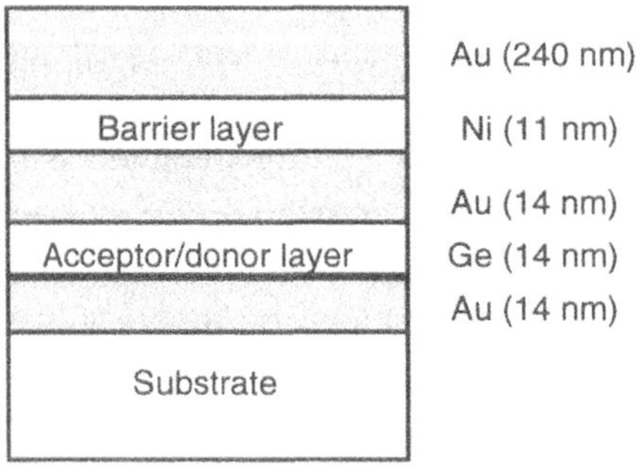

Figure 13. Five layer ohmic contact structure.

We have adopted the five layer ohmic contact structure proposed in [8]. In this structure (Future 13) the donor germanium layer is clearly separated from the barrier layer. The gold layer on the top is sufficiently thick to permit bonding. Only three metal evaporation sources are required to fabricate the whole five layer structure. The fabricated ohmic contacts are reproducible an have contact resistance in the range of 0.1 Ohm.mm.

Finally the SEM cross section of a fully completed PsHEMT is shown as an example in Figure 14. The device has 120 nm channel length, transconductance of 600 mS/mm and cut-off frequency of 120 GHz.

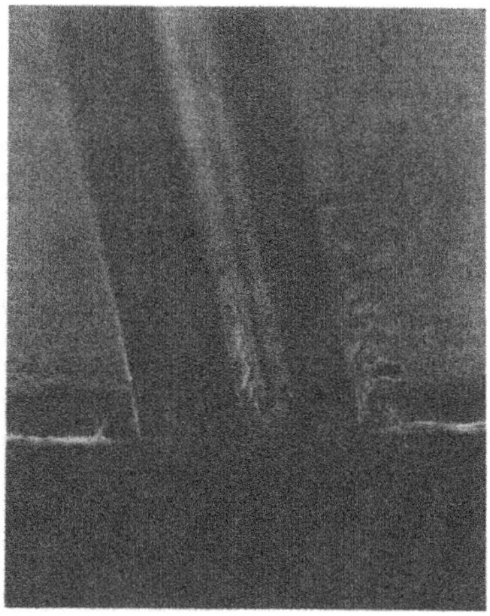

Figure 14 . SEM cross section of a fully fabricated PsHEMT with 120 nm gate length.

3. Numerical simulation

3.1. 1D NUMERICAL SIMULATION

One dimensional numerical simulations are widely used for optimisation of the vertical PsHEMT layer structure [9]. They are based on 1D solution of the Poisson equation in a direction normal to the PsHEMT surface, which may be coupled to 1D solution of the Schrödinger equation. The 1D simulations provide information for the band profile and the concentrations of mobile and trapped carries in all heterostructure layers. The differential gate capacitance can be extracted from two 1D solutions separated by a small increment in the gate voltage. Although the transport is excluded from the 1D PsHEMT simulations the gate voltage dependence of the mobile charge concentration in the channel can be used for rough estimation of the transconductance.

In the PsHEMT design process the 1D simulations can help to answer the following questions: What is the best place and distribution of the impurities? What is the optimum spacer thickness? What is the effect of the In fraction in the channel and the Al fraction in the supply layer on the carrier concentration in and out of the channel? How to avoid the parallel conductivity? What is the effect of the DX centres in the supply and the buffer layer? How the Fermi potential pinning in the impurity trap affects the 2DEG carrier concentration?

There is, however, some degree of uncertainty even in the relatively simple 1D simulations. The band structure in the compound layers and especially in the strained InGaAs is known only to some degree of approximation and the bandgap offset and the effective masses vary in different publications. The density and the energy position of the traps and DX centres in AlGaAs layers with high Al content are broadly unknown. The Schottky barrier height depends on the fabrication conditions, sometimes in uncontrollable fashion. The δ-doping may have a spatial distribution which extends in the channel through the spacer due to diffusion which is difficult to predict. The doping efficiency can be estimated through simulations but cannot be predicted.

3.1.1. *The Poisson Equation*

The 1D Posisson equation for a hetero-layer system is [10]

$$\frac{d}{dx}\left(\varepsilon(x)\frac{d\psi(x)}{dx}\right) = -q\left(N_D^+(x) - N_A^-(x) + p(x) - n(x)\right) \qquad (1)$$

where q is the electron charge, ψ is the electrostatic potential, $\varepsilon(x)$ is the dielectric permitivity which varies from layer to layer, N_D^+ and N_A^- are the ionised acceptors and donors which may include also deep acceptor and donor type states, and n and p are the electron and hole concentrations respectively.

In a classical approximation the electron concentration n in the 2DEG may be expressed in terms of Boltzmann or Fermi statistics. In the first case the concentration is given by the following expression

$$n(x) = N_c(x)\exp\left(\frac{E_F - E_C(x)}{kT}\right) \tag{2}$$

where k is the Boltzmann constant, T is the absolute temperature, N_c is effective density of states in the conducting band, E_F is the Fermi level and E_C is the conducting band minimum.

In the case of Fermi statistics the concentration is given by

$$n(x) = N_c(x)F_{1/2}\left(\frac{E_F - E_C(x)}{kT}\right) \tag{3}$$

where $F_{1/2}$ is a Fermi integral of order 1/2.

The 1D solution domain is normal to the PsHEMT's gate and extends usually to the impurity trap where the Fermi level may be pinned near the middle of the bandgap [11]. Under no current flow conditions the Fermi level is constant and its position is determined by the pinning. Without any restriction the potential may be set to zero at the bottom boundary of the solution domain. At the top boundary the potential should be equal to the sum of the built-in potential between the top and the bottom boundary, the Schottky barrier height and the applied gate voltage.

3.1.2. *Discretization*

In order to solve numerically the Poisson equation the 1D solution domain (Figure 15) is discretized into small intervals. The discretization is non uniform and the grid density is higher near the heterojunction interfaces and in the regions with heavy doping where the potential is expected to vary more rapidly. The lower density in the rest part of the layer structure simply saves processor time.

Applying the finite-difference approach and using the nomenclature from Figure 15, the first derivatives of the potential on the two sides of the node i can be approximated to

$$\left.\frac{d\psi}{dx}\right|_{i-1} = \frac{\psi_i - \psi_{i-1}}{h_{i-1}} \qquad \left.\frac{d\psi}{dx}\right|_{i} = \frac{\psi_{i+1} - \psi_i}{h_i} \tag{4}$$

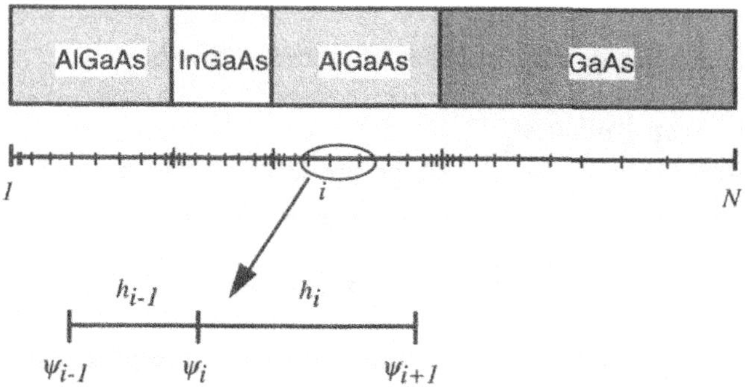

Figure 15. Non uniform 1D discretization in the direction normal to the PsHEMT gate used for numerical solution of the Poisson equation.

At the node the left hand side of the Poissons equation (1) can approximated as:

$$\frac{d}{dx}\left(\varepsilon\frac{d\psi}{dx}\right)\bigg|_i = \frac{\varepsilon_i\dfrac{d\psi}{dx}\bigg|_i - \varepsilon_{i-1}\dfrac{d\psi}{dx}\bigg|_{i-1}}{\dfrac{h_{i-1}+h_i}{2}} \tag{5}$$

Strictly speaking this approximation will preserve the continuity of the electrostatic displacement at the boundary between two layers with different permitivities only if the intervals h_{i-1} and h_i are equal. The substitution of the approximated first derivatives of the potential from equations (4) into equation (5) and then back in equation (1), leads to a discrete finite difference approximation of the Poisson equation at node i

$$\frac{\varepsilon_{i-1}}{h_{i-1}}\psi_{i-1} - \left(\frac{\varepsilon_{i-1}}{h_{i-1}} + \frac{\varepsilon_i}{h_i}\right)\psi_i + \frac{\varepsilon_i}{h_i}\psi_{i+1} = -\frac{h_{i-1}+h_i}{2}\rho_i(\psi_i) \tag{6}$$

Equation (6) is a non-linear algebraic equation since the charge on the right side depends on the potential through the carrier concentrations and the occupation of impurities and traps at the node i. Repeating the above procedure for all discretization nodes leads to a system of non-linear algebraic equations which can be presented in a matrix form

$$\mathbf{A\Psi} = \mathbf{b} \tag{7}$$

where Ψ is the vector of the potential at all discretization nodes, $\mathbf{b}$ is the right hand side vector and A is a thridiagonal coefficient matrix. $\mathbf{A}\Psi = \mathbf{b}$

3.1.3. *Linearization and Solution*

The Newton-Raphson method [12] can be applied to solve the non-linear system of equations (7). The system can be written in the general form

$$\mathbf{F} = \mathbf{A}\Psi - \mathbf{b} = \mathbf{0} \tag{8}$$

The expansion of $\mathbf{F}$ in terms of the difference between the approximate solutions at the nth and $n+1^{th}$ iterations gives

$$\mathbf{F}^{n+1} = \mathbf{F}^{n} + \left.\frac{\partial \mathbf{F}}{\partial \Psi}\right|^{n} \left(\Psi^{n+1} - \Psi^{n}\right) = 0 \tag{9}$$

The replacement in the above equation of $\mathbf{F}^n$ and the Jacobian $\left.\frac{\partial \mathbf{F}}{\partial \Psi}\right|^{n}$ from equation (8) leads to

$$\mathbf{A}\Psi^{n} - \mathbf{b}^{n} + \left(\mathbf{A} - \left.\frac{\partial \mathbf{b}}{\partial \Psi}\right|^{n}\right)\left(\Psi^{n+1} - \Psi^{n}\right) = 0 \tag{10}$$

which after rearrangement leads to the following linear system of equations

$$\left(\mathbf{A} - \left.\frac{\partial \mathbf{b}}{\partial \Psi}\right|^{n}\right)\Psi^{n+1} = \mathbf{b}^{n} - \left.\frac{\partial \mathbf{b}}{\partial \Psi}\right|^{n}\Psi^{n} \tag{11}$$

where $\left.\frac{\partial \mathbf{b}}{\partial \Psi}\right|^{n}$ is a diagonal matrix. The above linear system is thridiagonal and can be solved by Gauss elimination and back substitution using the fastl Thomas algorithm [13]. An initial approximation for the potential vector Ψ^0 which satisfies the conditions of local electroneutrality usually leads to a stable convergence of the Newton-Raphson method.

The PsHEMT has a well established confining potential in the channel and the classical 1D solution does not contain all the details and the accuracy of the quantum self consistent Posissn-Schrödinger approach. The classical 1D simulations, however, are much easier and faster and can be used for preliminary estimation of the potential of a wide range of structures. Figure 16 illustrates 1D classical results for three different

PsHEMT structures. In Figure 16 *a* acceptor δ-doping on the back side of the channel is used to improve the carrier confinement. In Figure 16 *b* the donor δ-layer is placed only at the back side of the channel and a uniform acceptor stopper is introduced in the buffer layer to reduce short channel effects. Figure 16 c illustrates a PsNEMT structure with a uniform channel doping and acceptor doped buffer.

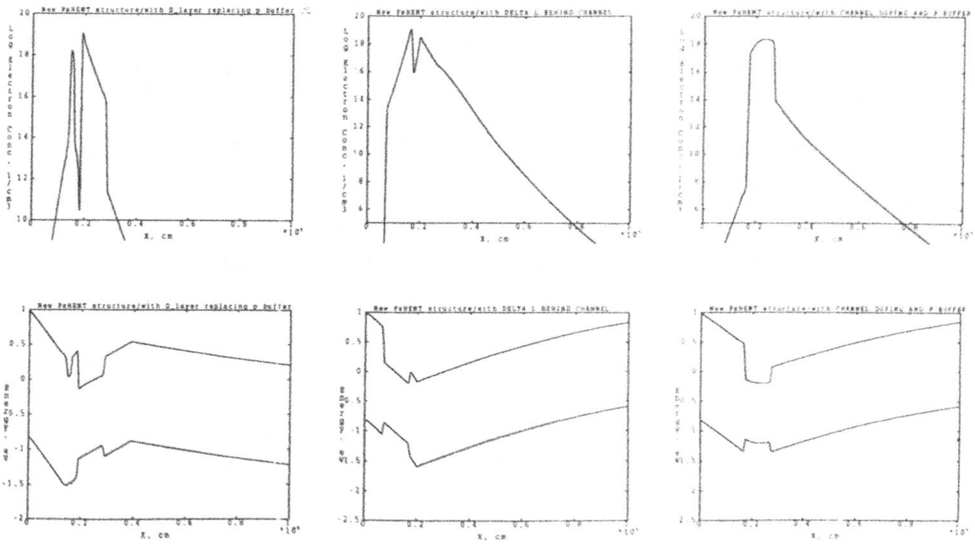

(a) PsHEMT structure with acceptor δ-doping on the back side of the channel is introduced to improve the carrier confinement.

(b) PsHEMT structure with donor δ-layer at the back side of the channel and a uniform acceptor doping in the buffer layer.

(c) Ps HEMT with uniform channel doping and acceptor doped buffer.

Figure 16. Examples of a classical 1D PsHEMT simulation.

3.1.4. *Quantum Mechanical 1D Simulation*

Two major effects lead to discrepancies between the classical and the quantum mechanical 1D simulation of the vertical PsHEMT structure. Firstly, the energy quantisation in the PsHEMT channel reduces the effective bandgap discontinuity at the AlGaAs/InGaAs interface. This affects the charge transfer efficiency and the 2DEG carrier concentration, and leads to smaller channel concentrations than what is predicted by the classical 1D simulations. Secondly, the classical simulation predicts that the carrier distribution in the channel peaks at the AlGaAs/InGaAs interface, although the real quantum mechanical distribution has a maximum several nanometers bellow the interface. This may lead to errors in the classically estimated gate capacitance which increases with the decreasing gate-to-channel separation in the modern shallow PsHEMT structures [14].

In the 1D quantum simulations the Poisson equation is solved self-consistently with the one dimensional Schrödinger equation in an effective mass approximation [15]

$$\left[\frac{-\hbar^2}{2}\frac{d}{dx}\frac{1}{m^*(x)}\frac{d}{dx}+q\psi^T(x)\right]\zeta_i(x)=E_i\zeta_i(x) \tag{12}$$

where m^* is the effective mass which varies in the different layers, E_i and ζ_i are the energy level and the envelope function of the ith subband, and ψ^T is the effective potential which is the sum of the electrostatic potential, the heterojunction potential and the local exchange-correlation potential [16]. The dependence of the envelope functions on the wave vector parallel to the heterointerfaces is usually neglected in such simulations. The electron concentration in the Poisson equation is calculated using the energy levels and envelope functions obtained from the solution of equation (12)

$$n(x)=\sum_i\frac{m^*(x)kT}{\pi\hbar^2}|\zeta_i(x)|^2\ln\left[1+\exp\left(\frac{E_F-E_i}{kT}\right)\right] \tag{13}$$

The differential operator in the left-hand-side of the Schrödinger equation (12) has exactly the same form as the left-hand-side of the Poisson equation (1) and can be discretized applying the finite-difference procedures (4) and (5) on the same non-uniform 1D grid. The energy levels E_i are then obtained by finding the eigenvalues of the discretization matrix. Different numerical eigenvalue methods may be employed for this purpose including the bisectional method [17], QR [18]. or QL [19]. algorithms. The wave functions ζ_i can then be calculated by substituting the corresponding eigenvalues in the discretized equation (12). This leads to a linear set of equations which has a thridiagonal matrix and can be solved using the Thomas algorithm [13].

Typical quantum mechanical PsHEMT simulation results [9] are illustrated in Figure 17. The energy levels, the potential and the charge distribution at $V_G = 0$ V are given in Figure 17 *a*. The ground energy level is approximately 100 meV above the bottom of the InGaAs conduction band. The separation between the ground and the first excited level is of the same order and for negative gate voltages almost all 2DEG electrons occupy the ground level. The corresponding envelope functions are illustrated in Figure 17 *b*. The gate voltage dependence of the total charge and the charge in the InGaAs channel, in the GaAs supply and in the GaAs buffer are given in Figure 17 *c*.

For negative gate voltages almost all electrons are in the InGaAs channel. When the gate voltage goes positive a parallel conductance in the supply layer develops, and the electron concentration in the buffer layer increases.

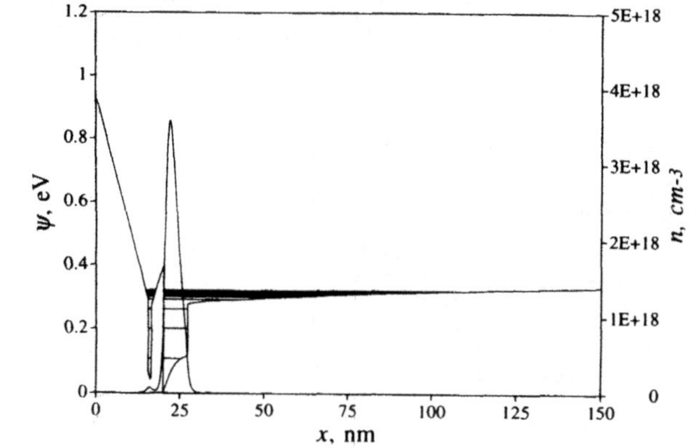

(a) Energy levels, potential, and charge distribution at $V_G = 0$ V.

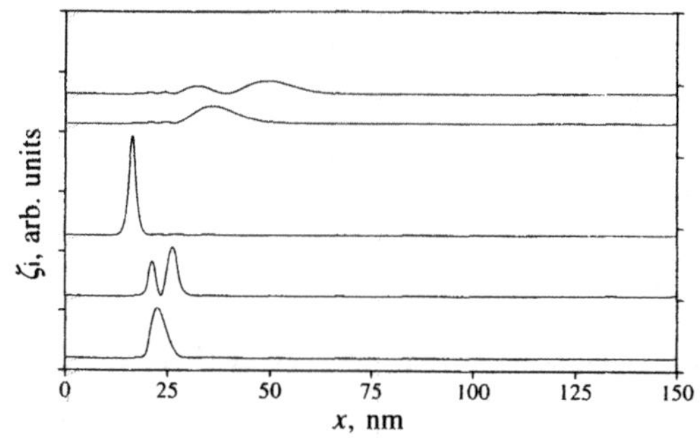

(b) The corresponding envelope functions

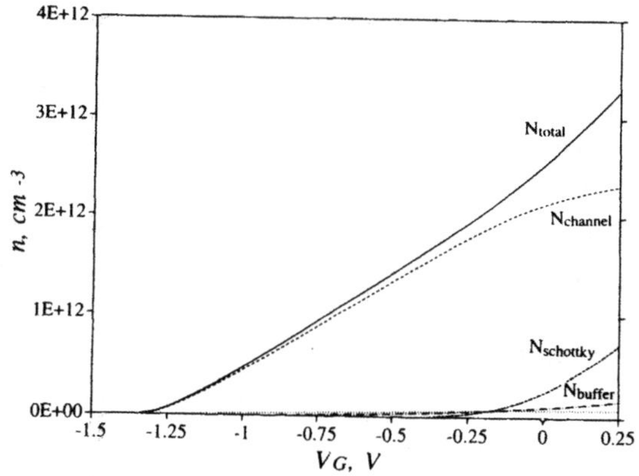

(c) Gate voltage dependence of the total charge and the charge in the InGaAs channel, in the GaAs supply and in the GaAs buffer.

Figure 17. Typical quantum mechanical PsHEMT simulation results

Although there are attempts to introduce quantum mechanics in the 2D numerical PsHEMT simulations they face severe conceptual and computational problems and most of the 2D simulation programs employ classical statistics. The 1D simulation can help to estimate the error which such classical approach introduce in the practical 2D device simulations. Such estimations for an ordinary HEMT structures are published in [20] and summarised in Figure 18.

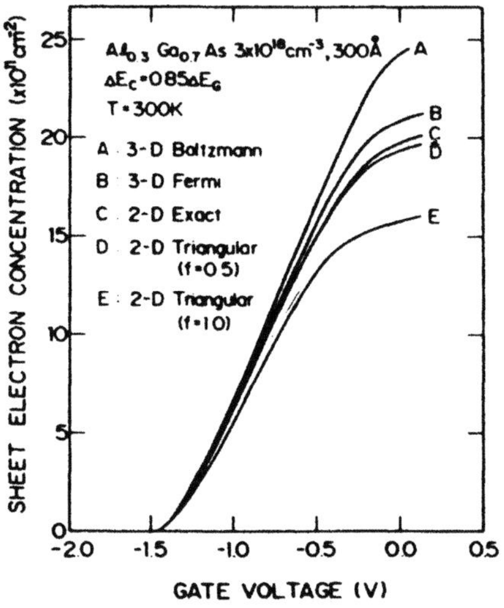

Figure 18. Comparison between the classical and the quantum mechanical simulation approach in 1D simulation of an ordinary AlGaAs/GaAs HEMT.

The error is almost negligible at low gate voltages and increases when the gate voltage approaches zero. It is surprising that the disagreement between the classical simulations with Femi statistics and the quantum simulations is less than 5% over the whole range of gate voltages. In PsHEMTs where the bandgap discontinuity at the channel interface is 2-3 times larger than in the ordinary HEMTs the discrepancy between the classical and the quantum mechanical simulations will be smaller which justifies the use of classical approach in the 2D PsHEMT simulations.

3.1.5. *PsHEMT Performance from 1D Simulations*

The PsHEMT performance can be estimated based on the results of the 1D simulations.. Such an estimation will be accurate for relatively long channel devices. Neglecting of the two dimensional, surface and transport effects in modern short channel PsHEMTs may introduce significant error. From the gate voltage dependence of the charge in the different device layers the overall gate capacitance C_g and the gate to channel capacitance C_{ch} can be calculated

$$C_g = q\frac{dn_{tot}}{dV_g}, \qquad C_{ch} = q\frac{dn_{ch}}{dV_g} \tag{14}$$

The transconductance can be approximate to $g_m \approx C_{ch}v_{av}$ and the cut-off frequency my be expressed as

$$f_T = \frac{C_{ch}}{C_g}\frac{v_{av}}{2\pi L_g} \tag{15}$$

3.2. 2D NUMERICAL SIMULATION

The modern PsHEMT are essentially two dimensional device and 2D numerical simulation is required to properly understand the device operation and to assist the device design. The 2D simulations are based on a self consistent solution of the Poisson and the Boltzmann transport equation in different approximations including DD, HD and MC approaches. Depending on their degree of complexity the 2D simulations can describe the transport effects in PsHEMTs including the velocity overshot [21], the geometry and fringing effects [22], the surface effects in the recess region and the bulk trapping and detrapping effects [23], the device breakdown [24].

In the PsHEMT design process the 2D numerical simulations can help to answer the following questions: How to scale down the device properly in order to avoid short channel effects? How to reduce the device parasitics like serial resistances and coupling capacitances? How the gate fringing effects influence the device performance and the breakdown voltage? To what extend the carrier deconfinement affects the transconductance? How to engineer better the device in order to obtain maximum advantage of the velocity overshoot.

The user, however, should be aware of the uncertainties involved in the 2D PsHEMT simulations. Everything said in Section 3.1. about the uncertainties in the 1D simulations is valid also in the 2D case. In addition the lack of accurate knowledge about the device geometry and the geometry fluctuations during the fabrication my affect the accuracy of the 2D simulation results. The PsHEMT ohmic contacts, which are sill not very well understood pose problems in the simulation. The surface states density and energy position in the recess region and the trap density in the buffer layer which are usually unknown and depend on the fabrication conditions affect the device performance and the accuracy of the simulation results. The lack of accurate knowledge about the band structure is even more important here because it affects the accuracy of the high field transport simulations.

3.2.1. *Geometry Effects on the PsHEMT Parasitics*

Three important factors determine the performance of the modern deep submicrometer PsHEMTs and should be included in the 2D simulations: the nonequilibrium electron transport in the high field regions which leads to a significant velocity overshoot, the device geometry which determines the parasitics and the surface conditions.

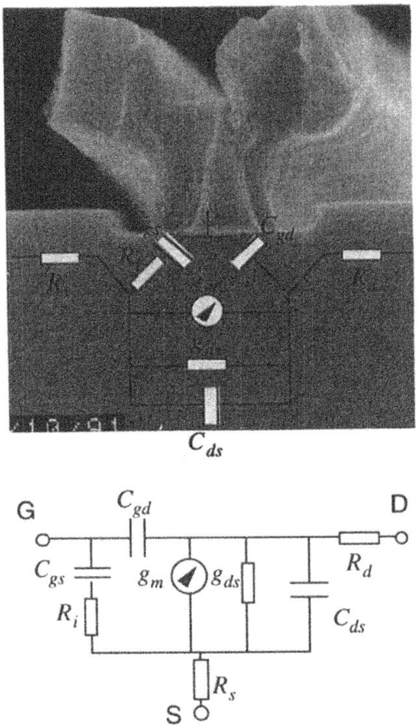

Figure 19. Intrinsic small signal equivalent circuit, superimposed on the schematic picture of a recess gate FET.

The impact of the device geometry and the surface conditions is illustrated in Figure 19 where the intrinsic small signal equivalent circuit is superimposed on the cross-sectional picture of a PsHEMT. The source and drain resistances R_S and R_D depend on the shape and the surface potential pinning in the recess region. The gate-to-source C_{gs} and gate-to-drain C_{gd} capacitances are determined by the shape of the gate and the recess and by the surface conditions. In most of the commercial devices these parasitic capacitances are influenced by a passivation filling the gap between the gate and the recess. In self aligned technologies additional capacitive coupling exists between the gate and the source and drain metal contacts. The reduction of the channel length increases the importance of the parasitics which becomes a major performance limiting factor.

3.2.2. *Velocity Overshoots*

Three types of velocity overshoot effects may affect the operation of the PsHEMT and should be distinguished.

The first type of overshoot, which we will call *static overshoot,* is typical for GaAs based compound semiconductors in which the electrons have small effective mass in the Γ valley and larger effective masses in the higher X and L valleys. It is manifested in the equilibrium drift velocity - field characteristics of an uniform sample with uniform electric field and is illustrated in Figure 20. At low fields almost all electrons occupy the Γ valley where they have a low effective mass, large mobility and velocity. If the field increases some of the electrons acquire enough energy to be transferred to the L valley where they have larger effective mass, lower mobility and velocity. The percentage of electrons occupying the highest valleys grows with the increasing field. The average velocity passes through a maximum value (overshoot) followed by a decrease in average velocity due to the high effective mass in higher valleys. At high fields the velocity saturates as a result of increasingly dissipative inter and intra-valley phonon scattering processes.

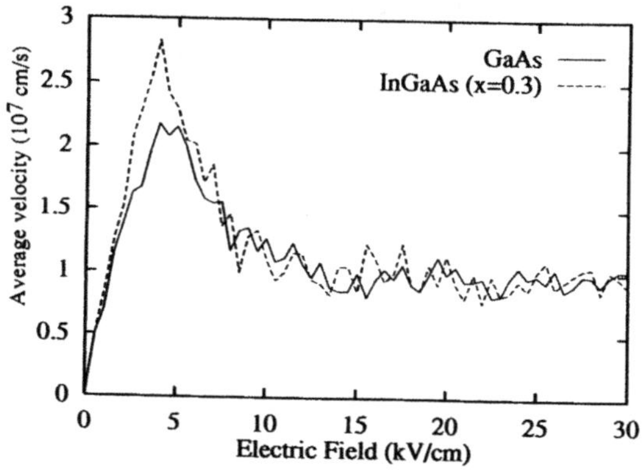

Figure 20. Field dependence of the drift velocity in GaAs and InGaAs

The second type of overshoot, which we will call *spatial overshoot* , may exist in any semiconductor but is more pronounced in semiconductors with small effective masses. It occurs when the carriers enter a region with a sharp increase in the electric field above the critical saturation field. The carriers will travel some distance at the beginning of the high field region with velocities higher than the saturation velocity.

This is related to the probabilistic nature of the scattering mechanisms and to the fact that the carriers travel some distance in order to acquire the energy necessary to trigger the dicipative scattering mechanisms responsible for the velocity saturation. The spatial overshoot region is comparable to the carrier mean free path and in GaAs it extends more than a quarter of a micron. In InGaAs the spatial overshoot region is even larger which means that the carries will travel with velocities higher than the saturation velocity all the way through the 200 - 100 nm channel of the modern PsHEMTs.

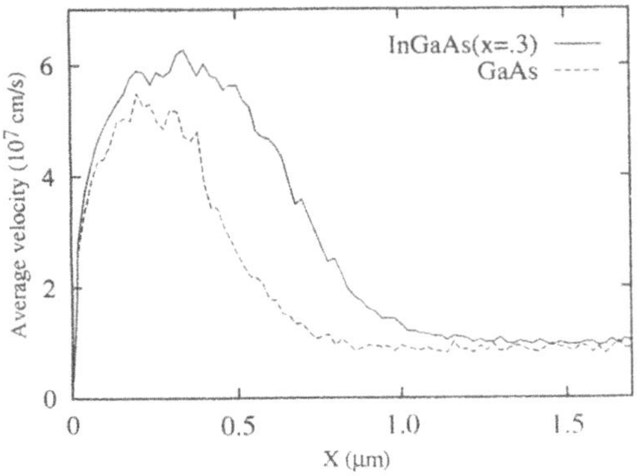

Figure 21. Spatial dependence of the drift velocity along 2 μm GaAs and InGaAs samples with uniform field E = 20 keV obtained form MC simulations. The carriers are injected in equilibrium at x=0.

The third type of overshoot, which we will call *dynamic overshoot*, occurs when the field is almost instantaneously switched on above the critical saturation level. For a short period the carriers will travel with average drift velocity which exceeds the saturation velocity (Figure 22).

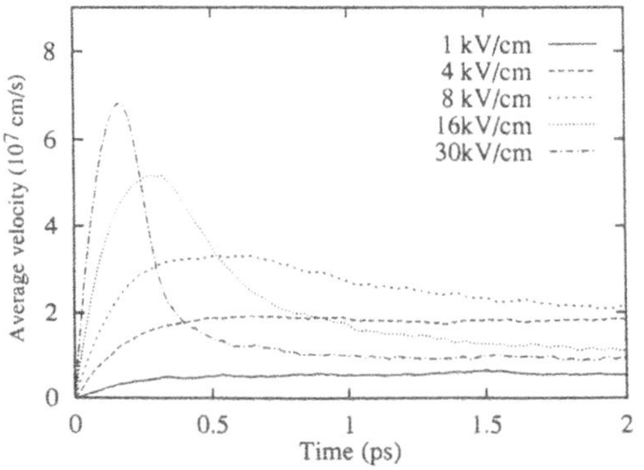

Figure 21. Evolution of the drift velocity with time for different strength of the electric field in a GaAs sample obtained form MC simulations.

This is again related to the probabilistic nature of the scattering and to the fact that the carries cannot acquire instantaneously the energy required to trigger the dissipative scattering processes leading to velocity saturation. In InGaAs all overshoot effects are stronger than in GaAs because the effective mass of the electrons in the Γ valley decreases with the increase of the In content, and the separation between the Γ and L valleys increases.

3.2.3. *The Drift-Diffusion Approach*

Formulation.. The simplest approximation to the Boltzmann transport equation (BTE) is the drift-diffusion (DD) approach. It arises from the zeroth and first moments of the BTE, which are basically statements of particle and momentum conservation. As the name implies the current is split into a diffusive component driven by the local density gradients and a drift component driven by the local electric fields. It is assumed that the carriers are in thermal equilibrium and in equilibrium with the driving field (*i.e.* moving with their equilibrium drift velocity).

In many cases of PsHEMT simulation it is sufficient to solve the Poisson equation self-consistently with only the current continuity equation for electrons in drift-diffusion approximation. At steady state the DD model for compound semiconductor devices is described by the following system of equations [25]

$$\nabla.(\varepsilon\nabla\psi) = -\rho \tag{16}$$

$$\nabla.\vec{J}_n = 0 \tag{17}$$

$$\vec{J}_n = -q\mu_n(E)\left(n\nabla\psi_1 - \frac{kT}{q}\nabla n\right) \tag{18}$$

where μ_n is the electron mobility and the pseudopotential ψ_1 is given by

$$\psi_1 = \psi + \frac{\chi}{q} + \frac{kT}{q}\ln(N_C) \tag{19}$$

Apart from the charge of the ionised acceptors and donors and the charge on deep acceptor and donor type states summarised in N^i_{bulk}, for PsHEMT simulation the charge density in the Poisson equation

$$\rho(\vec{r}) = q\left(N^i_{bulk} + \delta(\perp)N^i_{\text{int}} + p - n\right) \tag{20}$$

should include also the charge on the surface states in the recess region $\delta(\perp)N_{\text{int}}^{i}$ whose occupation depends on the surface potential distribution and the variation in the electron quasi-Fermi level.

In order to simulate impact ionisation related phenomena in PsHEMTs, including avalanche breakdown, the current continuity equation for holes and the generation/recombination terms should be included in the DD system of equation.

The mobility model reproduces the velocity-field characteristics of the different compound semiconductors and hence the static overshot. The local relationship between the mobility and the electric field, however, does not allow carriers to travel in high field regions with velocity higher than the saturation velocity. This means that the 2D drift diffusion simulations cannot predict the spatial overshot which is well pronounced and important in short PsHEMTs. However the DD simulations are much faster than the HD and the MC simulations, and after careful calibration can be used in the PsHEMT design.

Discretization.. For discretization of the 2D Poisson equation (16) standard finite difference and finite element methods can be applied.

The 2D finite difference discretization is a simple generalisation of the 1D finite difference discretization described in Section 3.1.2. The solution domain is covered with a rectangular grid (see Figure 29 *a*). The differential operators in the *x* and *y* directions in the Poisson equation are approximated with finite differences like equations (4,5). This leads to a five diagonal system of nonlinear algebraic equations.

The Galerkin [26] finite element discretization of the Poisson equation may be summarised as follows. The 2D solution domain is first divided into finite elements. which may have not only rectangular but triangular or quadrilateral shape [27]. An example of a quadrilatereal finite element grid is given in Figure 29 *b*. The potential is then approximated using the unknown values ψ_m in the grid nodes. The polynomial interpolation usually used for this approximation can be expressed in terms of shape functions N_m which depend on the type of the finite elements.

$$\psi \approx \hat{\psi} = \sum_{m=1}^{M} \psi_m N_m \tag{21}$$

This is illustrated in an example of 1D finite element interpolation using linear shape functions (Figure 22).

The shape function attached to the node *m* in this example is given by

$$N_m = \begin{cases} 1 - \dfrac{x - x_m}{x_{m+1} - x_m} & x_m \le x \le x_{m+1} \\ 1 - \dfrac{x_m - x}{x_m - x_{m-1}} & x_m \le x \le x_{m+1} \end{cases} \tag{22}$$

In general the shape function attached to a particular node of a finite element is equal to one at this node and is zero on all other nodes and on the element sides which do not belong to the node m.

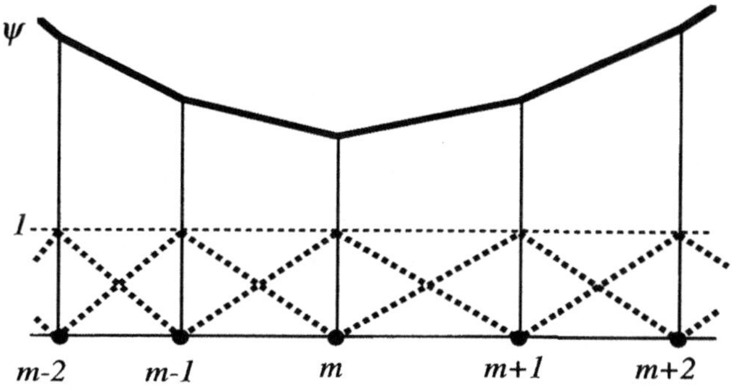

Figure 22. 1D finite element interpolation using linear shape functions.

The Galerkin method requires the residual of the Poisson equation weighted by the shape functions to vanish. Thus the unknown nodal values of the potential may be obtained by solving the system of algebraic equations

$$\int_{\Omega} \left[N_l \left(\nabla . \left(\varepsilon \nabla \hat{\psi} \right) + \rho \right) \right] dxdy = 0 \qquad l = 1, \dots N \tag{23}$$

where N is the number of grid nodes. Due to the properties of the shape functions each equation in the system (23) links usually only the values of the potential in the finite elements which condense in the node l.

The current continuity equation contains first and second order derivatives of the carrier concentration and the straight forward finite difference or finite element discretization described above are unstable. A stable finite difference approximation for the current was proposed by Scharfetter and Gummel [28] and used with some modification in all DD simulation programs. The current density between two points of the grid can be approximated to

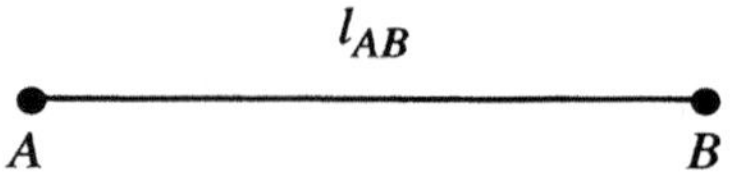

$$J_{AB} = \frac{D_{AB}}{l_{AB}}\left[\text{ber}\left(\frac{q\Delta\psi_{AB}}{kT}\right)n_A - \text{ber}\left(\frac{-q\Delta\psi_{AB}}{kT}\right)n_B\right] \tag{24}$$

where D_{AB} is the average diffusion coefficient, $\Delta\psi_{AB}$ is the potential difference and the Bernoulli function is defined as

$$\text{ber}(x) = \frac{x}{\exp(x) - 1} \tag{25}$$

The expression (24) can be used in a finite-difference approximation of the current continuity equation (17) but more frequently the control volume approach is adopted. It can be applied to a rectangular or triangular grid [29] (Figure 23). The nodes of the grid are surrounded by non overlapping control volumes. In the absence of generation and recombination the total current flowing in the control volume should be equal to zero. The current through each side of the control volume is approximated by the product of the corresponding nodal current density and the side length.

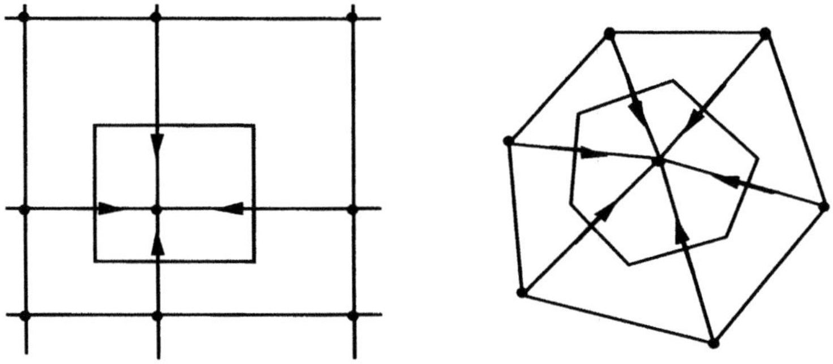

Figure 23. Control volumes in the case of rectangular and triangular grids.

Nonlinear Iterations. Different nonlinear strategies can be applied to solve self consistently the nonlinear set of DD equations (16-18). The most popular among them are the Gummel method [30] and the global Newton method [31].

In the Gummel method the Poisson and the current continuity equation are solved separately. First the discrete 2D approximation of the nonlinear Poisson equation is linearised and solved iteratively. The new approximation for the potential is then used in the discretization and solution of the current continuity equation. The electron quasi-

Fermi level calculated from this solution couples back the continuity equation to the Poisson equation. The algorithm is

Repeat until converge
  Solve the nonlinear Poisson equation
  Solve the current continuity equation

In the global Newton method the algebraic equations arising from the discretization of the Poisson and the current continuity equations are grouped in one nonlinear system of equations

$$\mathbf{F} = \begin{pmatrix} \mathbf{f}_\psi(\mathbf{w}) \\ \mathbf{f}_n(\mathbf{w}) \end{pmatrix} = \mathbf{0} \tag{26}$$

where $\mathbf{f}_\psi(\mathbf{w})$ and $\mathbf{f}_\psi(\mathbf{w})$ are the discrete approximations of the Poisson and the current continuity equations, and $\mathbf{w} = \begin{pmatrix} \Psi \\ \mathbf{n} \end{pmatrix}$ is a vector combining the vectors of the unknown nodal values of the potential and the electron concentration. Variants of the Newton method like the one described in Section 3.1.3 are then applied to solve $\mathbf{F} = \mathbf{0}$. The global Newton method provides faster convergence when the Poisson and the current continuity equations are strongly coupled at high current regime of PsHEMT operation.

Linear solvers. Direct or relaxation methods can be used to solve the linear systems of equations arising in both the Gummel or the Newton nonlinear approaches.

The direct methods are based on variants of Gaussian elimination. This commonly involves an LU factorisation of the coefficient matrix from which the solution can be calculated. The factorisation algorithms usually take advantage of the sparse structure of the discretization matrixes [32]. Ordering methods are used to number the nodes of the grid in such a way to reduce the bandwidth of the system. Pivoting maintains the numerical stability of the factorisation process. An appropriate scaling procedure can be applied to balance the size of the matrix elements prior to the factorisation.

Among the relaxation methods point and block Simultaneous Over-Relaxation (SOR) [33], Alternated Direction Implicit (ADI) [34], and Strongly Implicit Method (SIM) [35] can be used. They, however, are overperformed by the Bi Conjugate Gradient (Bi-CG) methods [36], which after careful preconditioning, using for example incomplete Cholesky decomposition, can deliver an extremely high rate of convergence. Among the modifications of the CG methods the BiCGSTAB algorithm is highly regarded [37].

3.2.4. *The Quasi-Hydrodynamic Approach*

A more sophisticated approximation to the BTE yields the hydrodynamic approach [38]. It can be derived from the BTE by retaining the full momentum conservation equation and by adding an equation for energy conservation, derived from the second moment of the BTE.

Neglecting the convection term in the momentum conservation equation leads to a relatively simple quasi-HD model which may applied in PsHEMT simulations [39] and is described by the following system of equations at steady state

$$\nabla.(\varepsilon \nabla \psi) = -\rho \tag{27}$$

$$\nabla.\vec{J}_n = 0 \tag{28}$$

$$\vec{J}_n = -q\mu_n(T_n)\left(n\nabla\psi_1 - \frac{1}{q}\nabla(kT_n n)\right) \tag{29}$$

$$\nabla.\vec{S} = -\vec{J}.\nabla\psi_1 - \frac{3}{2}nk\frac{(T_n - T_0)}{\tau_e} \tag{30}$$

$$\vec{S} = -\frac{5kT_n}{2q}(\vec{J} + \mu(T_n)nk\nabla T_n) \tag{31}$$

where T_n and T_0 are the electron and lattice temperatures, τ_e is the energy relaxation time which is energy dependent, and $\vec{S}$ is the energy flux. In the case of PsHEMT the charge in the Poisson equation (27) is given by the expression (20).

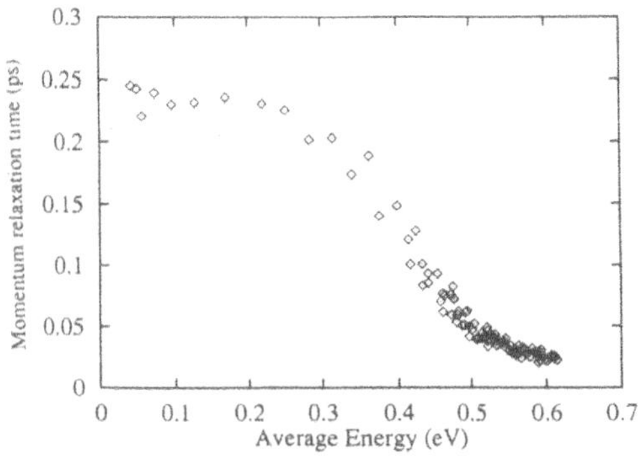

Figure 24. Momentum relaxation time vs average electron energy.

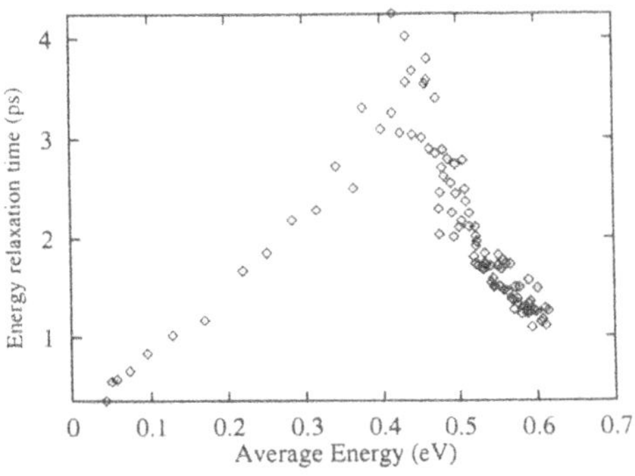

Figure 25. Energy relaxation time vs average electron energy.

In the momentum conservation equation (29) the electron momentum relaxation time and consequently the mobility are related to the average energy of carriers rather than to the local electric field. The average carrier energy lags the electric field and as a result, nonequilibrium effects like the spatial overshoot are included in the quasi-HD simulations. Phenomenological energy-dependent relaxation times are usually used, extracted from steady-state Monte Carlo simulations in uniform electric field. The energy dependence of the momentum relaxation time (resp. mobility) incorporates the change of the effective mass as a result of the $\Gamma \rightarrow L$ transfer. Typical energy dependence of the momentum and energy relaxation times for $In_{0.3}Ga_{0.7}As$ are shown in Figures 24 and 25.

In short channel devices the convection term in the momentum relaxation equation starts to play an important role and the full HD model should be used in the simulations. Recently quantum mechanical corrections have also been introduced in the HD equations [40].

The stable Scharfetter-Gummel discretization scheme (24) can be modified for the momentum conservation equation (29) [41]. Stable discretization schemes for the energy flux in the discretizarion of the energy conservation equation have also been proposed and implemented[42].

The Gummel iterative procedure, modified to solve the quasi-HD system of equations (27-31) can be described by the following algorithm

Repeat until converge
  Solve the nonlinear Poisson equation
  Solve the current continuity equation (28)
  Solve the energy balance equation (30)

Some simulators [43] solve the quasi-HD system in an external loop attached to the modified DD solver which incorporates equation (29). If for example the Newton method is used to solve the modified DD model the following algorithm can be used.

Repeat until converge
  Solve $\mathbf{F} = \begin{pmatrix} \mathbf{f}_{\psi}(\mathbf{w}) \\ \mathbf{f}_{n}(\mathbf{w}) \end{pmatrix} = \mathbf{0}$
  Solve the energy balance equation (30)

3.2.5. *The Monte Carlo Approach*

In contrast to the DD and HD methods which approximate the BTE to different degrees of accuracy, the ensemble Monte Carlo method [44] provides an exact solution to the BTE by following the phase-space trajectories of an ensemble of fictitious particles through the simulated device. The scattering processes are treated as discrete events separated by times-of-flight which are chosen at random using the total scattering rate. The trajectory of a single super particle in the real space is schematically illustrated in Figure 26.

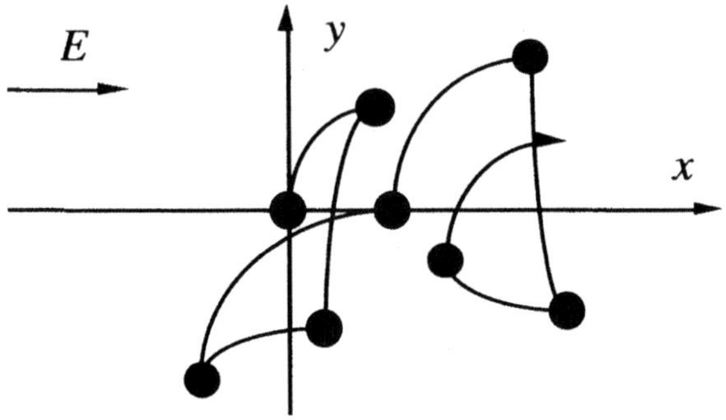

Figure 26. MC trajectory of a superparticle in the real space.

Many such flights for all fictitious particles in the device are used to build up a statistical picture of the device. This is a computationally intensive process because the statistical uncertainty after N measurements varies as $N^{-1/2}$. Thus a high density of data is required throughout phase space although variance-reducing techniques allow some effort to be saved. As well as its application to strongly non-equilibrium transport effects, MC is the only approach to provide *ab initio* information on intrinsic noise properties of a device.

The following scattering mechanisms play important roles in the simulation of PsHEMTs and expressions for some of the corresponding scattering rates may be found in [45]

- Phonons
  - intravalley
    - acoustic (ac)
    - polar optical (po)
      - absorbtion (a)
      - emission (e)
  - intervalley
    - absorbtion (a)
    - absorbtion (e)

Ionised impurity (ii)
Neutral impurity
Carrier-carrier
Impact ionisation
Alloy scattering
Remote impurity scattering

Usually the scattering mechanisms shown in italic above are not included in the conventional PsHEMT simulations. This, however, may lead to significant errors and discrepancy between the MC simulated and the measured device characteristics. Alloy and the remote impurity scattering for example determine the low field mobility in the InGaAs channel and neglecting them will underestimate the series resistance in the intrinsic device. The carrier-carrier scattering affects the mobility at high sheet concentrations typical for PsHEMTs. In order to treat properly the device breakdown in the MC simulations the impact ionisation should be included.

The accumulated scattering rates typically used in PsHEMTs simulations are illustrated in Figure 27. The area between two curves in this figure represent the contribution of any particular mechanism.

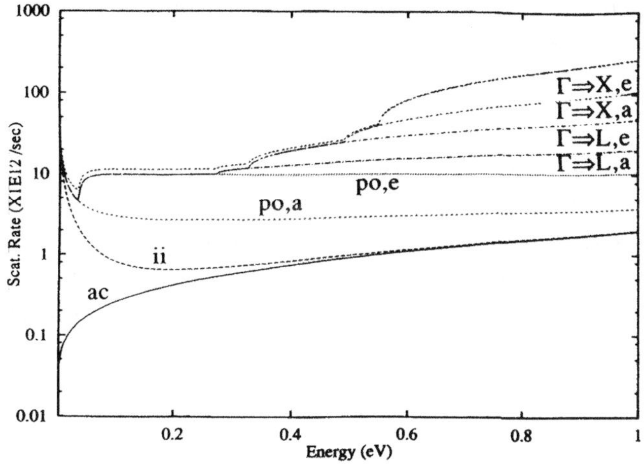

Figure 27. Accumulated scattering rates typically used in PsHEMTs simulations.

The total scattering rate which is the envelope curve in Figure 27 is a sum of the scattering rates of all scattering mechanisms included in the simulations $\Gamma = \sum_i \lambda_i[k(t)]$.

The numerical integration necessary to find the time of free flight from a flat random number r and the total scattering rate Γ is complicated and time consuming [46]. To simplify the procedure self scattering is introduced in most of the MC algorithms to keep the total scattering rate constant

$$\Gamma_0 = \sum_i \lambda_i[k(t)] + \lambda_0[k(t)] = const \tag{32}$$

In this case the time of free flight τ is given by a simple expression

$$\tau = -\ln(r) / \Gamma_0 \tag{33}$$

Once the time of free flight is determined the scattering mechanism terminating the free flight is selected by means of another random number. Finally the final state in the k-space is selected using a third random number and obeying the energy conservation principle in the case of non-elastic scattering mechanism.

3.2.6. *The Simulator H2F*

We have adopted a hybrid approach to the physical simulation of MESFETs and HEMTs which combines comprehensive treatment of the carrier transport and realistic description of the device geometry and the surface effects [47,48]. The corresponding **H**eterojunction **2D** **F**inite Element device simulator H2F which combines drift-diffusion and Monte Carlo engines in a same finite element simulation environment is a practical step towards the application of extensive physics based numerical modelling in the practical engineering of the new generation PsHEMTs.

Grid Generation. In the design of H2F a great deal of attention has been paid to the proper handling of the device geometry and the surface effects. For Poisson's equation, which is common for both the DD and the MC modules of the simulator the simulation domain includes the space above the semiconductor surface including the gate and the contacts. This provides a proper interaction between the charge on the surface states and the spreading surface potential and incorporates all parasitic capacitances. Quadrilateral finite elements have been used for discretization (Figure 28). The grid is generated by appropriate deformation of originally rectangular sub domains (Figure 29).

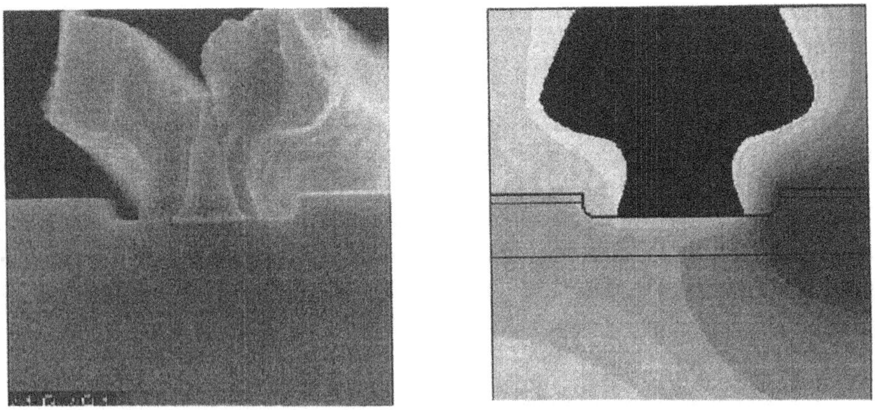

(a) 200 nm MESFET - SEM cross section

(c) Potential distributions at V_G=-0.4 V and V_D=2.5 V

Figure 28. The H2F solution domain.

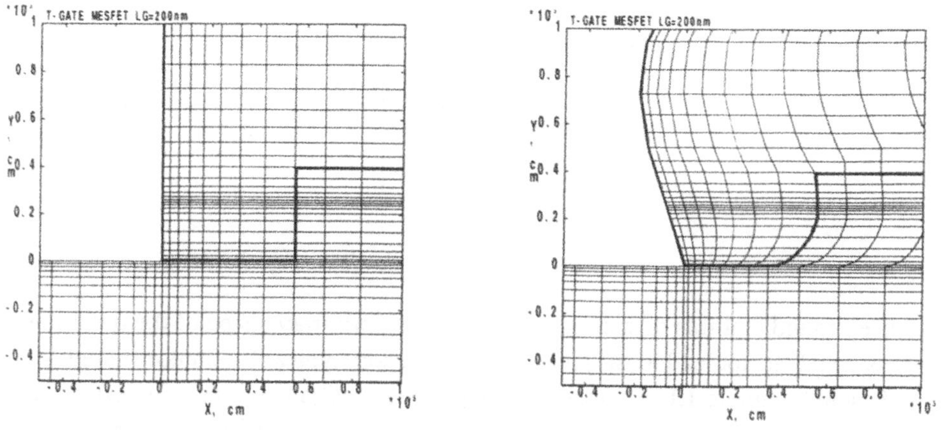

(a) The initial rectangular grid.

(b) The quadrilateral finite element grid obtained by deformation.

Figure 29. Generation of the quadrilateral finite element grid in the recess region.

Discretization and Solution. The Galerkin finite element method has been adapted to solve Poisson's equation. A linear isoparametric mapping (Figure 30) is used in the integration included in equations (23). A control volume method has been developed for the discretization of the current-continuity equation in the DD module [27].

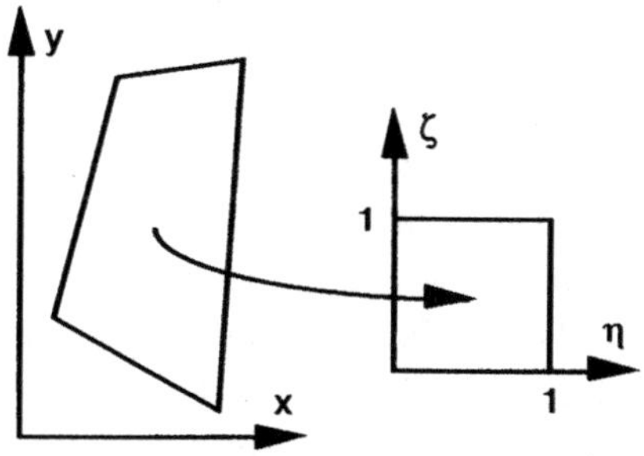

Figure 30. Linear isoparametric mapping.

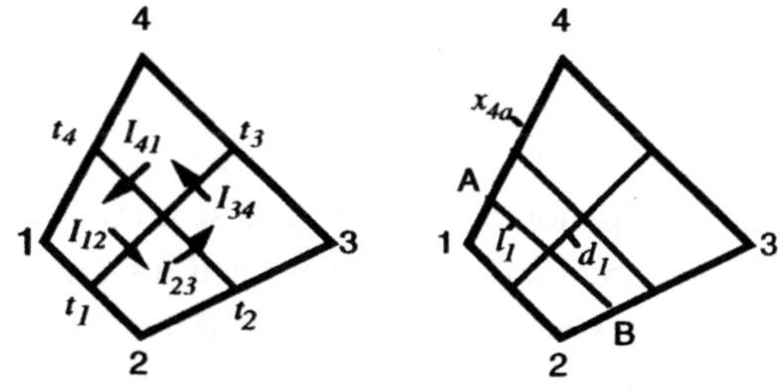

Figure 31. Discretization of the current continuity equation.

The electron current between the points A and B is approximated using the Gummel expression (34). The electron concentration at A and B can be obtained from the nodal values at the corners of the quadrilateral elements by interpolation. It takes into account the exponential variation of the electron concentration when the potential changes linearly. For point A the interpolation is given by

$$n_A = w_{A1}n_1 + w_{A4}n_4 \tag{34}$$

where

$$w_{A4} = \frac{x_{4A}\,\mathrm{ber}\left(\dfrac{q\Delta\psi_{14}}{kT}\right)}{\mathrm{ber}\left(\dfrac{q\Delta\psi_{14}x_{4A}}{kTl_4}\right)} \tag{35}$$

and

$$w_{A1} = 1 - w_{A4} \tag{36}$$

It has been found that this discretization is stable for arbitrary shape of the quadrilateral elements and does not lead to the spikes in the solution, typical for obtuse triangles.

A Fast Incomplete LU Factorisation Biconjugate Gradients (ILUBCG) solver is used for the numerically intensive iterations. The solution of the Poisson's equation involves only a few biconjugate gradient steps per Newton iteration that significantly reduces the total computation time. The convergence problems related to the strongly localised, potential dependent interface charge have been resolved by appropriate dumping. ILUBCG also solves without complication the discretized current continuity equation.

The Monte Carlo Module. A complete ensemble Monte Carlo procedure in a single quadrilateral element is the building block of the whole Monte Carlo module [49]. Apart form its FE implementation the MC procedure is standard. The time of free-flight and subsequent scattering mechanisms are chosen using the usual self-scattering scheme. The potential distribution is recalculated and the electric fields updated approximately every 1 to 5 fs depending on the minimum element size and the doping concentration in the heavily doped cap layers of our devices. The scattering mechanisms implemented include: ionised impurity, acoustic phonon, piezo-electric, optical phonon and polar optic scattering modes. A non-parabolic three valley (Γ,L,X) conduction band model was used for the III-V materials. The FE MC engine works in the same simulation domain as the drift-diffusion solver of H2F and uses the same FE discretization. Each finite element contains a single material and has uniform doping distribution and a local scattering table attached to it. The electric field also is considered uniform across the element which makes it possible to move particles inside each cell in a way similar to bulk material simulation where the trajectory of each particle is traced, both in momentum and real space. The boundary of the element may be homo-, hetero-interface, contact or line of symmetry, and particles reaching the boundary are transferred or reflected according to the local transition probability.

Typical H2F MC simulation domains and the superparticles distributions are given in Figure 32.

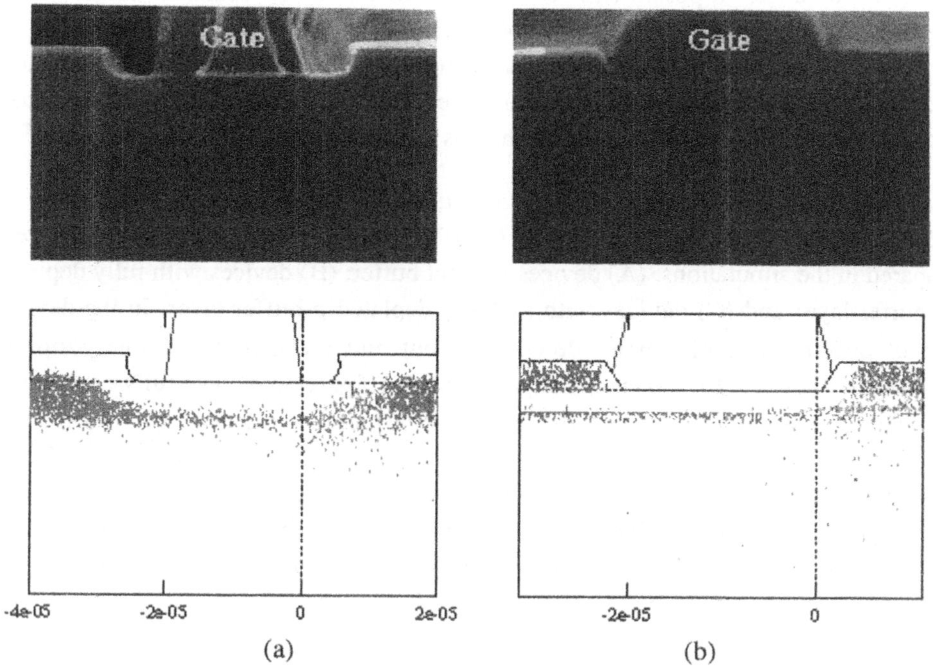

Figure 32. SEM cross sectional view and the H2F simulation domain with of (a) 200 nm gate length MESFET and (b) PsHEMT.

Advanced Monte Carlo ohmic contact model has been developed for the heterojunction multi-layer contacts in PsHEMTs. In the ohmic contact region the number of particles in each of the hetero-layers is kept constant and in thermal equilibrium.

The generalised surface trap model in H2F includes acceptor and donor like traps with an arbitrary energy position whose occupation depends on the quasi-Fermi level and the surface potential variation. The model is self-consistently coupled to the DD module in H2F. In the MC simulations fixed surface charge is used to emulate the surface potential pinning. The spatial variation of the surface charge which keeps the surface potential pinned to a certain value is estimated from a DD simulation at the same bias conditions. To have a uniform surface potential pinning in the recess region the spatial variation of the surface charge must be considered.

Implementing SPMD parallel approach [50] we are running H2F in parallel on Parsytec 64 Supercluster and an 8 node Power PC 601 Parsytec X-plorer (~ 0.6 MFlops). This makes it possible to use the MC simulations for practical design work, generating in parallel the steady state I-V characteristics. The MC simulation time to reach equilibrium starting from the initial conditions and to accumulate statistically acceptable results for the current is approximately 6 ps which corresponds to two hour processor time on a single Power PC. Eight bias points are calculated simultaneously.

Simulation Examples. The use of DD simulations with H2F are first illustrate in an investigation of the doping effects in the PsHEMT buffer. Three types of devices were compared in the simulations: (A) devices without buffer; (B) devices with fully depleted δ-*p*-buffer layer and (C) devices with partially depleted *p*-buffer layer. In the devices without buffer (A) both carrier de-confinement and electrostatic dipole coupling between the drain and the channel charge take place (Figure 33 *a*). The fully depleted δ-*p*-buffer dramatically improves the carrier confinement but cannot prevent the electrostatic drain influence on the potential and the charge control in the channel. Finally the partially depleted *p*-buffer improves the confinement and at the same time provides the necessary negative charge to terminate the drain field lines by expanding the depletion region from drain towards the source (Figure 33 *b*): this reduces the electrostatic coupling between the drain and the channel charge.

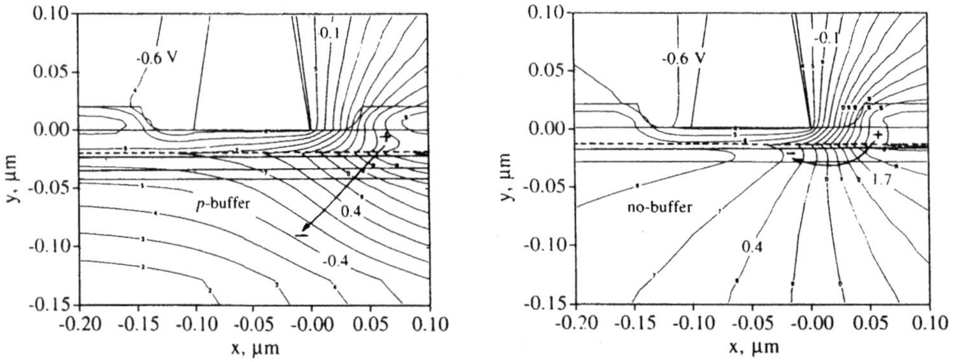

(a) No buffer doping. (b) Partially depleted buffer.

Figure 33. Potential distribution in PsHEMTs with different buffer doping.

The simulated dependence of the threshold voltage vs the channel length for the three types of devices (A), (B) and (C) is illustrated in Figure 34. It is clear that the confinement improvement in device (B) cannot prevent the drain-channel coupling and the threshold voltage reduction. The partially depleted *p*-buffer (C) is needed to stop the drain electrostatic influence and to reduce appreciably the short channel effects. The measured threshold voltage dependence in (A) and (C) type PsHEMTs is given in Figure 35.

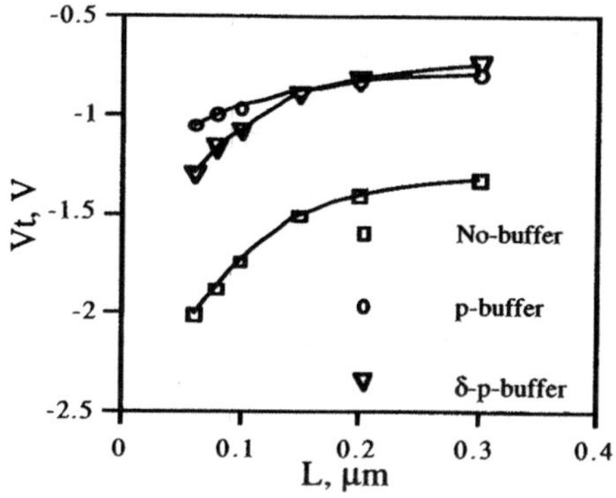

Figure 34. Calculated threshold voltage of the three types of devices (A), (B) and (C)

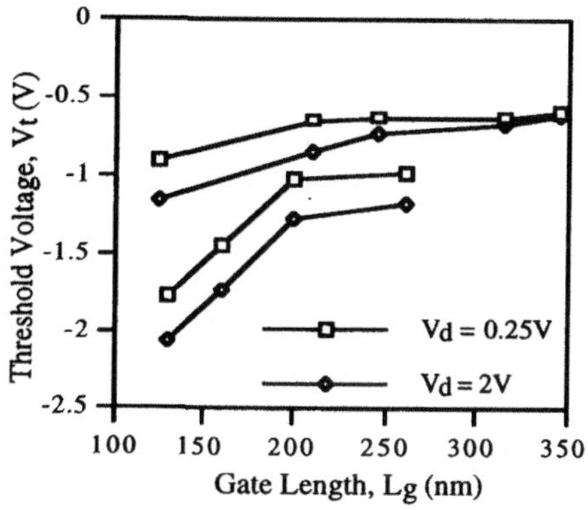

Figure 35. Measured threshold voltage of PsHEMTs without and with partially depleted *p*-buffer

Special care should be taken to incorporate the series resistances associated with the ohmic contacts and the RF probes of the measurement equipment in the MC simulation results [7]. This will allows a direct comparison with measurements and calibration of the MC simulation. Using such approach in combination with MC and DD simulations we have obtained a clear evidence for a large velocity overshoot ($v_{eff} > 3\text{x}10^7$ cm/s) in our latest design 200 nm PsHEMTs described in Section 2 and illustrated in Figure 32 *b*.

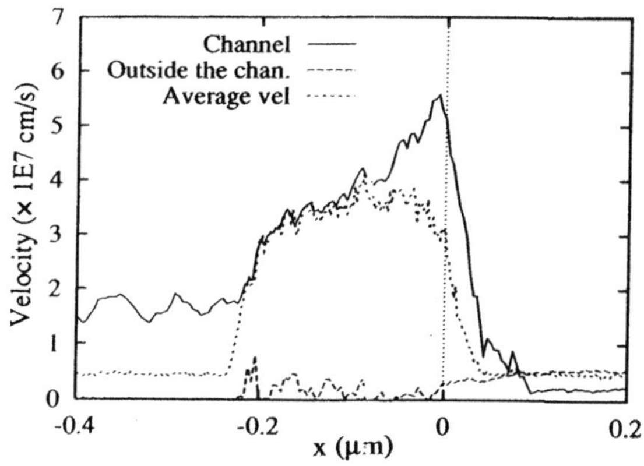

Figure 36. The average longitudinal velocity in the channel and outside the channel of the 200nm PsHEMT, V_G=0, V_D=1.5V

The calculated from the MC module average longitudinal velocity distribution in the channel of the 200 nm PsHENT is illustrated in Figure 36. Both the MC and the DD simulations with such high velocity in the channel lead to a current which is much higher than the experimentally measured (Figure 37). The agreement between the MC calculate I-V characteristics corresponding to such high velocity and the measured one is, however, good if the external to the MC simulations series resistances associated with the ohmic contacts and the measurement equipment are properly accounted for. The predicted high velocity in the channel was also reinforced in DD module which lead to a perfect agreement between the simulations and the measurements (Figure 37). It is clear that very low source series resistance is required in order to benefit from such large overshoot. This is an important challenge to the optimum PsHEMT design.

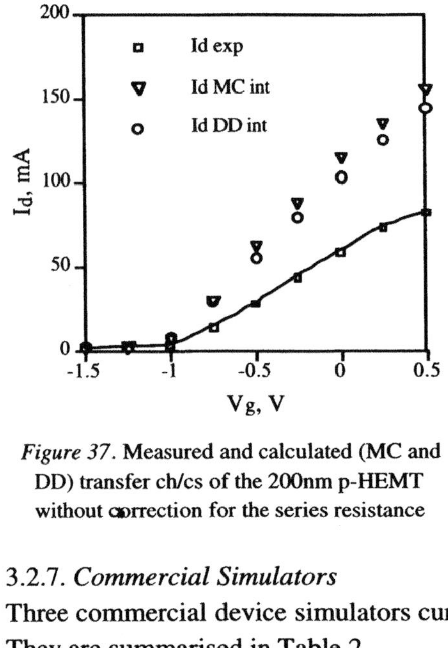

Figure 37. Measured and calculated (MC and DD) transfer ch/cs of the 200nm p-HEMT without correction for the series resistance

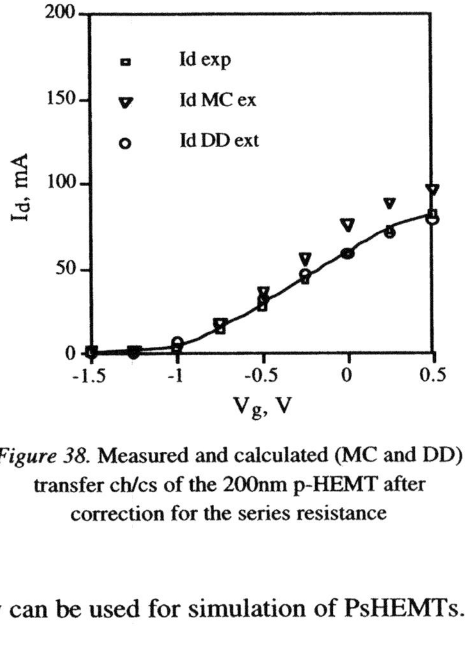

Figure 38. Measured and calculated (MC and DD) transfer ch/cs of the 200nm p-HEMT after correction for the series resistance

3.2.7. *Commercial Simulators*

Three commercial device simulators currently can be used for simulation of PsHEMTs. They are summarised in Table 2.

Table 3. Commercial simulators suitable for PsHEMT simulation

MEDICI (HD AAM)	Technology Modelling Associates (TMA)
ATLAS (BLASE)	SILVACO International
SEMICAD	DAWN Technologies, Inc.

All of them are general type finite element semiconductor device solvers, which can simulate an arbitrary semiconductor device. The different region of the device can consist of different semiconductor materials and may have different shapes. As many electrodes as necessary can be introduced in the solution domain representing the device. All of the listed commercial simulators are restricted to DD and quasi-HD models.

To illustrate the methodology of using commercial simulators for PsHEMT modelling we will present here some results obtained with the Heterojunction Device Advance Application Module (HD AAM) of the commercials simulator MEDICI [51]. To keep the input file short and readable a simple rectangular initial grid is used in this example and no gate recess offset is included The simulation domain and the corresponding MEDICI grid are illustrated in Figure 39.

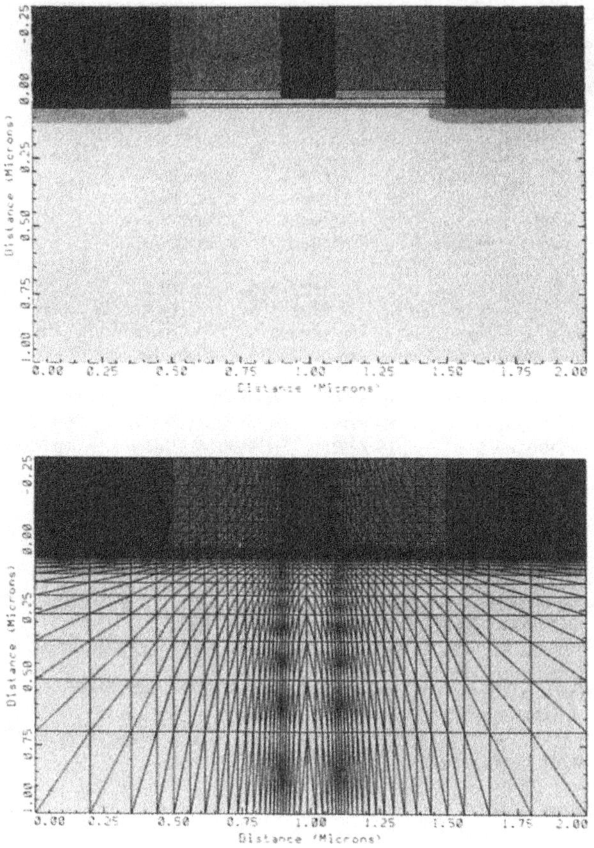

(a) PsHEMT solution domain

(b) MEDICI FE grid

Figure 39. PsHEMT simulation domain and MEDICI

Bellow is the MEDICI input file which generates the PsHEMT structure in Figure 39 used in the MEDICI simulation examples

```
TITLE      pHEMT Example

MESH

X.MESH     X.MAX=0.5     H1=0.2      H2=0.06
X.MESH     X.MAX=0.9     H1=0.06     H2=0.01
X.MESH     X.MAX=1.1     H1=0.01     H2=0.01     H3=0.03
X.MESH     X.MAX=1.5     H1=0.01     H2=0.06
X.MESH     X.MAX=2.0     H1=0.06     H2=0.2

Y.MESH     Y.MIN=-0.3    Y.MAX=0.0   H1=0.1      H2=0.03
Y.MESH     Y.MAX=0.03    H1=0.01     H2=0.005
Y.MESH     Y.MAX=0.0475 H1=0.005    H2=0.002
Y.MESH     Y.MAX=0.0495 H1=0.0005
Y.MESH     Y.MAX=0.052   H1=0.00125
Y.MESH     Y.MAX=0.062   H1=0.00125 H2=0.002    H3=0.005
Y.MESH     Y.MAX=1.0     H1=0.005    H2=0.3

REGION     NUM=1     Y.MIN=-0.3     Y.MAX=0.0      INSULATOR
REGION     NUM=2     Y.MIN=0.0      Y.MAX=0.03     ALGAAS     X.MOLE=0.0
REGION     NUM=3     Y.MIN=0.03     Y.MAX=0.052    ALGAAS     X.MOLE=0.3
REGION     NUM=4     Y.MIN=0.052    Y.MAX=0.062    ALGAAS     X.MOLE=0.0
REGION     NUM=5     Y.MIN=0.062    Y.MAX=1.0      ALGAAS     X.MOLE=0.0

ELECTR     NUM=1     Y.MIN=-0.3     Y.MAX=0.07     X.MIN=0.0     X.MAX=0.5     VOID
ELECTR     NUM=2     Y.MIN=-0.3     Y.MAX=0.07     X.MIN=1.5     X.MAX=2.0     VOID
ELECTR     NUM=3     Y.MIN=-0.3     Y.MAX=0.03     X.MIN=0.9     X.MAX=1.1     VOID
ELECTR     NUM=4     BOT

PROFILE    Y.MIN=0.0485     Y.MAX=0.0485     N-TYPE     DOSE=5E12     Y.CHAR=0.0005
PROFILE    Y.MIN=0.0        Y.MAX=0.03       N-TYPE     N.PEAK=4.0e18     UNIFORM
PROFILE    Y.MIN=0.03       Y.MAX=1.0        P-TYPE     N.PEAK=1.0e14     UNIFORM
PROFILE    Y.MIN=0.0        Y.MAX=0.1        X.MIN=0.0     X.MAX=0.5
+          N.PEAK=1e20      N-TYPE           UNIFORM
PROFILE    Y.MIN=0.0        Y.MAX=0.1        X.MIN=1.5     X.MAX=2.0
+          N.PEAK=1e20      N-TYPE           UNIFORM

MATERIAL   REGION=1   PERMITTI=1.0
MATERIAL   REGION=2   PERMITTI=12.9    EG.MODEL=0    EG300=1.425    AFFINITY=4.070
+          NC300=4.35E17    NV300=8.21E18
MATERIAL   REGION=3   PERMITTI=12.03   EG.MODEL=0    EG300=1.805    AFFINITY=3.747
+          NC300=1.00E18    NV300=9.51E18
MATERIAL   REGION=4   PERMITTI=13.58   EG.MODEL=0    EG300=1.011    AFFINITY=4.318
+          NC300=3.38E17    NV300=7.60E19
MATERIAL   REGION=5   PERMITTI=12.9    EG.MODEL=0    EG300=1.425    AFFINITY=4.070
+          NC300=4.35e17    NV300=8.21e18

CONTACT    NUMBER=3    WORKFUNC=4.670

MODELS     FLDMOB

MOBILITY   REGION=2   MUN0=2000   FLDMOB=1   VSATN=1.0E7    BETAN=2.0
MOBILITY   REGION=3   MUN0=2000   FLDMOB=1   VSATN=1.0E7    BETAN=2.0
MOBILITY   REGION=4   MUN0=5000   FLDMOB=1   VSATN=3.5E7    BETAN=2.0
MOBILITY   REGION=5   MUN0=2000   FLDMOB=1   VSATN=1.0E7    BETAN=2.0
```

The grid generation reproduces the device structure of the Glasgow PsHEMTs described in Section 2. Since InGaAs is not available in the material library of our current version of MEDICI AlGaAs is used instead in all active layers of the device in the REGION input statement. Later in the MATERIAL statement the parameters of each layer including the InGaAs channel are redefined changing the default AlGaAs values.. A simple velocity saturation model is used in the simulations [*] but in order to include the spatial overshoot in the channel the saturation velocity in the InGaAs layer is set to 3.5x10^7 cm/s. The latest value represents the average velocity in the channel estimated from the MC simulation of the same device. The δ-doping is introduced by a Gaussian distribution in the PROFILE statement. The ohmic contacts which extends to the buffer through all active layers are heavily doped.

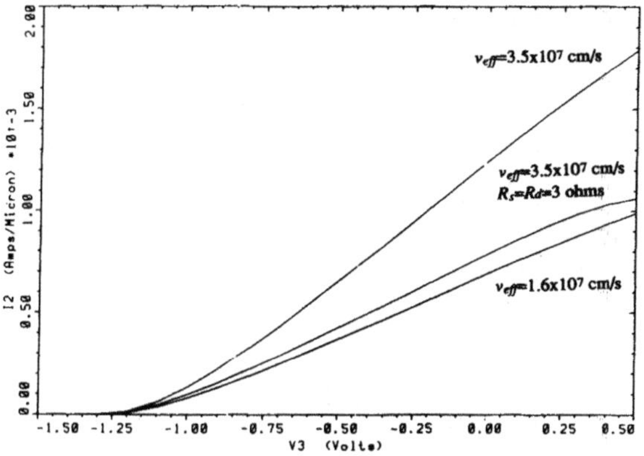

Figure 40. MEDICI generated PsHEMT ID-VG characteristics with and without contact resistance included in the simulations.

The I-V characteristics generated with MEDICI are illustrated in Figure 40. They are very similar to those obtained with our simulator H2F (see Figures 37 and 38). If only the intrinsic device is simulated the current obtained exceeds almost twice the actual measured current in the device. If, however, the contact and the measurement equipment resistances are introduced in the simulations the current drops to the proper level. The third curve in Figure 40 represents simulations without the contact and the measurement equipment resistances but with much lower average velocity in the channel of 1.6x10^7 cm/s. This result is very instructive and shows how neglecting the series resistances in the PsHEMT simulation can lead to apparently low average velocities in the device channel.

3.2.8. *Transient 2D Simulations.*

Background. The transient DD, HD and MC simulations provide information about the switching waveforms and the small signal RF performance of the PsHEMT. The transient MC simulations are the only source providing *ab initio* information on intrinsic noise properties of a device. In the PsHEMT design process the transient simulations answer the questions how the vertical layer structure and the device geometry affect the current gain G_c and the cut off frequency f_T, the power G_p gain and the frequency at which it fall to unity f_{max} , the frequency dependence of the minimum noise figure F_{min}. Parameters of the small-signal PsHEMT equivalent circuit may be extracted from transient simulations for the purposes of circuit design [52].

It should be noted however that all uncertainties related to the 2D steady state simulation will affect also the 2D transient simulations. Further to those at high frequencies the trapping and detrapping dynamics of the impurity and the defect states may affect the results of the transient simulations but very little research has been done in this area. For a device operating in nearly Terahertz regime the electromagnetic effects may have significant impact and should be included in the simulations through the Maxwell equations.

The time dependent version of the simple single carrier DD PsHEMT model described in Section 3.2.3 includes a time dependent version of the current continuity equation

$$\nabla.\vec{J}_n - q\frac{\partial n}{\partial t} = 0 \tag{37}$$

The time dependent HD simulations require more rigorous treatment [53] and the different variants of time dependent HD models are beyond the scope of this chapter. The MC simulations are inherently time dependent following the trajectory of the super- particles within the device.

Examples. Typical MC simulated switching characteristics of a the PsHEMT described in Section 2. (see also Figure 32 *b*) are shown in Figure 41. The transient simulations are carried out by first establishing the steady-state conditions for V_D = 1.5 V and V_G = -0.5 V. This is then followed by changing the gate voltage to V_G=-0 V. The evolution of the drain current following the gate perturbation is shown for three different cap layer donor concentrations. Terachertz oscillations are clearly observed in the transient region. The amplitude of the current swings and the frequency of the oscillations were found to be independent of the mesh size, the simulation time step and the total number of particles employed in the simulation. The Fourier transform of the drain current

(Figure 42) indicates the presence of secondary peaks which may correspond to modes generated at regions of different charge densities and different effective masses.

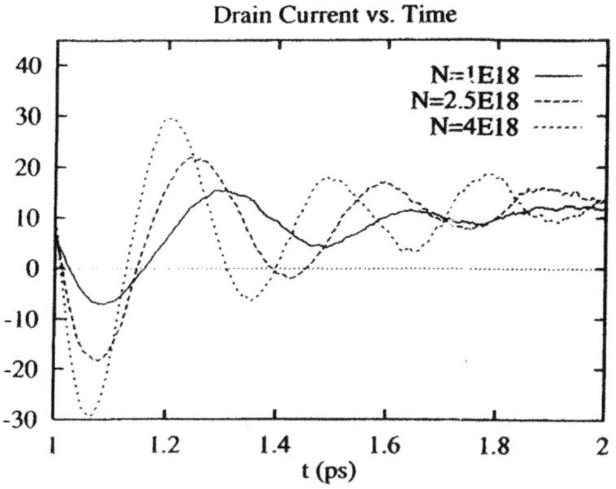

Figure 41. Switching behaviour of PsHEMT obtain with MC simulation.

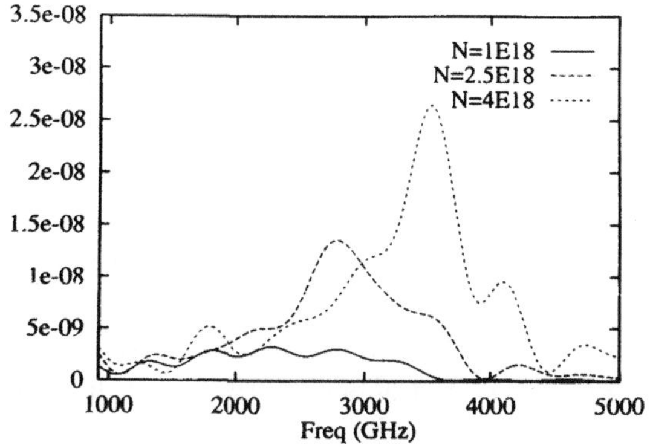

Figure 42. Fourier transforms of the transient currents from Figure 41.

The physics of the plasma oscillations is not included in the DD simulations and the fine details of the transient can not be resolved. However the DD simulations include the charging and discharging effects and can be a good approximation from which the small signal PsHEMT parameters can be extracted. It is also easy to introduce external circuit element like series resistances and coupling capacitances as a boundary conditions in the DD simulations. Typical switching characteristics simulated with MEDICI are given in Figures 43 and 44 for the PsHEMT from above MC simulation.

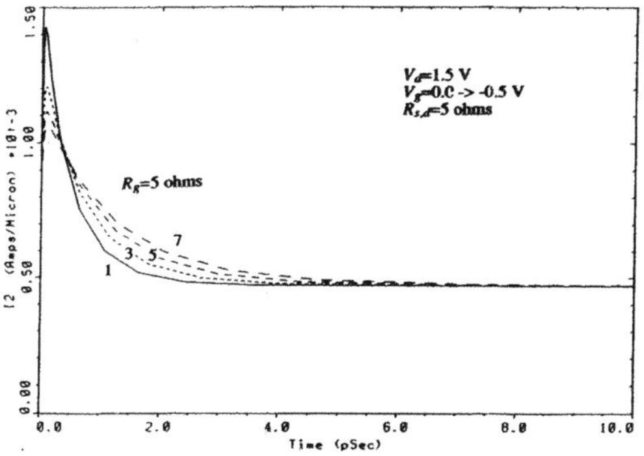

Figure 43. Switching behaviour of PsHEMT obtain with MEDICI simulation for step change in the gate voltage.

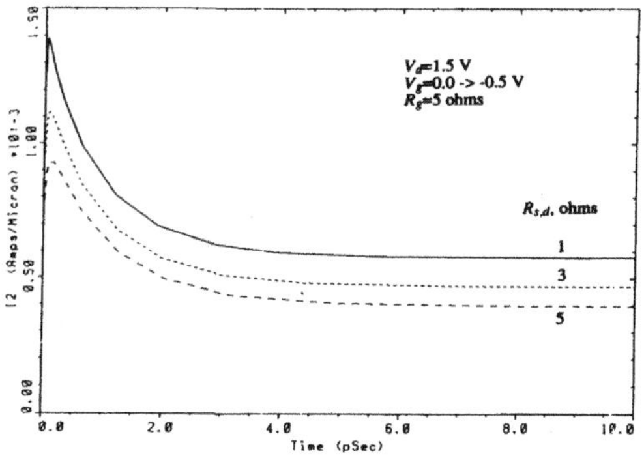

Figure 44. Switching behaviour of PsHEMT obtain with MEDICI simulation for step change in the drain voltage.

Extraction of Equivalent Circuit Parameters. The two port Y PsHEMT parameters can be extracted from both MC and DD transient simulations for a small step change in the drain and the gate voltages [54] using the following relationships

$$\mathrm{Re}\left[Y_{ij}(\omega)\right] = \frac{I_i(\infty) - I_i(0)}{\Delta V_j} + \frac{\omega}{\Delta V_j}\int_0^\infty \left[I_i(t) - I_i(\infty)\right]\sin\omega t dt \tag{38}$$

$$\mathrm{Im}\left[Y_{ij}(\omega)\right] = \frac{\omega}{\Delta V_j}\int_0^\infty \left[I_i(t) - I_i(\infty)\right]\sin\omega t dt \tag{39}$$

The Y parameters extracted from transient simulations near the maximum gain operating point of our PsHEMTs are illustrated in Figure 45.

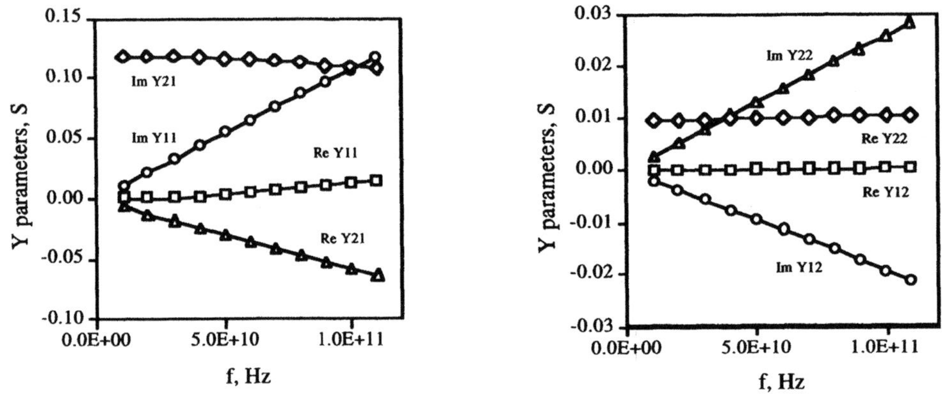

Figure 45. Y PsHEMT parameters extracted from DD MEDICI simulations.

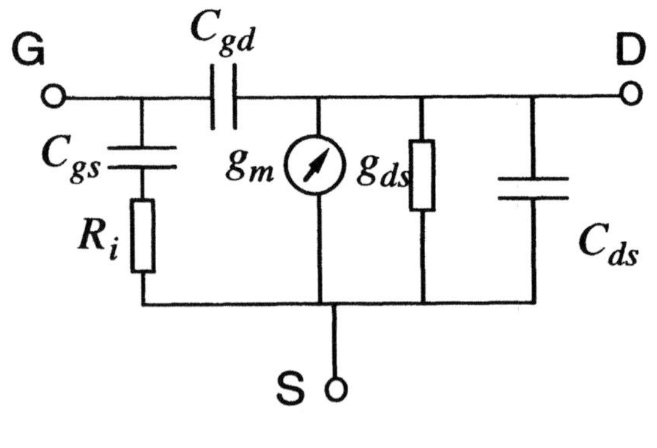

Figure 46. Small signal intrinsic equivalent circuit of a PsHEMT.

The Y parameters are related to the equivalent circuit parameters of the simple intrinsic device equivalent PsHEMT circuit shown in Figure 46 by the following relationships

$$Y_{11}(\omega)=\frac{R_i C_{gs}^2 \omega^2}{D}+j\omega\left(\frac{C_{gs}}{D}+C_{gd}\right) \tag{40}$$

$$Y_{12}(\omega)=-j\omega C_{gd} \tag{41}$$

$$Y_{21}(\omega)=\frac{g_{mo}\exp(-j\omega t)}{1+j\omega R_i C_{gs}}-j\omega C_{gd} \tag{42}$$

$$Y_{22}(\omega)=g_{ds}+j\omega(C_{ds}+C_{gd}) \tag{43}$$

where $D=1+\omega^2 C_{gs}^2 R_i^2$ .The system of equations (40 - 43) can be solved to provide analytical expressions for all intrinsic circuit parameters [55]. This approach has been used to extract the equivalent circuit parameters shown in Figure 47 from the Y parameters shown in Figure 46. The extracted parameters are almost frequency independent up to 100 GHz indicating that the intrinsic equivalent circuit is still valid.

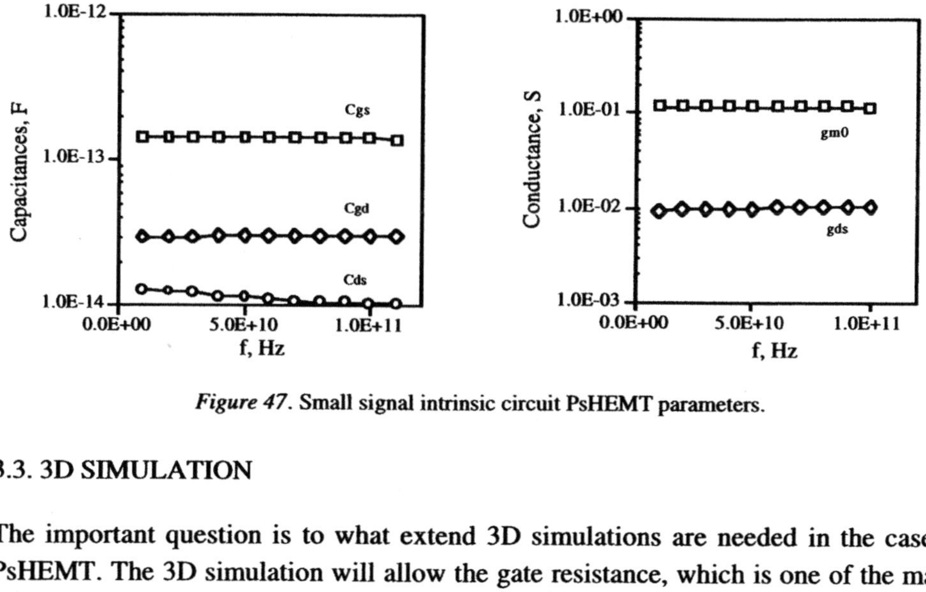

Figure 47. Small signal intrinsic circuit PsHEMT parameters.

3.3. 3D SIMULATION

The important question is to what extend 3D simulations are needed in the case of PsHEMT. The 3D simulation will allow the gate resistance, which is one of the major RF performance limiting factors, to be incorporated properly in the simulations. The influence of the gate resistance dictates the fabrication of narrow gate multifinger

PsHEMTs and 3D simulation is require the to study effect of the channel width on the performance of multifinger devices. The 3D simulation can also be used to study the effect of lithography and etching fluctuation in the channel length and the recess on the device performance. The random distribution of a countable number of discrete impurities, which is now believed to be the major restriction in the device miniaturisation, may be included only in 3D simulations [56]. This is schematically illustrated in Figure 48 in the case of a simple MESFET. The potential distribution in such a device is illustrated in figure 49.

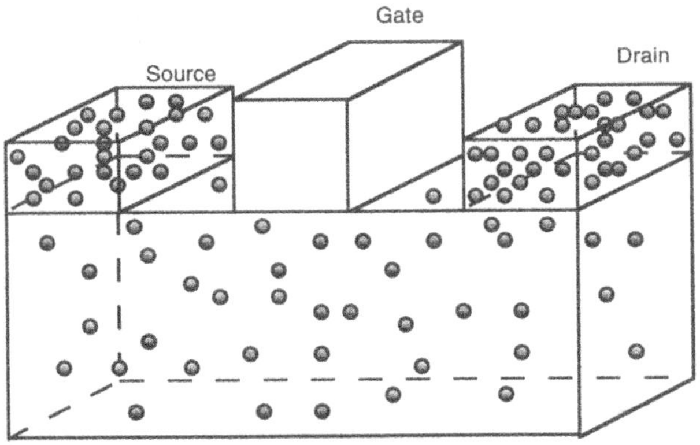

Figure 48. Schematic illustration of the discrete nature of electric charges in a very small MESFET.

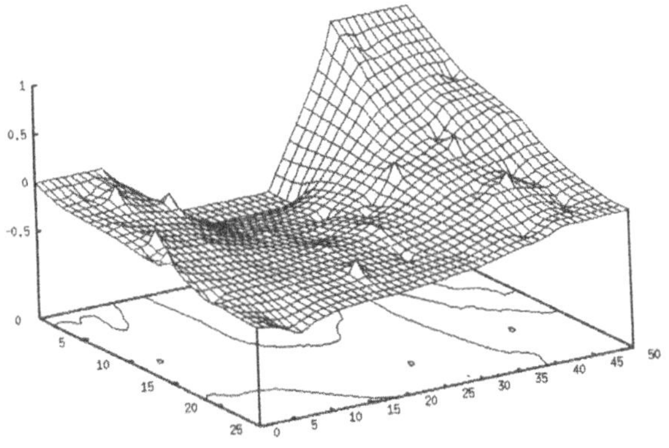

Figure 49. The potential distribution in the middle of the device illustrated in Figure 48.

The practical 3D DD, HD simulations are, however, computationally and memory expensive and are feasible only on relatively large parallel processing systems.

References

1. Mc.Elhnney, M. and Stanley, C.R. (1993) *Electronics Letters* **14**, 1302-1303.
2. Cameron, N.I., Taylor, M.R.S., McLelland, H., Holland, M., Thayne, I.G., Elgaid, K, and Beaumont, S.P. (1995) *IEEE Microwave Theory and Techniques Symposium Digest*, Orlando, FL 435-443.
3. Tedesco, C., Zononi, E., Canali, C., Bigliardi, S., Manfredi, M., Streit, D.C., Anderson, W.T. (1993) *IEEE Trans. Electron Dev.* **40,** 1211-1214.
4. Asenov, A., Babiker, S., Cameron, N., and Beaumont, S.P. (1995) IEE Colloquium on Terahertz Technology Digest, Ref. No: 1995/229, 2/1-2/6.
5. Cameron, N.I., Ferguson, S, Taylor M.R.S., Beaumont, S.P. and Holland, M. (1993) *J. Vac. Sci. Technol.* **B 11**, 2244-2248.
6. Murad, S. K., Cameron, N. I., Beaumont, S. P., and Wilkinson C. D. W. (1995) *MicroelectronicEnigineering* **27**, 439-444.
7. Babiker, S., Cameron, N., Asenov, A., and Beaumont, S.P. (1995), *Proc. ESSDERC'95*, Eds. H.C. de Graaf, H. van Kraneburg, Editions Frontiers, 173-176.
8. Verlangieri, P.A., Kuznetsow, M., and Shneider, M.U. (1991) *IEEE Microwave and Guided Wave Let.* **3**, 51-53.
9. Singh, R. and Snowden, C. M. (1993) *Proc. Int. Workshop on Computational Electronics*, ed. C. M. Snowden, University of Leeds Press, 173-176.
10. Snowden, C.M. (1986), *Introduction to Semiconductor Device Modelling*, World Scientific.
11. Krantz, R.J., Mayer, D.C., and Bloss W.L. (1990) *Solid-State Electron.* **33**, 1189-1195.
12. Greenfield, J.A. and Dutton R.W. (1980) *IEEE Trans. Electron Dev.* **ED-27,** 1520-1528.
13. Thomas, L. (1949) *Elliptic Problems in Linear Differential Equations Over a Network*, Columbia Univ. Press.
14. Deuchmann, R., Fisher, C., Sala, C., and Selberherr, S. (1993) in *Simulation of Semiconductor devices and processes*, **5** Eds. Selberherr, S., Stippel, H., Strasser, E., Springer Verlag, 461-464,
15. Ando. T. (1982) J. Phys. Soc. Japan **44**, 3895-3899.
16. Stern. F. Phys. Rev. B **30**, 840-848.
17. Wilkinson, J.H. (1971) *The Algebraic Eigenvalue Problem*, Springer-Verlag.
18. Golub, G.H. and Van Loan, C.F. (1983) *Matrix Computations*, John Hopkins University Press.
19. Peters, G. NPL Algorithms Library, Document No. F2/03/A.
20. Yoshida, J. *IEEE Trans. Electron Dev.* **36**, 154-156.

21. Park, D.H. and Brennan, K.F. (1989) *IEEE Trans. Electron Dev.* **36**, 1254-1263.
22. Asenov, A., Cameron, N.I., Taylor, M.R.S., Holland M., and Beaumont, S.P.(1995) in *Preparation, Properties and Applications of Low Dimensional Semiconductors*, Eds. M. Balkanski and I. Yanchev, Kluver Academic Publishers, 101-103.
23. Horio, K., Satoh, K., and Yamada, T. (1995) *Simulation of Semiconductor Devices and Processes* **6**, Eds. H. Ryssel and P. Pichler, Springer Verlak 78-81.
24. Butel, Y., Hedoire, J., De Jaeger, J.C., Lefebvre, M. and Salmer, G. (1995) *Simulation of Semiconductor Devices and Processes* **6**, Eds. H. Ryssel and P. Pichler, Springer Verlak 360-363.
25. Lundstrom, M.S. and Shuelke, R.J. (1983) *IEEE Trans. Electron Dev.* **ED-30,** 1151-1159.
26. Zienkiewich, O.C. and Morgan, K (1982) Finite Elements and Approximation, John Wiely & Sons.
27. Asenov, A., Reid, D., Barker, J.R., Cameron, N., and Beaumont, S.P.(1993) *Proc. Int. Workshop on Computational Electronics*, ed. C. M. Snowden, University of Leeds Press, 45-49.
28. Scharfetter, D.L., and Gummel, H.K. (1969) *IEEE Trans. Electron Dev.* **ED-30,** 64-77.
29. Buturla, E.M., Cottrell, P.E., Grossman, B.M., and Salsburg, K.A. (1981) *IBM J. Res. Develop.* **25**, 218-331.
30. Gummel, H.K. (1964)*IEEE Trans. Electron Dev.* **ED-30,** 455-465.
31. Ortega, J.M. and Rheinboldt, W.C. (1970) Iterative *Solution of Nonlinear Equations in Several Variables*, Academic Press.
32. Duff, I.S. (1977) *Proc. IEEE 65*, 500-535.
33. Varga, R.S. (1962) *Matrix Iterative Analysis*, Printice Hall.
34. Young , D.M. (1971) Iterative Solution of Large Linear Systems, Academic Press.
35. Stone, H.L. (1968) *SIAM J. Numer. Anal.* **5**, 530-558.
36. Fletcher, R. (1976) *Lecture Notes in Mathematics*, Springer-Verlag.
37. Van de Vorst, H.A. (1992) *SIAM J. Sci. Stat. Comput.* **13(2)**, 631-644.
38. Blotekjaer, K. (1970) *EEE Trans. Electron Dev.* **ED-30**, 38-47.
39. Meinerzhagen, B. and Engl, W.L. (1988) *IEEE Trans. Electron Dev.* **35**, 689-697.
40. Zhou, J-R., Ferry, D.K (1992) *IEEE Trans. Electron Dev.* **39**, 473-478.
41. Rudan, M. and Oden, F. (1986) COMPEL 5, 149-183.
42. Forghieri, A., Guerieri, R., Ciampolini, P., Ghui, A., Rudan, M., and Baccarani, G. *IEEE Trans. Computer-Aided Design* **7**, 231-242.

43. Axelrad, V, Rollins, G, Motzny, S and Jaczynski, M. (1991) *Proc. SISDEP'91.*
44. Jacoboni, and Lugli, P (1989) *The Monte Carlo Method for Semiconductor Device Simulation* , Springer-Verlag.
45. Jacoboni, C. and Reggiani, L. (1983) *Rewiews of Modern Physics* **55**, 645-705.
46. Moglestue, C. (1993) *Monte Carlo Simulation of Semiconductor Devices*, Chapman & Hall.
47. Asenov, A., Reid, D., Barker, J. R., Cameron, N., and Beaumont, S. P. (1993) in *Simulation of Semiconductor Devices and Processes*, Selberherr, S., Stippel, H., Strasser, E., eds.,Springer-Verlag, **5**, 265-268.
48. Babiker, S., Asenov, A., Barker, J.R., and Beaumot (1995) In *Simulation of Semiconductor Devices and Processes* Eds. H. Ryssel and P. Pichler, Springer Verlag **6**, 226-229.
49. Babiker, S., Asenov, A., Barker, J.R., and Beaumont, S. P (1996) *Solid State Electronics*, in press.
50. Barker, J.R., Asenov, A., Brown, A.R., Cluckie, J., Babiker S., and Atrokianathan, C.R. (1994) in *Massively Parallel Processing, Applications and Development*, Eds. L. Dekker, W. Smith, J.C. Zuidervaart, Elsevier, 683-690.
51. MEDICI, Technology Modelling Associates, Palo Alto CA.
52. Laux, S.E. (1985) *IEEE Trans. Electron Dev.* **ED-32** 2028-2037.
53. Yoganathan, S. and Banerjee, S.K. (1982) *IEEE Trans. Electron Dev.* **39,** 1578-1587.
54. Gonzalez, T and Padro, D. *IEEE Trans. Electron Dev.* **42,** 605-611.
55. Shirakava, K., Oikava, H., Shimura, T., Kawasaki, Y., Ohashi, Y., Saito, T., and Daido, Y. (1995) *IEEE Trans. Microwave Theor. and Techniques* **43**, 504-510.
56. Wong, H.-S. and Taur, Y. (1993) *IEDM Tech. Digest* 29.2.1-29.2.4.

Subject Index

GPSR Compliance

The European Union's (EU) General Product Safety Regulation (GPSR) is a set of rules that requires consumer products to be safe and our obligations to ensure this.

If you have any concerns about our products, you can contact us on ProductSafety@springernature.com

In case Publisher is established outside the EU, the EU authorized representative is:

Springer Nature Customer Service Center GmbH
Europaplatz 3
69115 Heidelberg, Germany

Batch number: 10755284

Printed by Printforce, the Netherlands